Advancing Personalized Medicine With AI-Driven Microbiome and Nutrigenomics:

Innovations and Future Perspectives

Reyed M. Reyed
https://orcid.org/0000-0001-5027-3478
Genetic Engineering and Biotechnology Research Institute (GEBRI), Alexandria, Egypt & City of Scientific Research and Technological Applications (SRTA-City), Alexandria, Egypt

Pranav Kumar Prabhakar
Department of Biotechnology, School of Engineering and Technology, Nagaland University, India

A.K. Haghi
https://orcid.org/0000-0002-7259-6718
University of Coimbra, Portugal

IGI Global
Scientific Publishing
Publishing Tomorrow's Research Today

Vice President of Editorial | Melissa Wagner
Director of Acquisitions | Mikaela Felty
Director of Book Development | Jocelynn Hessler
Production Manager | Mike Brehm
Cover Design | Jose Rosado

Published in the United States of America by
IGI Global Scientific Publishing
701 East Chocolate Avenue
Hershey, PA, 17033, USA
Tel: 717-533-8845 | Fax: 717-533-7115
Website: https://www.igi-global.com E-mail: cust@igi-global.com

Library of Congress Cataloging-in-Publication Data

(CIP Data Pending)
ISBN13: 979-8-3373-3121-8
Isbn13Softcover: 979-8-3373-3122-5
EISBN13: 979-8-3373-3123-2
British Cataloguing in Publication Data
A Cataloguing in Publication record for this book is available from the British Library.

To the cherished memories of my beloved father and mother, may Allah's mercy and blessings be upon them. Their enduring guidance and love continue to illuminate my path every day.

To my devoted wife and children, whose unwavering support, patience, and encouragement inspire me endlessly and give meaning to every endeavor.

This work is a humble tribute to the pillars of my life — the source of my strength, motivation, and purpose.

Table of Contents

Foreword xiii

Preface xvii

Acknowledgment xlvi

Introduction 1

Chapter 1
Advancing Diabetes Reclassification Using the NexDi AI System With Multi-Omic Insights: Egypt's Living Laboratory 1
Reyed M. Reyed, Genetic Engineering and Biotechnology Research Institute, Egypt & City of Scientific Research and Technological Applications, Egypt

Chapter 2
Decoding the Human Polybiome Through AI and Nutrigenomics in Next Generation Personalized Health: Integrating Multi-Omics for Precision Medicine 93
Hina Bansal, Amity Institute of Biotechnology, India
Arshita, Amity Institute of Biotechnology, India
Kalidindi Divya Srija, Amity Institute of Biotechnology, India

Chapter 3
Biointelligent Reframing of Colorectal Cancer: From Somatic Mutation to Systemic Coherence 173
Mona A. Esawy, Pharmaceutical and Drug Industries Research Institute, Egypt
Reyed M. Reyed, Genetic Engineering and Biotechnology Research Institute, Egypt & City of Scientific Research and Technological Applications, Egypt

Chapter 4
AI-Guided Monoclonal Antibody Design Integrating Polybiome and Nutrigenomic Insights: Integrating AI, Polybiome, and Nutrigenomics in Therapy ... 219
Ashraf Tabll, Egypt Centre for Research and Regenerative Medicine, Egypt

Chapter 5
Algorithmic Nutrient Engineering via Dynamic Microbial Bioprocesses: Omics Translation for Precision Nutrient Design ... 263
Irem Kılınç, Tokyo University of Marine Science and Technology, Japan
Sherein Ismail Abd El-Moez, National Research Centre, Egypt

Chapter 6
Microbiome-Driven Food Allergies and Precision Therapeutic Innovations ... 309
Shweta Thusoo, Dr. D.Y. Patil School of Allied Health Sciences, Dr. D.Y. Patil Vidyapeeth, Pune, India
Rohini Vipin Adepwar, Dr. D.Y. Patil School of Allied Health Sciences, Dr. D.Y. Patil Vidyapeeth, Pune, India

Chapter 7
Using AI to Profile Egyptian Lactobacillus for Next Generation Functional Foods: Integrating AI for Probiotic Development ... 357
Sherein Ismail Abd El-Moez, Veterinary Research Institute, Egypt
Mohamed Gomaa Seadawy, Biodefense Center for Infectious and Emerging Diseases, Egypt
Mostafa Fetoh El-Hosseny, Biodefense Center for Infectious and Emerging Diseases, Egypt
Mohamed Abd Elhakim Kelany, Central Lab of Residue Analysis of Pesticides and Heavy Metals in Food, Egypt
Amr Mohamed Ageez, Independent Researcher, Egypt
Miral Gamal AbdElwahab, Faculty of Technological Industry and Energy, Thebes Technological University, Luxor, Egypt
Ahmed Zaki Ghareeb, Pharmaceutical and Fermentation Industries Development Canter, Egypt
Marwa Ashraf Zayed, Faculty of Veterinary Medicine, Damanhour University, Egypt
Doaa A. Ghareeb, Independent Researcher, Egypt
Abdelmohsen M. Soliman, Pharmaceutical Research Institute, Egypt

Chapter 8
Functional Profiling and Therapeutic Potential of Egyptian Probiotics: Food Safety and Viral Defense Applications 395
Sherein Ismail Abd El-Moez, Veterinary Research Institute, Egypt
Mohamed Gomaa Seadawy, Main Chemical Warfare Laboratories, Egypt
Mostafa Fetoh El-Hosseny, Biodefense Center for Infectious and Emerging Diseases, Egypt
Doaa A. Ghareeb, Independent Researcher, Egypt
Osama Ibrahim El Batawy, Independent Researcher, Egypt

Chapter 9
Ethical and Regulatory Considerations in AI-Driven Healthcare 439
Ashok Kumar Sah, A' Sharqiyah University, Oman
Rahul Kumar, Vidya University, India
Sakshi Patel, Meerut Institute of Engineering and Technology, India
Prem Shankar Mishra, Vidya University, India
Hiren Doshi, Vidya University, India
Rakhi Mishra, Noida Institute of Engineering and Technology Pharmacy Institute, India

Conclusion 495

About the Contributors 499

Index 511

Detailed Table of Contents

Foreword xiii

Preface xvii

Acknowledgment xlvi

Introduction 1

Chapter 1
Advancing Diabetes Reclassification Using the NexDi AI System With Multi-Omic Insights: Egypt's Living Laboratory 1
Reyed M. Reyed, Genetic Engineering and Biotechnology Research Institute, Egypt & City of Scientific Research and Technological Applications, Egypt

NexDi is introduced as an advanced framework that reconceptualizes diabetes as a fluid continuum of β-cell vulnerability shaped by genetic, nutritional, and environmental pressures. Type 1 diabetes is a state where autoimmune activity increasingly overlaps with insulin resistance pathways, whereas type 5 diabetes emerges from maternal anemia, hidden hunger, and early-life undernutrition. This spectrum reflects the dual burden observed in Egypt, where rising childhood diabetes coexists with persistent nutritional deficits. Environmental stressors, including exposure to endocrine-disrupting agents and microplastic, further disturb metabolic and polybiome stability. By integrating nutrigenomics, AI-enabled prediction, and polybiome analytics, NexDi provides a scalable and culturally grounded pathway toward predictive, preventive, and personalized diabetes care.

Chapter 2
Decoding the Human Polybiome Through AI and Nutrigenomics in Next Generation Personalized Health: Integrating Multi-Omics for Precision Medicine 93
Hina Bansal, Amity Institute of Biotechnology, India
Arshita, Amity Institute of Biotechnology, India
Kalidindi Divya Srija, Amity Institute of Biotechnology, India

This chapter investigates the integration of artificial intelligence, nutrigenomics, and the human polybiome to enhance precision healthcare. It outlines how deep learning and multiomics approaches decode host–microbe interactions influencing nutrition, gene expression, and disease risk. This chapter introduces advanced analytical frameworks linking microbial metabolites, epigenetic regulation, and metabolic balance. By merging computational biology with clinical insight, AI-driven strategies for predictive nutrition, individualized therapy, and early disease intervention can be developed. Furthermore, this chapter discusses the ethical and interpretability dimensions of AI in biomedical data, emphasizing transparent algorithms and privacy-preserving models that ensure reliability, reproducibility, and responsible innovation in personalized healthcare systems.

Chapter 3
Biointelligent Reframing of Colorectal Cancer: From Somatic Mutation to Systemic Coherence 173
Mona A. Esawy, Pharmaceutical and Drug Industries Research Institute, Egypt
Reyed M. Reyed, Genetic Engineering and Biotechnology Research Institute, Egypt & City of Scientific Research and Technological Applications, Egypt

This chapter presents a critical analysis of the existing textual paradigm of linear somatic mutation in colorectal cancer (CRC) and suggests reinterpretation as a failure of distributed biological intelligence in the holobiome. It presents the idea of the cognitive oncology system (COS) as an ontological requirement, conceptualizing cancer as an adaptive process that is caused by the loss of systemic coherence along microbial, nutrigenomic, and metabolic axes. The framework integrates polybiome–nutrigenome analytics, biointelligent modeling and neuroecological organoid interfaces. Together, these factors form the basis of circular medicine, which assumes the systematic recycling of biological intelligence to attain therapeutic efficacy and secure national autonomy. This view is in line with Egypt Vision 2030, which promotes the development of scientific infrastructure at the forefront of a biointelligent civilization. The chapter ends with a translational blueprint connecting disease reframing to a sustainable national health strategy.

Chapter 4
AI-Guided Monoclonal Antibody Design Integrating Polybiome and Nutrigenomic Insights: Integrating AI, Polybiome, and Nutrigenomics in Therapy 219
Ashraf Tabll, Egypt Centre for Research and Regenerative Medicine, Egypt

This chapter explores how the microbiome, the broader concept of the polybiome, and nutrigenomic factors converge within precision immunotherapy. Monoclonal antibodies (mAbs), once confined to infectious and oncological diseases, are now being designed to correct dysbiosis and nutrigenomic disturbances, restoring immune and metabolic balance. Artificial intelligence (AI)-driven discovery refines antibody specificity and predicts therapeutic outcomes. Case studies demonstrate how mAbs interact with microbial metabolites and individual nutrigenomic profiles. By integrating AI, polybiome profiling, and nutritional genetics, this chapter outlines a new personalized framework for chronic and autoimmune disease management.

Chapter 5
Algorithmic Nutrient Engineering via Dynamic Microbial Bioprocesses: Omics Translation for Precision Nutrient Design 263
Irem Kılınç, Tokyo University of Marine Science and Technology, Japan
Sherein Ismail Abd El-Moez, National Research Centre, Egypt

This chapter describes the mechanistic approach used to translate nutrigenomic data into clinically meaningful nutritional formulations through artificial intelligence (AI) and precision fermentation (PF). This highlights the central challenge of personalized nutrition: converting complex genomic and metabolic signals into consistent, actionable nutrient specifications. This chapter introduces an AI-driven inverse-design system that generates digital nutrient specifications (DNSs) that define the molecular composition and production parameters. PF is a biomanufacturing platform capable of executing these specifications through engineered microbial systems and real-time process control to produce genotype-aligned peptides, vitamins, and metabolic cofactors. The chapter also outlines the analytical pipeline linking multiomic interpretation to fermentation optimization and provides examples relevant to metabolic and neurodevelopmental applications, concluding with key regulatory and ethical considerations for implementing AI-based personalized nutrition.

Chapter 6
Microbiome-Driven Food Allergies and Precision Therapeutic Innovations ... 309
Shweta Thusoo, Dr. D.Y. Patil School of Allied Health Sciences, Dr. D.Y. Patil Vidyapeeth, Pune, India
Rohini Vipin Adepwar, Dr. D.Y. Patil School of Allied Health Sciences, Dr. D.Y. Patil Vidyapeeth, Pune, India

Food allergies have become a growing concern in the general population, and an increase of evidence is pointing to gut microbiome disbalances as the cause of immune malfunctions and allergic reactions. Altered filtering between the microbes may cause defects in immune tolerance and elevated allergicity. Conventional therapies tend to deal with symptoms and not the cause. The novel developments in the metagenomics, microbiome sequencing, and AI derived analytics have further elucidated that the gut microbes have the ability to regulate immunity. Microbial biomarkers can now be detected with the help of these technologies, and personalized interventions involving probiotics, prebiotics, fecal microbiota transplantation (FMT) and engineered microbial consortia can be developed. In spite of the promise, there are still challenges of the variability in the individual microbiome, regulatory issues, and long-term safety. In this chapter of the book, the author offers a detailed account of the role of the microbiome in food allergies, the development of AI-enhanced research instruments, and the development of microbiome-based therapeutics. It also bridges the gap between science, technology, and clinical implementation by pointing out the possibility of precision medicine changing the way allergy is treated into a more focused and sustainable method of healthcare provision.

Chapter 7
Using AI to Profile Egyptian Lactobacillus for Next Generation Functional Foods: Integrating AI for Probiotic Development 357
Sherein Ismail Abd El-Moez, Veterinary Research Institute, Egypt
Mohamed Gomaa Seadawy, Biodefense Center for Infectious and Emerging Diseases, Egypt
Mostafa Fetoh El-Hosseny, Biodefense Center for Infectious and Emerging Diseases, Egypt
Mohamed Abd Elhakim Kelany, Central Lab of Residue Analysis of Pesticides and Heavy Metals in Food, Egypt
Amr Mohamed Ageez, Independent Researcher, Egypt
Miral Gamal AbdElwahab, Faculty of Technological Industry and Energy, Thebes Technological University, Luxor, Egypt
Ahmed Zaki Ghareeb, Pharmaceutical and Fermentation Industries Development Canter, Egypt
Marwa Ashraf Zayed, Faculty of Veterinary Medicine, Damanhour University, Egypt
Doaa A. Ghareeb, Independent Researcher, Egypt
Abdelmohsen M. Soliman, Pharmaceutical Research Institute, Egypt

In response to the growing demand for sustainable, health-promoting foods, this chapter details an AI-powered pipeline for whole-genome sequencing and functional annotation of six Egyptian Lactobacillus isolates selected from 50 dairy-derived candidates. High-coverage Illumina MiSeq reads were quality-filtered (FastQC), assembled (PATRIC), and annotated via RASTtk, antiSMASH, BAGEL4, ResFinder, and CRISPRCasFinder. Analyses revealed four bacteriocin biosynthesis clusters, two Type I-E CRISPR arrays, prophage elements, and no antibiotic resistance or virulence genes. Phylogenomic analysis confirmed membership in the *L. paracasei* and *L. rhamnosus* groups. This *in silico* approach accelerates safety assessment and guides the selection of bacteriocin-rich cultures for next-generation functional foods, with alignment with Egyptian regulatory standards and recommendations for digital-twin modeling and pilot-scale validation.

Chapter 8
Functional Profiling and Therapeutic Potential of Egyptian Probiotics: Food Safety and Viral Defense Applications 395
Sherein Ismail Abd El-Moez, Veterinary Research Institute, Egypt
Mohamed Gomaa Seadawy, Main Chemical Warfare Laboratories, Egypt
Mostafa Fetoh El-Hosseny, Biodefense Center for Infectious and Emerging Diseases, Egypt
Doaa A. Ghareeb, Independent Researcher, Egypt
Osama Ibrahim El Batawy, Independent Researcher, Egypt

This chapter presents a practical follow-up to earlier AI-guided genomic investigations, examining the functional characterization and therapeutic benefits of Egyptian *Lactobacillus* strains for food safety and viral defense. Metabolomic and phenotypic assays were used to confirm bacteriocin production and viral inhibition in vitro. The link between probiotic features—adhesion, immunomodulation—and host genetics has been explored to inform personalized nutrition. Key regulatory and ethical frameworks, including the Codex, EFSA, and WHO-Food Safety, are integrated. The chapter concludes with a roadmap—from laboratory validation to commercial products, preclinical trials, and cultural adaptation—promoting scalable, safe, and precision-oriented probiotic applications consistent with the thematic logic of the book.

Chapter 9
Ethical and Regulatory Considerations in AI-Driven Healthcare 439
Ashok Kumar Sah, A' Sharqiyah University, Oman
Rahul Kumar, Vidya University, India
Sakshi Patel, Meerut Institute of Engineering and Technology, India
Prem Shankar Mishra, Vidya University, India
Hiren Doshi, Vidya University, India
Rakhi Mishra, Noida Institute of Engineering and Technology Pharmacy Institute, India

Artificial Intelligence (AI) is revolutionizing clinical practice by providing previously unheard-of capabilities in patient monitoring, diagnosis, treatment personalization, and healthcare system optimization. Even though artificial intelligence (AI) promises to improve accuracy, efficiency, and scalability, its quick adoption also brings with it difficult moral, legal, and social issues. Using a critical lens based on medical ethics, legal frameworks, and practical implications, this chapter examines the dual reality of AI-driven healthcare—its transformative potential and its associated risks. In the context of algorithmic decision-making, fundamental ethical concepts like beneficence, autonomy, justice, and transparency are reviewed. Data bias, health disparities, patient consent, data privacy, clinical deskilling, and accountability in the case of AI system failures are among the urgent issues covered in this chapter. It assesses new ethical frameworks from international organizations and medical institutions and offers a comparative overview of regulatory environments across the globe. The chapter highlights the importance of collaborative, multidisciplinary approaches to ensure AI remains a tool for equitable and human-centered care by highlighting the roles of key stakeholders, including developers, healthcare professionals, regulators, and patients. In order to integrate technological innovation with the ethical underpinnings of healthcare, it concludes with recommendations for the future that include adaptive regulation, ethical design practices, and ongoing vigilance. This chapter is a call to action as well as a guide for using AI responsibly for the benefit of humankind and health.

Conclusion 495

About the Contributors 499

Index 511

Foreword

Cancer immunology has reached a defining stage in its scientific evolution. Despite substantial progress in molecular oncology, immune checkpoint therapies, and targeted treatment strategies, cancer continues to reveal limitations in reductionist models that isolate genetic, immunological, or metabolic dimensions. The persistence of therapeutic resistance and interpatient variability underscores a fundamental reality: cancer is not governed by singular biological axes, but by integrated systems shaped across time, context, and environment.

This volume is positioned precisely within this emerging systems-based understanding. By extending the microbiome paradigm into the more comprehensive Polybiome framework, the book introduces a constitutional model that integrates microbial ecology, host genetics, nutrigenomic modulation, metabolic signaling, and environmental exposure. From an immunological standpoint, this integration is both timely and essential. Immune surveillance, tolerance, inflammation, and exhaustion represent dynamic outcomes of sustained biological interactions rather than isolated immune events. These processes are continuously influenced by microbial metabolites, diet-derived epigenetic regulation, and long-term exposomic pressures.

A distinguishing feature of this work is its disciplined integration of Artificial Intelligence (AI) as a methodological tool for biological interpretation rather than a substitute for scientific reasoning. The chapters collectively demonstrate how AI enables the analysis of multidimensional biological data, supports the identification of latent immunological and metabolic patterns, and facilitates predictive modeling across complex host–microbe– environment interfaces. In cancer research, such approaches contribute to the emergence of predictive and preventive immunology, allowing earlier identification of immune dysregulation and therapeutic vulnerability before clinical manifestation.

The interdisciplinary scope of this volume enhances its relevance across multiple domains. Researchers and practitioners in immunology, oncology, microbiology, nutrition science, and biomedical data science will find structured methodological approaches, applied case perspectives, and integrative analytical frameworks. This breadth positions the book as both a scholarly reference and an educational resource for advanced academic training and translational research initiatives.

Equally important, the volume maintains a clear commitment to ethical responsibility and scientific governance. By addressing data integrity, transparency, and biological sovereignty, the editors recognize that innovation in AI-driven healthcare must be aligned with equitable access, regulatory accountability, and societal trust. Such considerations are indispensable as data-intensive biomedical models increasingly influence clinical decisionmaking and public health strategies.

The editor and contributing authors are to be commended for delivering a coherent and forward-looking contribution that bridges theoretical modeling with applied biomedical relevance. This book offers a structured vision for how multi-dimensional biological systems, immunology, and AI can be responsibly integrated to advance Personalized and Precision Medicine. It represents a meaningful addition to the global scientific literature and will be of lasting value to researchers, clinicians, and policymakers alike.

Mohamed Labib Salem
Faculty of Science, Tanta University, Egypt

Mohamed Labib Salem, PhD, is a Distinguished Professor of Immunology at the Faculty of Science, Tanta University, Tanta, Egypt. He is internationally recognized as a leading expert in Immuno-oncology, with an extensive career distinguished by groundbreaking contributions to both fundamental immunology and translational cancer research. Over several decades, Prof. Salem has conducted pioneering studies and collaborative projects at renowned institutions in Egypt, Japan, and the United States. His tenure as an Associate Professor at the Medical University of South Carolina, spanning ten years, was particularly influential, during which he significantly advanced the scientific understanding of immune system mechanisms involved in cancer development and therapy. Prof. Salem is the Founding *Director of the Center of Excellence in Cancer Research (CECR) and the President of the Egyptian Association for Cancer Research (EACR)*, based in Tanta, Egypt. Through these leadership roles, he has been a driving force in advancing cancer immunotherapy research both nationally and across the Middle East. His efforts have fostered vital collaborations that integrate laboratory discoveries with clinical applications, thereby enhancing therapeutic approaches and improving patient outcomes. His vision and strategic initiatives have substantially contributed to the development of Egypt's research infrastructure and capacity in oncology. In addition to his scientific achievements, Prof. Salem holds key positions influencing science policy and innovation. He serves on *the Board of Directors at Zewail City of Science and Technology in 6th of October City, Giza, Egypt*, and chairs *the Council for Scientific Culture at the Academy of Scientific Research and Technology (ASRT) in Cairo, Egypt*. Through these influential roles, he actively shapes national strategies to promote scientific excellence, foster interdisciplinary research, and accelerate technological innovation. His leadership helps bridge the gap between scientific discovery and policy-making, thereby driving sustainable development in Egypt's science and technology sectors. Prof. Salem's academic impact is exemplified by an *H-index of 45* and more than 10,000 citations, reflecting his broad influence and respect within the global scientific community. His prolific body of work effectively bridges molecular immunology with clinical oncology, highlighting the translational potential of immunotherapeutic strategies. His research continues to push the boundaries of personalized medicine, emphasizing the importance of understanding immune regulation in cancer treatment. His endorsement of this volume reflects a deep alignment with the transformative potential of Precision Medicine, particularly through the integration of complex biological networks as conceptualized by the *Polybiome framework*. Prof. Salem recognizes that innovations driven by artificial intelligence and systems biology represent a paradigm shift in healthcare. He highlights the critical importance of these approaches in supporting *national bio-sovereignty*, advancing personalized therapeutic interventions, and ultimately improving health outcomes on a global scale. His support underscores the scientific rigor and visionary scope of this work, positioning it as a foundational contribution to the future of medicine.

Preface

THE EDITORIAL JOURNEY AND SCIENTIFIC RIGOR

The development of this scientific volume unfolded through a highly competitive and methodologically rigorous process that engaged scholars from a broad international landscape. The initial stage began with the submission of *60 chapter proposals* from researchers representing multiple countries, academic traditions, and disciplinary backgrounds. This large and diverse pool demonstrated the global interest in the book's central theme and underscored the need for a strong, structured editorial and peer-review system.

Following an in-depth evaluation of these 60 proposals, which assessed scientific relevance, methodological suitability, thematic alignment, and the potential contribution of each topic to the intellectual arc of the volume, *40 proposals were formally accepted* for chapter development. The remaining *20 proposals were respectfully declined* due to issues such as thematic misalignment, insufficient conceptual grounding, or limited visibility within the framework of the book's scientific objectives.

Following the formal acceptance of 40 chapter proposals, the second stage of the editorial process involved the submission of full manuscripts. However, only *34 full chapters were submitted* within the prescribed editorial deadlines. The remaining *6 chapters were excluded and marked as "Rejected: Incomplete Submission"* due to the authors' failure to submit the full manuscripts despite the initial acceptance of their proposals. This attrition underscores the importance of strict deadline adherence in maintaining the integrity and schedule of the publication process.

Each of these submitted chapters entered a rigorous and multi-layered review process involving several expert reviewers whose collective number significantly exceeded the number of submitted manuscripts. This approach ensured diverse scientific perspectives, minimized bias, and strengthened the methodological and conceptual quality of the assessments.

The peer-review cycle for many of these chapters included multiple rounds of revision. Authors were asked to address detailed scientific critiques, refine their methodology, restructure specific arguments, correct technical issues, and align their work with the book's overarching analytical model. As Lead Editor, I provided extensive guidance, including conceptual clarifications, structural recommendations, and iterative feedback aimed not only at improving individual manuscripts but also at enhancing the overall coherence of the volume. Authors were given multiple opportunities to revise their work when their manuscripts demonstrated potential for contributing meaningfully to the book's aims.

At the conclusion of this comprehensive process, *9 chapters were approved for publication*, representing the strongest scientific contributions across the entire submission pipeline. These chapters demonstrated originality, methodological rigor, thematic relevance, and clear alignment with the book's conceptual framework. The remaining *25 chapters were declined* after failing to meet the necessary scientific or structural requirements despite opportunities for revision.

The resulting volume reflects a meticulously curated body of work shaped by disciplined editorial oversight, robust international peer review, and sustained scientific dialogue between authors, reviewers, and the editorial office. These processes collectively ensured that the final content maintains the highest standards of academic integrity and advances the scholarly objectives of the field.

The following table summarizes the geographic distribution of contributing authors based on their institutional affiliations

Table 1.

Rank	Country	Number of Authors
1	India	52
2	Egypt	27
3	United States of America	5
4	Morocco	3
5	Canada	2
5	Bangladesh	2
5	Turkey	2
8	Greece	1
8	Ireland	1
8	Oman	1
8	Spain	1
8	United Kingdom	1
8	Japan	1

Statistical Analysis and Insights

- The contribution is markedly concentrated, with the top two countries, *India (52 authors, 43.7%) and Egypt (27 authors, 22.7%)*, together accounting for approximately *66.4%* of all contributing authors. This highlights strong regional leadership and scholarly engagement from South Asia and North Africa in the fields of microbiome, AI, and nutrigenomics.
- The *United States*, although representing a global research powerhouse, contributed 5 authors (4.2%), reflecting a more selective but significant involvement.
- Countries ranked 4th to 7th (Morocco, Canada, Bangladesh, Turkey) collectively contribute 9 authors (7.6%), indicating moderate international reach extending into Africa, North America, and Asia.
- The remaining 7 countries share a combined total of 7 authors (5.8%), demonstrating a broad but thinly distributed global participation, which enriches the volume's diversity and cross-cultural perspectives.
- This wide international participation not only strengthens the scientific rigor but also ensures that the book covers a variety of geographic contexts, healthcare systems, and technological applications, essential for a volume aiming to impact global personalized medicine.

SYSTEMS MEDICINE: A METHODOLOGICAL CROSSROAD POLYBIOME-DRIVEN FRAMEWORK

Contemporary research in biomedical science has reached an unprecedented crossroad. Highthroughput sequencing, metabolomics, proteomics, and digital health instrumentation have developed rapidly, and their use has provided humans with data of theoretically revolutionary value in scale. But even with all this data, there is a dearth of translation into actual clinical benefits. This is mainly because of a lack of integrative models that can transform complex, heterogeneous data into actionable information. The classical paradigms of reductionism that lead to the isolation and focus on specific molecular pathways or genetic determinants do not reveal the emergent nature of living systems. Such models frequently fail to consider the interactions between genetics, metabolism, immunity, environmental exposures, and microbial ecosystems. Consequently, the concept of health and illness, particularly chronic and multifactorial illnesses, is only partially understood. Such issues must be met with a shift in thinking towards systems-level thought, which encompasses biological complexity and incorporates several streams of data into realistic, plausible models of human physiology. The book presents the Polybiome

Systems Medicine (PSM) model, a model that views the host as a self-regulated, dynamic biological system. In contrast to traditional microbiome studies, PSM combines host genetic variations, nutritive-genomic interactions, immune physiology, metaclomic pathways, and environmental exposures into a unified network. In this context, health and disease are conceptualized as conditions of systemic order or disorder, including disease heterogeneity, personal variability, and longitudinal biological variation-important for precision medicine in the face of single-gene genomic frameworks.

Conceptual Documentation and Terminological Framework

The theoretical constructs and methodological terminologies introduced in this volume are part of the broader Polybiome Systems Medicine framework. These concepts have been formally documented in publicly accessible scholarly repositories (Zenodo), providing persistent digital object identifiers (DOIs) for transparency, citability, and proof of intellectual provenance. Researchers can examine these archives to understand the methodological foundations and extend them in future interdisciplinary studies on AI-enabled precision health and nutrigenomic systems science. Representative references include:

- Reyed, R. M. (2026a). Polybiome Systems Medicine: Conceptual Architecture, Methodological Foundations, and Translational Applications — Volume I. Zenodo. https://doi.org/10.5281/zenodo.18294835
- Reyed, R. M. (2026c). Polybiome Systems Medicine (PSM): A Systems-Level Biodiversity Conversion Framework for Precision Nutrition and Predictive Health 2040. Zenodo. https://doi.org/10.5281/zenodo.18727664
- Reyed, R. M. (2026e). Polybiome Systems Medicine – Volume III: AI Nutrigenomics, Precision Fermentation, and Genotype-Aligned Biomanufacturing. Zenodo. https://doi.org/10.5281/zenodo.18871656

By directly citing these archives, this book ensures intellectual attribution, ethical rigor, and protection of original ideas for future scholars. The terminologies and hierarchical system presented here are unique to this volume, uniting human biology, multi-omics analytics, and ethical AI within a consistent, practical, and novel framework of precision medicine.

The Urgency of the Big Data Challenge in Medicine

Despite the proliferation of biomedical data, the potential of Big Data to improve patient outcomes has not been fully realized—a problem often referred to as the Big Data Crisis. Several interconnected factors contribute to this challenge:

1. Data Heterogeneity: Biomedical datasets are enormous and highly variable, spanning genomics, transcriptomics, metabolomics, proteomics, clinical imaging, electronic health records, environmental sensors, and patient-reported outcomes. Each domain uses distinct technologies and standards, complicating interoperability. Without cohesive frameworks and shared ontologies, constructing a unified view of health or disease is exceedingly difficult.
2. Single-omic Prioritization: Genomics has historically dominated precision medicine efforts, often at the expense of crucial domains such as the microbiome, nutrition, immune function, and environmental factors. This narrow approach fails to reflect the multidimensional nature of chronic disease.
3. Population Variability: Differences in genetics, culture, diet, environmental exposures, and socioeconomic conditions challenge model generalizability. Models derived from small cohorts may underperform when applied to diverse populations, risking inequitable health outcomes.
4. Data Quality and Reproducibility: Variability in sample handling, analytical workflows, and data cleaning introduces biases that undermine confidence in predictive models and clinical translation.

Collectively, these factors produce a significant gap between data generation and actionable clinical knowledge, limiting individualized therapy, escalating healthcare costs, and obstructing precision medicine implementation.

A Paradigm Shift: Systems Integration

Overcoming these challenges requires integrative, systems-oriented methodologies that merge multidimensional data across biological, environmental, and social levels. Polybiome Systems Medicine addresses this need by framing health as an emergent property of interconnected biological networks encompassing genetics, microbial communities, metabolism, immunity, nutrition, and environmental exposures.

This approach enables dynamic, context-sensitive models that better reflect real-world complexities, enhance predictive accuracy, and tailor interventions to individuals and populations. The framework is scalable and adaptable, capable of integrating novel data types and emerging scientific knowledge, while promoting equitable healthcare by considering population diversity and context.

ARTIFICIAL INTELLIGENCE AS AN INTEGRATIVE METHODOLOGICAL KEYSTONE

Artificial Intelligence (AI) is not merely a computational tool within this framework; it serves as a central methodological integrator. AI analyzes vast, heterogeneous datasets to uncover complex, often nonlinear relationships and emergent patterns that traditional statistical or reductionist methods cannot detect.

Chapter 1 exemplifies AI's role through the NexDi system, which reclassifies diabetes along a continuum of beta-cell vulnerability rather than a binary diagnosis. NexDi integrates nutrigenomic phenotypes, polybiome signals, and environmental stressors, tailored to the unique nutritional and socio-environmental realities of Egypt. This demonstrates AI's capacity to contextualize disease knowledge within population-specific frameworks.

Importantly, AI complements rather than replaces clinical judgment. It serves as a decisionsupport system that enhances hypothesis generation, risk stratification, and individualized intervention planning. The book emphasizes ethical considerations including algorithmic bias, transparency, data privacy, and equitable access. Applications of AI follow international standards of reproducibility, interpretability, and governance, establishing a foundation of trust and responsible use in healthcare.

By turning individual data silos into interconnected systems, AI supports the identification of hidden biological correlations, enables dynamic representations of health states, and defines new standards for precision health that are both scientifically rigorous and ethically sound.

Intellectual Ownership Note

The concepts and frameworks presented in this book, including the Polybiome Systems Medicine (PSM) model, are the original work of Dr. Reyed M. Reyed. They have been formally documented in publicly accessible repositories (Zenodo) to ensure transparency and proper attribution. Readers are kindly asked to acknowledge the author when referencing or building upon these ideas.

THE ETHICAL FRONTIER: FROM MICROBIOME TO POLYBIOME SOVEREIGNTY

This methodological integration advances a new paradigm of Polybiome Ethics. Unlike traditional microbiome ethics, which primarily emphasized individual privacy, polybiome ethics frames the human holobiont as an active, context-dependent network operating across biological, environmental, and nutritional dimensions.

At the heart of this volume is the principle of Ethical AI Sovereignty. We advocate for locally trained, culturally grounded AI systems that counter Western-centric data biases—biases that routinely misclassify, underrepresent, or pathologize microbiome profiles from the Global South. By aligning the methodological backbone of this work with national ethical frameworks—specifically the Egyptian Charter for Responsible AI (MCIT, 2023) and the Indian Ethical Guidelines (ICMR, 2023)—we ensure that technological innovation is directed toward collective benefit and equitable scientific advancement.

This ethical repositioning is fully aligned with the One Health framework and Egypt's Vision 2030, transforming nutribiome research from a "data-rich exercise" into an Ethical AI Science that promotes global health equity and reinforces the scientific sovereignty of emerging nations.

CHAPTER-BY-CHAPTER OVERVIEW AND REFLECTIVE INTEGRATION

This book offers an integrated and visionary analysis of systems medicine within the framework of the polybiome approach, making it a revolutionary element of modern biomedical practice. The chapters develop sequentially on the principles of biological integration, utilizing more sophisticated instruments like artificial intelligence and multi-omics data to address the complexity of human health and disease. As illustrated in Figure 1, the framework follows a structured progression from theoretical foundations to translational applications, ethical considerations, and future perspectives. This integrated module ensures a coherent and holistic understanding for readers across disciplines.

Figure 1. Integrated Module of Polybiome Systems Medicine: From Diagnostics to Ethical Governance.

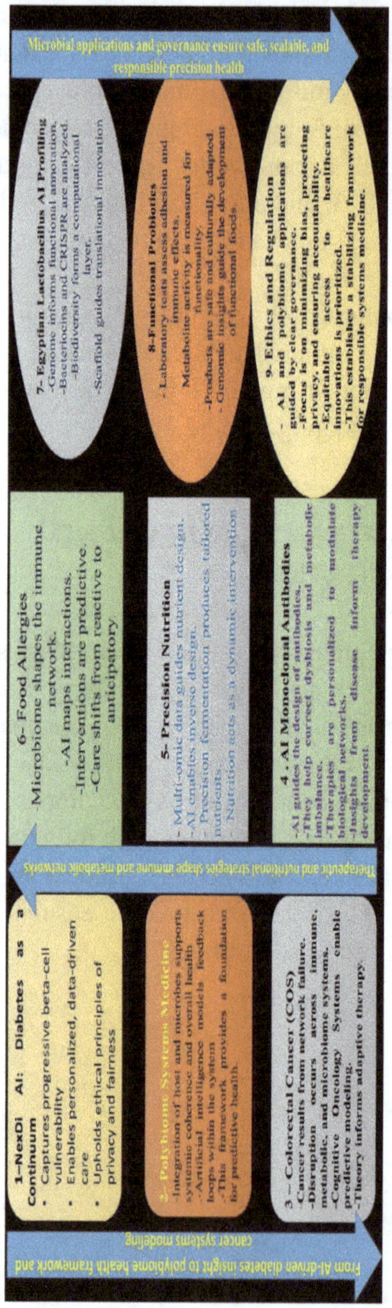

This figure takes the nine chapters of the book and consolidates them into an integrated polybiome-minded systems medicine framework. It emphasizes the development that is based on AI-powered diagnostics and dynamic disease modeling, predictive and adaptive therapeutic strategies, precision nutrition, probiotic innovation, and ethical governance. The focus of the module on systemic coherence, interventions built on feedback, and translational applicability, and a critical view of barriers to interpret inter-individual variability, regulatory complexity, and ethically appropriate AI integration. The figure illustrates a novel, holistic vision of next-generation healthcare by mapping every domain as an interdependent part and not separate silos; combining computational understanding with practical, patientcentered action.

Chapter 1. Advancing Diabetes Reclassification Using the NexDi AI System with Multi-Omic Insights: Egypt's Living Laboratory

A Diabetes Reclassification Advantage. The book begins with a paradigm shift, questioning conventional clinical views on diabetes through an original approach to its classification. The dynamic framework offered by the NexDi AI system replaces the traditional models that divide diabetes into strict categories mainly based on glucose readings. In this case, the conceptualization of diabetes is presented as a continuum that involves the progressive vulnerability of beta cells determined by a complex interaction of nutrigenomic attributes, polybiome cues, and environmental influences. This sophisticated model recognizes that metabolic diseases are multifactorial and the significance of context-specific information should be acknowledged. The case study that centers on Egypt demonstrates how local nutritional shifts and socio-environmental situations can be incorporated to increase the relevance and predictive potential of the model. The fact that urbanization and dietary changes are occurring rapidly in Egypt presents a unique metabolic challenge, which the NexDi system captures to enable personalized intervention strategies. The values of ethics are placed as the highest priority, and the principles of data fairness, patient confidentiality, and equitable access are enshrined in the AI framework. As a result, this chapter establishes a precedent in the responsible utilization of AI in the intricate healthcare environment.

Chapter 2. Decoding the Human Polybiome Through AI and Nutrigenomics in Next Generation Personalized Health

Chapter 2 serves as the intellectual and methodological axis of the volume. It moves the discussion from disease reclassification toward systemic interpretation.

While Chapter 1 challenges conventional diagnostic boundaries, this chapter constructs the analytical architecture that makes such reclassification scientifically coherent and computationally actionable.

The chapter presents an integrative exploration of personalized health through the convergence of artificial intelligence and nutrigenomics within a host–microbe systems framework. Its development marked a decisive transition from a predominantly microbiome-centered vocabulary toward a broader Polybiome Systems Medicine paradigm. Under the scientific Page 10 of 31 guidance of Reyed M. Reyed, the conceptual lens was expanded to encompass cross-kingdom microbial dynamics, host genomic architecture, immune modulation, metabolic fluxes, nutritional signaling, and environmental exposures as interacting regulatory strata rather than parallel variables.

This expansion was not cosmetic. It required restructuring the chapter's analytical flow, harmonizing terminology, refining modeling assumptions, and re-aligning its computational logic. The finalized version reflects this deeper systems orientation. Visual frameworks, modeling diagrams, and multi-layer analytical pathways were recalibrated to ensure that the host–microbe dialogue is represented not as a linear cascade but as a feedback-governed network with bidirectional regulatory loops.

At its core, the chapter argues that the Polybiome is not simply an ecological extension of the microbiome. It is a co-regulated dynamic network in which host genetics, immune signaling, metabolism, diet-derived molecular inputs, and environmental exposures continuously interact. Health is framed as systemic coherence within this network. Disease emerges when coordination across these layers becomes unstable or fragmented.

This shift from component analysis to systemic coordination is central. Traditional microbiome research often emphasizes compositional shifts. Chapter 2 instead emphasizes relational architecture. It explores how microbial metabolites influence host epigenetic states, how nutritional patterns modulate microbial gene expression, and how immune tone mediates crosskingdom communication. These interactions are modeled as dynamic states rather than static measurements.

The chapter provides theoretical grounding in systems biology. Mathematical and computational modeling approaches are introduced to formalize these multilayered interactions. Graph theory, network topology, and multi-omic integration pipelines are described as essential tools for preserving structural coherence in interpretation. Without such formalization, complex biological data risks fragmentation into isolated findings.

Multi-omics integration occupies a central methodological position. Genomic, transcriptomic, proteomic, metabolomic, and exposomic datasets are not treated as additive layers. They are integrated within unified analytical scaffolds that allow cross-layer inference. Artificial intelligence is positioned as the infrastructure enabling this synthesis. Machine learning models are employed for pattern recognition, stratification, and predictive simulation. Explainable AI strategies are emphasized to ensure interpretability and epistemological consistency.

This level of methodological rigor is essential to avoid conceptual dilution. The chapter maintains strict terminological precision. It differentiates between microbiome variability and Page 11 of 31 polybiome coordination. It distinguishes compositional dysbiosis from systemic incoherence. It clarifies that prediction without mechanistic plausibility risks oversimplification. Such clarity preserves scientific integrity while enabling translational application.

Importantly, Chapter 2 establishes the epistemological foundation for the entire volume. By defining health and disease as emergent properties of systemic coherence or incoherence, it provides a unifying conceptual ground for subsequent chapters. Whether addressing metabolic vulnerability, oncological reframing, antibody engineering, nutrient design, probiotic authentication, or ethical governance, each applied domain operates within the systems logic articulated here.

The chapter therefore functions as more than a theoretical exposition. It is a structural anchor. It translates the Polybiome from a conceptual proposal into an operational analytical framework. It demonstrates how AI-enabled nutrigenomics can transform heterogeneous biological signals into predictive health architectures aligned with next-generation personalized medicine.

In doing so, Chapter 2 establishes the methodological continuity required for the volume to function as an integrated scientific project rather than a collection of independent thematic contributions. It defines the shared scaffold upon which the translational ambitions of Polybiome Systems Medicine are constructed.

Chapter 3. Biointelligent Reframing of Colorectal Cancer: From Somatic Mutation to Systemic Coherence

The submitted chapter represents a conceptually grounded and systems-oriented academic contribution that advances a bold analytical thesis for the redefinition of colorectal cancer (CRC). Rather than interpreting oncogenesis as the cumulative outcome of linear somatic mutations occurring within isolated cellular compartments, the chapter advances a reframed systems-level interpretation in which carcinogenesis is understood as an emergent consequence of distributed biological intelligence failure and the progressive loss of systemic coherence within the holobiome. In this context, tumorigenesis is not treated as a localized pathological endpoint, but as the macroscopic manifestation of network-level regulatory disruption spanning microbial metabolism, host immunity, nutritional signaling, and genomic stability. During the course of editorial and conceptual development, the chapter's initial tumor-centric explanatory model was further expanded through the methodological and analytical contribution of Reyed M. Reyed in his capacity as co-author. This contribution was not Page 12 of 31 confined to structural refinement or terminological harmonization; rather, it involved the introduction of a systems-oncological interpretive layer that repositions colorectal cancer within the broader Polybiome Systems Medicine framework. This reframing enabled the integration of microbial fermentative dynamics as measurable regulatory inputs participating in host-level immunogenomic modulation. As a result, disease emergence is modeled not as a stochastic genetic event, but as a bioinformational breakdown across metabolically mediated host–microbe signaling networks.

The chapter's core conceptual propositions may be summarized as follows:

1. **Ontological Repositioning of Cancer**

   Cancer is redefined as a self-organizing adaptive response occurring within a dysregulated biological network rather than as a terminal failure of cellular control. This perspective emphasizes compensatory system behavior under conditions of metabolic and ecological incoherence.
2. **Expansion Toward the Polybiome Domain**

   The analytical scope extends beyond bacterial microbiota to incorporate viral, fungal, archaeal, and metabolically active consortia operating within cross-kingdom ecological gradients. Within this expanded domain, the polybiome is conceptualized as a form of extended biological intelligence capable of dynamically interacting with host signaling architectures.
3. **Fermentobiometric Interpretability Layer**

   A key methodological development introduced in the finalized version of the chapter is the application of fermentobiometric parameters—including short-chain fatty acid ratios (e.g., butyrate/acetate balance) and secondary bile acid flux—as informational intermediates linking microbial metabolism with host genomic and immunological regulation. This interpretive layer permits the detection of pre-neoplastic states through metabolically anchored bioinformational gradients rather than through delayed histopathological markers.
4. **Cognitive Oncology Systems (COS)**

   The chapter further outlines a predictive technological framework incorporating organoidsimulated tumor environments, biologically informed machine learning models, and systemslevel inference architectures. These tools support the anticipation of tumor behavioral trajectories and enable the design of adaptive therapeutic interventions aligned with individual host–microbe–metabolic configurations.
5. **Circular Medicine as Diagnostic Ecology**

Metabolic by-products traditionally regarded as physiological waste are repositioned as diagnostically valuable signaling residues. Within this circular medical logic, fermentation- Page 13 of 31 derived metabolites are reinterpreted as actionable indicators of systemic resilience or instability, thereby linking ecological sustainability with clinical precision.

Methodologically, the chapter adopts a transdisciplinary synthesis integrating theoretical systems biology, ecological oncology, and bioinformatic inference. Drawing upon a curated body of literature spanning the period from 2021 to 2025, the objective is not statistical generalization but conceptual stabilization—namely, the construction of a coherent explanatory scaffold capable of supporting predictive disease modeling across heterogeneous biological contexts.

Strategically, this reframed oncological model is aligned with national priorities articulated under Egypt's Vision 2030, particularly in relation to the development of biologically intelligent health infrastructure, digital clinical ecosystems, and sustainability-oriented medical innovation. In doing so, the chapter situates CRC modeling within a forward-looking framework that integrates systems medicine, artificial intelligence, and context-aware biomedical governance.

Collectively, Chapters 2 and 3 introduce and operationalize the Polybiome Systems Framework as an applied systems-analytic paradigm for next-generation precision oncology. Within the specific context of colorectal cancer, Chapter 3 advances this paradigm by enabling the transition from descriptive microbial taxonomy toward analytically grounded modeling of disease emergence through distributed bioinformational disruption, thus reinforcing the methodological trajectory of systems medicine approaching the 2040 horizon.

Chapter 4. AI-Guided Monoclonal Antibody Design Integrating Polybiome and Nutrigenomic Insights: Integrating AI, Polybiome, and Nutrigenomics in Therapy

This chapter represents the culmination of a tightly interwoven framework that begins with Chapter 1's foundational case study and is theoretically scaffolded by Chapter 2 and conceptually applied in Chapter 3. In this integrated narrative, diabetes reclassification via the NexDi AI system demonstrates how high-dimensional clinical and omics data can be operationalized to capture the continuum of beta-cell vulnerability, shaped by polybiome interactions, nutrigenomic variability, and environmental determinants. Chapter 2 then formalizes this polybiome-centric framework, situating the host as a co-regulated, dynamic system in which microbial, genetic, immunologic, metabolic, and nutritional inputs interact in non-linear, context-specific patterns. Chapter 3 applies this lens to colorectal cancer, reframing

Page 14 of 31 carcinogenesis as a failure of systemic coherence and demonstrating that immune dysregulation, metabolic perturbation, and microbial imbalance are not isolated events but interconnected phenomena that define disease emergence and progression.

Building on these foundational insights, Chapter 4 translates the theoretical and pathological understanding into therapeutic strategy. AI-assisted monoclonal antibody design is positioned not merely as a precision intervention but as a method to modulate distributed biological intelligence within the host–microbe–nutrition network. Computational modeling anticipates antibody-binding affinity while integrating polybiome signals, allowing therapy to be dynamically adapted to an individual's biological and microbial context. Nutrigenomic data further refine the treatment, informing dosage, targeting, and intervention timing, thereby operationalizing systems medicine into actionable clinical strategies. In this sense, the chapter presents monoclonal antibodies as tools that intervene in complex signaling networks rather than solely neutralizing pathogens or tumor markers, embodying a shift toward biointelligent, ecologically contextualized therapeutics.

Throughout the editorial development of this chapter, Reyed M. Reyed provided critical scientific oversight and integrative guidance, ensuring that the polybiome systems medicine paradigm remained coherently applied across conceptual, pathological, and therapeutic domains. This contribution focused on enhancing the analytical depth of the chapter, linking fermentobiometrics, metabolomic readouts, and immune modulation pathways to clinically relevant intervention strategies without infringing upon authorship boundaries or misrepresenting primary contributions. By coordinating the conceptual alignment with the earlier chapters, this work reinforces the book's internal logic, demonstrating a seamless progression from theory (Chapter 2) through disease modeling (Chapter 3) to therapeutic translation (Chapter 4), illustrating the book as a coherent, integrative school of thought rather than a series of discrete contributions.

Furthermore, the chapter situates these developments within the future-oriented PSM 2040 vision, emphasizing adaptability, predictive analytics, and sustainability in therapeutic design. It anticipates the convergence of ethical AI, polybiome-informed modeling, and precision nutrition to create interventions that are not only personalized and contextually grounded but also scalable across diverse healthcare settings. By embedding these forward-looking perspectives, the chapter exemplifies how advanced computational and biological integration can reshape clinical practice, informing both immediate patient care and long-term biomedical strategy. In doing so, it underscores the book's overarching goal: to provide a methodological and ethical blueprint for next-generation systems medicine that harmonizes high-dimensional Page 15 of 31 data, translational insights, and actionable therapeutic innovations into a unified framework for precision health.

Chapter 5. Algorithmic Nutrient Engineering via Dynamic Microbial Bioprocesses: Omics Translation for Precision Nutrient Design

This chapter represents a strategic extension of the polybiome-informed systems framework introduced in Chapter 2, applied through the pathological modeling of Chapter 3, and therapeutically translated in Chapter 4. It focuses on precision nutrition as a multidimensional, data-driven intervention, demonstrating how AI-enabled inverse design algorithms can integrate genomic, metabolomic, and microbial profiles to generate individualized nutrient formulations. High-precision microbial bioprocesses are leveraged to systematically design functional nutrients that dynamically interact with host biology, creating adaptive nutritional interventions that surpass conventional dietary guidelines in specificity and effectiveness. The methodology emphasizes adaptive, context-specific nutrient engineering, wherein microbial fermentation outputs are quantitatively mapped to host metabolic and genomic profiles, enabling predictive, responsive, and iterative interventions. AI-driven modeling translates complex multi-omic datasets into actionable strategies, operationalizing the concept that nutrition can be both personalized and dynamically responsive to the polybiome state of the individual. By doing so, nutrient design becomes an integral component of a broader therapeutic continuum, linking the theoretical infrastructure of Chapter 2, the systemic understanding of disease emergence in Chapter 3, and the adaptive therapeutic modeling outlined in Chapter 4.

The chapter highlights translational innovation and methodological precision, demonstrating a coherent trajectory from foundational systems principles to applied nutritional therapeutics. This integrated approach positions personalized nutrition not merely as a supportive adjunct, but as a proactive, clinically relevant modality capable of modulating host–microbe interactions, metabolic pathways, and systemic homeostasis within the polybiome context. By bridging AI analytics, multi-omic intelligence, and bioprocess engineering, the chapter operationalizes a forward-looking vision for precision nutrition as part of holistic health management aligned with emerging global standards and sustainable health strategies.

Throughout the chapter, conceptual leadership and analytical orchestration guide the design, refinement, and integration of complex datasets into actionable outcomes. This leadership ensures that all methodologies, models, and translational insights maintain rigor, coherence, Page 16 of 31 and applicability, reinforcing the book's unified vision as a school of thought that moves from theoretical infrastructure to predictive pathology and adaptive therapeutics. The framework exemplifies a future-ready paradigm for precision nutrition, demonstrating how innovative scientific synthesis can translate multi-layered biological intelligence into sustainable, patientcentered solutions.

The metabolic coordination patterns examined in Chapter 5 invite a necessary reframing of immune dysfunction not as an isolated pathological event, but as a downstream manifestation of disrupted microbial–host regulatory synchrony. This shift in perspective becomes especially relevant in conditions such as food allergies, where immune hypersensitivity may reflect not an overactive defense system, but a misaligned interpretive response emerging from ecological incoherence within the gut microbiome. Accordingly, the transition toward microbiomeinformed immunological modeling, as developed in Chapter 6, represents a critical expansion from metabolic adaptability to immune interpretability within precision health frameworks.

Chapter 6. Microbiome-Driven Food Allergies and Precision Therapeutic Innovations

Chapter 6 examines the complex, multiscale dynamics of the gut microbiome and its direct influence on the development, modulation, and long-term regulation of immune tolerance, with a specific focus on food allergies—an increasingly prevalent and multifactorial global health challenge. It critically analyzes how variations in microbial composition, functional gene expression, metabolite signaling—such as short-chain fatty acids, tryptophan derivatives, and bile acid intermediates—shape mucosal immune education, epithelial barrier function, and systemic homeostasis. These microbial–host interactions are not static; they represent dynamic networks whose perturbations can precipitate hypersensitivity reactions, offering both mechanistic insight and actionable clinical targets. By integrating high-resolution immune profiling with AI-enhanced analytics, the chapter elucidates pathways that are observable, quantifiable, and potentially modifiable through precision interventions.

While Chapter 6 focuses on the microbiome as a subsystem, these observations are interpreted within the Polybiome Systems Medicine framework—a conceptual model that situates biological, computational, and institutional layers as interconnected, emergent subsystems. Within this framework, immune-network instability is viewed not as an isolated biological anomaly but as a reflection of systemic fragility across scales. Patterns of dysregulation at the microbiome–immune interface can inform anticipatory strategies for clinical intervention and Page 17 of 31 reveal potential vulnerabilities in translational and regulatory processes. This perspective reframes allergy not merely as a symptomatic disease but as an indicator of distributed systemic resilience or fragility, linking mechanistic biology with institutional governance.

AI-assisted analytics are employed as interpretive, decision-support tools rather than prescriptive solutions. They allow clinicians and researchers to map probabilistic relationships among microbial taxa, metabolites, and immune networks, identify functional bottlenecks, and design adaptive interventions. Precision strategies—including metabolite-guided immune recalibration, targeted microbiome restoration, and tolerance induction protocols—are conceptualized as interventions to restore coherence across dynamic host–microbial networks. Such interventions move allergy management from reactive treatment toward anticipatory, predictive, and preventative frameworks, laying the groundwork for systems-informed clinical pathways that can evolve alongside technological, environmental, and societal changes through 2040.

The chapter also anticipates future challenges and opportunities in AI-enabled clinical translation. As microbiome-informed interventions increasingly depend on advanced computational infrastructures, issues such as algorithmic transparency, data integrity, governance alignment, ethical stewardship, and equitable access will become intrinsic requirements rather than post hoc considerations. The same distributed vulnerabilities that create immune-network fragility biologically may propagate if institutional and technological layers are not rigorously aligned. Thus, the chapter situates clinical innovation within a broader ethical and operational architecture, emphasizing resilience, adaptability, and sustainability.

By connecting microbiome-level phenomena to systemic and institutional dimensions, this chapter establishes a conceptual bridge to Chapter 9, which explores the governance, ethical, and regulatory frameworks necessary for the safe, responsible, and equitable deployment of AI-enhanced healthcare interventions. This dual framing ensures that biological insights are translated into actionable strategies without compromising systemic coherence or patient safety.

In summary, Chapter 6 provides a critical, forward-looking analysis of microbiome-mediated food allergies, integrating mechanistic, clinical, and systems-level perspectives. By situating microbiome insights within the Polybiome Systems Medicine framework, it not only clarifies current translational pathways but also anticipates the challenges and opportunities for AIdriven, integrative healthcare systems by 2040. The chapter presents a defensible, rigorous, and innovative academic narrative that bridges detailed mechanistic understanding with future- Page 18 of 31 proof strategies for clinical, technological, and institutional integration, emphasizing coherence, resilience, and sustainability across scales.

Chapters 7 and 8. Integrated Preface: AI-Guided Microbial Biodiversity within a Systems Precision-Health Architecture

The accelerating convergence of artificial intelligence and biodiversity science represents a defining inflection point in contemporary biomedical research. Computational sophistication alone, however, does not guarantee epistemological coherence or translational responsibility. As high-resolution sequencing, strain-level analytics, and machine-learning tools increasingly permeate microbiome science, the central challenge lies in architectural integration— structuring data, inference, and application within a disciplined systems framework that is both scientifically rigorous and ethically grounded.

Chapters 7 and 8 are conceived as interdependent components within this broader scientific architecture. Together, they operationalize a structured approach to AI-guided microbial profiling that moves beyond descriptive taxonomy toward analytically grounded, translationally relevant modeling. Their design reflects a deliberate systems-level approach in which microbial biodiversity is treated not merely as ecological variation, but as a computationally navigable and clinically interpretable layer within evolving precision-health ecosystems projected toward 2040.

The conceptual direction guiding these chapters represents a coordinated effort to situate strainlevel microbiological research within an integrative precision-health paradigm. Rather than isolating genomic discovery from translational application, the chapters articulate a continuum: genomic profiling informs functional annotation; functional annotation guides phenotypic validation; phenotypic validation supports translational and regulatory pathways. This progression transforms microbial analytics from static datasets into structured inputs for longterm, evidence-based health strategy.

Chapter 7. AI-Driven Profiling of Egyptian Lactobacillus Strains

Chapter 7 establishes the analytical foundation of this integrated framework by presenting a comprehensive AI-driven methodology for profiling Egyptian Lactobacillus strains in the context of next-generation functional food development. The chapter integrates genomic sequencing, functional annotation pipelines, safety assessment parameters, and regulatory Page 19 of 31 considerations into a reproducible, decision-support evaluation pathway. Rather than presenting strain identification as a mere endpoint, the chapter positions profiling as a structured process linking computational inference with cultural, nutritional, and public-health contexts specific to Egypt.

The methodological design emphasizes reproducibility, transparency, and regulatory awareness. AI-based comparative genomics, predictive functional mapping, and safety screening are embedded within a disciplined analytical structure that avoids speculative extrapolation. Tables, figures, and analytical matrices are meticulously constructed to clarify inference pathways, ensuring interpretive coherence across genomic, metabolic, and safety domains.

Reyed M. Reyed provided conceptual direction, structural organization, and analytical design oversight for the chapter. His leadership ensured methodological rigor, alignment with the overarching systems perspective of the book, and structural clarity. All contributing authors reviewed and approved the final content, ensuring collective scholarly responsibility while accurately representing individual contributions.

Chapter 8. Functional Profiling and Translational Potential of Egyptian Probiotics

Chapter 8 advances the analytical trajectory established in Chapter 7 by transitioning from genomic profiling to applied functional validation and translational modeling. While Chapter 7 constructs the computational and genomic scaffold, Chapter 8 interrogates functional capacity, therapeutic plausibility, metabolomic characterization, and safety evaluation within a structured translational framework.

The chapter integrates laboratory phenotyping, metabolomic profiling, AI-supported inference modeling, and regulatory alignment into a roadmap for culturally contextualized probiotic innovation. It resists premature therapeutic claims and emphasizes evidence layering, linking genomic indicators with experimentally validated traits prior to suggesting health applications. This approach stabilizes methodological standards in AI-guided probiotic research, balancing innovation with scientific restraint, and supports responsible translational decision-making. Analytical coherence across Chapters 7 and 8 reflects a systems-oriented architectural logic. Microbial biodiversity is interpreted through computational analytics, ecological adaptation models, and translational feasibility assessments as interconnected elements rather than isolated investigative silos. This integrative design strengthens the epistemic foundation of AI- Page 20 of 31 assisted microbial science and contributes to resilient, biodiversity-informed precision health strategies.

Reyed M. Reyed's oversight ensured consistency across genomic and functional layers, disciplined integration, and alignment with responsible translational principles. His guidance in drafting analytical tables, figure architecture, and interpretive modeling structures reinforces coherence while maintaining full collaborative authorship integrity

Forward Orientation Toward 2040

Looking toward 2040, the decisive question is not whether microbial biodiversity will shape precision medicine, but how systematically and responsibly that influence will be engineered. Chapters 7 and 8 collectively argue that sustainable progress depends on structured integration—linking ecological intelligence, computational inference, experimental validation, and regulatory accountability within ethically grounded frameworks.

These chapters serve a dual purpose:

1. As a scholarly reference for AI-guided strain-level microbial profiling.
2. As a strategic template for interdisciplinary collaboration across computational biology, microbiome science, food innovation, and translational medicine.

By emphasizing architectural coherence over fragmented discovery, they contribute to the maturation of biodiversity-informed precision health, transforming microbial analytics into structured components of resilient and culturally contextualized health infrastructures.

As biological research advances toward increasingly complex multi-omic integration, responsibility shifts from simply generating high-resolution data to designing interpretive frameworks capable of applying that data with rigor and accountability. These chapters embody that responsibility, offering a disciplined systems-level pathway for future research, innovation, and collaborative advancement in computational biodiversity medicine.

Note on contribution: Reyed M. Reyed provided conceptual direction, analytical guidance, and editorial oversight for Chapters 7 and 8, ensuring structural and methodological coherence throughout, while all contributing authors reviewed and approved the final manuscripts, reinforcing collective scholarly responsibility. This phrasing recognizes the leadership role without conferring contractual authorship rights on the chapters outside the scope of the original agreement.

Chapter 9. Ethical and Regulatory Considerations In AI-Driven Healthcare

Chapter 9 serves as the integrative normative framework that consolidates the scientific, analytical, and translational pathways progressively developed across the volume. Its placement at the conclusion of the book reflects a deliberate structural logic: the preceding chapters collectively establish what AI-enabled, polybiome-informed systems medicine can achieve across diagnostic, therapeutic, nutritional, immunological, and microbial domains, while this final chapter examines the ethical, regulatory, and governance conditions under which such capabilities must responsibly operate.

The intellectual trajectory of the volume begins in Chapter 1 with the NexDi AI system, which reclassifies diabetes not as a fixed metabolic disorder but as a dynamic continuum influenced by nutrigenomic variability, microbial ecology, and environmental context. This redefinition initiates a broader methodological transition toward multi-omic disease modeling. Chapter 2 provides the theoretical systems infrastructure necessary for this transition by formalizing the polybiome paradigm as an integrative analytical lens through which host–microbe–environment interactions can be computationally interpreted.

Building upon this conceptual foundation, Chapter 3 extends the systems framework into oncology by reframing colorectal carcinogenesis as a disruption of systemic biological coherence rather than an isolated mutational cascade. This pathological reinterpretation is subsequently translated into adaptive immunotherapeutic strategy in Chapter 4, where AIguided monoclonal antibody design is informed by polybiomic and nutrigenomic signaling dynamics. Chapter 5 further operationalizes the systems paradigm within the domain of precision nutrition through algorithmic nutrient engineering, demonstrating how microbial bioprocesses and genomic data can inform the inverse design of adaptive dietary interventions. Chapter 6 introduces an additional clinical dimension by examining microbiome-driven food allergies through the lens of host–immune–microbial interaction. Using AI-enhanced analytics, the chapter elucidates mechanistic pathways linking microbial dysbiosis to immune dysregulation and supports a transition from symptomatic allergy management toward preventative, personalized immune modulation strategies grounded in multi-omic profiling.

Chapters 7 and 8 extend this translational continuum into microbial biodiversity science by presenting AI-guided frameworks for strain-level profiling of Egyptian probiotic organisms. Chapter 7 establishes a genomic and computational scaffold for evaluating culturally contextualized Lactobacillus strains, while Chapter 8 advances toward functional validation, metabolomic characterization, and therapeutic plausibility assessment. Together, these Page 22 of 31 chapters transform microbial biodiversity from descriptive ecological data into computationally navigable inputs for scalable, precision-health innovation within food safety and therapeutic domains.

Against this multidimensional backdrop—spanning metabolic disease reclassification, oncological systems modeling, immunotherapy design, algorithmic nutrition, immunemediated allergy prevention, and biodiversity-informed probiotic development—Chapter 9 interrogates the governance architectures required to sustain such integrative innovation. It advances the position that ethical reasoning must be embedded within AI-system design from inception, rather than appended retrospectively as regulatory compliance.

Issues of data stewardship, transparency, algorithmic accountability, bias mitigation, patient autonomy, and equitable access are examined as operational design imperatives that directly influence the clinical legitimacy of AI-guided interventions. As international policy frameworks evolve to address algorithmic healthcare systems, the chapter emphasizes the need for governance models capable of accommodating cross-domain biological data integration, predictive inference, and culturally contextualized implementation strategies.

By explicitly linking technological capability with societal trust, Chapter 9 reframes the question from whether AI-enabled polybiome systems medicine is scientifically feasible to how responsibly it can be institutionalized. Its concluding position underscores that sustainable progress toward 2040 will depend not solely on advances in computational biology or microbial analytics, but on the disciplined alignment of innovation with regulatory accountability and public values.

In doing so, the final chapter reinforces the conceptual unity of the volume. The systems-level thinking introduced in the biological modeling of earlier chapters is extended into governance modeling at the institutional level. Just as host–microbe networks require systemic coherence for physiological stability, AI-driven healthcare ecosystems require ethical and regulatory coherence for long-term societal legitimacy. The chapter therefore functions not only as a conclusion, but as the stabilizing framework through which the book's integrated vision is rendered both scientifically ambitious and operationally responsible.

INTEGRATED NARRATIVE AND FORWARD-LOOKING ROADMAP FOR CHAPTERS 1–9

This volume presents a cohesive, sequential, and conceptually unified framework for understanding, analyzing, and applying systems-informed biomedical knowledge. Each chapter contributes to a deliberate intellectual architecture that situates the polybiome and AIenabled analytics at the center of a holistic precision-health paradigm, with an explicit trajectory from theoretical foundations to clinical translation, technological innovation, and ethical governance. The chapters are not discrete units but interconnected nodes within a broader scientific and strategic continuum, reflecting the vision of 2040 for integrative, sustainable, and ethically responsible systems medicine.

Chapter 1. Advancing Diabetes Reclassification Using the NexDi AI System with Multi-Omic Insights

Chapter 1 establishes the paradigm-shifting case study that frames the book's methodology. By integrating polybiome cues, nutrigenomic data, and environmental determinants, it reconceptualizes diabetes as a dynamic continuum rather than rigid categories defined solely by glucose metrics. This chapter lays the foundational logic for the system-wide analyses in subsequent chapters, demonstrating how AI can synthesize complex multi-omic data to support personalized interventions in real-world clinical contexts. Ethical principles, data fairness, and patient-centered design are embedded as integral components, setting the tone for the volume's methodological and ethical rigor.

Chapter 2. Polybiome Systems Foundation

Building on the clinical foundation of Chapter 1, Chapter 2 formalizes the Polybiome Systems Framework (PSM) as an integrative platform that links microbial, metabolic, immunological, and genomic networks. It introduces analytical and computational infrastructures that enable mechanistic modeling, predictive simulation, and cross-layer integration. By establishing the core systems lens, this chapter provides a unifying conceptual architecture that supports both disease modeling (Chapter 3) and therapeutic translation (Chapters 4–5).

Chapter 3. Biointelligent Reframing of Colorectal Cancer

Chapter 3 operationalizes the polybiome lens within oncology, presenting cancer as a disruption of systemic coherence rather than a simple accumulation of somatic mutations. By integrating immune, metabolic, and microbial dysregulation, it demonstrates a systems-level understanding of carcinogenesis. The chapter introduces cognitive oncology and circular medicine, framing therapeutic strategies that restore systemic balance rather than targeting isolated tumors, thereby providing a conceptual bridge to adaptive immunotherapy and precision nutrient design in Chapters 4 and 5.

Chapter 4. AI-Guided Monoclonal Antibody Design

This chapter translates the theoretical and pathological insights from Chapters 2 and 3 into therapeutic applications, demonstrating how AI can design monoclonal antibodies that modulate dysregulated microbial-metabolic signaling circuits. The chapter positions antibodies not merely as neutralizing agents but as adaptive modulators of distributed biological intelligence, exemplifying the therapeutic arm of the biointelligent systems paradigm and illustrating how multi-omic data can be operationalized for individualized immunotherapy.

Chapter 5. Algorithmic Nutrient Engineering via Dynamic Microbial Bioprocesses

Chapter 5 extends the therapeutic paradigm into precision nutrition, integrating AI-driven inverse design with polybiome and nutrigenomic analytics. Through high-precision microbial bioprocesses, it develops customized functional nutrients tailored to individual metabolic and microbial profiles. This chapter links the mechanistic understanding of disease emergence (Chapter 3) with adaptive dietary interventions, highlighting the convergence of computational modeling, experimental biology, and clinical applicability. Together, Chapters 4 and 5 illustrate the full translational spectrum from systems-level pathology to actionable therapeutic and nutritional interventions.

Chapter 6. Microbiome-Driven Food Allergies and Precision Therapeutic Innovations

Chapter 6 bridges the therapeutic applications of Chapter 5 with immune-mediated conditions, demonstrating how AI-enhanced analytics reveal mechanistic pathways of allergy formation. By situating microbiome dysbiosis at the center of immune regulation, it supports personalized immune modulation plans and transitions allergy management from reactive treatment to Page 25 of 31 preventative, predictive care. This chapter emphasizes multi-omic integration and provides a clear example of how systems medicine can reshape emerging public health challenges.

Chapters 7 and 8. AI-Guided Microbial Biodiversity within a Systems PrecisionHealth Architecture

Chapters 7 and 8 constitute the operational core for microbial profiling and translational application. Chapter 7 develops a comprehensive AI-guided framework for profiling Egyptian Lactobacillus strains, while Chapter 8 applies these insights to functional validation, therapeutic potential, and scalable probiotic interventions. Together, they exemplify interconnected data architecture, embedding ecological intelligence, computational modeling, and translational feasibility within a single continuum. These chapters illustrate how strainlevel microbial analytics feed into evidence-based functional foods, clinical interventions, and regulatory pathways, reinforcing the book's systems-level design.

Chapter 9. Ethical and Regulatory Considerations in AI-Driven Healthcare

As the final chapter, Chapter 9 functions as the capstone of the volume, integrating the lessons, methodologies, and innovations of all preceding chapters. It situates the entire polybiomeinformed, AI-driven systems medicine initiative within a framework of ethical oversight, governance, and societal accountability. By addressing data stewardship, transparency, bias mitigation, patient autonomy, and compliance with international standards, this chapter ensures that the technical and therapeutic innovations described in Chapters 1–8 remain aligned with ethical imperatives and sustainable practice. Its placement as the concluding chapter underscores the central premise that responsible integration of AI and systems medicine requires ethical foresight, making it both a reflective and actionable endpoint for the book.

Cohesive Vision Toward 2040

Together, Chapters 1–9 form a sequential, integrated narrative:

- Chapters 1–2 establish foundational theory and infrastructure,
- Chapters 3–6 demonstrate pathological, therapeutic, and nutritional applications,
- Chapters 7–8 operationalize microbial profiling and translational frameworks, and
- Chapter 9 consolidates these efforts under ethical and governance oversight.

This design ensures that the book is not a collection of independent chapters but a coherent "school of thought" in systems medicine, where AI, polybiome science, multi-omic analytics, Page 26 of 31 and translational innovation converge to define precision healthcare as a holistic, ethical, and future-ready discipline. By interweaving scientific, computational, therapeutic, and ethical dimensions, the volume establishes a forward-looking roadmap for 2040, guiding research, clinical practice, policy development, and educational frameworks in multi-disciplinary, integrative biomedical science.

INTENDED AUDIENCE

The book will appeal to a multidisciplinary audience, including biomedical researchers, clinicians, bioinformaticians, nutrition scientists, policymakers, and health governance stakeholders. It offers conceptual richness and practical models that support cross-disciplinary insight and cooperation, which are crucial to the development of systems medicine.

Scholarly Contribution and Future Outlook

The scientific value of this book is conceptually deep and methodologically radical, covering the gaps in biomedical research and professional practice that have long existed.

Firstly, it solidly lays down the polybiome as one of the organizing principles of systems medicine. The polybiome model incorporates a broad range of biological fields (genetics, immunology, metabolism, nutrition, and environmental exposures) into a dynamically coregulated system by extending far beyond the microbiome focus. This holistic view conceptualizes health and illness as an emergent process that exists based on the orderliness or disorderliness of this interrelated biological system. This reconceptualization facilitates better modeling of the heterogeneity of diseases, variability across people, and time, thus leading to personalized healthcare.

Second, the volume transforms the role of artificial intelligence from a more peripheral and analytic supplement to a more integrative methodology. AI enables the generation of highly multi-omic data into coherent and interpretable models that can be directly applied to clinical decision-making and customized intervention development. The volume encourages responsible AI innovation by focusing on such crucial aspects as reproducibility, model transparency, and mitigation of bias in line with changing international ethical and regulatory guidelines. This combination of methodologies plays a central role in addressing the disintegration and interpretive dilemmas that have traditionally hindered the translation of precision medicine into scalable, equitable healthcare solutions.

These contributions, in combination, address systemic barriers to progress, including biological complexity, cultural diversity, and moral urgency, which are needed to achieve sustainable and inclusive innovation. This book therefore offers a rigorous blueprint to inform future Page 27 of 31 biomedical research, clinical practice, and health policy with the further growing volume, diversity, and complexity of biomedical data.

TOWARD 2040: FUTURE PERSPECTIVES ON INTELLIGENT SYSTEMS MEDICINE

While the strategic implementation of these innovations aligns with immediate national priorities such as Egypt's Vision 2030, the foundational frameworks established in this volume are designed to catalyze a long-term global transition toward the 2040 horizon.

As we look toward 2040, the trajectory of systems medicine is poised to undergo a profound transformation, driven by the convergence of AI-enabled analytics, polybiome science, and integrative multi-omic data. The forthcoming era will demand more than incremental improvements in diagnostics and therapeutics—it will require a redefinition of health and disease as emergent, context-dependent phenomena, interpreted through the lens of systemic coherence across molecular, microbial, metabolic, immunological, and environmental networks.

AI-enhanced polybiome systems will enable precision interventions that are not only individualized but contextually adaptive, accounting for genetic background, microbial composition, environmental exposures, diet, and socio-cultural determinants of health. By 2040, such systems are expected to close the critical gap between data generation and actionable insight, allowing clinicians, researchers, and policymakers to navigate the increasing complexity of biomedical information with unprecedented clarity. This integrative approach will foster predictive, preventive, personalized, and participatory medicine, moving beyond reductionist paradigms to a holistic understanding of human biology.

Moreover, the maturation of intelligent systems medicine will depend on innovative governance frameworks that embed ethical responsibility, equity, and transparency at every stage. Active stewardship will ensure equitable access to advanced diagnostics and therapeutics, promote societal trust, and safeguard against bias or misuse of sensitive biomedical and omics data. By proactively designing governance models alongside technical solutions, the field can prevent the widening of global health disparities and ensure that precision health strategies are both scientifically rigorous and socially responsible.

From a research perspective, the 2040 vision addresses critical gaps in current knowledge, particularly the integration of microbial ecology, host physiology, and environmental systems into actionable clinical frameworks. Systems medicine will leverage multi-omic intelligence and AI-driven predictive modeling to identify emergent disease patterns, reveal mechanistic interactions previously obscured by siloed approaches, and generate interdisciplinary Page 28 of 31 therapeutic innovations that are both robust and scalable. Functional foods, microbiometargeted therapeutics, and adaptive nutritional interventions will increasingly be designed using algorithmic frameworks that translate complex omics data into clinically relevant strategies, ensuring interventions are dynamic, precise, and culturally contextualized.

Importantly, this future perspective emphasizes a proactive, innovation-driven approach: rather than reacting to diseases as they occur, healthcare will anticipate systemic dysregulation, leveraging predictive modeling and integrative analytics to prevent pathology before it manifests. AI-assisted simulations of host–microbiome–environment interactions will guide intervention design, policy formulation, and public health strategies, fostering a global health ecosystem that is resilient, adaptive, and ethically grounded.

In this context, the book serves as a foundational reference for 2040-oriented systems medicine, offering both conceptual scaffolding and practical frameworks for integrating polybiome insights with AI-driven analytics. It provides scholars, clinicians, and policymakers with a methodologically rigorous, ethically informed, and future-ready roadmap to navigate the complexities of human health, address research gaps, and design innovative interventions. By combining analytical depth with translational foresight, the volume positions intelligent systems medicine not merely as a technological endeavor but as a holistic paradigm shift, setting the stage for sustainable, equitable, and actionable precision healthcare worldwide.

The 2040 vision outlined here is both aspirational and operational: it articulates a path toward integrative, adaptive, and ethically responsible medicine, ensuring that the next generation of healthcare innovation is informed by robust science, cultural awareness, and societal accountability, closing existing gaps and opening unprecedented avenues for research, clinical practice, and policy.

INTEGRATIVE CONCLUSION: TOWARD A SYSTEMS MEDICINE FUTURE

From the pioneering reconceptualization of diabetes classification through the NexDi AI system to the formal development and operationalization of the polybiome framework, this volume consistently emphasizes three interdependent pillars: systemic coherence, methodological rigor, and ethical accountability. These pillars collectively underpin a transformative vision for biomedical science, one that transcends fragmented, reductionist approaches and embraces the dynamic, multi-layered complexity of human biology within its full environmental, microbial, nutritional, and cultural contexts.

Across the chapters, the volume systematically demonstrates how polybiome-informed systems medicine can integrate genomics, metabolomics, microbiome profiling, and clinical Page 29 of 31 phenotyping into cohesive analytical and translational frameworks. The work highlights the convergence of AI-driven analytics and mechanistic modeling, showing how sophisticated computational tools can generate actionable insights while maintaining stringent standards of reproducibility, interpretive clarity, and regulatory compliance. In doing so, it provides a concrete blueprint for translating multi-omic complexity into practical, individualized, and population-aware health strategies.

Crucially, the volume situates ethical responsibility at the core of innovation. By embedding data governance, transparency, bias mitigation, and patient-centered principles into each stage of analysis and application, it addresses one of the most pressing challenges of contemporary biomedical research: ensuring that technological sophistication aligns with societal trust and equitable healthcare delivery. This integration reflects a deliberate move beyond technical problem-solving toward a principled, accountable framework for the responsible deployment of AI-enhanced systems medicine.

Moreover, the work establishes a future-oriented, 2040-ready vision of healthcare. It provides scholars, clinicians, and policymakers with tools and conceptual scaffolds capable of addressing rising data complexity, predicting emergent pathological states, and designing interventions that are both personalized and contextually adaptive. By demonstrating the interplay between computational intelligence, experimental validation, and translational innovation, the volume serves as a strategic guide for interdisciplinary collaboration, enabling the development of therapies, functional foods, and public health strategies that are resilient, culturally contextualized, and scientifically sound.

The integrative conclusion of the preface emphasizes that this book is more than a compendium of chapters: it represents a methodologically coherent, ethically grounded school of thought. It fosters a paradigm shift in which health and disease are understood as emergent phenomena shaped by complex regulatory networks, rather than isolated or purely symptomatic events. Through this lens, systems medicine is positioned as a forward-looking, adaptive discipline that equips the next generation of researchers, practitioners, and policymakers to anticipate challenges, innovate responsibly, and advance healthcare in ways that are holistic, sustainable, and equitable.

In essence, the volume provides a landmark synthesis: it bridges theory, computational modeling, clinical application, and ethical governance into a cohesive narrative, offering a rigorous, actionable, and visionary framework for the future of precision health. It establishes the intellectual and operational foundation for a systems-informed, AI-empowered, and ethically accountable approach to medicine, ensuring that advances in knowledge are matched Page 30 of 31 by social responsibility, translational feasibility, and a global commitment to inclusive healthcare.

Reyed M. Reyed

Genetic Engineering and Biotechnology Research Institute, Egypt & City of Scientific Research and Technological Applications (SRTA-City), Egypt

Acknowledgment

This volume emerges from a sustained intellectual commitment to rethinking human health beyond reductionist and compartmentalized paradigms, toward integrative, systems-oriented models capable of capturing biological, nutritional, technological, and contextual complexity. *Advancing Personalized Medicine with AI-Driven Microbiome and Nutrigenomics: Innovations and Future Perspectives* was conceived as a coherent scholarly framework rather than a thematic aggregation, articulating a forward-looking vision for precision healthcare grounded in *polybiome-driven systems medicine*, enabled by *artificial intelligence* and *integrative nutritional sciences.*

I extend my sincere appreciation to all individuals whose direct engagement contributed to the academic rigor, critical refinement, and scholarly integrity of this work. The development of this volume reflects the collective discipline required to *advance knowledge within complex, multi-layered biological systems, where host biology, nutrition, microbial ecology, and computational intelligence converge.*

Foremost, I wish to acknowledge with deep respect the expert reviewers whose careful evaluations, methodological scrutiny, and constructive critiques were essential to strengthening the scientific and conceptual foundation of this book. Their assessments extended beyond technical review toward safeguarding coherence, epistemic clarity, and intellectual responsibility within an emerging *polybiome-centered framework.* Such rigorous peer engagement remains indispensable to advancing credible and ethically grounded systems medicine.

The reviewers, listed alphabetically, are:

- *Ashraf Tabll* Egypt Centre for Research and Regenerative Medicine (ECRRM), Cairo, Egypt; and Microbial Biotechnology Department, Biotechnology Research Institute, National Research Centre (NRC), Dokki, Giza, Egypt. Email: aa.tabll@nrc.sci.eg | ORCID: https://orcid.org/0000-0001-7041-5445

- *Hina Bansal* Amity Institute of Biotechnology, India. Email: hinabansal@gmail.com | ORCID: https://orcid.org/0000-0003-1683-1581
- *Mona A. Esawy* Pharmaceutical and Drug Industries Research Institute, National Research Center (NRC), Egypt. Email: mona_esawy@hotmail.com | ORCID: https://orcid.org/0000-0001-5306-4946
- *Sherein Ismail Abd El-Moez* Microbiology and Immunology Department, Veterinary Research Institute, National Research Center (NRC), Egypt. Email: si.abdelmoez@nrc.sci.eg | ORCID: https://orcid.org/0000-0003-4413-7589
- *Shweta Thusoo* Dr. D.Y. Patil School of Allied Health Sciences, Dr. D.Y. Patil Vidyapeeth, Pimpri, Pune, Maharashtra, India. Email: shweta.thusoo@gmail.com | ORCID: https://orcid.org/0000-0002-0316-3193

Professional acknowledgment is also extended to *Pranav Kumar Prabhakar* (Department of Biotechnology, School of Engineering and Technology, Nagaland University, India) and *A. K. Haghi* (Honorary Research Associate, Chemistry Centre, University of Coimbra, Portugal) for their association within the broader academic and publication framework supporting this volume.

I gratefully acknowledge the institutional environment provided by the *Genetic Engineering and Biotechnology Research Institute (GEBRI), City of Scientific Research and Technological Applications (SRTA-City)*, Borg El Arab New City, Alexandria, Egypt. The institute's research culture and interdisciplinary orientation offered a stable foundation for sustained inquiry and conceptual integration throughout the development of this work.

Finally, I recognize the essential yet often understated role of family members, colleagues, and institutional staff whose patience, encouragement, and trust enabled the continuity and focus required for long-term scholarly endeavor. Their support represents the human dimension underlying every rigorous academic contribution.

This volume embodies a distinctive and pioneering scholarly vision, thoughtfully developed under my conceptual guidance and editorial leadership. It represents a sustained intellectual commitment to advancing the frontiers of personalized medicine through an integrative, polybiome-driven systems medicine paradigm. This framework harmoniously unites host biology, nutritional genomics, microbial ecology, and artificial intelligence to offer novel insights and practical methodologies that address the complex, multi-layered nature of human health.

However, the realization of this ambitious academic endeavor would not have been possible without the invaluable contributions of a diverse group of expert authors, diligent reviewers, and collaborative scholars. Their deep expertise, rigorous analytical input, and constructive critiques have been indispensable in refining the conceptual clarity, methodological rigor, and scientific coherence that underpin this volume. Each chapter benefits from their dedication to academic excellence and ethical scholarship, collectively elevating the quality and impact of the work.

I deeply appreciate the open and collaborative spirit of all contributors, whose insights and intellectual generosity helped shape a cohesive and future-oriented framework. Their involvement exemplifies the true essence of scholarly partnership, where shared knowledge and critical dialogue drive innovation and responsible progress in biomedical research.

In sum, this volume stands as a testament to the power of collective effort guided by clear leadership, uniting diverse perspectives into a coherent and transformative approach. It is my sincere hope that this work will serve as a foundational reference and catalyst for ongoing research, clinical practice, and policy development in personalized medicine and beyond.

ACKNOWLEDGMENTS TO IGI GLOBAL AND PRODUCTION TEAM

I extend my deepest gratitude to *IGI Global Scientific Publishing* for providing an exceptional platform that fosters innovation, rigor, and global dissemination of cutting-edge academic research. Their longstanding commitment—spanning over 35 years—to excellence in scholarly publishing has created an invaluable environment for authors and editors to share impactful knowledge.

My special thanks go to Ms. Mikaela Felty, Director of Acquisitions, whose guidance, openness, and collaborative spirit were essential in facilitating the review process and finalizing both the manuscript and the book cover design.

I am also grateful to DJ Kreiss, Assistant Development Editor, and Ms. Rebekah Rocha, Assistant Production Manager – Journals, whose diligent follow-up, attention to detail, and communication ensured a seamless production process and timely publication.

I extend appreciation to Ms. Elizabeth Barrantes, Managing Editor of Book Development, and Ms. Melissa Wagner, Vice President of Publishing Ethics & Research Integrity, for their guidance and oversight in maintaining publishing integrity.

The entire IGI Global team's expertise and responsiveness have been fundamental in navigating modern academic publishing, making this volume accessible to an international audience and advancing personalized medicine and AI-driven sciences.

It is through such dedicated partnerships that academic works can realize their fullest potential, and for this, I am deeply appreciative.

This volume was developed under the conceptual leadership of Prof. Reyed M. Reyed. The integrative Polybiome Framework articulated in this book provided the thematic foundation that shaped the interdisciplinary dialogue across chapters. Individual chapter authors retain full responsibility for the scientific content and conclusions presented in their respective contributions. The editor's role was to ensure conceptual coherence, methodological alignment, and translational integration across the volume.

Introduction

THE POLYBIOME FRAMEWORK: SYSTEMS INTELLIGENCE, ETHICAL AI, AND PRECISION MEDICINE

Biomedical research is undergoing a structural transformation. The classical view of disease as a localized molecular defect is no longer sufficient to explain the complexity observed in chronic disorders, immune dysregulation, metabolic instability, or cancer progression. Health itself can no longer be reduced to the absence of symptoms. It is increasingly understood as a dynamic state emerging from continuous interactions across multiple biological and environmental layers. Genetics, epigenetics, microbial ecosystems, nutrition, metabolism, immune regulation, environmental exposures, and digital health infrastructures operate as interconnected components of a living system.

Within this expanded understanding, precision medicine must evolve. It cannot remain confined to pathway-specific targeting or isolated biomarker detection. Instead, it must aim to restore systemic coherence, reinforce adaptive capacity, and preserve biological resilience. This requires a conceptual and analytical shift from fragmentation to integration.

This volume, *Advancing Personalized Medicine with AI-Driven Microbiome and Nutrigenomics: Innovations and Future Perspectives*, is situated within this evolving paradigm. It introduces Polybiome-Driven Systems Medicine as a structured framework designed to interpret complexity rather than simplify it prematurely. By integrating microbiome science, nutrigenomics, multi-omics analytics, and artificial intelligence, the book advances a coherent architecture for translating biological data into clinically meaningful strategy. The objective is not merely descriptive synthesis. It is methodological clarity, analytical rigor, and translational impact.

From Big Data Expansion to Conceptual Coherence

The rapid expansion of high-throughput technologies has generated vast biomedical datasets. Genomics, transcriptomics, proteomics, metabolomics, exposomics, clinical imaging, wearable sensors, and electronic health records collectively produce volumes of information unprecedented in medical history. Yet, despite this abundance, clinical outcomes have not improved at a proportional rate.

The difficulty lies not only in computational scale. It lies in fragmentation. Data sources operate under different technical standards. Terminologies vary across disciplines. Analytical pipelines often remain isolated within specialized silos. As a result, integration becomes technically possible but conceptually unstable. The challenge is therefore structural: how can heterogeneous biological signals be organized into interpretable, reproducible, and clinically relevant models?

This tension is frequently described as the translational gap in Big Data medicine. Without shared conceptual architecture, data remain descriptive rather than actionable. Biological complexity becomes cataloged rather than understood.

Artificial intelligence emerges in this context as a necessary interpretive infrastructure. Machine learning models can process nonlinear interactions and high-dimensional datasets. Deep neural networks can identify patterns not immediately visible through conventional statistical methods. Explainable AI approaches attempt to preserve interpretability while enhancing predictive power.

However, artificial intelligence alone cannot resolve conceptual ambiguity. Algorithms require structured ontologies, defined variables, and biologically meaningful categories. Without a coherent language, computational outputs risk remaining abstract correlations. Therefore, the integration of AI must be grounded in clearly articulated biological frameworks.

Conceptual Scope of the Polybiome

The Polybiome is presented in this volume as a composite biological ecosystem governing individual health trajectories. It extends the traditional microbiome concept by situating microbial communities within a broader systemic matrix. The framework integrates:

- Host genomic and epigenomic architecture
- Microbial communities across anatomical sites and biological kingdoms
- Nutritional inputs and metabolic flux networks
- Environmental and chemical exposures
- Dynamic host–microbe–environment feedback loops

Health, within this model, depends on coordinated interaction across these layers. Stability emerges when regulatory processes maintain balance. Disruption occurs when feedback mechanisms weaken or when cumulative stress exceeds adaptive capacity.

Chronic disease is therefore interpreted not as a single defect but as a network-level disturbance. A perturbation in one biological domain can propagate through metabolic, immune, or microbial pathways. Resilience diminishes when compensatory systems fail. This systemic interpretation allows prevention and intervention strategies to focus on restoring coordination rather than suppressing isolated symptoms.

The Role of Artificial Intelligence in Systems Interpretation

Biological systems generate dense, overlapping, and time-sensitive signals. Nutritional fluctuations influence microbial metabolites. Environmental exposures modulate immune thresholds. Epigenetic markers respond to both diet and stress. These interactions occur simultaneously and often nonlinearly.

Traditional linear models are limited in capturing this dynamic complexity. Within the Polybiome framework, artificial intelligence serves as an analytical engine capable of integrating multi-layered inputs into predictive and adaptive models.

Machine learning facilitates stratification and risk modeling. Deep learning enables recognition of complex interaction patterns. Explainable AI contributes transparency, allowing clinicians and researchers to understand model rationale rather than relying on opaque predictions.

This integration supports a shift in clinical philosophy. Instead of reacting to established disease, medicine can move toward anticipation, risk mitigation, and personalized modulation of biological systems. AI becomes a tool for enhancing human decision-making rather than replacing it.

METHODOLOGICAL ORIENTATION

The chapters within this volume are unified by shared methodological commitments:

- **Multi-Omics Integration:** Coordinated analysis across genomics, transcriptomics, proteomics, metabolomics, and exposomics.
- **Structured AI Analytics:** Development of interpretable predictive models grounded in biological plausibility.
- **Translational Validation:** Alignment between experimental findings, computational outputs, and clinical feasibility.

- **Systems Modeling:** Use of digital twins and simulation frameworks to test intervention scenarios before implementation.

These principles are applied across diverse domains, including diabetes reclassification, colorectal cancer reframing, monoclonal antibody design, probiotic development, functional food innovation, and disease stratification. Each chapter demonstrates how systems thinking can be operationalized rather than merely theorized.

Intellectual Architecture and Terminological Precision

To prevent conceptual ambiguity, this volume introduces a defined lexicon supporting Polybiome-Driven Systems Medicine. These terminologies were developed specifically to articulate relationships between biological integration, computational analytics, and translational application within the scope of this work.

They are not presented as universal replacements for established biomedical language. Nor are they imposed uniformly across all chapters. Instead, they function as analytical instruments. Their purpose is to clarify systemic relationships, enhance methodological reproducibility, and provide precision where interdisciplinary ambiguity might otherwise arise.

By explicitly documenting these conceptual structures, the volume establishes intellectual attribution while reinforcing scientific accountability. Terminological clarity supports not only academic recognition but also reproducible application in future research.

Ethical Governance and Regulatory Alignment

The integration of biological data, environmental profiling, and computational modeling introduces significant ethical responsibilities. Genomic sequences, microbiome signatures, and health records are deeply personal forms of information. Their misuse can lead to privacy violations, inequitable access, or algorithmic discrimination.

This volume treats ethical governance as foundational rather than supplementary. Core considerations include:

- Data privacy and secure infrastructure
- Algorithmic transparency and bias mitigation
- Reproducibility and auditability
- Equitable benefit-sharing

Bio-Sovereignty is articulated as a guiding principle. It emphasizes responsible stewardship of biological resources, traceability of data use, and alignment with international governance frameworks such as FAIR data principles and global regulatory standards. Egyptian case studies illustrate localized implementation while maintaining international scientific consistency.

CONTRIBUTORS AND ROLE CLARIFICATION

The development of this volume reflects structured collaboration:

- **Editor:** Coordinated methodological coherence, ensured regulatory alignment, and supervised translational direction.
- **Lead Authors:** Developed chapter analyses and synthesized domain-specific evidence.
- **Conceptual Architects:** Formulated core frameworks and terminological precision.
- **Technical Contributors:** Implemented multi-omics integration and AI modeling pipelines.
- **Ethics Consultants:** Ensured adherence to governance standards and bio-sovereignty principles.

This delineation of roles promotes transparency, accountability, and interdisciplinary rigor.

Intended Audience and Practical Contribution

This book is written for advanced researchers, clinicians, computational biologists, and policy-oriented professionals working at the intersection of systems biology and artificial intelligence. It assumes foundational knowledge of omics technologies and machine learning methodologies.

Its contribution is practical as much as conceptual. Researchers will find structured models for integrating heterogeneous data. Clinicians will encounter frameworks for stratification and personalized intervention. Computational scientists will engage with biologically grounded modeling targets. Policy professionals will observe how governance principles can be embedded into innovation pathways.

The central contribution of this work lies in translating systems-level thinking into operational precision medicine strategy. By integrating polybiome science, computational intelligence, and ethical governance, this volume offers a cohesive and forward-looking paradigm for the future of personalized healthcare.

TRANSFORMATIVE APPLICATIONS ACROSS THE POLYBIOME–AI FRONTIER

The conceptual and methodological foundations outlined in this introduction are not intended to remain abstract. They are translated, tested, and refined across the chapters of this volume. Each contribution applies the Polybiome–AI framework within a distinct biomedical context, demonstrating how systems thinking, multiomics integration, and computational intelligence can converge into measurable translational outcomes.

The chapters collectively move from disease reclassification and mechanistic reframing to therapeutic engineering, functional nutrition design, and ethical governance. This progression reflects the central thesis of the book: precision medicine must integrate biological systems coherence, analytical depth, and regulatory responsibility within a unified architecture.

The structure of the volume unfolds as follows:

1. Advancing Diabetes Reclassification Using the NexDi AI System With Multi-Omic Insights: Egypt's Living Laboratory
2. This chapter reconsiders diabetes beyond conventional typologies. By integrating multi-omic datasets through the NexDi AI system, it proposes a stratified model grounded in biological heterogeneity and population-specific context.
3. Decoding the Human Polybiome Through AI and Nutrigenomics in Next Generation Personalized Health: Integrating Multi-Omics for Precision Medicine Here, the polybiome is analyzed as a dynamic regulatory ecosystem. Artificial intelligence supports interpretation of nutritional–genomic interactions, enabling more refined personalization strategies.
4. Biointelligent Reframing of Colorectal Cancer: From Somatic Mutation to Systemic Coherence This chapter challenges mutation-centered paradigms by situating colorectal cancer within broader systemic imbalance, emphasizing networklevel disruption rather than isolated oncogenic events.
5. AI-Guided Monoclonal Antibody Design Integrating Polybiome and Nutrigenomic Insights Therapeutic development is approached through computational modeling that incorporates host–microbe–nutrition interactions, demonstrating how biologically informed AI can guide immunotherapeutic innovation.
6. Algorithmic Nutrient Engineering via Dynamic Microbial Bioprocesses: Omics Translation for Precision Nutrient Design Nutrient formulation is reframed as a systems engineering challenge. Multi-omics data inform adaptive nutrient architectures aligned with individual metabolic states.

7. Microbiome-Driven Food Allergies and Precision Therapeutic Innovations Food allergy is examined as a dysregulated ecological interface. Systems-level interpretation informs targeted, microbiome-aware therapeutic approaches.
8. Using AI to Profile Egyptian Lactobacillus for Next Generation Functional Foods Indigenous microbial resources are analyzed through AI-assisted profiling, linking regional biodiversity to functional food innovation.
9. Functional Profiling and Therapeutic Potential of Egyptian Probiotics: Food Safety and Viral Defense Applications This chapter extends probiotic research beyond digestion, examining immune modulation, safety architecture, and antiviral resilience within a systems framework.
10. Ethical and Regulatory Considerations in AI-Driven Healthcare The final chapter anchors innovation within governance. It addresses algorithmic accountability, data sovereignty, and regulatory harmonization to ensure responsible implementation of AI-integrated medicine.

Together, these chapters do not function as isolated scientific reports. They form a structured continuum. Foundational systems theory informs analytical modeling. Analytical modeling informs therapeutic design. Therapeutic innovation remains aligned with ethical governance. The result is a cohesive and future-oriented framework for precision medicine—one that integrates polybiome science, artificial intelligence, and regulatory stewardship into a coherent translational paradigm.

Reyed M. Reyed

Genetic Engineering and Biotechnology Research Institute, Egypt & City of Scientific Research and Technological Applications, Egypt

Chapter 1

Advancing Diabetes Reclassification Using the NexDi AI System With Multi-Omic Insights: Egypt's Living Laboratory

Reyed M. Reyed
http://orcid.org/0000-0001-5027-3478
Genetic Engineering and Biotechnology Research Institute, Egypt & City of Scientific Research and Technological Applications, Egypt

ABSTRACT

NexDi is introduced as an advanced framework that reconceptualizes diabetes as a fluid continuum of β-cell vulnerability shaped by genetic, nutritional, and environmental pressures. Type 1 diabetes is a state where autoimmune activity increasingly overlaps with insulin resistance pathways, whereas type 5 diabetes emerges from maternal anemia, hidden hunger, and early-life undernutrition. This spectrum reflects the dual burden observed in Egypt, where rising childhood diabetes coexists with persistent nutritional deficits. Environmental stressors, including exposure to endocrine-disrupting agents and microplastic, further disturb metabolic and polybiome stability. By integrating nutrigenomics, AI-enabled prediction, and polybiome analytics, NexDi provides a scalable and culturally grounded pathway toward predictive, preventive, and personalized diabetes care.

DOI: 10.4018/979-8-3373-3121-8.ch001

1. INTRODUCTION: REDEFINING THE DIABETES SPECTRUM THROUGH THE NEXDI FRAMEWORK

1.1. The Expanding Landscape of Precision Medicine: Rethinking Host-Environment Complexity

Figure 1. Graphical Abstract

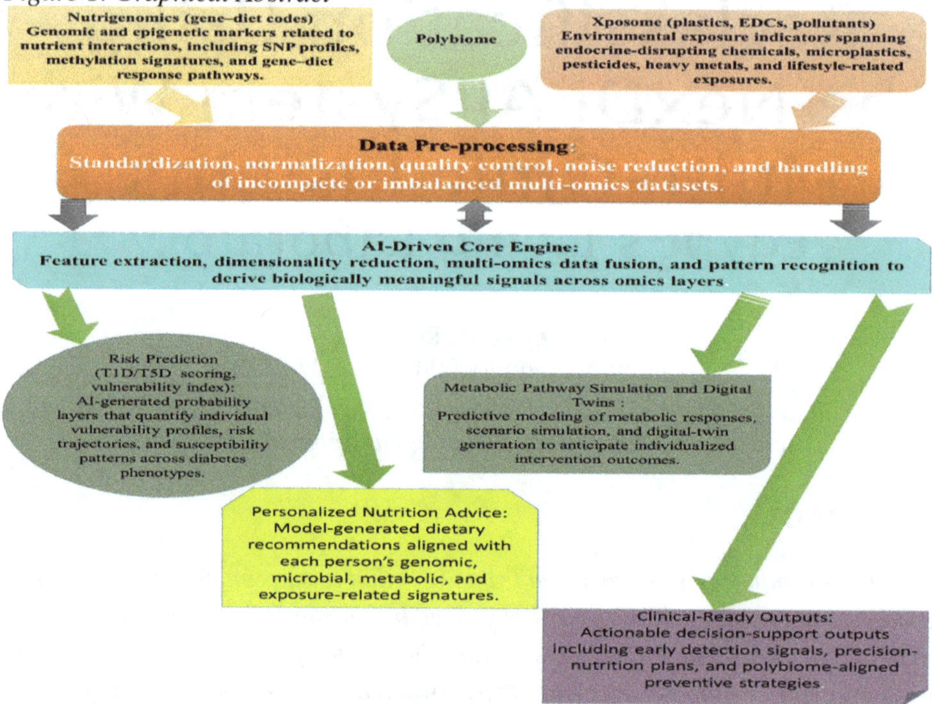

Graphical Abstract : NexDi Framework Overview

A graphical scheme diagram that illustrates the fundamental architecture of the NexDi Framework, in which nutrigenomics, multi-kingdom polybiome science, environmental exposures, and AI-based multi-omics inference are integrated into a single predictive model throughout the T1D-T5D spectrum.

The graphical abstract presents the central architecture of the NexDi Framework in a schematic form using the integration of nutrigenomic signals, multi-kingdom polybiome dynamics, early-life nutritional determinants, and environmental exposures through AI-based multi-omics analytics. The figure shows that all these data streams can intersect to produce personal risk maps and personalized nutrition paths, which redefine diabetes as a predictive, preventive continuum in the Egyptian setting (Author-generated figure).

Precision medicine has developed at an extraordinary rate in the last ten years, radically changing our understanding of the multifaceted interactions among our host biology, resident microbial communities, environmental exposures, and nutritional status. This crossroad is the most complex, especially in chronic metabolic diseases, in which immunological, epigenetic, genomic, and metabolic processes are colocalized and function nonlinearly. Conventional microbiome-based approaches, in which the primary technique is taxonomic profiling personal space, only shed light on a small portion of this complexity. The small scale of their coverage, which often overlooks fungi, viruses, archaea, protists, and unculturals, has limited our knowledge about our system-wide interactions of microbes (Reyed, 2023a; Reyed & Prabhakar, 2025a).

These restrictions are further compounded by expedited dietary changes, the advent of population-specific genomic vulnerability, the spread of endocrine disrupting chemicals (EDCs), and increased metabolic stress (El-Degwi et al., 2024; Bora et al., 2024). This complex ecosystem requires an analytical system that can unlock multiple kingdom interactions and cross-functional cross-talk as well as how microbial and environmental stresses affect host vulnerability, resilience and disease manifestation (Ahn & Jeung, 2023). To address these conceptual and methodological limitations, a broader ecological framework becomes essential—one that captures the full spectrum of biological and environmental determinants shaping metabolic health.

1.2. The Creation of the Polybiome: A Foundational Multi-Kingdom Ecosystem Framework.

To mitigate these conceptual and methodological shortcomings, this chapter uses the polybiome as a reference framework. Polybiome. The polybiome, initially described by Reyed (2023a) and defined by Reyed and Prabhakar (2025a), redefines microbial ecology as an entirely multikingdom ecosystem. They include bacteria, fungi, viruses, archaea, protists, and uncultivable organisms, all of which constantly interact with the host metabolic, immunological and epigenomic circuits.

In this systems-based paradigm, microbial entities are not described as categories of taxa but as functional hub nodes, the activities of which change and adapt throughout the lifespan to mediate variations in β-cell responsiveness, inflammatory tone, metabolic efficiency and pathogenesis. As a result, the polybiome serves as a translational prism, allowing us to understand how metabolic susceptibility, environmental exposure, and nutrition signals all coalesce through similar biological pathways to establish risk and resilience. Building on this explanation, the Polybiome–NexDi framework operates through four interdependent analytical pillars that structure how metabolic vulnerability is understood and predicted (Box 1).

Box 1. Basic Elements of Analysis on which the Polybiome-NexDi Framework Relies

Polybiome science, included in the NexDi model, is based on four interdependent pillars of analysis that maintain the structure of the question:

1. Multikingdom Microbial Ecosystems: An understanding of bacteria, fungi, viruses, archaea, protists, and uncultivable organisms as inherently cooperative interdependencies with direct impacts on metabolic, immune, and endocrine activity.
2. Nutrigenomic Modulation of Host Physiology: Diet-gene interaction is acknowledged to be an active regulatory process that results in lifetime metabolic programming, epigenetic signatures, inflammatory set points and disease disposition.
3. Artificial Intelligence-Assisted Systems Inference and Predictive Modeling: The use of machine learning and multiomics analysis schemes to identify patterns, stratify risk, and predict early perturbation before clinical disease.
4. A Continuum-Based Model of Diabetes Physiology: Reconceptualization of diabetes as a spectrum of β-cell weakness that develops through early-life programming, environmental toxicants, nutritional exposures, and microbial-host interactions.

1.3 Coherence of the Structure of Chapters: The Polybiome as the Analytical Model.

The Polybiome definition is the constitutional point of reference to all analyses connecting nutrition, genomics and microbial ecology because they follow this paradigm to guarantee conceptual continuity and terminological consistency throughout the chapters holding to this scheme. Chapter 2, Decoding the human polybiome via AI, Nutrigenomics in the next generation of personalized health, offers the first detailed and paradigmatic methodological comparison between the microbiome and polybiome paradigms on the basis of the methodological development of Reyed (2023a). It will also present methods of analysis, AI-integrative approaches and case-based implementation that bring this definition to operationalization in the context of precision medicine. Accordingly, any further chapters that Reyed might write are furnished with this systems-level point of view as their analytical framework. Building on this systems-level definition, the polybiome framework becomes directly relevant for reinterpreting complex metabolic diseases—most notably diabetes—where multisystem interactions shape vulnerability and clinical expression.

1.4. The NexDi Perspective: Redefining Diabetes as a Sequence of β -cells Fragility.

This preliminary section contextualizes the polybiome in a wider reassessment of the pathophysiology of diabetes. Modern trends in epidemiological studies of diabetes mellitus show that dual metabolic classification has given rise to a more comprehensive, systems-based conceptual model (Agrawal et al., 2024; Chaudhary et al., 2025). Recent international and regional evidence indicates that the traditional differences between autoimmune type 1 diabetes and lifestyle-related type 2 diabetes cannot explain the multidimensional interactions that produce the foundation of metabolic risk. Some of the critical determinants include early-life adversity, environmental toxicology, epigenetic programming, polybiome architecture and intergenerational nutritional stress.

Growing evidence suggests that type 1 diabetes (T1D) can actually be seen as an insulin resistance system problem involving multiple failures instead of a breakdown of β-cells, making conventional treatment models softer (Apostolopoulou et al. 2025). According to new data, T1D can also be associated with wide-scale insulin resistance superimposed on autoimmunity, which makes customary clinical histories more complicated. At the other end of the vulnerability continuum, which is determined by metabolic programming during development, type 5 diabetes (T5D) is a disease linked to maternal anemia, hidden hunger, and early childhood undernutrition. This animal model highlights the long-term implications of nutritional programming and metabolic stressors during early life stages in permanently reconfiguring physiology, resulting in the generation of unique diabetic phenotypes in populations with socioeconomic and environmental determinants.

The current condition is expanded with supplementary evidence from Egypt that reports changes in the polybiome in pediatric patients with type 1 diabetes (T1D). It has been reported that physiological development is negatively impacted by undue exposure to endocrine-disrupting agents, including bisphenol A (BPA), as well as pervasive contamination with nanoplastics and socioeconomic determinants (El-Degwi et al., 2024). A purported association between BPA exposure and T1D incidence has been reported in Egyptian children (El-Degwi et al., 2024). The overall degree of environmental hazard increases with increasing nanoplastic ingestion, which has been described as a genetic and epigenetic mediator of endocrine pathways (Aloisi & Poma, 2025). The pathogenesis of chronic diseases involves dysbiotic changes in the gut polybiome caused by microplastics (Bora et al., 2024). In the Egyptian context, social issues of plastic waste and environmental pollution also increase the urgency of discussing these determinants (Reyed & Prabhakar, 2025 a; Abdu et al., 2025; Reyed, 2023a).

Taken together, these differing tendencies highlight the insufficiency of the dichotomous paradigm and emphasize the need for an integrative paradigm. These findings prove that diabetes cannot be viewed solely as an autoimmune or lifestyle-induced insulin-resistant disease; instead, the disease must be considered a spectrum of susceptibility in which biological and environmental factors interact. The importance of developing a comprehensive scientific approach that can capture this spectrum of metabolic fragility is highlighted by these convergent outcomes.

1.5. NexDi Framework appearance: The rationale of a synthesizing framework.

The knowledge gap that remains is described as the lack of a homogenizing theoretical paradigm that would allow the integration of these multifactorial determinants into a blueprint that can be implemented in clinical practice. The current models are not sophisticated enough to allow the provision of exact preventive maneuvers. Additionally, the scientific corpus continues to show the lack of an inclusive systemic model that can be used to combine the segmented biological, environmental, and socioecological determinants into a logicalized predictive, preventive, and personalized intervention model. This gap is filled by the next-generation diabetes (NexDi) framework.

NexDi is a multiomics platform that integrates nutrigenomic regulation of human metabolic pathways, intelligent predictive algorithms, multikingdom microbiome analytics and environmental and developmental factors into a unified continuum platform. It reconfigures diabetes as an evolving and dynamic metabolic vulnerability that is experienced throughout the lifespan but not as a categorical entity. This strategy uses nutrigenomics, which is the study of how food and other nutritional constituents regulate gene expression (Almoghrabi et al., 2025; Chaudhary et al., 2025), takes advantage of the power of artificial intelligence to conduct sophisticated predictive modeling (Brugger et al., 2025; Ahmed et al., 2024) and uses polybiome science as a translational interface between diet, host genomics, and environmental exposures. Soon after its intellectual underlying principles were developed, larger epidemiological investigations and geophysical scenarios have made even clearer why the NexDi framework is urgently needed.

1.6. Convergent Moment: National and Global Imperative.

The need for NexDi is not limited to incremental growth; it is acutely needed at a critical historical crossroad of epidemiological emergencies, technological breakthroughs, and environmental demands. Egypt is a special intersection point where the increasing incidence of T1D in children, the ubiquitous presence of environmental toxins, changing nutritional patterns, and maternal malnutrition intersect, transforming the metabolic risk landscape in the country. The prevalence of T1D is now a concerning trend, especially in pediatric and adolescent populations (American Diabetes Association, 2024). The combination of BPA, nanoplastics, and endocrine-disrupting chemicals is assumed to increase the risk of disease, redefining the biological determinants of diabetes (El-Degwi et al., 2024; Ahn & Jeung, 2023; Aloisi et al., 2025; Bora et al., 2024). The emergence of advanced technologies, such as AI-based modeling, nutrigenomics, and polybiome science, which provide the possibility of integrating these multidimensional factors into effective preventive measures of public health, increases this urgency (Almoghrabi et al., 2025; Brugger et al., 2025; Ahmed et al., 2024). The convergence therefore provides a historical opportunity, as well as an urgent challenge: unless new ways of doing things are initiated, the issue of creating health inequities and economic strains can only worsen, but, on the contrary, by introducing the NexDi trajectory, Egypt would become a global pioneer in precision public health innovation. That, hence, makes NexDi not just an innovation but also a necessity in this sense.

To achieve a comprehensive conceptualization of the NexDi framework, it is necessary to place it within the wider path of diabetes typology. The identification of T1D and T5D as salient extremes not only cuts across the rigidity of traditional categories but also requires rethinking the definition of diabetes as well as its epistemological foundations. To entrench this imperative within a contextual setting, it becomes essential to initially examine the changing global and regional environment of diabetes, moving beyond the classical subtypes to new phenotypes, and in the process illuminating the gaps in the existing classification models. The following chapters of this text outline the conceptual framework of NexDi, provide an example of the application of the concept to a real-world case in the Egyptian public health context, and map out an indisputable roadmap to precision nutrition and regenerative medicine research.

2 DIABETES MELLITUS IN TRANSITION: CLASSICAL FORMS TO EMERGING FORMS.

The current global and regional dynamics in the epidemiology of diabetes also dictate reclassification. These chronic conditions have become more widespread, thereby questioning the traditional therapeutic paradigm, and diabetes mellitus has become a global epidemic in the twenty-first century. The International Diabetes Federation estimates that around 589.5 million adults (20-79 years) across the world are already living with diabetes, and it is estimated that this figure will increase to 853.5 million in 2050 (IDF, 2025a). According to the International Diabetes Federation (IDF, 2025a), the total number of deaths linked to diabetes reached about 3.4 million people, which is a fatality rate of 1 person every 9 seconds, and the total health expenditures worldwide related to the disease were USD 1 trillion; this is a 338% increase over the last 17 years.

Type 2 diabetes (T2D) is still the most prevalent type, but type 1 diabetes (T1D) has increased at an average rate of 3-4% per year, which is evidence of dramatic changes in the epidemiology of the disease (Allakany et al., 2023). T1D is currently experienced by more than 9.5 million individuals around the world.

Egypt is illustrative of these tendencies in the world, although there are regional peculiarities. Based on the estimation of the IDF, Egypt is one of the top ten countries most impacted by diabetes, and the number of adults living with diabetes is more than 13. 2 million (IDF, 2025b (. Childhood diabetes is increasing in large urban areas like Cairo and Alexandria, but in rural regions, numerous issues are encountered, including late diagnosis, nutritional insecurity, and access to specialized care (El - Ziny et al., 2014). This two-polar landscape of urban metabolic overload on the one hand and rural undernourishment on the other supports the thesis of the redefinition of diabetes as a continuum that is predisposed genetically, modulated by diet, and subject to changes in the gut microbiome and environmental determinants.

3. RETHINKING THE EVOLUTION OF DIABETES SUBTYPES

Historically, diabetes mellitus has been subdivided into a series of distinct clinical types: type 1 diabetes (T1D), which is an autoimmune disease, with the bulk of evidence linked to the destruction of pancreatic β-cells and a complete lack of insulin; type 2 diabetes (T2D), a multifactorial metabolic condition with a high likelihood of obesity and lifestyle habits; and gestational diabetes mellitus (GDM), which is impaired glucose tolerance during pregnancy, with enduring metabolic consequences. In addition, some clinical practices, research priorities, and global health guidelines have been influenced by this taxonomic framework for decades (Hasin et al., 2017; Verma et al., 2021).

However, the phenotypic heterogeneity observed among different populations cannot be ascribed entirely to their structural rigidity. The questions that clinical experience and emerging evidence pose are critical: why are there phenotypes of diabetes-in-children who do not show the traditional autoimmune phenotype of T1D? Why do specific patterns of β-cell failure seem to be caused by developmental or nutritional disturbances rather than through the metabolic mechanisms to which T2D is linked? These discrepancies highlight the conceptual shortcomings of current classification systems and the need to revise them in support of an enhanced etiological system (Akhlaq et al., 2024).

As of 2025, type 5 diabetes (T5D) is recognized as a developmental phenotype. This classification breakthrough underscores the need to include nutritional and developmental determinants in the diabetes taxonomy, as well as to emphasize that autoimmune mechanisms are not the only ways of predisposing the β-cell population to vulnerability; other factors of malnutrition of the fetus and maladaptive metabolic programming can also predispose the body (IDF, 2025a).

4. T1D AND T5D: INTEGRATIVE PERSPECTIVE IN THE FRAME OF NUTRIGENOMICS AND AI.

Despite resembling each other clinically in terms of severe insulin deficiency, the biological basis, pathology of development and therapeutic consequences are inherently different between T1D and T5D (malnutrition-related diabetes). T1D is associated with genetic predisposition, which is encoded by HLA and autoimmune devastation of β-cells, and T5D is caused by disruptions in normal pancreatic development and insufficient endowment of β-cells due to persistent undernutrition in early life (IDF, 2025a; El-Sayed & Saleh, 2024). These two conditions are characterized by hyperglycemia and chronic complications, but the difference in their origins requires distinct approaches to research and specific clinical measures.

Clinical and developmental data indicate that undernutrition during gestation and the initial phase of early life may irreversibly reduce the mass and functional strength of β-cells, which leads to a lifetime predisposition to metabolic frailty (Pammi et al., 2025). This pattern of development typically presents with severe insulin insufficiency in lean people, with or without exocrine pancreatic dysfunction but no autoimmune markers of T1D. By comparison, T1D is characterized by a well-defined immunopathology of autoantibodies, autoreactive T cells, and cytokine-mediated destruction of β-cells that require lifelong insulin therapy. Although the two disorders are etiologically different, numerous downstream amplifiers of β-cell stress-oxidative damage, mitochondrial dysfunction, and exposure to environmental pollutants, including endocrine disrupting chemicals and plastic-derived toxins, are common in both disorders and increase the susceptibility of the cells to attack (El-Sayed & Saleh, 2024). The gut polybiome is central to this spectrum, with dysbiosis facilitating immune activation and loss of tolerance in T1D patients and dysbiosis inhibiting metabolic homeostasis-promoting short-chain fatty acid (SCFA)-producing microbial communities in T5D patients. The restoration of SCFA-promoting microbial communities is therefore an attractive treatment goal in both diseases to balance nutritional rehabilitation in T5D and immune-modulatory approaches in T1D.

This reframing requires a change between a categorical classification and mechanism-focused care, clinically. The management of T1D is based on immune moderation, insulin replacement and the avoidance of autoimmune comorbidities. Conversely, T5D management has taken nutrition-first interventions (such as protein and micronutrient replacement, malabsorption restoration and polybiome restoration) as the core of management. Most low-cost adjuncts, mechanistically based, such as nutraceuticals lowering oxidative stress and increasing SCFA production, including dietary fiber, berberine, and polyphenols, can offer accessible therapeutic advantages, especially in resource-limited environments (Pammi et al., 2025; Reyed & Prabhakar, 2025 d, e).

The future of enhancing the results of the T1D-T5D range involves AI-intelligent personalization and integrative nutrigenomics. AI-based risk stratification models are able to combine prenatal developmental predictors and polybiome phenotypes (the microbial networks in compartments of the body) with genetic risk factors to detect individuals at risk of T5D significantly before clinical onset. Nutritional and microbiome-targeted interventions can be simulated by digital twin models and help design trials in a cost–benefit and culturally reflective way. AI may improve immunomodulatory and personalized insulin therapy in T1D patients. AI in T5D patients has unparalleled prospects for creating nutrition-based interventions to promote the resilience of the pancreas. Nevertheless, such systems need to be trained on datasets locally curated so that the algorithms become less biased and equitable with clinical benefit (El-Sayed & Saleh, 2024; Pammi et al., 2025).

There are also significant gaps, such as the absence of longitudinal cohorts of early nutrition, pancreatic imaging, and polybiome outcomes; the unavailability of regional, nutrition-first, randomized studies; and the affordability of cheaper nutraceuticals in local food systems. The solutions to these gaps would need a concerted policy and research investment, i.e., maternal–child nutrition interventions combined with diabetes prevention, open-source multiomics databanks to low- and middle-income cohorts, and open and ethical AI solutions.

A promising translational approach in Egypt, as well as similar settings, would be to combine screening of high-risk pediatric groups, enhanced nutritional recovery, the restoration of the polybiome with locally produced fermented products and dietary fibers, and AI-supported individualized care. The strategy transforms diabetes management into a nonreactive, insulin-dependent care model to create a proactive, nutrition-based, digitally empowered ecosystem. These findings concur with global endeavors to frame diabetes as a life-long disease that is influenced by the interplay between nutrition, immunity, and environmental exposures. A summary of the etiologies of T1D and T5D and some common vulnerabilities downstream are given in Box 2, and Table 1 provides a comparative view of the etiologies.

Box 2. Branches: Autoimmunity versus Malnutrition.

Understanding Type 5 Diabetes: A Silent Epidemic.

In type 1 diabetes mellitus (T1DM) insulin-dependent diabetes, the destruction of the pancreatic β-cells is mediated by immune mechanisms, and the disorder is mainly genetically predetermined in children, culminating in absolute insulin insufficiency.
Conversely, Type 5 diabetes (T5DM) develops as a result of undernutrition during early life stages which limit pancreatic development. Losses in vital proteins, vitamins and minerals decrease the mass of β- cells and long-term metabolic ability. Because children with T5DM are typically shown as having normal body weight, they may be hiding an underlying condition of hidden hunger. Aging affected individuals are characterized by a growing burden of ultraviolet foods increasing oxidative stress and disrupting the Polish biome, as well as by accelerating β-cell depletion. This synergy of childhood malnutrition and subsequent food stress makes T5DM a silent epidemic of many low- and middle-income nations.

Table 1. Comparative features of type 1 diabetes (T1D) and type 5 diabetes (T5D).

Feature	T1D (Type 1 Diabetes)	T5D (Type 5 Diabetes(	References
Biological origin)Pathogenesis(	Autoimmune β - cell decreasing energetic by genetic risk	Pancreatic underdevelopment from chronic undernutrition (critical first 1000 days)	Ikegami et al., 2021; Awadalla et al., 2017; El-Sayed & Saleh, 2024
Age of appearance	Childhood or adolescence	Childhood & adolescence, often after poor early-life nutrition	Allakany et al., 2023; Prentice, 2017
β-cell function)Insulin Production(	Complete insufficiency	Reduced β-cell mass & weakened secretory capacity	Ikegami et al., 2021; Prentice, 2017; El-Sayed & Saleh, 2024
Primary Driver)Key factor(	Genetic predisposition & immune triggers	Nutritional deprivation, hidden hunger, & later ultra-processed diet exposure	Awadalla et al., 2017; Prentice, 2017; IDF 2025
Gut microbial impact	Dysbiosis amplifies immune activation	Dysbiosis from malnutrition & processed food reduces SCFA support	Radwan et al., 2020; Gruttola et al., 2016; Buonocore, 2024
Therapeutic approach	Lifelong insulin therapy immune modulation	Nutritional therapy, polybiome restoration, supportive insulin when needed	Podobnik & Prentice, 2025; Kassem et al., 2023 ;Pammi et al., 2025
Research outlook	Immune-targeted therapies, AI-optimized insulin treatments	AI-driven nutrition-first interventions, low-cost nutraceuticals, polybiome support	Lundberg et al., 2020; Podobnik & Prentice, 2025
Epidemiological pattern	Rising worldwide, especially in children	Emerging in LMICs, linked to food insecurity & rapid dietary transition	Allakany et al., 2023; IDF 2025; El-Ziny et al., 2014

This table compares and explains the mechanistic and etiological differences between Type 1 diabetes (T1D), stemming from autoimmunity, and T5D, attributed to malnutrition, as two mechanistically distinct forms of diabetes. This finding highlights major distinctions in what causes them (ranging from autoimmunity to insufficient pancreas development as a result of early-life malnutrition), the efficiency of β-cell function, major contributing factors (genetic predispositions and immune system dysfunctions versus insufficient dietary intake), and their influence on the intestinal microbiota (gut bacteria dysregulation due to immune responses versus gut bacteria dysregulation induced by malnutrition). The data also demonstrate that therapies and investigation protocols vary (insulin/immune changes versus better nutrition and fixing the polybiome), highlighting the requirement for customized, individualized strategies such as NexDi. This approach may be beneficial in managing the growth of T1D on a global scale and T5D in economically disadvantaged countries.

6. MOVING TO FUTURE-FOCUSED ARCHITECTURE: NEXDI.

There is an overlap between type 1 diabetes (T1D) and type 5 diabetes (T5D), which is why developing a more coherent and cohesive conceptual framework that goes beyond traditional dichotomous disease categories is essential. The NexDi principle, which refers to next-generation diabetes, redefines diabetes as a spectrum of immune-metabolic diseases induced by the interplay between genetic predisposition, nutritional status, polybiome composition, environmental stressors, and developmental chronology (Grewe-Salfeld, 2021; Sharmar et al., 2024).

From this integrative perspective, the gut microbiota acts as a focal point that connects nutritional deprivation and immune dysregulation. Dysbiosis can increase oxidative stress, disrupt immune tolerance and impair β-cell survival, which is a credible explanation for why these mechanistic weak points are shared in both T1D and T5D. Similarly, the development of nutrigenomics has helped explain how micronutrient deficiencies affect metabolic processes to shape disease progression. In this context, the future of diabetes studies and clinical applications depends on the implementation of integrative models such as NexDi that combine microbiology, nutrigenomics, natural health sciences, and artificial intelligence. The concept transforms diabetes as a fixed categorical disorder into a life-course disorder that is affected by nutritional, immunological and environmental factors. This strategy can assist in more consistent prevention, diagnosis, and treatment plans, thereby simplifying clinical outcomes and possibly transforming the modern scientific knowledge of diabetes.

7. ENVIRONMENTAL FACTORS AND IMMUNOGENIC TRIGGERS.

The long-standing suggestion of the role of environmental triggers has been to map environmentally induced autoimmune cascades underlying type 1 diabetes. However, many studies have explored the role of early-in-life determinants, such as maternal obesity, cesarean section, rapid weight gain, length of breastfeeding, vitamin D supplementation, and omega-3 intake, but no studies have shown an unmistakable association with disease occurrence (Kotsiri et al., 2025; Tilahun et al., 2025; Vasilev et al., 2025; Ouabbou et al., 2020). This discrepancy highlights the shortcomings in the reductionist paradigms of research and highlights the need for an integrative concept that explains the relationships among immunity, nutrition and the polybiome.

When type 5 diabetes is considered, nutritional deprivation and environmental stressors tend to play out each other to make the discussion more critical. More than a quarter of children under five years of age have iron deficiency anemia, and more than three-quarters of Egyptian adolescents have an absence of vitamin D (Hassan & Bayomi, 2024). An increasing number of children with hyperglycemia along with undetectable insulin secretion without typical T1D autoantibodies are being labeled with emerging T5D phenotypes (Allakany et al., 2023). These are some of the cases that can culminate in hidden hunger where they are calorie-rich foods but lack the necessary micronutrients, such as essential micronutrients, causing impaired pancreatic development and weakened β-cell functioning.

The rapid transformation to ultra (ultra-processed) foods in urban environments causes considerable acute oxidative and inflammatory loads and is part of this silent epidemic (Reyed & Prabhakar, 2025 d, e). An additional example of how such trends emerge through the impacts of the real world is provided by a clinical vignette observed in a rural Egyptian setting (Box 3).

Box 3. Egypt as a Case Study

Case Study 1: National Nutrition Paradox.
The Egyptian case shows that there is a collision between undernutrition and overnutrition that form a twofold nutritional burden. There are high rates of obesity and great deficiency of micronutrients that are documented by national surveys. Iron deficiency anemia is found in between one-quarter of children below the age of 5, and Vitamin-D deficiency in adolescents is more than one-sixth (Hassan et al., 2023). Meanwhile, the dietary intake of ultra-processed foods is increasing aggressively in urban areas at the expense of the traditional diets that are high in legumes, whole grains, and fermented foods.
This nutritional paradox, the overconsumption of caloric and a lack of micronutrients, sets the best stage to develop type 5 diabetes. T5D national data are not available yet, but the clinical evidence of increasing numbers of young patients reporting with diabetes-like symptoms without autoimmune-typical T1D markers is being seen with a common history of nutritional deprivation (Misra et al., 2025).

Case Study 2: Hidden Hunger in rural Egypt.
A 13-year-old rural upper Egyptian boy presented with the problem of recurrent hyperglycemia and weight loss. Engineering outlined that tests conducted in the laboratory showed disrupted insulin secretion without autoantibodies that are associated with T1D (GAD65, IA -2). The dietary assessment revealed a trend of high consumption of refinement bread, sweetened tea, and processed snacks, with low intake of protein, fresh fruits, or vegetables. Although the patient had a normal BMI, micronutrient deficiency (iron-deficiency anemia and Vitamin-D deficiency) was evident based on the clinical evaluation. The presentation was described by endocrinologists as a non-autoimmune diabetic phenotype, which aligns with new definitions of Type 5 Diabetes (Hassan et al., 2023).
The current case is a good example of how impaired development of the pancreas due to hidden hunger (sufficient calories, but inadequate micronutrients) results in insufficient growth of β-cells and exhaustive functional performance (Nurkolis et al., 2025).

8. NEXDI BIOLOGY, PLASTICS, ENDOCRINE-DISRUPTING CHEMICALS.

Besides the conventional nutritional determinants, plastics and the endocrine-disrupting chemicals (EDCs) (e.g., bisphenol A, phthalates, PBDEs) that accompany them are becoming active agents in the environment that trigger metabolic and immune maladjustment. They have a significant mechanistic effect, overlapping with β-cell fragility, metabolic inflexibility, and autoimmune priming pathways (Huang et al., 2025).

A conceptual framework, as shown in Figure 1, suggests that these environmental exposures are causally involved in the process of accelerating disease progression within the nutrigenomic-AI model of NexDi.

Figure 2. Pathophysiological cascade linking plastics and endocrine-disrupting chemicals (EDCs) to autoimmune and metabolic diabetes. The figure illustrates the step-by-step mechanism by which EDC and plastics exposure triggers oxidative stress, inflammation, and subsequent barrier damage, ultimately leading to disease initiation (Type 1 and Type 5 Diabetes).

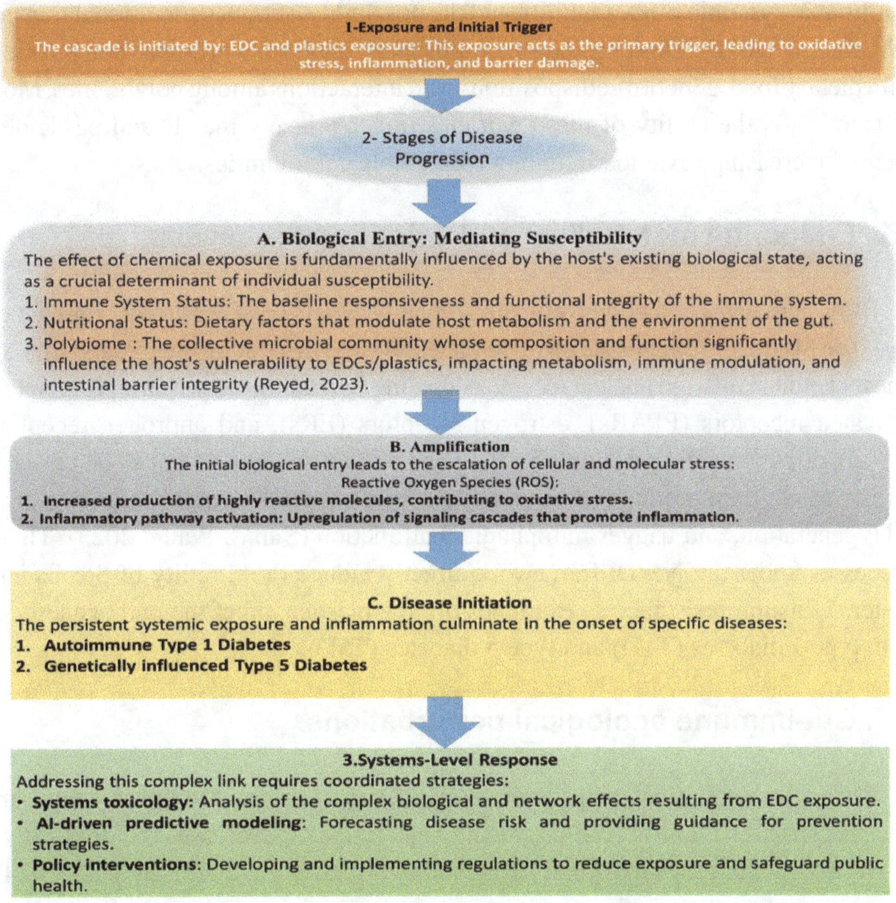

The figure illustrates the sequential damage that EDCs and plastic exposure cause to develop disease. The cascade initiates at the level of chemical absorbance and cellular sensory responses, followed by the development of oxidative stress (increased reactive oxygen species, ROS) and protracted inflammatory communication (Reyed & Prabhakar, 2025 d, e) . These mechanisms disrupt epithelial/gut-barrier integrity and polybiome interactions at multikingdom boundaries, disturbing immune-metabolic steady states. The resultant autoimmune stimulation is pertinent to Type-1 diabetes (T1D), and comorbid micronutrient deprivation, endocrine imbalances, and stresses on mitochondrial functions augment β -cell fatigue, typical of Type-5 diabetes (T5D). Combined, the figure highlights environmental toxicants as the upstream agents of disease initiation within the NexDi continuum. This model emphasizes how this integration can enhance precision in the prevention and management of the whole system through systems toxicology, AI-based predictive modeling, and policy intervention at a systemic level. Within this model, the sequential pathway is consistent with NexDi artificial intelligence analytics, where toxicological, microbial, and metabolic biomarkers are co-processed to discover the initial disruption in immune-metabolic balance. The AI system links exposome-generated signatures of stress to eventual β-cell functional programs, thus enabling the early prediction of autoimmune mobilization or metabolic overload along the T1D-T5D continuum (Author-generated figure).

8.1. Extended exposure and systemic intake.

Microplastics and nanoplastics have already developed beyond the status of environmental pollutants and have already become traces in the human body. Blood, placenta, breast milk, adipose tissue and cerebral-spinal fluid are disassociated, resulting in systemic bioavailability and the possibility of transgenerational transfer (Nihart et al., 2025). This omnipresence creates a background presence that interacts with dietary food, genetic predispositions, and interactions among polybiomes. Most concerning is the ability of these particles to carry heavy metals and pesticides, thereby increasing toxic loads among susceptible communities.

8.2. Endocrine metabolism and oxidation processes.

BPA, phthalates and polybrominated diphenyl ethers (PBDEs are all compounds based on plastics and are made up of potent endocrine-disrupting chemicals (EDCs). These chemicals disrupt nuclear receptor signaling (including peroxisome proliferator-activated receptors (PPARs), estrogen receptors (ERs), and androgen receptors (ARs)), disrupt insulin receptor signaling and disrupt the thyroid hormone balance. Furthermore, they produce reactive oxygen species (ROS), inhibit mitochondrial ATP generation, and trigger mitophagic malfunction (Sun & Wang, 2023). These processes under the NexDi framework directly induce the fragility of β-cells and systemic insulin resistance as examples of how plastics affect the pathogenesis of both type 1 diabetes (T1D) and type 5 diabetes (T5D).

8.3. Gut-immune ecological perturbations.

Plastics have significant impacts on gastrointestinal ecology. Nanoplastic exposure depletes short-chain fatty acid-producing bacterial taxa, decreases the diversity of microbes and increases intestinal permeability (Abdu et al., 2025; Lu et al., 2024). These disruptions help in the translocation of microbial metabolites, including lipopolysaccharide (LPS) and plastic-associated antigens, into the systemic circulation, which subsequently triggers innate immunity and reduces the autoimmune priming threshold. The gut–immune axis is an essential gateway in the NexDi paradigm, where exposure to the environment recreates metabolic and immune health via the gut–immune axis.

8.4. Egypt as a high-exposure living laboratory.

The mechanisms are especially relevant in Egypt, where biomonitoring surveys indicate the presence of high exposure to BPA and phthalates, especially in vulnerable groups such as children, adolescents, and lactating mothers (Gharanizadeh et al., 2025; El Wu et al., 2024; Shalaby et al., 2020). Moderate levels of BPA found in the urine have been associated with early-onset diabetic phenotypes, which means that environmental contaminants are not just hypothetically dangerous; indeed, they are active agents of disease burden. However, considering that Egypt is highly dependent on plastic-packed foods and its increased air pollution rate in urban areas, the nation is the perfect real-life test setting of NexDi plans, especially those aimed at exposure alleviation and environmental health measures.

8.5 Research policy implications

Taken together, these results highlight the need to integrate environmental exposure science into multiomic NexDi frameworks. Key priorities include the following:

A. *Predicting exposomic metrics*, such as nanoplastic deposition and EDC burden, into nutrigenomic and AI-based dietary modeling to model exposure–diet–gene interactions.
B. *High-risk clinical studies*, such as exposure-sensitive studies in Egyptian populations, are needed to define the links between plastic loads, EDC biomarkers, glycemic indices, immune profiles, and structural polybiomes.
C. *Standardization of plastic metabolite assays* in biofluids to enable risk assessment tools that are reproducible and compatible with AI.
D. *Regulatory and nutrition-focused policies*, including safer food-packaging regulations, maternal–child nutrition programs, and country-wide plastic-reduction strategies, have been adopted (Ikegami et al., 2021; Hassan and Bayomi, 2024; Bao et al., 2025; Rizaldi et al., 2025; Jiao et al., 2025).

Overall, plastics and EDCs are not peripheral but central immune-metabolic risk factors in the NexDi continuum. Their integration into mechanistic frameworks, computer modeling, and preventive health initiatives is critical to bringing environmental susceptibility into actionable high-precision nutrition and policy inventions.

9. DYSFUNCTION OF THE GUT BARRIER: THE BIOLOGICAL BRIDGE.

Barrier dysfunction is an important mechanistic mediator of the relationship between environmental exposure and immune misrecognition by β-cells. Under physiological conditions, the intestinal barrier has selective permeability, allowing nutrients to be absorbed but barring potentially harmful antigens. A leaky gut, as is sometimes said, in people predisposed to NexDi, nevertheless allows microbial fragments, endotoxins and dietary antigens to pass through into systemic circulation when permeability is excessive. All of these factors elicit innate immunity, weaken immune tolerance, and favor the development of autoreactive T cells (Macura et al., 2024; Martinez-Augustin et al., 2025).

Dysbiosis also enhances this susceptibility by decreasing butyrate-producing commensals, increasing pathobionts, and increasing inflammatory signaling. Viral triggers: Transient or chronic destabilization of the mucosal barrier is caused by viruses such as enteroviruses or rotaviruses. Together, these disruptions carry risk beyond classical T1D and determine the developmental β-cell vulnerability of T5D. Environmental toxicants, such as plastics and EDCs, worsen this imbalance by affecting mucosal integrity and distorting the structure of polybiomes, thus reducing the threshold of immune activation.

The intersection of such mechanisms and other disorders, such as celiac disease, where peptidogluten takes advantage of inadequate barrier integrity to trigger metabolic autoimmunity, indicates the similarity between diet, the environment, and immune maladaptation (Valitutti et al., 2025; Johnson-Pitt et al., 2024). Celiac disease within the framework of NexDi provides a strong example of moving forward to explore the interplay between dietary antigens and environmental exposures and the gut–immune pathways that make an individual susceptible to diabetes.

10-CELIAC DISEASE A MODEL OF DIET-IMMUNE-DIABETES INTERVENTION.

A case in point is celiac disease, which demonstrates the multifactorial relationship between diet and genetic factors, immune tolerance and barrier function in the development of diabetes risk. Intestinal inflammation, villous atrophy, and chronic barrier dysfunction: gluten-derived peptides trigger chronic inflammation of the intestines, villous atrophy, and unregulated entry of microbial and dietary antigens into the systemic circulation in genetically susceptible individuals (HLA-DQ2 or HLA-DQ8) (Daley et al., 2025). This cascade contributes to β-cell autoimmunity and explains the high co-occurrence of celiac disease and T1D. In that context, contemporary international practices suggest periodical screening of celiac disease among children newly diagnosed with T1D (Ruera et al., 2025). Biomarker surveillance has also emerged as a vital tool to characterize this comorbidity (Alzahrani, 2024).

Other dietary antigens, such as bovine milk proteins, high-glycemic-load foods, and novel food additives, can further destabilize gut-immune homeostasis beyond gluten (Naredi Scherman et al., 2025). As a result, celiac disease is not just a comorbidity but can serve as a paradigmatic example of how gut barrier integrity coordinates immune-metabolic effects.

This theory is core to the concept of NexDi, which views T1D and T5D as two poles of a combined spectrum of immunometabolic vulnerability in response to dietary antigens, impaired mucosal barriers, and susceptibility of the pancreas to external challenges (Ikegami et al., 2021) (Table 2). This model is combined into a consistent immune–metabolic nexus that integrates nutrition programming in the first 1,000 days, nutrigenomics-enabled risk stratification, and polybiome modulation inversely into a coherent whole (Figure 2A–C).

These interactions are amplified in Egypt, where nutrient deficiency in rural areas and nutrient excess in urban areas coexist in the form of dietary imbalance. Celiac disease thus demonstrates the combination of autoimmunity, malnutrition, toxicants in the environment, and unfair treatment access that increases the probability of diabetes. The NexDi paradigm offers a source of combined roadmaps of genomics, nutrition science, environmental epidemiology, polybiome ecology, and AI analytics together to form a next-ready model of precision prevention.

Figure 3. (A, B, C) Immune–metabolic nexus: Unifying genetics, the gut barrier, and the AI in the diabetes spectrum, these boards show, in a strong and associated way, how immune activity and metabolic pressure network along the T1D–T5D continuum.

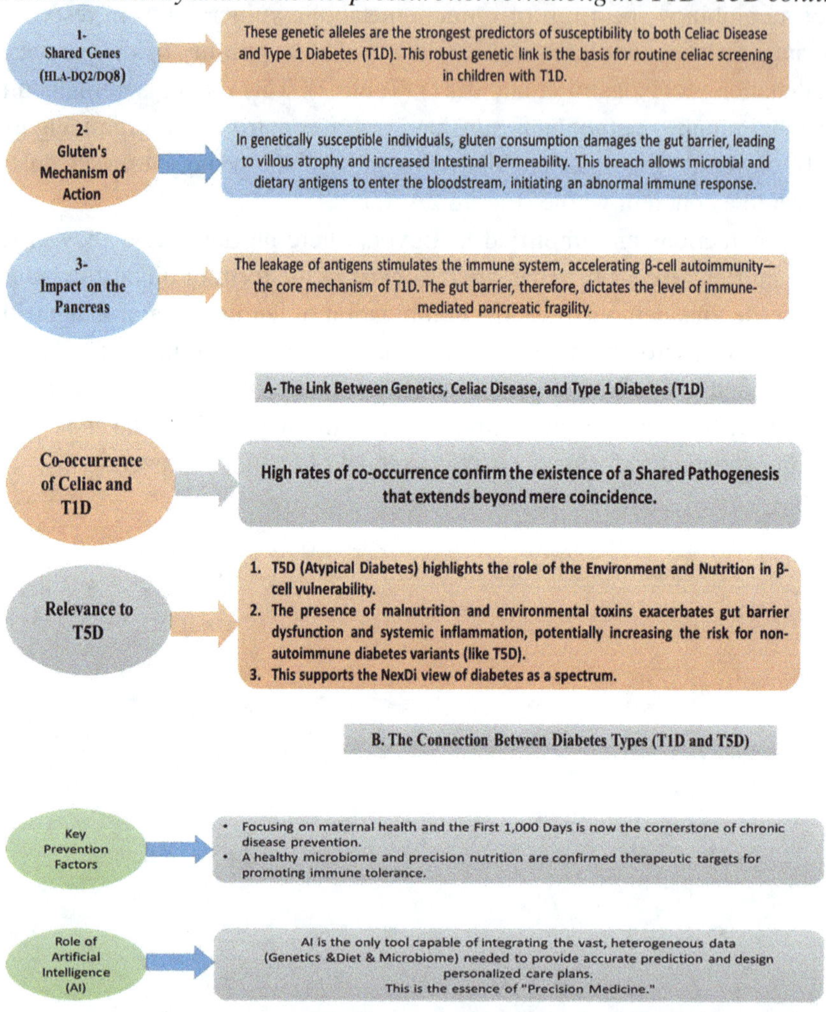

Panel A describes genetic vulnerability and gut-barrier integrity as initial sentinel mechanisms that determine baseline immune preparedness and inflammatory tone. The interaction between multi-kingdom polybiome dynamics and these signals, either amplifying or buffering them, affects β-cell stress and metabolic compensation, as announced in panel B. Panel C incorporates AI-derived metabolic signatures that project these interacting pathways into early predictive paths along the spectrum of diabetes. Collectively, the three panels indicate that diabetes development represents a coherent immune-metabolic circuit, not defined by individual disease categories, and common mechanistic forces operationalize susceptibility through autoimmune to nutrition-induced metabolic inflammation. In all the panels, the figure shows the convergence of nutrigenomic markers, microbial diversity indicators, and exposomic stress indicators pursued by the AI to build individualized susceptibility profiles. With deeper insights into these multi-omic inputs coupled with patterns of immune-metabolic responses, the AI system identifies earlier signals of autoimmune activation (T1D) or environmentally mediated β-cell overload (T5D), thus contributing to proactive, individualized intervention (Author-generated figure).

Table 2. Unifying the Diabetes Spectrum: An Integrated Analysis of Pathogenesis, Genomic Susceptibility, and AI-Driven Intervention within the NexDi Framework.

Basic Scientific Knowledge	Precise Mechanism /Example	NexDi Implication
1. Genetics & Gut Barrier (T1D/Celiac Link)	The HLADQ2/DQ8 alleles have been linked to increased susceptibility to type 1 diabetes mellitus (T1D) and celiac disease. Eating gluten causes intestinal hyperpermeability, which allows the translocation of antigens, thus promoting β-cell autoimmunity (Ahmadzadeh, et al., 2025; Apostolopoulou et al., 2025).	Integrating Pathogenesis: Diabetes is not a pancreatic pathology but a disease instigated by the failure of gut-immune barriers in genetically predisposed persons (Łukowski, 2025)
Disease Spectrum (T1D vs. T5D)	T1D and celiac disease have a comorbid manifestation which supports a common pathogenic framework. Environmental toxins and malnutrition worsen gut dysfunction in susceptible contexts thus altering the risk profile toward nonautoimmune forms such as type 5 diabetes (atypical diabetes) (Vincentini et al., 2025)	Spectrum Management: The T1D-T5D spectrum should be viewed as a spectrum that is regulated by the environmental and nutritional determinants that influence the phenotypic outcome (Gómez-Peralta, et al., 2024).
Prevention & Modifiable Targets	Immune tolerance is a critical programming window in the early part of life, during the first 1,000 days along with maternal health. A polybiome and a healthy diet are the main priorities in promoting a more robust metabolism (Bankole & Li 2025).	Proactive Intervention: Therapeutic approaches must be shifted toward proactive, lifelong prevention, with a high focus on early-life nutrition and microbiota wellbeing (Hsu et al., 2025) .
AI as the Translational Tool	Artificial Intelligence is the only tool that can combine multifaceted, heterogeneous data, genetic, dietary, and microbiomic data, to accurately forecast risk and create highly personalized care plans (Agrawal et al., 2025)	Precision Implementation: Artificial intelligence operationalizes the complexity, transforming disaggregated scientific knowledge into action and adaptive strategies regarding personalized public health (Panteli, et al., 2025).

This table summarizes the essential scientific opinions that explain the NexDi paradigm, demonstrating the junction of biological risk factors (genetics and the gut barrier) and the vital role of technology (AI) in translating complex data into actionable public health strategies across the diabetes spectrum.

Generally, the example of celiac disease demonstrates a wider scientific paradigm: the intestines are not an inert pathway but rather an active determinant of the fate of metabolism. Its vulnerability to both dietary and environmental pressures combined with genetic predisposition preconditions a range of diabetes phenotypes. The way forward is a commitment to an integrative, futuristic business model that entails starting with precision nutrition, polybiome regulation, and AI-informed personalization, so the way diabetes is now conceptualized as an avoidable and responsive consequence in the interrelationship of biological and environmental systems.

11. THE 1,000DAY JOURNEY: PROGRAMMING THE FATE OF Β -CELLS AND THE METABOLIC HEALTH OF MOTHERS AND CHILDREN.

One thousand days of life—between conception and the second birthday—are a critical developmental period in which immune programming, gut polybiome colonization (Reyed & Prabhakar, 2025a), metabolic resilience and the future of pancreatic β-cells are determined. It is not a passive developmental stage but rather an active biological process in which maternal health, prenatal exposure and early nutrition intersect to determine intergenerational vulnerability to immune-metabolic pathology. In this context, NexDi risk is written well in advance of clinical manifestations, which supports the inseparable connection between maternal health and childhood metabolic conditions.

11.1 Maternal nutrition and micronutrient deficiency

The nutritional status of the mother has a background role in the programming of β-cell resilience. Iron, folate, and vitamin D deficiency disrupt immune development in the fetus, maladaptive placental functionality, increase oxidative stress, and alter the developmental pathway of pancreatic β-cells (Hassan et al., 2023; Bezie et al., 2025).

Maternal anemia is very common in Egypt, and this leads to intergenerational hidden hunger, which even makes offspring weaker in supporting metabolic processes. Poor intake of high-quality proteins, essential fatty acids and antioxidants amplifies this predisposition, resulting in infants with impaired immuno-metabolic systems and a low ability to cope with postnatal nutritional demands.

11.2 Crosstalk between maternal infections and the immune system

There is an extrabiological stressor of maternal infections. Maternal–fetal immune tolerance dysregulation may result in exposure to viruses and microorganisms, manipulation of regulatory T-cell networks, and predisposition of the infant to heightened autoimmune reactions (Kalam et al., 2025). Latent or subclinical infections during pregnancy can also reprogram fetal immune responses, increasing their vulnerability to β-cell autoimmunity or developmental insufficiency in adulthood. These mechanisms demonstrate how maternal immune imbalance may pendulum through generations.

11.3 Sociocultural and healthcare inequalities

In addition to biological factors, structural and sociocultural determinants play an enormous role in success during the 1,000-day window. The lack of maternal nutrition literacy, insufficient antenatal care, and programs addressing gut health and prevention of infection decrease the likelihood of early intervention (Calvez et al., 2024). In Egypt, healthcare disparities such as inadequate antenatal education and irregular micronutrient supplementation contribute to intergenerational vulnerability and create unaddressed critical gaps, which further increase the cumulative burden on the metabolic health of mothers and infants.

11.4 Polybiome and early postnatal determinants

Metabolic programming is further solidified in the neonatal and infant stages by the creation of gut microbial ecosystems. The mode of delivery, duration of breastfeeding, exposure to antibiotics, and quality of the infant diet are the determinants of early gut colonization (Aquino -Domonguez et al., 2025).

A- Cesarean birth inhibits the exchange of microbial taxa in mothers that are vital to immune tolerance.
B- Reduced duration of breastfeeding avoids exposure to bioactive oligosaccharides and immunomodulatory microbes.
C- Age-related antibiotics interfere with the diversity of microbes and weaken barriers.
D- Ultra-processed and low-fiber infant diets decrease the generation of SCFs, decreasing tolerance and suggesting increased inflammatory capacity (Tilahun et al., 2025). These mechanisms jointly influence the polybiome in a manner that can precondition children with β-cell vulnerability via the NexDi pathway.

11.5 Case Study: Egypt, the Living Laboratory.

Egypt suggests a noteworthy synergistic compounding of maternal and infant determinants that increase developmental risks in metabolism. NexDi is exposed to a high-risk environment due to high rates of maternal anemia, extensive cesarean sections, excessive antibiotic use in the early life of children, and a reduced period of breastfeeding, among other widespread environmental exposures.

Recent evidence documents:

- High levels of BPA in the urine of Egyptian children with T1D (El-Degwi et al., 2024).

- Phthalate metabolites have been identified in the breast milk of Egyptian mothers (Shalaby et al., 2020). These results support the conclusion that maternal and infant exposure to EDCs ultimately coincides with the metabolic vulnerability of the first 1,000 days. International findings also suggest that the rate of maternal exposure to phthalates during pregnancy is associated with gestational diabetes mellitus (GDM), which is highly likely to be associated with pregnancy in Egypt but is under researched locally (Salem et al., 2019). Together, these determinants can be grouped into pregnancy-related and early postnatal determinants, which are summarized in Boxes 4 and 5. They emphasize the combination of environmental toxicants, nutritional inadequacies, and health care disparities to form long-term β-cell dysfunction, immune tolerance, and metabolic resiliency.

Box 4. Pregnancy Determinants and Maternal Health.

1. Micronutrient deficiency (iron, folate, vitamin D) impairs immune system development, increases oxidative stress, and lowers tolerance of β-cells (Hassan et al., 2023).
2. Maternal anemia and nutritional deficiency disrupt protective immuno platforms, disrupt placental activity, and expose newborn to permanent vulnerability to lifelong weakness (Uta et al., 2022).
3. Autoimmunity of newborns is caused by the impairment of immune tolerance between mothers and offspring due to maternal infections including viral and microbial infections (Hussain et al., 2022).

The gap in healthcare that is the most common is linked to the lack of prenatal education, and the ineffective use of prenatal care that delays the implementation of the preventative interventions (Calvez et al., 2024).

Box 5. Early Postnatal Exposures.

- Mode of Delivery (C -section): does not involve the transfer of maternal microbes, changes colonization of the polybiome (Mohamed et al., 2025).
- Reduced Breastfeeding Length of Duration: decreases the protective oligosaccharide and immune-modulating factors transfer (Purkiewicz et al., 2025).
- The first wave of antimicrobials disrupts the microbial communities of the intestines, establishing a sustainable bewilderment of immunity (Tilahun et al., 2025).
- Early-Childhood Feeding: when diets have low microbial diversity, the establishment of gut protective measures and immunological defenses is impeded (Aquino-Dominguez et al., 2025).

To highlight these significant determinants, important maternal and early postnatal determinants are plotted in Figure 3.

11.6 In the Direction of an Integrated Framework

As described in Box 5, case studies in Egypt demonstrate patient evidence of the relationship between early-life determinants, such as environmental exposure to bisphenol 1,2-dichlorobenzo p-dioxin (BPA), and negative β-cell function and the risk of developing diabetes. These maternal influences and early-life factors mechanistically converge with the environmental factors mentioned above, such as plastics, endocrine-disrupting chemicals, and food antigens, such as gluten, involved in celiac disease. All these factors mediate the aggravation of β-cell resiliency by oxidative stress, barrier dysfunction, immune imbalance, and polybiome instability, thus highlighting the importance of comprehending the fact that the danger of diabetes has been predetermined long before the emergence of clinical manifestations. The NexDi paradigm proposes that the first 1,000 days of life need to be re-conceptualized as a prime intervention period in which beneficial micronutrient supplementation of mothers, advocacy for breastfeeding, minimization of unnecessary cesarean births, and minimization of exposure to environmental risks can change patterns of disease at the population level.

Figure 4. Mechanisms of metabolic programming in NexDi.

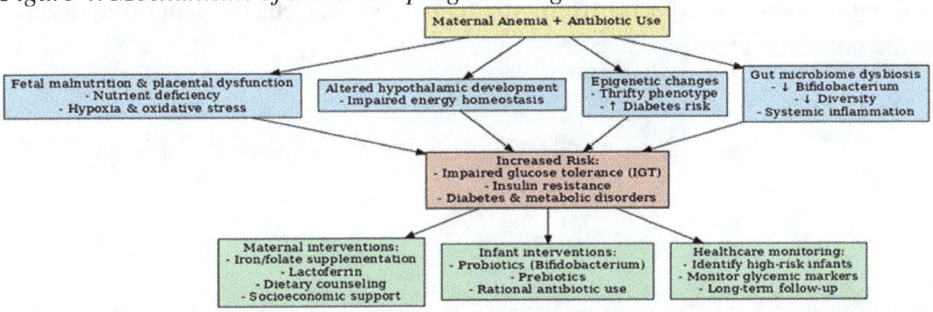

This figure shows that early-life metabolic programming is a result of the interaction of maternal nutritional health, placental activity, and postnatal ecological experiences. Mother anaemia, micronutrient deficiency, and suboptimal feeding habits compromise the placental delivery, undermine the fetal β cell stability, and propel the newborn to an increased metabolic vulnerability. These susceptibilities, after birth, interact with infant exposures such as cesarean delivery, shortened breastfeeding exposure, use of antibiotics, and dietary low-diversity to interrupt polybiome colonization, destabilize immune tolerance, and increase oxidative stress. This interplay between these pathways enhances the loss of metabolic flexibility, glucose dysfunction, and predisposes individuals to diabetes early in their lives. The first 1000 days highlighted in the model are a critical period where nutritional, microbial and environmental signals influence the pathway of metabolic health in the NexDi framework. In the NexDi framework, this figure combines the impacts of a nutritional condition early in life, microbiome setup, and environmental exposures on metabolic programming. It highlights the initiation of a process of β-cell exposures by maternal deficiencies of micronutrients and placental stressors, followed by the aggravation of the problem by infant biases in the neonatal period, including limited breastfeeding, delivery by cesarean section, the use of antibiotics, and monolithic diets. Within this integrative framework, the figure explains how the overlap of these convergent pathways undermines metabolic flexibility and places infants on early predictive pathways toward T1D and T5D risk, consistent directly with the life-course perspective of NexDi (Author-generated figure).

11.6 In the Direction of an Integrated Framework

As depicted in Box 6, case studies in Egypt demonstrate patient evidence of the relationship between early-life determinants, such as environmental exposure to bisphenol 1,2-dichlorobenzo p-dioxin (BPA), and negative β-cell function and the risk of developing diabetes. These maternal influences and early-life factors mechanistically converge with the environmental factors mentioned above, such as plastics, endocrine-disrupting chemicals, and food antigens, such as gluten, involved in celiac disease. All these factors mediate the aggravation of β-cell resiliency by oxidative stress, barrier dysfunction, immune imbalance, and polybiome instability, thus highlighting the importance of comprehending the fact that the danger of diabetes has been predetermined long before the emergence of clinical manifestations. The NexDi paradigm proposes that the first 1,000 days of life need to be reconceptualized as a prime intervention period in which beneficial micronutrient supplementation of mothers, advocacy for breastfeeding, minimization of unnecessary cesarean births, and minimization of exposure to environmental risks can change patterns of disease at the population level.

Box 6. Case Study: Early-Life Determinants in Egypt.

Recent clinical and epidemiological evidence in Egypt highlights the interplay between maternal nutrition, early-life exposures, and risk of diabetes:

1. Bisphenol A (BPA) and Type 1 Diabetes (T1D):

In a case–control study carried out in Dakahlia, significant increases in urinary BPA levels in the children with T1D were found compared with their controls. BPA-induced pancreatic architectural changes were supported by complementary animal modeling studies, thus suggesting a mechanistic process that mediates endocrine-disrupting chemicals (EDCs) to β-cell damage (El-Degwi et al., 2024).

2. Phthalates in Breast Milk and Maternal Exposure:

Research conducted in Helwan, Cairo, found monosaccharides, which included mono-n-butyl phthalate (MnBP), in quantifiable amounts in maternal breast milk and in mother urine. Domestic plastic use, such as PVC flooring, plasticizing wall finishes, and keeping food or water products in plastic containers, were the most salient predictors of exposure, highlighting the imminent presence of transgenerational plastic exposure during the crucial phase of breastfeeding (Shalaby et al., 2020).

3. Gestational Diabetes Mellitus (GDM):

Although there are limited native-origin data associating phthalate exposure with gestational diabetes mellitus (GDM), local research studies elucidated the increasing incidence of GDM in women in Egypt (Salem et al., 2019). However, international cohort studies consistently have demonstrated that maternal exposure to phthalates, particularly di-2-ethylhexyl phthalate (DHC) metabolites, correlates with poor glucose tolerance during pregnancy (Mitra et al., 2025). This proposes a hypothetical but underresearched causal mechanism through which the use of plastics can cause GDM in Egypt.

Key Insight:

The existing Egyptian evidence base, albeit limited, demonstrates that plastics and their objects of metabolism may penetrate maternal-child health during the pivotal time of the first 1,000 days, potentially affecting autoimmune T1D in children, as well as GDM in mothers.

Future Direction:

There is an urgent need to develop large-scale, prospective Egyptian birth cohorts of deaths to numerically determine the relationships between BPA and phthalate exposures and longitudinal health outcomes. Introducing exposure biomarkers, polybiome proficiency, and AI-based risk assessments can make Egypt one of the leading centers of exposure science and diabetes prevention in the region.

12. INTEGRATION OF ENVIRONMENTAL AND BIOLOGICAL PATHWAYS.

Plastics and EDCs no longer behave in isolation while within the body; instead, they interact with core biological systems—particularly the gut-immune axis (Kiran et al., 2025). This process results in a two-way cycle: environmental toxicants impair gut protective capabilities and modify the polybiome, and the consequent microbial infiltration and immune maladaptation increase the susceptibility of the body to very toxicants (El-Degwi et al., 2024). By doing so, the external exposures and internal dysfunctions are compounded in themselves and each other in a self-perpetuating risk cycle.

12.1 Drivers of NexDi as Environmental Protection Toxicants

The presence of environmental toxicants, especially plastics and monomers, is active in transforming the immunometabolic landscape (Keskesiadou et al., 2024). Micro- and nanoplastics not only enter food chains but also bioaccumulate in tissues and enter epithelial barriers (Faja et al., 2022). Recent research on bottled water reported that hundreds of thousands of nanoplastics, also known as tiny plastics, are present in bottled water and can find their way into human cells and tissues. Their additives, e.g., BPA and phthalates, are endocrine disruptors that cause oxidative stress, dissociate and disrupt mitochondrial activity and amplify systemic inflammation (Gambino et al., 2022). Comprehensive correlations between metabolic disorders and endocrine-disrupting chemicals, especially those found in plastics, have been established. An extensive umbrella meta-analysis involving 52 systematic reviews and 759 meta- analyses established uniform relationships among bisphenol A (BPA), phthalates, PFAS, and various adverse metabolic effects, such as insulin resistance, obesity, and type 2 diabetes (Symeonides et al., 2024).

BPA exposure in children has been specifically associated with emerging autoimmune or metabolic imbalances due to biomonitoring evidence in Egypt (Hassan & Bayomi, 2024). Being a country with a high diabetes burden and significant exposure to chemicals related to plastics, Egypt is a highly informative case for the NexDi framework that combines environmental, metabolic, and epidemiological determinants (Awadalla et al., 2017; El-Ziny, 2014).

12.2 Gut–Immune Interface as a Mechanistic Bridge

Mechanistically, the interface between external stimuli and β-cell autoimmunity is a bridge between barrier dysfunction and external triggers at the gut–immune interface. The normal intestinal barrier filters what is absorbed and what is taken out of the intestines. Heightened permeability—or leaky gut—permits microbial pieces, endotoxins, and dietary antigens to enter circulation. These antigens stimulate innate immunity, tolerance interference, and autoreactive T-cell responses.

Dysbiosis contributes to this effect by decreasing protective butyrate-producing microbes and permitting the proliferation of pathobionts that increase proinflammatory signaling (An et al., 2022). Viral infections also increase instability in the system: enteroviruses may infect islets directly, and rotavirus is associated with temporary dysfunction of the barrier (Obaied et al., 2025).

A good example is celiac disease. In genetically predisposed individuals, gluten peptides cause chronic inflammation of the intestines, augment the translocation of antigens, and accelerate βcell autoimmunity. Therefore, T1D and T5D are not separable but are related products of similar mechanistic pathways.

12.3 Beyond Insulin-Centric Models: Translational Strategy

This approach highlights the shortcomings of insulin-based therapies. In Egypt, insulin pumps and continuous glucose monitoring (CGM) are limited and are only maintained by elite groups in urban settings, rendering rural citizens underserved. Unless efforts are made to convert these toxins into upstream drivers, such as environmental toxicants, microbial imbalances, and dietary antigens, contemporary therapies are inactive (Ghazi et al., 2025).

In this case, NexDi contributes value by harmonizing mechanistic research with translational work:

1. *AI-based analytics and systems toxicology* combine genomics and metabolomics with exposomics to stratify patients and age-related risk outcomes (Sarigiannis et al., 2025; Pammi et al., 2023).
2. **Nutrigenomics**: An accurate diet that increases gut barrier capacity and polybiome stability.
3. *Polybiome-targeted correlates*, comprising probiotics, prebiotics, and fecal microbiota transplantation, have the ability to restore immune equilibrium.

12.4 Policy implications as a natural extension

The research roadmap does not make policy implications peripheral:

1. High-fiber, antioxidant diets are nutritional countermeasures.
2. Maternal child involvement includes micronutrient supplementation, maternal vaccination, increased breastfeeding, and a reduction in excessive cesarean section.
3. Plastic controls want tougher BPA guidelines and investment in waste management frameworks.

These interventions can rewrite risk pathways at the population level (Alum et al., 2025; Weiner et al., 2025).

12.5. Integrated Summary

Taken together, NexDi redefines the concept of diabetes as a disorder of the system rather than a dichotomy between T1D and T2D. It characterizes the existence of causal convergence between immunity, metabolism, and the environment and puts this intricacy forward in translational conceptualization. Egypt is a bright example of the urgency of the crisis in question and the possibility of becoming one of those who introduce pioneering solutions.

NexDi represents a new approach to predict, prevent, and personalize diabetes treatment; through integrated approaches, factoring in precision nutrition, and polybiome control along with AI-based modeling and evidence-based policymaking, NexDi promises a new paradigm of the treatment process.

Although environmental triggers and policy responses are important, a nutritional and polybiome-based approach offers the most effective and culturally applicable facilities for NEXDI prevention and interception. On the basis of the Egyptian context and the framework of the first 1,000 days, the subsequent sections discuss how dietary, microbial, and nutrigenomic interventions have the potential to transform disease paths.

13. NUTRITIONAL AND POLYBIOME-BASED OPPORTUNITIES

In a NexDi context, nutritionally based and polybiome-oriented opportunities represent promising similarities between prevention and management, particularly within populations in which any significance of advocated dietary customs collides with the modern influx of metabolic and autoimmune illnesses.

Traditionally and historically, the Egyptian meal called *laban rayeb*, *kishk*, or mish is full of probiotics and, therefore, is a culturally grounded nudging that takes into account both its accessibility and its environmental friendliness (Roshdy, 2024; Zhong et al., 2022; Tian et al., 2025). In addition to harboring microbial components (*Lactobacillus* and *Bifidobacterium* species), these foods enrich their bioactive metabolites, which promote the production of short-chain fatty acids, reinforce the health of epithelial barriers, and strengthen mucosal immune tolerance (Reyed & Prabhakar, 2025 a,b,c; Thakur et al., 2025; Reyed 2007).

There is a possibility of addressing the microbial deficiencies evidenced among newly diagnosed NexDi patients by reconciling this culinary legacy with modern nutritional regimens. In fact, Adel et al. (2025) reported that disease onset in Egyptian children was associated with reduced abundances of *Bifidobacterium* and *Lactobacillus*, accompanied by a resultant increase in the growth of pathogenic *Enterobacteriaceae*. These dysbiotic patterns highlight that dietary reconstitution of microbial diversity and presentation is a fundamental, rather than a peripheral, component of disease prevention procedures (Enache et al., 2025; Buonocore et al., 2024).

14. NUTRITIONAL STRATEGIES WITHIN THE NEXDI FRAMEWORK

In addition to the usual diets, specific nutraceuticals form the second level of polybiome-based intervention that may be added to the paradigm of precision nutrition. The anti-inflammatory actions of omega-3 fatty acids found in fish oils have been shown to modulate eicosanoid signaling pathways and increase the production of resolvin and protectin. β-cell autoimmunity is based on this dampening of chronic immune activation (Povaliaeva et al., 2020). Similarly, polyphenolic compounds derived from turmeric, such as curcumin, have been shown to inhibit NF-like activation, reduce oxidative damage, and facilitate the proliferation of butyrate-producing bacterial taxa, which strengthens anti-inflammatory and barrier-supportive actions (Servida et al., 2024).

Vitamin D plays a very specific role in the Egyptian context. It is also regularly and significantly deficient in both pediatric and adult populations (Hafez et al., 2017). In addition to disrupted regulatory T-cell proper operations, deficiency in vitamin D has been linked to increased exposure to autoimmune reactions, making supplementation an immediate approach to restoring immunoglobulin sugar (El-Sayed et al., 2019). In addition, polyphenols found in pomegranate, hibiscus, and green tea have to be mentioned because of their dual action as both antioxidants and prebiotics that discriminate in favor of beneficial microbiota living in the gut.

Fermentation, as a longstanding preserved food technique, can also serve as a powerful nutrient-enhancing instrument. According to Annunziata et al. (2020), fermentation significantly increases the concentration of bioactive compounds, especially short-chain fatty acids (SCFAs), that have established health benefits. This process is based on microorganisms that convert food matrices to health-promoting substances. Bell et al. (2018) also highlighted the importance of fermented food in the diets of children and its contribution to the establishment of the early-life gut polybiome. Since the colonization of the gut microbial community starts at infancy, there is promising potential for early lifestyle modifications during childhood in terms of long-term health.

Another important scientific aspect that the authors noted, owing to the combination of fermented foods and fungal biomass, is the synergistic effect achieved, an important element that must be considered when explaining the probiotic constituents of the gut polybiome.

These interventions utilizing nutrition and the polybiome as a component should be viewed through a systems-based approach by considering maternal and early-life exposures. The nutrition of the human immune system predisposes individuals to nutritional challenge at the earliest stages of embryonic development, when deficiencies in macro- and micronutrients or surpluses of attributes to a health system are detrimental. In Frontiers in Immunology, Prentice (2017) mentioned that nutrition during key periods of immune system formation—conception through the entire gestation process and at the initial stages of postnatal life—has a significant effect on infant immune system maturation. She emphasized that nutritional status, particularly during these developmental stages, has a significant influence on the developmental path of the human immune system. Through her review, she discovered that a lack of nutrients and an excess of nutrients can trigger negative consequences in immune development, thus presenting a partial explanation for the increased susceptibility of neonates to infectious diseases, which can be explained in part by immunological immaturity. Finally, strategic nutritional modulation was also suggested by the Prentice as a key to centralizing defense against infants through immunotransplantation. In agreement with this assumption, insufficient micronutrient levels during maternal intake, including vitamin D, zinc, and omega-3 fatty acids, may continue to mediate aberrant fetal immune programming and increase the risk of postnatal dysbiosis (Reyed & Prabhakar, 2026; Park & Manna, 2025). As a result, the incorporation of culturally approved fermented foods, as well as a specific approach in terms of micronutrient supplementation when working with maternal and infant nutrition issues, would help to strengthen immunological stability in the initial stages of immune development. Additionally, the development of excessive changes in dietary traditions based on high-sugar and processed foods may be somewhat suppressed by public health measures such as the preservation and modernization of traditional diets, which further support the suppression of the worsening of dysbiosis and the discharge of physiological balance.

In summary, the overlap between conventional Egyptian dietary assets, modern nutraceuticals, and specific interventions with microbiocytes provides a multidimensional scope into which to apply the NexDi frame. With the inoculation of cultural heritage and biomedical advancements, cost-effective, scalable, and context-specific measures can be designed that directly address the microbial and immunological miscoordination underlying the NexDi.

These measures emphasize that nutritional interventions are not beneficial, but they are core trusted, as they have the potential to remodel the gut–immune relationship, help to recover metabolic homeostasis, and reduce the intergenerational morbidity of diabetes in Egypt and other countries.

15. NUTRIGENOMICS, THE FIRST 1,000 DAYS, AND THE PATH TO NEXDI: MECHANISMS, EVIDENCE, AND RESEARCH PRIORITIES

The first 1,000 days—conception to the second year of the child—are a programming window where dietary exposures generate long-lasting molecular signals that influence immune development, metabolic set points, and β-cell resilience. Food bioactives and nutrients are epigenetic modulators: methyl donors (folate, vitamin B12, and choline) provide one-carbon metabolism and DNA-methylation pathways; polyphenols and other phytochemicals manipulate histone modifications and miRNA networks; and fatty acids remodel the membrane composition and downstream signaling. Such molecular impacts during prenatal and early postnatal development impose traces in the placenta and subsequent footage of the offspring that can continue for decades and alter hazard (maternal nutritional modulation of metastable epialleles and epigenetic aging have been duplicated and duplicated) (Hsu et al., 2025). These dietary signals are translated into systemic biology that is applicable in NexDi by two mediators that are interlocking.

First, epigenetic reprogramming of fetal and infant tissues can change the expression of genes regulating immune tolerance, antioxidant defenses, and β-cell development, altering the autoimmune activation threshold or developmental β-cell insufficiency threshold (Qin et al., 2025). For example, there is a correlation between maternal deficiencies in methyl donors and hypomethylation of immune-regulatory loci, which, reasonably, suppresses regulatory (Treg)-based immunity and predisposes individuals to autoimmunity. On the other hand, maternal nutritional excess (high sugar/high fat) facilitates proinflammatory transcriptomic programs and metabolic stress pathways in offspring hepatocytes and adipocytes, predisposing them to insulin resistance (Prentice, 2017; Hsu et al., 2025).

Second, the gut polybiome is the key biochemical intermediary between diet and host epigenetic regulation. Dietary fiber fermented by microbes produces short-chain fatty acids (SCFAs), including butyrate and propionate, which are histone deacetylase (HDAC) inhibitors and agonists through G protein-coupled receptors. Butyrate enhances epithelial barrier homeostasis, affects Tregs, and programs anti-inflammatory measures; its loss (on low-fiber diets or during early dysbiosis) eliminates a critical epigenetic check of inflammation and compromises mucosal tolerance (Valitutti et al., 2025). On the other hand, dietary patterns that produce harmful microbial metabolites, including lipopolysaccharide (LPS) translocation and trimethylamine N-oxide after excess consumption of animal choline and carnitine, support oxidative stress and inflammatory signaling and act in complex with genetic vulnerability to drive β-cell damage. The interdependence between host–microbial epigenomic interactions and microbial metabolites has been comprehensively reviewed in both mechanistic and translational research (Stein & Riber, 2023).

Plastic particles and their chemical additives are environmental toxicants that have a cross-section across the two pathways. Micro- and nanoplastics can be found in human blood, placenta, and tissues, indicating their bioavailability at the systemic level. Endocrine-disrupting chemicals such as plastic, BPA, phthalates, and PBDEs disrupt endocrine signaling, activate mitochondrial dysfunction, and increase the production of reactive oxygen species (ROS). ROS directly harm pancreatic β cells and destroy epigenetic stressors by modifying DNA methylation and histone marks and harm the integrity of the polybiome and barrier, which feeds a loop of amplified dietary and infectious harm, establishing a feed-forward loop that amplifies dietary and infectious insults. Human biomonitoring and mechanistic reviews support the plausibility of causal relationships among plastics/EDCs, ROS formation, and destabilization of the polybiome, as well as metabolism/immune dysregulation; however, causal human cohort studies revealing causal pathways remain limited (Ullah et al., 2023).

These intersecting pathways have been mapped onto NexDi, which is the continuum to which autoimmune type 1 diabetes (T1D) and developmental type 5 diabetes (T5D) belong but is conceptually easy but not empirically well tested. Early nutrition deficiencies lower β-cell mass or functional reserve, a central hallmark of T5D (Misra et al., 2025); simultaneous epigenetic and microbial disturbances during early development may also impair immune tolerance, leading to increased susceptibility of immune assets to autoimmune freezing attacks typical of T1D. In many of these environments, maternal micronutrient deficiencies (e.g., vitamin D insufficiency being widespread), the early establishment of processed diets, high infectious burden, and high exposure to environmental chemicals combine to meet the clinical tipping point require the developmental or immunological compromise of β-cells (or both) in a child.

Experimental evidence of the polybiome in newly diagnosed children revealed a decline in the abundance of advantageous taxa, and supplementation with pathobionts supported an ecological interaction among diets, microscreening amenities, and disease expression (Radwan et al., 2020).

15.1 Research gaps and an action agenda to address them:

1- **Longitudinal causal inference**: There are imminent requirements for prospective birth cohort studies that concurrently evaluate maternal and infant exposure measures (such as endocrine-disrupting chemicals and micro- and nanoplastics), extensive dietary records, sequential polybiome and metabolomic sampling, assays of oxidative stress and mitochondrial performance, epigenetic scrutinies in placental tissue and neonatal blood, immunophenotypic analyses of regulatory T cells and Th17 cells, and clinical biomarkers, including islets. The inquiry is that such cohorts should be properly powered and stratified in terms of urban versus rural environments to identify nutritional transitions. The lack of multiomic data that are resolved in terms of time makes it impossible to move beyond association to causation (Sarigiannis et al., 2025; Rukavina Mikusic et al., 2025).

2- **Harmonization of micro- and nanoplastic measurements**: Quantitative techniques of micro- and nanoplastic measurements (e.g., m-FTIR, pyrolysis-GC/MS) should be harmonized, biofluids should be measured in validated panels of endocrine-disrupting chemicals, and standard biomolecules of ROS/oxidative-stress reactions (e.g., 8-hydroxy-2-deoxyguanosine, F2-isoprostanes, and glutathione redox pairs) should be adopted. This harmonization of assays will simplify data pooling and meta-analyses across sites (Leslie et al., 2022).

3- Arcaded interventional human studies: Having embedded short-duration randomized controlled trials as part of the cohort framework should confirm whether specific maternal nutritional interventions (e.g., methyl donors, vitamin D, omega-3 fatty acids), early-life dietary fiber and fermented food supplement supplementation, or household exposure mitigation (e.g., plastic-free food storage) have effects on intermediate biomarkers such as short-chain fatty acids, proxies of histone deacetylase activity, and intestinal permeability markers.

4- Multiomic causal analytics and interpretable AI: The use of causal inference procedures such as marginal structural models and mediation uses explainable machine-learning tools to detail a sequence of stages of increasing risk of autoimmunity and telesustainable pancreatic-β dysfunction spanning the increasing exposure through the obligate polybiome, metabolome, and epigenetic changes. Transparency in discovering and developing testable intervention hypotheses should be prioritized in the deployment of AI (Rooks et al., 2016).

15.2 Contextualized translation:

Interventions need to be culturally practical, including increasing the conventional production of traditional fermented foods such as *laban rayeb* and *kishk* to reestablish butyrate-producing microbiota and introducing low-cost maternal food supplement programs in regions where deficiencies of vitamins and methyl donors have been reported. These interventions should be paired with biomonitoring guidelines that are able to measure reductions in exposure to endocrine-disrupting chemicals, including BPA and phthalates, thus establishing the relationship between exposure trends and biomarker gains and eventually resulting in a reduction in NexDi (Allakany et al., 2023).

Synthesis conclusion (practical implication): A dietary approach based on nutrigenomics implemented within the first 1,000-day framework redefines diet as a key alterable factor in which ideal maternal and infant nutrition and intervention based on the restoration of the mucosal barrier could recalibrate relative to epigenetic-based mechanisms, reeducate immune regulatory mechanisms, and hamper biologically permissive conditions of plastic exposure, infectious agents, and immune-imbalanced subsystems of epigenetics, all of which contribute to NexDi. The scientific frontier is thus without a doubt the production of longitudinal, multiomic, intervention-ready findings in high-burden settings—including in Egypt—in addition to normalized exposure measurement and causal artificial intelligence analytics, thus furnishing the evidentiary foundation required in population-wide translational policies. These might include policies addressing maternal nutrition, policies focused on polybiomes, and policies addressing plastic-related endocrine-disrupting chemicals (Sarigiannis et al., 2025; Brügger et al., 2025; Trasande &Sarigis, 2024).

Although mechanistic explanations about nutrigenomics, polybiome metabolites, and dietary acclimation provide a biological framework for understanding NexDi, the lived experience of patients can never be encircled by a single framework that may be enclosed in molecular pathways. Persistent metabolic fluctuations, control over food consumption, and immunological imbalance lead to significant psychological strain, which, in turn, affects adherence, stability, and the overall course of the disease. This psychological aspect is not foreign to biology but has an extraordinarily strong connection: stress and depression may be able to regulate immune responses, remodel polybiome structure and nutritional behaviors, and thus create a vicious cycle between metabolic malfunction and psychological well-being. It is crucial to recognize this two-way crosstalk when developing integrative approaches where functional nutrition, nutrigenomics, and computationally optimized diets should serve not only as metabolic interventions but also as tools to strengthen mental resilience. This line of thinking preconditions the examination of how the concepts of psychological well-being, functional foods, and AI-based precision nutrition can intertwine to create sustainable change in type 1 and type 5 diabetes within the framework of the NexDi.

Although nutrigenomics and lifelong programming explain the biological factors behind NexDi, it is also crucial to discuss the psychological aspects and their intersection with functional nutrition. The reciprocal relationship between mental health, dieting behavior, and polybiome dynamics composes a bidirectional network that stunts resilience and susceptibility, making integration essential to a fully comprehensive NexDi framework (Reyed & Prabhakar, 2026; Wachamo & Gaultier, 2025; Weiner, 2025).

16. PSYCHOLOGICAL AND NUTRITIONAL STRATEGIES IN THE NEXDI FRAMEWORK

Mental health is an outcome and an effector of poor outcomes in individuals with diabetes who develop diabetes in childhood. The data consistently revealed that the prevalence rates of depression, anxiety, and diabetes-specific distress are higher in children and adolescents with type 1 diabetes than in nonill children. These psychiatric comorbidities reduce compliance with treatment protocols, increase glucose fluctuations, and increase the risk of complications in the future. In the conceptual framework of NexDi, the field of mental health cannot be pushed to the marginal plane; it represents a core factor that contributes to immune dysregulation, areas of fragility in the β-cell determination of immune functions, and metabolically unstable conditions. Likewise, this psychosocial aspect should be seen as a top priority in preventive and therapeutic interventions (Beverly & Gonzalez, 2025).

Human nutrition, which was formerly driven by the utilitarian interest of survival, has transformed into an advanced scientific field to increase health via individual and integrative approaches. This change has been driven by progress in defining how dietary and environmental factors likely contribute to the initiation and development of chronic noncommunicable diseases (NCDs), including obesity, type 2 diabetes, and an emerging neurometabolic syndrome, NexDi. These conditions are genetically preconditioned, but their development is an elaboration of a set of interactions between nutrition, the environment, and lifestyles. Since NCDs have escalated globally, outcomes have become more premature because of their economic burden, and given the unsinisterly known export on the subject in question, nutrition-related solutions can be viewed as one of the most cost-effective and sustainable means of reducing disease risk and improving outcomes (Agrawal et al., 2024; Singh, 2023; Sikalidis, 2019).

This paradigm shift has spawned the emergence of functional nutrition—a migrant rigorous approach to nutrition that aims to reestablish metabolic, immune, and psychological homeostasis via a focused emphasis on whole foods, traditional fermented foods, and specifically targeted nutraceuticals. The NexDi context specifically concerns the concept of functional nutrition since this is where psychiatry collides with metabolism and the polybiome sciences reveal themselves to be like these bridges between fields.

Laban rayebs and other products have been shown to maintain gut microbiota diversity and increase the generation of short-chain fatty acids (SCFAs). SCFAs, especially butyrate, have direct epigenetic and immunoregulatory actions, such as stimulating the growth of regulatory T cells, decreasing systemic inflammation, and consolidating mood and metabolic control. These polybiome-driven pathways offer convincing evidence that dietary programs can concurrently increase psychological resilience and enhance metabolic endpoints (Rukavina Mikusic et al., 2025; Sharma et al., 2024; Stein & Riber, 2023).

In addition to functional nutrition, nutrigenomics provides a fine-tuning layer by matching dietary interventions with genomic and epigenomic profiles. Nutrients and bioactive compounds can directly alter gene expression patterns, modify DNA methylation, and control histone modifications, thereby controlling pathways that are core to metabolism, immunity, and neurobehavioral control. This interaction in the genome forms the foundation behind individualized dietary plans that are based on the genetic composition of an individual, and the paradigm of broad-based dietary recommendations is replaced by the specifics of gene-based intervention (Wachamo &Gaultier, 2025).

Importantly, nutrigenomic effects are enhanced during a pivotal period of life, i.e., the first 1,000 days, during which maternal and infant diets have the potential to influence epigenetic programming, gut polybiome ecology, and future vulnerability to β-cell fragility (Stein &Riber, 2025).

Through a combination of genomic readouts and maternal–infant functional nutrition, it is feasible to suppress the two risk pathways of developmental NexDi (T5D) and autoimmune NexDi (T1D), thus providing a novel platform to intercept disease at the earliest life stages. In addition to psychosocial and nutritional lines, precision nutrition and nutrigenomics provide a mechanistic basis for how diet reprograms gene expression, immune tolerance, and β-cell resilience in the NexDi spectrum.

17. THE NEED FOR PRECISION NUTRITION IN NEXDI

A new working group dedicated exclusively to type 5 diabetes, a subtype that previously received little attention, was officially created by the International Diabetes Federation (IDF, 2025). Within the context of accepted types of diabetes, including type 1 diabetes (T1D), and the emerging form of diabetes, which is the fifth type (T5D or NexDi), precision nutrition has ceased to be an optional practice but has evolved into an inseparable part of the disease management procedure. Traditional dietary protocols are generally very broad-based and often tend to prescribe broad guidelines on carbohydrate intake, glycemic regulation, and micronutrient supplementation while paying little attention to dramatic intersubjective variability in metabolic response.

Surrogate biomarkers such as HbA1c, fasting glucose, and C-peptide levels provide partial information but cannot capture how all these interfaces dynamically interact to influence immune function, β-cell resilience, gut polybiome activity, and host metabolic processes. Such policy inconsistency between the established dietary guidelines and the changing analysis desperately introduces a hole to compromise management and prevention interventions in the long term (Selim, 2025). The literature indicates that two people with similar genetic preconditions might have different reactions: one might show a positive reaction to high-fiber or probiotic-enriched diets, and the other might show an imperceptible or even negative phenomenon. Such discrepancies highlight the shortcomings of the one-size-fits-all paradigm and indicate the necessity of adaptive systems that are able to process metabolic, genomic, and polybiome-derived data. Without these types of precise tools, dietary interventions risk remaining underpowered or mistimed, especially during the immature stage of the first 1,000 days of lifespan, during which nutritional signals have long-term population-wide effects on β-cell stability and immune regulation (Bacaloni & Agrawal., 2025; Shaffer et al., 2017). This integrative pathway is conceptually illustrated in Figure 4, which visualizes how nutrigenomic signals—particularly during the first 1,000 days—interact with epigenetic, metabolic, and polybiome-mediated mechanisms to shape β-cell fragility and define the NexDi spectrum.

Figure 5. The NexDi Concept: Linking Nutrigenomics, β-Cell Fragility, and Precision Nutrition.

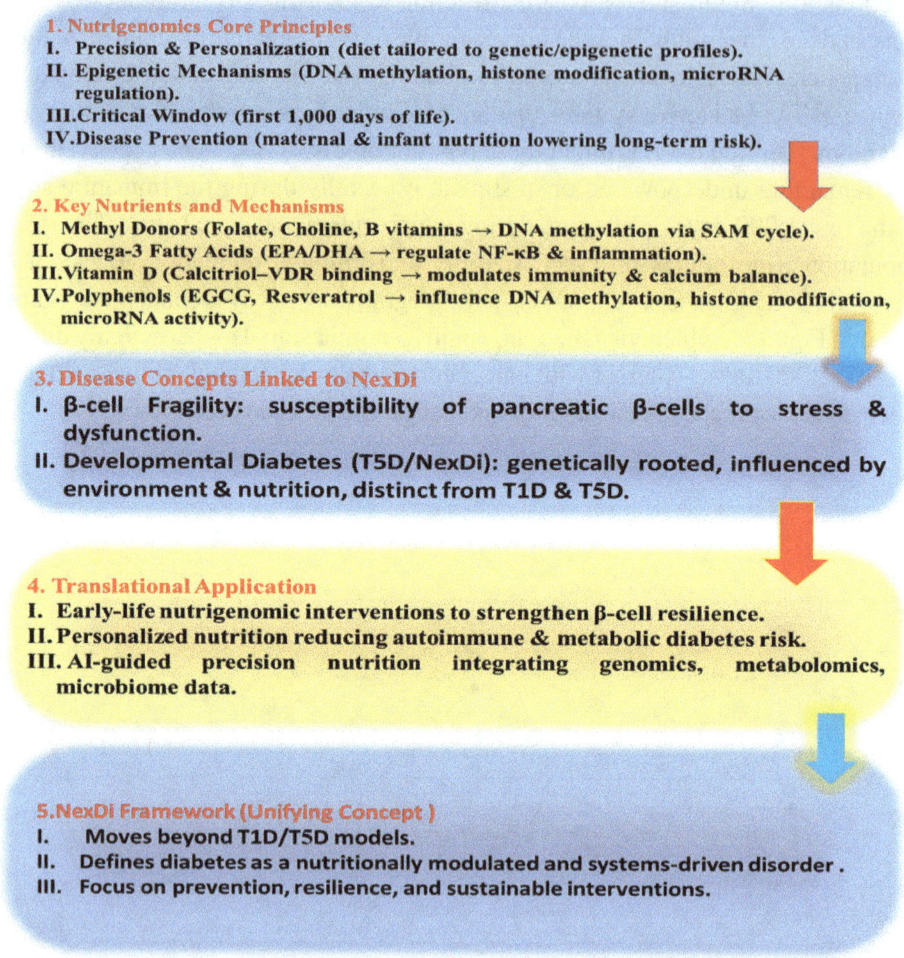

This figure illustrates the core integrative logic of the NexDi framework, beginning with nutrigenomic and epigenetic mechanisms that are most active during the first 1,000 days of life. Key nutrient-dependent pathways--such as methyl-donor availability, omega-3-mediated inflammatory regulation, vitamin D-driven immunomodulation, and polyphenol-related antioxidant systems--shape immune tone, metabolic flexibility, and long-term β-cell robustness. These processes converge on the central NexDi hypothesis of β-cell fragility and developmental diabetes (T5D/NexDi), distinguishing this pathway from classical autoimmune (T1D) and insulin-resistant (T2D) disease courses. The downstream translational implications include early-life nutrigenomic optimization, personalized dietary modulation, and AI-assisted integration of multiomics signals spanning genomics, metabolomics, and the polybiome. The framework underscores diabetes as a systemic, nutritionally modulated condition and shifts the clinical paradigm from glucose control toward prevention, resilience building, and metabolic sustainability. This value puts in perspective the integrative overlap of nutrigenomic pathways, early-life epigenetic governance, and micronutrient-specific biochemical pathways in regulating β-cell stability across life. The figure provides explanations of the mechanistic principles of the NexDi reclassification model by demonstrating the interactions between DNA methylation balance, inflammatory control, antioxidant defenses, and microbial-nutrient interactions. It highlights the fact that diabetes vulnerability, particularly T5D/NexDi, arises through the combination of cumulative biological strains rather than singular defects, which supports the paradigm shift toward anticipatory, nutrition-based precision prevention (Author-generated figure).

17.1 Genetic Foundations of NexDi and MODY as a Model

Monogenic types of diabetes, especially maturity-onset diabetes of the young (MODY), provide a specific tool that allows us to view the effects of specific genetic perturbations on metabolic pathways. In contrast to type 1 and type 2 diabetes, which are complex and involve the interaction of polygenic architectures with the environment, MODY separates the impact of single-gene mutations. There are currently at least fourteen well-known genes implicated in normal defects, including those involved in diabetes, such as GCK, HNF1A, HNF1B, HNF4A, PDX1, NEUROD1, KLF11, CEL, PAX4, INS, BLK, ABCC8, KCNJ11, and APPL1 (Riddle et al., 2020; Naylor et al., 2024). These genes establish specific functional realms of β-cell biology, such as glucose sensing (GCK), transcriptional regulation (HNF1A, HNF4A), developmental programming (PDX1, PAX4), and ion-channel modulation (ABCC8, KCNJ11).

Clinical heterogeneity is particularly strong in MODY, which highlights an important principle associated with precision therapeutics: genetics determines different responses to treatment and nutrition. As an example, patients with a harmful mutation in GCK typically do not require pharmacological treatment; rather, management through changes in diets balancing basal glucose levels could provide metabolic protection in the long term. On the other hand, HNF1A-MODY patients show high sensitivity to sulfonylureas, although the response to such treatment is effectively controlled by the dietary carbohydrate content. These findings demonstrate that genomics could be converted directly into nutrition-drug banks, which would represent a viable paradigm of precision medicine (Naylor et al., 2024).

However, MODY has been neglected as a research paradigm, as it has been confined and relegated to a rare disease category instead of discussing its advantages as a natural laboratory to investigate the weakness of β-cells and their interactions with nutrients. Mechanistic clarity intrinsic to MODY should therefore be utilized not only to test and perfect therapeutic measures during the management of affected families but also as an example of possible intervention during polygenic and environment-driven diabetes forms. Connecting to the theme of NexDi, MODY is a demonstration of principled evidence that the same environmental exposures, be they a dietary load, metabolic stress, or a pharmacologic challenge, may produce different responses depending on the underlying genetic environment.

The difficulty is how to scale this paradigm: to convert the GD of MODY into large-scale nutrigenomic models that can integrate genomics and polybiome data, as well as environmental exposures. If this effort yields results, there is an opportunity that MODY can be recast as a clinical niche into a paradigm-setting model of individualized metabolic management that would seal the gap between the sporadic genetic understanding and conventional treatment of diabetes.

17.2 Nutrigenomic cycle: Nutrition as a gene modulator

Nutrigenomics has had some fundamental impact on our theoretical understanding of nutrition and has shifted the debate so that we can start seeing food as something that engages in the regulation of gene expression and cellular activity instead of treating it merely as a source of energy. This metacognitive transition can be summarized as what can be called the nutrigenomic cycle, a system highlighting the bidirectional relationship that exists dynamically between nutrients and the genome (Archibald & Joffe, 2021). Genetic variation, on the one hand, determines the structure of nutrient absorption, metabolism, and use; on the other hand, nutrients and dietary bioactive compounds are molecular instructors capable of adjusting the circumference of the epigenome, transcriptional state, noncoding RNA functions, and mitochondrial activity (El-Sayed & Saleh, 2024). In the context of the NexDi framework, where diabetes is perceived as a nutrition-sensitive systemic disorder involving both genetics and the environment, the nutrigenomic cycle is not the peripheral concept but the core of the formation of actionable interventions (Hassan & Bayomi, 2024).

17.3 Epigenetic Message: Scripting and Remodeling the Metabolic Code.

A key mechanism by which diet can shape the genome is epigenetic changes, especially the methylation of DNA and histone modifications. Methyl donors such as betaine, choline, and folate can lead to the provision of substrates that directly deposit methylation patterns on the promoters of crucial immune and metabolic genes (Aloisi, & Poma, 2025). Systematically, the extent of hypermethylation of loci that regulate β-apoptosis can prolong cell survival, and modulated methylation of insulin receptor signaling can improve insulin sensitivity (Leesang et al., 2024). It has also been reported that exposure to environmental disruptors such as nanoplastics and endocrine-disrupting chemicals (EDCs) can lead to the destabilization of these methylation patterns and thus predispose humans to these conditions. This means that nutrients are important inputs not only as corrective factors but also as protection boards in response to environmentally induced epigenomic drift (Abdu et al., 2025).

17.4 MicroRNA Modulation: Nutrients as Noncoding RNA Regulators

In addition to epigenetic modifications, diet has a far-reaching effect on post-transcriptional regulation, as diet can induce changes in microRNAs (miRNAs).

Antioxidants such as bioactive polyphenols (resveratrol and quercetin) and long-chain omega-3 fatty acids (EPA and DHA) have been demonstrated to increase the expression of certain miRNAs that regulate insulin signaling cascades, inflammatory agents, and oxidative-stress responses. For example, miRNAs that promote endothelial and metabolic activities, including miR126, and miRNAs that promote proinflammatory activities, including miR21 and miR34a, have been linked to interventions with high levels of polyphenols (Liu et al., 2024). Likewise, preclinical and omics-based studies indicate that omega-3 supplementation inhibits proinflammatory miRNA vectors and increases insulin sensitivity (Zhou & Pang, 2025). Interestingly, nanoplastics, as environmental pollutants, can disrupt miRNA patterns in gastrointestinal and metabolic tissues to increase inflammatory and barrier dysfunction signaling, which dietary bioactives mitigate (Chen et al., 2024). Under NexDi, miRNA modulation provides a crucial, diet-forgiving lever in which nutrition can mediate the switching of the immune response to either tolerance or pathology under simultaneous conditions of metabolic stress and environmental exposure.

17.5 Transcription factor activation: Nutrients as molecular switches

Another branch of the nutrigenomic cycle is characterized by the stimulation of transcription factors by particular nutrients. Short-chain fatty acids, or short-chain fatty acids such as butyrate produced by a normal gut polybiome, activate peroxisome proliferator-activated receptors (PPARs) to regulate lipid metabolism, glucose uptake, and inflammatory homeostasis. Similarly, vitamin D, through its receptor (VDR), has been reported to alter not only calcium homeostasis but also the expression of genes that play critical roles in β-cell survival and immune tolerance. On the other hand, disruptions in the contents of polybiome-generated nutrient pools due to high-fat diets, nanoplastics, or dysbiosis induced by antibiotics weaken these regulatory cues, further disrupting the balance of metabolic homeostasis. The nutrigenomic cycle, therefore, provides the connection among the quality of dietary intake, polybiome resilience, and regulation of gene levels to a singular axis where health or disease outcomes are determined.

17.6 Mitochondrial Bioenergetics: Protecting the Powerhouse

Mitochondrial function is one of the points of intersection in the cellular pathogenesis of diabetes, where the ability of bioenergetic compounds and the redox balance together determine whether mitochondria will operate optimally or lead to oxidative damage. An examination by Iheagwam et al. (2025) revealed that the accumulation of reactive oxygen species (ROS), including vitamins C/E, polyphenols, and carotenoids, continues to be suppressed by antioxidant-rich diets and preserves mitochondrial integrity, which is especially important with respect to β-cells, which have a comparatively low antioxidant capacity. Equally supportive of this finding, Ge et al. (2023) suggested that disruptions in redox homeostasis act as a direct causal mechanism, rendering β-cells more vulnerable and initiating insulin resistance. Conversely, Zhang et al. (2024) demonstrated that Mino reserves can be hindered by certain enzymatic deficiencies in diabetic kidney disease, indicating that mitochondrial disturbances can be facilitated by endocrine-disrupting chemicals (EDCs) and nanoplastics. In the nutrigenomic context, dietary antioxidants consequently become not auxiliary performers but rather part and parcel molecular defense mechanisms against environmental maladies.

17.7 Bidirectionality: Genes Shape Diet, Diet Shapes Genes

The most fundamental feature of the nutrigenomic cycle is its appreciation of reciprocity. Genetic variation dictates individual metabolic response to nutrients--a concept vividly seeing application in monogenic diabetes such as MODY when mutations in a gene, such as GCK or HNF1A, cause an underlying change in the metabolic reaction to dietary carbohydrates, as illustrated by Ge et al. (2017) in their original review of the genetics of MODY. However, the cycle is reversible: nutrients affect real-time gene expression patterns, and cellular resistance adapts itself under metabolic, immune, and environmental challenges. An example is the way in which Zhou and Pang (2025) smaller entities (polyphenols) regulate transcriptional and epigenetic programs and how Hamadou (2025) identified the ability of bioactive lipids to tune immune-metabolic signaling. This bidirectionality bridges the old dichotomy of genetic versus environmental causes of diabetes, making nutrition a determinant of genetic risk and a modulator of genetic risk. Computational inventiveness and situational awareness are essential for translating the nutrigenomic cycle into practice. Recent progress in the fields of artificial intelligence and systems biology has yielded a novel opportunity to simulate gene–diet interactions that have never been available. As a complement to this idea, Ahn et al. (2023) incorporated environmental exposure science as a part of nutrition pathways, and localized studies, including those conducted by Almoghrabi et al. (2025) in Egypt, emphasize that local dietary and environmental contexts are key in running precision nutrition programs.

18. ARTIFICIAL INTELLIGENCE IN NUTRIGENOMIC DIET DESIGN

Artificial intelligence (AI) has become the engine of translating the discoveries of nutrigenomics into individually tailored and personalized dietary models. AI has enabled the combination of high-dimensional omic data, real-time physiological observations, and even contextual lifestyle data to create accurate dietary prescriptions instead of depending on population averages and has typically been used to encapsulate these factors into accurate dietary guidelines. This move makes nutrigenomics more of an operational science than a descriptive science, which is the power of sustained intervention that can be directed and personalized (Symeonides et al., 2024; Allakany et al., 2023).

At the predictive level, individualized glycemic and metabolomic responses can be predicted via supervised machine-learning algorithms, including gradient-boosted trees and deep neural networks. These models combine various streams of data, such as continuous glucose monitoring (CGM), gut polybiome sequencing, and host genomic variants. Indicatively, TCF7L2 or FTO genomic variability can interact with polybiome-generated metabolites, and AI software can estimate such nontrivial interactions and predict a particular individual's response to carbohydrate load or fiber supplementation (Lalani, et al., 2025).

In addition to dead-eye forecasts, AI is most successful in dynamic adaptation. Patterns of glycemic variability and immunometabolic stress in CGM data can be detected by recurrent neural networks with a specific focus on long short-term memory (LSTM) units and transformer-based architectures. These time series models dynamically regulate the timing, macronutrient ratios, and probability of probiotic supplementation even in response to changes in stress hormones, inflammation, or circadian variation (Lim et al., 2025). In this way, diet becomes a living prescription, changing nearly in real time in response to the physiological condition of the patient.

Predictive models are also more specifically refined using causal inference frameworks that discriminate correlation as opposed to causation. Systematic methods, such as targeted maximum likelihood estimation and causal forests, help researchers determine whether certain interventions, such as the intake of butyrate-inducing fiber or omega-3 supplementation, have a direct effect on β-cell function or whether they coincide with better results (Powell, 2018). As a result, the scientific basis of AI-based suggestions is strengthened because the interventions must indicate not only predictive associations but also mechanistic plausibility.

These insights are then converted into measurable meal plans via optimization algorithms. Techniques such as integer programming and reinforcement learning can be used to build meal patterns that meet and surpass a variety of goals simultaneously: glycemic stability, production of short-chain fatty acids, micronutrient adequacy, affordability, and cultural acceptability. Reinforcement learning, in more specific terms, allows us to refine meal plans through repeated optimization, as users follow the advice, and the metabolic response approximates the solutions yielding the most effective and sustainable solutions (Hassan et al., 2023).

To be able to explain things is paramount to clinical adoption. The counterfactual analysis and SHAP (SHapley Additive exPlanations) values are also considered tools that help explain the logic behind a particular suggestion, e.g., replacing barley with rice. This transparency increases trust among clinicians, patients, and policymakers, thus making the computational complexity–human understandability gap less pronounced (Lundberg et al., 2020). As a part of this hybrid framework, AI does not act as an additional auxiliary, but its integration of multiomic data, environmental exposures, and individual patient preferences creates a unified scheme channeling them toward consistent, practical, and adaptive nutritional plans (Symeonides et al., 2024; Allakany et al., 2023).

19. NUTRIGENOMICS AND ENVIRONMENTAL DRIVERS OF NEXDI: INSIGHTS FROM GLOBAL AND EGYPTIAN EVIDENCE

AI-based diet modeling continues to evolve, as does the emergence of environmental stressors as mechanisms to regulate metabolic health. Specifically, Ahn and Jeung (2023) highlighted the disruption of hormonal and immune pathways with endocrine-disrupting chemicals (EDCs) carried by plastics, but Hamadou (2025) elucidated the actions of bioactive peptides in some cases, which alleviate the abuse of insulin sensitivity and inflammation. Similarly, Chaudhary et al. (2025) noted that nutrigenomics has appeared as a source of mechanistic prediction of the connection between dietary elements and genetic susceptibility, which allows nutrition to be viewed not only as a gift of observation but also as a gift of precise intervention.

Egypt is an eloquent example of such interactions. Ahmed et al. (2025) reported the potential of low-carbohydrate nutritional approaches, including their effectiveness and representativeness of Egyptian culture, in Egyptian patients with type 2 diabetes. Moreover, Farag et al. (2024) reported that genetic polymorphisms such as TCF7L2 rs7903146 are considerable predictors of diabetes in Egyptian populations, indicating the importance of the localization of dietary recommendations in the context of genetic variations. To add to these observations, Ahmed et al. (2024) demonstrated that the perception of artificial intelligence in healthcare is still contradictory among Egyptian professionals, whose trust and explainability are key concerns in digital nutrition systems.

Zhang et al. (2023) exemplify how dietary modulation of the gut ecology can delay or even prevent diabetes onset, whereas Zhou and Pang (2025) illustrate the idea that dietary polyphenols reorganize miRNA networks, providing an additional molecular dimension in which diet shapes immune and metabolic strength. Similarly, Brügger et al. (2025) confirmed that real-time predictive modeling of postprandial glucose responses allows both personalized and real-time dietary interventions.

Vodovotz et al. (2020) reported that lifestyle-based interventions are a key component of prevention and reversal of chronic diseases in the framework of the translational research paradigm, but Fatumo et al. (2022) reported increased genomic heterogeneity in studies to ensure applicability throughout the world, which is highly pertinent in the Egyptian context since full data are scarce and are most often urgently needed. These studies all converge around a single metabolic state. Within the NexDi framework, diet, genes, and environmental stressors are not distinct determinants but rather operating (or malfunctioning) components of one unified metabolic regulation cycle (El-Sayed & Saleh, 2024; Archibald & Joffe, 2021; Kassem et al., 2023).

20. ARTIFICIAL INTELLIGENCE AS A TRANSLATIONAL BRIDGE

Although the global potential of these computation systems is indicated by their relevance in AI-powered nutrigenomics, their conceptual relevance can be further considered once personalized to a particular population, and in Egypt, the interplay of genetic polymorphisms and culturally embedded eating patterns requires location-specific consideration and adaptation (Ahmed et al., 2025). The variety instigated by these polymorphisms highlights the necessity of a computational apparatus that uses multilayered data. Artificial intelligence (AI) is thus a translational channel through which genetic variation can be connected with dietary response patterns and environmental exposures (Topol, 2019).

Predictive models can incorporate PPARγ variants, the nature of TCF7L2, ongoing glucose metabolism patterns, metabolomic profiles, and microbiome-derived short-chain fatty acids.

Potential educational implications in Egypt: Egyptian research has recently revealed that TCF7L2 polymorphisms affect the risk of gestational diabetes, demonstrating the importance of including local data (Farag et al., 2024). Similarly, the positive gains of carbohydrate-heavy diet methods in Egyptian groups have demonstrated significant metabolic advantages, highlighting the importance of culture-specific predictive models (Ahmed et al., 2025).

Reinforcement learning models are able to examine the impacts of small adjustments to the dietary compounds of fat or fiber on diet eaters with particular genotypes; this will generate adaptive food intake that is biologically specific (Vodovotz et al., 2020; Brügger et al., 2025). Additionally, explainable AI models such as SHAP values can clarify the interaction between genes and diets, leading to improved clinical outcomes and contributing to the development of an atmosphere of trust between patients and clinicians and helping to relieve so-called black-box skepticism (Topol, 2019). This interpretability is especially crucial in Egypt, where dietary changes must be scientifically sound and still socially acceptable (Ahmed et al., 2024).

In conclusion, universal genomic research highlights the importance of cross-ancestry validation, which demonstrates that refining risk scores across populations reduces bias and increases clinical applicability (Fatumo et al., 2022). In this way, nutrigenomics in this global–local framework can be applied to Egypt as an example of how AI-mediated nutrition can be implemented in culturally specific but scientifically sound settings (Reyed, 2023b; Guevara--Ramirez et al., 2023; Almoghrabi et al., 2025).

20.1 Addressing Scientific Gaps and Cultural Barriers

The practical application of this multifaceted promise faces three critical challenges that limit its real-world feasibility. First, there are no genomic or nutrigenomic datasets available in Egypt at present that would potentially allow the establishment of allele frequencies and their dietary associations among local cohorts. Without such datasets, predictive models run the risk of using biases generated by studies covering Europeans or Asians, thus failing to document the genetic architecture unique to Egyptians. Second, cultural eating patterns, such as a strong dependency on *Baladi bread* (an ancient Egyptian flatbread, see Hussein et al., 2024), fried legumes, and a limited variety of vegetables, affect the applicability of hierarchical dietary prescriptions. Third, the computational and communication framework to unify genetic, nutritional, and clinical data is immature, with suboptimal research potential and clinical implementation.

A combination of three interconnected strategies is suggested to correct such shortfalls. First, investigations of a population-specific nature concerning the context of the prevalence and functional implications of the sections of PPAR7L2, TCF7L2, and other candidate polymorphisms in Egyptian cohorts, including maternal–infant dyads, are needed. Second, nutritionists should co-design culturally adapted guidelines with stakeholders in the community to ensure that food recommendations resonate with local food availability and culture. Finally, AI-based platforms need to be built to combine such datasets and provide personalized nutrition suggestions that respond dynamically, remain adaptive, and be based on Egyptian realities (Ge et al., 2022; Almoghrabi et al., 2025).

20.2 Broader scientific and clinical implications

Recent studies worldwide support the information presented on the presence of gene–diet interactions by showing that individuals with TCF7L2 risk alleles exhibit more pronounced improvements in their glycemic regulation in response to high-fiber, low-glycemic dietary regimens than to high-carbohydrate regimens (Rein, et al., 2022).

Extrapolating these findings to the Egyptian setting has two benefits: it can help refine individualized interventions in addition to enhancing the global understanding of how genes and the environment interact in diverse populations. With the integration of nutrigenomics into the extensive NexDi framework, Egypt has the potential to serve as a hub where genomic, nutritional, and environmental exposures intersect and provide insights that can be extrapolated to other Middle Eastern and African regions (Riddle et al., 2020; Sivri & Gezmen-Karadag, 2024).

20.3 Agenda Research and Implementation.

On the basis of a mechanistic understanding of the nutrigenomic cycle, the computing power made possible by AI, and the identification of environmental perturbations as indirect initiators of immune–metabolic pathology, an articulate roadmap develops on how to put NexDi into practice. Priority should include adaptive clinical trials, longitudinal cohorts, standardized assays of nanoplastics and metabolites, federated learning frameworks, and policy translation strategies that incorporate maternal nutrition subsidies, the regulation of endocrine-disrupting chemical (EDC) exposures, and equitable access to digital health help (Aloisi & Poma, 2025; Chen et al., 2024; García-García et al., 2024; Leesang et al., 2024; Qin et al., 2025; Zhang et al., 2024; Li et al., 2025; Iheagwam et al., 2025),

21. INTEGRATED NEXDI MODEL: NUTRIGENOMICS, MITOCHONDRIAL DYNAMICS, AI, AND ENVIRONMENTAL DRIVERS IN EGYPT.

Together, these priorities underscore a continuum that spans the molecular underpinning of NexDi as to its wider societal implications. Mody is a classic example of a perturbation at the single-gene level, revealing the principle of nutritionally modifiable genetic susceptibility. The nutrigenomic cycle extends this concept, demonstrating how dietary interventions repattern gene expression trajectories through both epigenetic and transcriptional rearrangements associated with mitochondrial remodeling. AI is an operational integrator that transforms multiomic data into adaptive, personalized interventions. This cycle is complicated by environmental disruptors, i.e., nanoplastics and endocrine-disrupting chemicals, but in any case, it stresses the need to curtail genetic risk with the help of protective nutritional interventions (Ahn & Jeung, 2023; Chen et al., 2024).

Located at the intersection of genetic predispositions, fast food shifts, and increased pollutant exposure, Egypt will provide an informative trial of the NexDi framework, which could be converted into a beneficial outcome of the proposed study. Its problems, including nutrient deficits, socioeconomic inequalities, and environmental factors, resemble those faced by many countries in the Global South, thus making Egyptian-based interventions globally relevant (Ahmed et al., 2025; Farag et al., 2024).

In next-generation diabetes (NexDi), the interaction between genetic predispositions and dietary patterns and polybiome metabolite-derived metabolites modulates epigenetic and transcriptional networks, which determine immune-metabolic trajectories, as shown in Figure 5. This balance is disturbed by environmental stressors (nanoplastics and endocrine disruptors), and artificial intelligence offers adaptive means to design precision nutrition plans. This framework highlights the convergence of genomics, nutrigenomics, polybiome science, and AI to produce personalized interventions that have translational effects.

Figure 6. The Nutrigenomic–AI Pathway from Genetic Susceptibility to Public Health Policy.

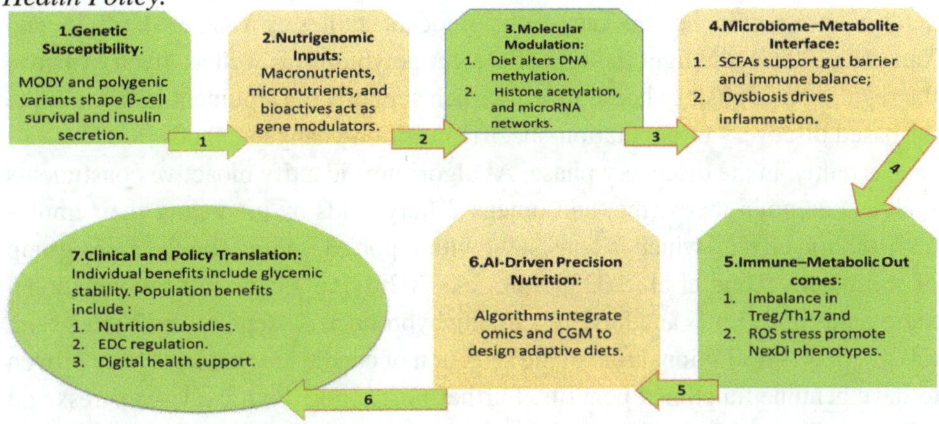

This graphic shows the integrated systems framework that forms the basis of NexDi, demonstrating the interaction of nutrigenomic pathways, mitochondrial resilience, polybiome-derived metabolites, and environmental stressors to determine the level of vulnerability to β- cells in the diabetes continuum. The diagram emphasizes nutrient-sensitive processes that control mitochondrial signaling, the effects of EDCs and microplastics on oxidative balance, and the destabilization of polybiome-host crosstalk in early and late metabolic transitions. These multidimensional biological signals are all integrated using AI to create predictive patterns and stratify the risk of metabolic problems, allowing the use of precision nutrition interventions during specific windows. The model is mirrored by Egypt, with its unique environmental and nutritional context, highlighting how a mixture of genomic, ecological, and dietary stressors accelerates the transition into the T1D-T5D range within the NexDi framework. Nutrigenomic signals, mitochondrial stress pathways, and polybiome-host interactions with environmental toxicants are combined to marker-trace pathways that define β-cell vulnerability on the spectrum of diabetes. By applying these biological layers to AI-based pattern recognition, the model shows how early mechanistic perturbations scale to population-level risk profiles and insights in policy. The integrative pathway indicates that the effects of the ecological and nutritional exposures in Egypt on such mechanisms are unique, thus supporting the necessity of anticipatory, precision-designed, and publicly targeted health strategies as part of the NexDi framework (Author-generated figure).

22. ARTIFICIAL INTELLIGENCE IN NUTRACEUTICAL DISCOVERY: EGYPT AS A LIVING LABORATORY

Artificial intelligence (AI) is set to transform the sphere of research on nutraceuticals. It includes the combination of genomics, metabolomics, microbiota profiling and their interactions, dietary intake, and environmental exposure evaluation, thus helping to provide accurate and individualized therapy.

In an Egyptian setting, changes in environmental factors produce a unique metabolic environment. Interactions with indigenous dietary patterns are conditioned by genetic polymorphisms, including TCF7L2 and PPAR/indigenous dietary interactions. These interactions are subsequently modulated by environmental stress factors, particularly microplastics and endocrine-disrupting chemicals.

These links have been highlighted by Martínez-Pinna et al. (2023) and He et al. (2024) and are examples of how environmental and genetic factors overlap to determine metabolic health. The results of recent studies by Yang et al. (2024) and Yan et al. (2025) highlight the role of the gut polybiome as well as dietary fiber in the regulation of glucose homeostasis, which supports the argument in favor of the AI-based discovery of population-specific nutrition-change methods.

Currently, in the discovery phase, AI algorithms identify bioactive constituents such as curcumin, berberine, and omega-3 fatty acids by predicting their immunometabolic effects, which is consistent with reported findings on the same group of substances (Ding et al., 2025; Xu et al., 2025). An analysis of the nutritional ingredients of legumes and food-processing byproducts, as reported by Nartea et al. (2023), plays an additional role in the selection of candidates that have been proven to have genuine functional benefits. Furthermore, under both oxidative stress and inflammatory stress conditions, molecular interactions are modeled via simulation and optimization via deep learning to reinforce learners and enhance such synergies between probiotics, prebiotics (Baimakhanova et al. (2025) and antioxidants, as demonstrated in studies on longevity-promoting compounds by Santos et al. (2025) and Kiran et al. (2025).

Within the translational step, AI supports virtual trials and predictive modeling for capturing population-specific dynamics in urban and rural environments, where real-time data assessed via wearable sensors, glucose monitors, and digital health platforms are now available to enable adaptive interventions, as reported by Ding et al. (2025) and El-Sayed & Saleh (2024). Figure 6 represents this three-level layer cake of discovery, simulation, and translation: it allows Egypt to serve as a living laboratory where AI gathers molecular insights and suggests nutrigenomic and polybiome-based approaches to magnificence, which are culturally embedded, as has been argued by D'Ambrosio et al. (2022), Radwan et al. (2020), Kassem et al. (2023), and Riddle et al. (2020) on precision nutrition and personalized medicine.

Figure 7. Deep Learning an AI-based Nutraceutical Pipeline to Translate Genomic and Nutritional Insights into Egyptian Health Strategies.

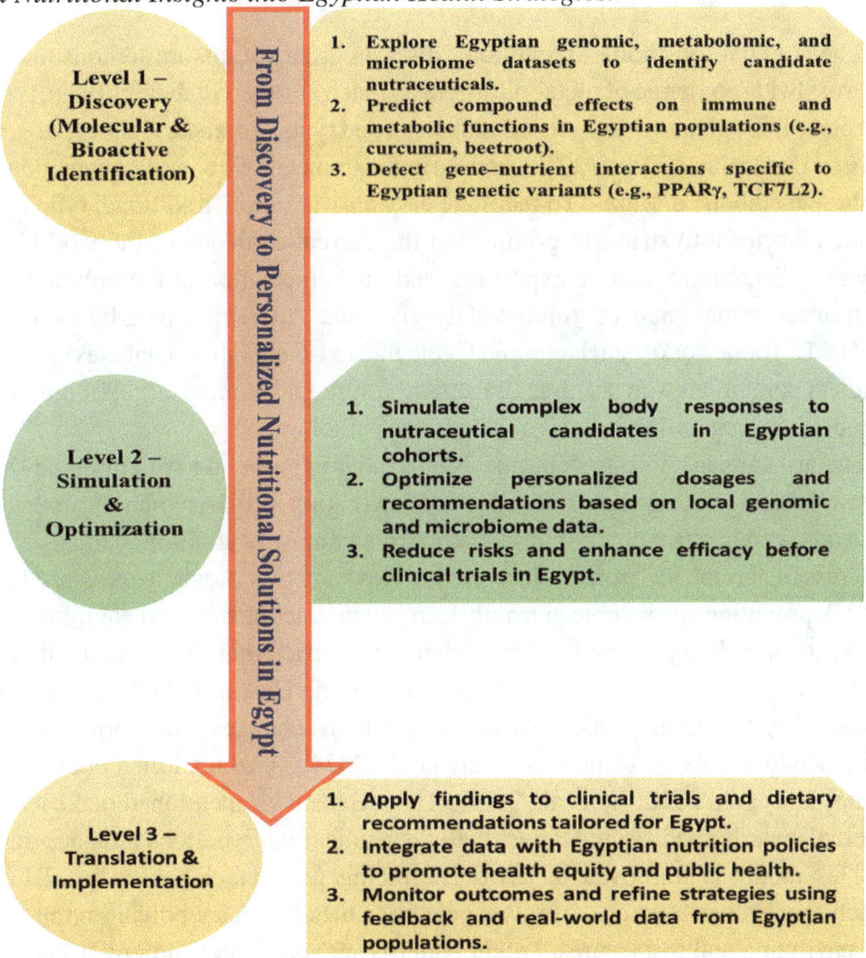

The figure visualizes the three layers of the AI pipeline, namely discovery, simulation, and translation, that propel nutraceutical innovation under the NexDi ecosystem. The discovery layer combines genomics, metabolomics, and polybiome signatures to find candidate bioactive compounds and nutrient-gene pairs applicable to Egyptian metabolic susceptibility. The simulation layer is an expression of oxidative, inflammatory, and immunometabolic pathways through deep learning optimization, enabling the prediction of synergistic effects between probiotics, prebiotics, antioxidants, and culturally grounded dietary components. The translational layer is a system that integrates real-life data using wearables, online platforms, and environmental exposures in order to create population-specific nutrition plans that are tailored to the unique ecological and genomic environment of Egypt. Together, these layers place Egypt squarely as a living laboratory where AI-directed nutraceutical development can bring molecular knowledge to bear on context-relevant precision nutrition solutions. The figure represents the application of the AI-guided nutraceutical discovery approach designed to bridge the gap between multi-omic information and culturally specific nutritional treatments at NexDi. The figure explains how AI can predict metabolic responses before clinical manifestation by illustrating how genomic and polybiome markers are incorporated into simulation models to refine antioxidant, probiotic, and nutrient synergies. The concluding translational layer highlights the potential of population-specific data, based on the unique exposomic and dietary landscape of Egypt, to be applied into practical, precision-crafted nutrition plans. Altogether, the model illustrates that Egypt can be used as a real-world testbed to implement AI-enabled nutraceutical innovation (Author-generated figure).

23. RETHINKING DIABETES BEYOND CONVENTIONAL TYPOLOGIES: THE NEXDI PARADIGM

The essential constraints of traditional disease classifications are demonstrated by the worldwide epidemic of metabolic diseases, such as type 1 diabetes (T1D), type 5 diabetes (T5D, malnutrition-related diabetes), and type 3c diabetes (pancreatogenic diabetes) Vonderau & Desai (2022), and the immune-metabolic complications induced by these metabolic diseases. To facilitate easy classification, traditional typologies divide diabetes into stringent groups, but the current evidence of the world links genetics, diet, environmental exposures, and the composition of the polybiome to form an active malignant continuum of dysglycemia. Alternative paradigms such as the NexDi framework (Nutrigenomic Cycle in Next-Generation Diabetes) propose a radical point of view in defining diabetes as a systemic manifestation of immune–metabolic dysregulation instead of a separate pathology.

Dietary antigens, intestinal permeability, and polybiome-derived immune signs have a resilient effect on autoimmune β-cell destruction in T1D, which illustrates the inadequacy of exclusively immunological explanations. In addition, damage to the pancreas or chronic inflammation, as well as exposure to ecological toxicants, leads to T5D, providing an example of a multifactorial interaction between environmental toxins and hereditary strengths. Researchers are worried that short-chain fatty acid (SCFA) depletion and dysbiosis, along with the increase in pathobiont numbers promoted by pollutants or microplastics, substantially change immune tolerance and resilience to metabolic responses (Yang et al., 2024; Martínez-Pinna et al., 2023; Campanale et al., 2020; He et al., 2024). The mechanisms mentioned make it clear that molecularly specific and contextually dependent interventions are necessary.

This model is improved by artificial intelligence, which incorporates high-dimensional information on the basis of genomics, metabolomics, polybiome profiles, and environmental disclosures. Latent patterns across clinical and omics datasets, such as machine learning and deep learning models, can be used to predict the development of T1D in genetically predisposed humans who have been exposed to endocrine-disrupting chemicals, nanoplastics, or nutrient depletion (Kiran et al., 2025; Ding et al., 2025). Digital twin models can also simulate personalized behaviors that optimize nutritional and nutraceutical treatments, such as the use of nitrate-dense beetroot supplementation by patients with T5D and an intact polybiome-mediated nitrate reduction ability (Yan et al., 2025; Xu et al., 2025).

Nutrigenomics also provides intricate granularity by determining the effects of polymorphisms in loci such as TCF7L2, PPARG, and HLA on how individuals respond to bioactive compounds such as curcumin, berberine and adipol (El-Sayed & Saleh, 2024; Guevara-Ramirez et al., 2023). Findings assist in the process of informing personalized nutrition strategies, which are beyond the limits of a one-size-fits-all paradigm. In Egypt and other populations located in the Global South, where complete genomic datasets are limited, these frameworks can shed light on the most applicable research gaps and open up avenues for the generation of region-specific insights that are often overlooked by global frameworks. The case analysis of comparative cases highlights the clinical implications of the NexDi paradigm:

On the basis of the former clinical contrasts (Table 2), the following analysis develops an assessment of the molecular and translational dimensions that define the NexDi framework.

In patients with type 1 diabetes mellitus (T1DM), high levels of environmental pollutants accelerate β-cell attrition, which is enhanced by a lack of short-chain fatty acid-producing microbiota; in patients with type 5 diabetes (T5D), as they present with pancreatic damage, the response to probiotic and antioxidant supplementation is heterogeneous and depends on the composition of the polybiome and diet. These observations are pooled together in Table 2, as genotype, environmental exposure, polybiome profile and responses to bespoke interventions are compared, and at the same time, the current gaps in therapeutic addressing are outlined. This comparative framework foreshadows the necessity of adaptive, integrative therapies that have a synergistic combination of artificial-intelligence-based predictive modeling, nutrigenomic stratification, and cost-effective local nutraceutical modalities (Pammi et al., 2025; Nartea et al., 2023; Li et al., 2025).

Despite this potential, the NexDi framework faces several essential challenges. The predictive power of AI depends on the lack of data on the underrepresented population. The design of research protocols should therefore be informed by ethical issues related to privacy, equity, and informed consent. Moreover, there are economic and logistic barriers that could hinder the transfer of computational models into clinical practice. To alleviate these difficulties, the combination of high-tech innovations such as patented nutraceutical compositions and AI-based discovery systems with low-cost, cultural consonant policies that can be incorporated into existing health care systems is needed.

In the future, the NexDi paradigm will envision a holistic, integrated, and fair future of diabetes research and care. This approach will synthesize immunometabolic science, artificial intelligence, nutrigenomics, and polybiome ecology, along with localized nutraceutical innovation, to establish a scalable model applicable to both individualized patient care and population-wide interventions. In the Egyptian setting, this positioning was successful in transforming national health systems into vibrant laboratories and subsequently making it possible to develop solutions that are scientifically rigorous, culturally sensitive, and economically feasible (Riddle et al., 2020; Naylor et al., 2024; Archibald & Joffe, 2021; Kassem et al., 2023).

Table 2. Comparative features of type 1 and type 5 diabetes in the NexDi framework: Molecular, environmental, and translational insights from the Egyptian context

Characteristic	T1D	T5D	Translational significance	Reference
Genetic susceptibility	Immune and metabolic gene variation (HLA, TCF7L2)	Metabolic and insulin-regulating gene (PPARγ, TCF7L2)	Guides personalized nutrigenomic strategies	(Almoghrabi et al., 2025)
Environmental influence	Hormone-disrupting chemicals and nanoplastics	Chronic inflammation, environmental toxins	Machine learning-based disease forecasting	(El-Degwi et al., 2024)
Polybiome profile	Loss of beneficial metabolites and microbial imbalance (SCFA depletion, dysbiosis)	Altered intestinal microbial diversity	Probiotic/prebiotic optimization for immune–metabolic balance	(Allakany et al., 2023)
Nutraceutical Response	Anti-inflammatory bioactives/Plant-derived modulators)Curcumin, Berberine)	Nitric oxide precursors and oxidative stress protector (Beetroot nitrate, antioxidants)	Customized nutrient support aligned with individual biology	(Chaudhary et al., 2025
Knowledge limitations/Evidence deficit	Scarce local genetic datasets	Few population-based studies in developing regions	Emphasize context-specific research and flexible strategies	(Reyed & Prabhakar, 2025a)

This table provides an integrative comparison of type 1 diabetes (T1D) and type 5 diabetes (T5D) as conceptualized through the framework of NexDi. It outlines critical genetic, environmental, and polybiome-related dissimilarities that influence disease manifestations and intervention responses and relationship designs, which are based on the unique nutritional changes in Egypt, environmental contact, and information gaps, support nutrigenomic stratification, AI-supported prediction, and the generation of a locally tailored precision nutrition strategy.

The systematic approach outlined in this comparative framework highlights the potential of the NexDi model as an integrative, evidence-based, and regionally contextualized model. It integrates genetic vulnerability, environmental exposure, and polybiome-mediated mechanisms, which predict a realistic roadmap of research, policy development, and clinical practice that is directly related to the changing metabolic environment in Egypt. This combination, therefore, supports the need to uphold the power of fine-tuning nutrition and AI-informed interventions as the cornerstone of addressing the dual burden of T1D and T5D among high-risk groups.

As a continuation on this regional basis, the next frontier is to integrate NexDi initiatives in Egypt with international research and development in nutraceuticals.

24. GLOBAL INNOVATIONS IN NUTRACEUTICALS FOR DIABETES PATIENTS: AN ANALYTICAL PERSPECTIVE

The nutraceutical developmental environment has changed over the last twenty years, moving away from the traditional approach to nutraceutical products, which is more of an herbal-based remedy, toward consistently characterized, patented products that combine pharmacological, nutritional, and biotechnological properties. These products are characterized by standard bioavailability, mechanistic validation, and clinical testing, which puts them in a position as acceptable and persuasive supplements or substitutes for traditional pharmacotherapy in diabetes and associated metabolic conditions. Global innovations are characterized by scientific complexity but accessibility, and their applicability to resource-income-restricted cases, including those in Egypt, depends on population-worthy affirmation, price, and customization (El-Sayed & Saleh, 2024; Ding et al., 2025; Nartea et al., 2023). 23.1. Comparison and analysis of major patient nutraceuticals.

24.1.1 Curcumin and its derivative curcumin: a major bioactive component of turmeric (Curcuma longa), are examples of polyphenolic compounds that have pleiotropic impacts on metabolism and immune-modulatory effects. It makes AMPK active and NF-kB inactive and causes changes in the composition of the gut microbiota, which suppresses systemic inflammation and enhances insulin sensitivity. It is clinically effective in both type 1 diabetes (T1D) and atypical type 5 diabetes (T5D) because its increased bioavailability is based on two factors: the use of nanoformulations and the use of phospholipid complexes. In the NexDi paradigm, the immunomodulatory potential of curcumin especially applies to the process of modulating the loss of β-cells into autoimmune destruction while using it as an optimal source of metabolic resistance (Ding et al., 2025; El-Sayed & Saleh, 2024).

24.1.2 Berberine: An example of the transformation of traditional medicines into drug-like nutraceuticals is the translation of berberine, a drug that is an isoquinoline alkaloid found in Berberis species. Sustained-release pills, probiotic coformulations, and artificial intelligence-controlled dosage optimization are patented formulations. Mechanistically, berberine does not promote the proposed antiglycemic actions in GHG accumulation but modulates bile acid metabolism, affecting the restructuring of the gut microbiota to flourish SCFA-producing taxa, improving glucose homeostasis, and bypassing the immune response (Yang et al., 2024; Yan et al., 2025). Clinical trials have shown that, similar to metformin, metformin reduces HbA1c, fasting glucose, and triglyceride levels. However, population-specific responses vary and depend on genetic polymorphisms (TCF7L2, PPARγ), the diversity of the polybiome, and environmental exposures, including pesticides and microplastics, which are specific factors that are especially relevant to Egypt (Radwan et al., 2020; Martínez-Pinna et al., 2023).

24.1.3. Omega-3 fatty acids: Their relevance in cardiovascular settings is expanded to metabolic and immune settings by supporting omega-3 fatty acids in the form of patented EPA and DHA formulations. They stimulate resolvin and protectin signaling and facilitate anti-inflammatory and immunoresolving interactions involving immune-nutrient overlap with NexDi 's current user interests. Artificial intelligence-based network pharmacology models have been shown to play a role in the regulation of β-cells and T-cell differentiation, expanding their potential uses in both T1D and T5D (Ding et al., 2025; Kiran et al., 2025).

24.1.4. Probiotics and Synbiotics Distinctive modulation of the polybiome, particularly in populations, is now the central subject of patented interventions. Next-generation synbiotics (e.g., *Lactobacillus rhamnosus* GG and *Bifidobacterium longum* BB536) and engineered probiotics aim to accomplish the reestablishment of gut barrier integrity, augment the production of SCFAs, and reduce autoimmune activation. Metagenomic profiling using AI will increase the rate of discovering strains optimized for the local polybiome, which offers possibilities for personalized probiotic therapies (Pammi et al., 2025; Kassem et al., 2023).

24.1.5. Adipol: Adipol is a new generation of patentable bioactive agent that is an adipose tissue signaling agent. Its two-pronged action, where it improves insulin sensitivity and regulates adipokine profiles, is representative of how AI-assisted compound discovery and digital twin modeling can lead to quicker innovations. At a preliminary stage, evidence suggests that Adipol can be tailored to fit local adaptations in Egypt in combination with nutrigenomic profiling and culturally applicable dietary approaches (El-Sayed & Saleh, 2024; Ding et al., 2025).

24.2 Analytical Synthesis and Comparative Perspective

This section outlines the mechanistic, clinical, and translational nature of the major nutraceuticals under investigation, highlighting how each of the analyzed agents functions via specific biological pathways and how their efficacy at any moment depends on the circumstances. These nutraceuticals, including curcumin, berberine, omega-3 fatty acids, probiotics, and emerging compounds such as adipol, precipitate their actions through a combination of these mechanisms, which include metabolic regulation, immune modulation, and gut–polybiome crosstalk. However, these methods are not always successful; individuals in different populations respond sharply differently because of variations in their genetic composition, nutrition, polybiome composition, and exposure to the environment. These contextual variables cannot be understood without an understanding of the ultimate usefulness of interventions to develop approaches that are effective and satisfactory. The following is an example: TCF7L2 or PPARG genetic polymorphisms can influence responsiveness to berberine or omega-3 supplements, but the local diet and polybiome should predetermine nutrient availability and polybiome personnel. Similarly, the fact that curcumin is a natural compound with low bioavailability underscores the need to develop technological solutions, e.g., the creation of nanoparticles, to ensure therapeutic effects. In addition to biological variability, new optimization opportunities arise from the combination of artificial intelligence (AI) and multiomic analytics. AI-based models can identify latent patterns of response and geometries of metabolic interactions and individualize nutraceutical combinations, mitigating responses at the individual or population level. This methodology is especially relevant in resource-rich environments, where adaptive AI systems have the potential to enhance clinical outcomes with the assistance of locally available dietary and nutraceutical sources. Under these circumstances, the discussion establishes a conceptual space between nutrigenomics, polybiome research, and computational precision nutrition, which are designed to place nutraceuticals in the context of a living, progressive, data-driven strategy that analyzes the broader spectrum of human health within the framework of NexDi.

Curcumin accomplishes its functions through stimulation of AMP-activated protein kinase (AMPK) pathways, repression of NF-κB signaling, and gut polybiome regulation. The effect of a combination of these mechanisms is thought to be better glycemic control and the mitigation of systemic inflammation. The natural bioavailability of the compound is, however, restricted, and to improve this absorption, the design of nanoparticle delivery systems has been developed. In addition, artificial intelligence platforms have the potential to further streamline the formulation and dosage decisions of curcumin by predictively modeling metabolic and inflammatory responses (Servida et al., 2024; Tian et al., 2025; Sandeep et al., 2024).

Berberine exerts its pharmacological effects via AMPK activation, bile acid metabolism, and short-chain fatty acid (SCFA) synthesis. Decreases in hemoglobin A1c, fasting glucose, and triglyceride levels have been recorded through the use of randomized clinical trials. Nonetheless, the therapeutic response seems to be dependent on genetic polymorphisms, such as differences in TCF7L2 and PPARG, in addition to differences in gut polybiome composition and region-specific environmental exposures that are common in Egypt. Digital twin simulations and virtual clinical trials based on artificial intelligence should be taken as opportunities to demarcate favorable therapeutic regimens and adaptive dosing plans (Almoghrabi et al., 2025; Brügger et al., 2025).

The effects of omega-3 polyunsaturated fatty acids, i.e., eicosapentaenoic and docosahexaenoic acids, on the execution of anti-inflammatory effects involve defined, mechanically specific, proresolving mediator pathways, i.e., protectins and resolvins, that support the physiological maintenance of β cells and balance T-cell differentiation. The relative clinical performance of these fatty acids is tempered by local eating patterns and the presence of concerned foodstuffs. The interaction of omega-3 fatty acids and the flexibility of predicting personal benefits cannot be achieved without the use of AI-based network pharmacology technologies because these approaches require the combination of molecular and dietary information (Singh, 2023; Agrawal et al., 2024).

Probiotics and synbiotics increase the integrity of the intestinal barrier and the production of short-chain fatty acids (SCFAs). These actions significantly suppress the activation of the immune system and promote metabolic balance. Considering the interpopulation heterogeneity in terms of polybiome diversity, there is a critical need to optimize strains. Thus, combining metagenomic profiling with personalized artificial intelligence models enables the development of specific probiotic interventions to restore the immuno-metabolic balance (Thakur et al., 2025; Gonzalez et al., 2025).

Adipol is a novel adipokine-derived compound that improves insulin sensitivity by regulating adipokine signaling pathways. Although existing data are mainly of a preclinical nature, the compound has a high degree of translational potential. Discovery pipelines and models of adaptation specific to the population, which are facilitated by artificial intelligence, could hasten its development and validation in regional cohorts (Ding et al., 2025).

24.3 Critical Reflection and Future Directions

The assumption of tension between global innovation and contextual applicability can be observed through patented nutraceuticals. The fact that such products are highly sophisticated cannot be denied, but they cannot be easily accessed in Egypt owing to low affordability, reliance on imports, and regulatory failures. Predictive analytics via artificial intelligence, digital twin models, and nutrigenomic profiling are powerful solutions to fill this gap, allowing them to be validated and optimized to suit certain population groups. The NexDi framework reiterates the combination of these innovations with foods that are locally available, culturally consistent dietary practices, and economical modes of production, thus creating a hybrid model with a recap of global scientific developments and regional viability (El-Sayed & Saleh, 2024; Nartea et al., 2023; Ding et al., 2025).

23.4 Future research priorities:

1. Population-specific genomic and polybiome data should be created to inform and utilize individual nutraceutical interventions.
2. Long-term effectiveness, safety, and sustainability of patented compounds in cohorts of Egyptians.
3. - Local production pipelines of high-evidence nutraceuticals will be developed so that they do not rely on imported patented products.
4. Using AI-powered virtual trials and predictive modeling to streamline formulations and doses.
    - Addressing knowledge divides in regulation and education to adopt nutraceuticals as part of the population health agenda.

To balance global patented innovations with locally hierarchically adjusted, evidence-based strategies through synthesizing the former into locally pragmatic and scalable tools that can alleviate the burden of diabetes and metabolic disorders arising from it, Egypt and other rising economies can turn nutraceuticals from expensive, experimental interventions into scalable and long-term decisions (Radwan et al., 2020; Pammi et al., 2025; Kiran et al., 2025). This unifying vision not only focuses on the gaps in research but also looks deeply into the future needs of the health system, harmonizing the understanding of mechanisms, AI-inspired discovery, and tendencies during implementation within its cultural context.

24. THE NEXDI FRAMEWORK: INTEGRATING AI, NUTRIGENOMICS, AND THE POLYBIOME FOR PRECISION PUBLIC HEALTH IN EGYPT

The NexDi framework defines a breakthrough intersectional synthesis of complex scientific, technological, and populist mechanisms aimed at changing the approach to the prevention and management of diabetes in Egypt. Fundamentally, NexDi represents a combined effort of artificial intelligence (AI), nutrigenomics, and polybiome studies into intensive and data-driven means of nutrition and metabolic well-being. This integration facilitates highly personalized interventions that forecast and optimize reactions to nutraceuticals and take into account the particular genetic, dietary, and environmental circumstances of populations in Egypt (Akhlaq et al., 2024; Verma et al., 2021; Hasin et al., 2017). It provides molecular insights into the context of community life and advances the field of precision public health by positioning it between molecular knowledge and a conventional scale-up strategy based on a one-size-fits-all model.

24.1 Artificial Intelligence as the Analytical Engine.

The most critical component of NexDi is AI, which processes large, heterogeneous data, including genomics, transcriptomics, metabolomics, polybiome sequencing, and detailed lifestyle and dietary data. According to recent research, AI and machine learning have the potential to identify discrete trends within nutrigenomic and metabolic data, thereby making it possible to guide interventions with precision (Topol, 2019; Akbar et al., 2024). An example of this is the use of AI models to simulate interactions of genes and nutrients, including the effects of common Egyptian polymorphisms in PPAR7 or TCF7L2 on glycemic responses to nutraceuticals. These polymorphisms are linked to both insulin insensitivity and variable responses to dietary factors intrinsic to the Middle Eastern population (Santos et al., 2025; Abd El Razek et al., 2016).

Reinforcement learning further adjusts intervention strategies by using real-time data from wearable and continuous glucose monitors to adapt a combination of functional foods and nutraceuticals continuously. This adaptive model is increasingly validated via digital health platforms, which provide a dynamic process to maximize their effectiveness (Xu et al., 2025). AI-based clinical trials will be less expensive, eliminate transportation-related effects, and appreciate the diversity of urban and rural populations in Egypt, which will be critical in embodying equitable health care (Sandeep et al., 2024). Table 3 affords a short overview of the crucial moral, public, and controlling thoughts that figure the safe and responsible adoption of the NexDi AI-enabled outline.

Table 3. Ethical, Social, and Regulatory Dimensions Relevant to the NexDi Framework

Aspect	Main Issues	References
Ethical	The integrity of personal genomic and microbial data cannot be compromised. Artificial intelligence models should not be biased and must be transparent. Patients must have an explicit understanding of how their data are used.	(WHO, 2024a, 2024b); Britannica, 2024)
Social	Dietary and health choices are influenced by social habits. The availability of artificial intelligence tools should be equal across demographic groups. Families must be given instructions on how to use digital health tools safely	(Haroen et al., 2025); Singh et al., 2025)
Regulatory	Countries impose strict regulations regarding health data management. Clinical deployment requires that artificial intelligence models be safe and meet performance benchmarks. The environmental exposure information must comply with national environmental and chemical safety regulations.	(Boege et al. 2024); (Monlezun et al. 2025)

These scopes highpoint the wider household tasks related with AI-driven precision-health tools and offer a real-world lens for estimating NexDi's willingness for real-world organization.

24.2 Nutrigenomics: Bridging the Gene-to-Nutrition Phenotype

NexDi is anchored by nutrigenomics, which, without clinching a preliminary link, directly connects genetic variation to dietary responsiveness. Polymorphisms in the TCF7L2 and MTHFR genes and PPARγ are common in Egypt and have already been proven to be closely related to vulnerability to or protection against type 2 diabetes and individual responses to nutrients (Maghraby et al., 2022; Shawki et al., 2022). For example, carriers of specific PPAR7 variants might gain greater advantages from berberine supplementation or adipokine regulators such as Adipol, which enhance insulin responsiveness by targeting major glucose metabolism pathways (Chen, et al., 2023). TCF7L2 polymorphisms, which are the strongest predictors of type 2 diabetes, govern how individuals respond to dietary fiber interventions, underscoring the importance of incorporating local functional foods, such as beetroot and carrot, into individual diets (Muruganantham & Veerabathiran, 2025). The integration of these nutrigenomic concepts into AI programs allows interventions to be more precise and to account for polybiome, genetic, and dietary diversity simultaneously to generate a multidimensional approach (D'Ambrosio et al., 2022).

24.3 Polybiome as a Translational Bridge.

The gut polybiome interacts with diet, genetics, and physiological health. Egyptians have high levels of second messenger carbohydrates with low intake of prebiotic fibers, which are correlated with lower microbial diversity, decreased SCFA synthesis, and increased inflammatory activity (Radwan et al., 2020). NexDi uses polybiome research to scale by providing a mapping framework of baseline ecological phenotypes in the Egyptian population and feeds predictive models to understand microbial interactions with nutraceuticals, such as beetroot fibers or berberine (Muguerza et al., 2025; Li et al., 2025). With the assistance of AI-based platforms, the combination of probiotics, prebiotics, and plant-based bioactive compounds has been optimized, improving the population-specific effect and providing culturally acceptable translation (Baimakhanova et al., 2025; Yan et al., 2025; Yang et al., 2024).

24.4 Dynamic Cycle of NexDi

As a continuous, adaptive loop, NexDi guarantees that the interventions are enhanced through new data and adapts to the needs of the population:

1. **Discovery**: Multiomic datasets are used to identify candidate nutraceuticals and biodaries, with beetroot, carrot, and fenugreek currently located in proximity (Nartea et al., 2023; Dima et al., 2020).
2. **Simulation**: AI-based digital twins simulate chemical reactions, metabolic behaviors, and polybiome behaviors to provide intervention testing, digital twin simulations, and scalable low-cost ethical simulation programs (Pammi et al., 2025).
3. **Translation:** Updated implementations are used in practical settings, and performance is utilized to transmit responses into the machine to improve predictive accuracy and form a cycle of continuous enhancement (Kiran et al., 2025).

This cycle will make interventions scientifically sound, culturally aware, and cost effective to propel Egypt toward a vision of precision-guided public health.

24.5. Egyptian strategy implications.

NexDi is an avenue for integrating low-cost, domestically sourced nutraceuticals into national health policy. The advantages include lower reliance on external supplements, programs scaled according to cultural dietary habits, and the opportunity to become the regional center of AI-based nutrigenomics. Integration of NexDi into national initiatives makes it possible to detect infections at an early stage with genomic and polybiome profiling, while AI-based registries make real-time measurements possible, and low-cost functional foods facilitate equal intervention (El - Sayed & Saleh, 2024; Nartea et al., 2023). Accessibility, equity, and long-term sustainability are also reinforced through policy measures, which include crop subsidies, school and workplace nutritional programs, and educational campaigns.

Societal and environmental aspects are also incorporated into NexDi. Sustaining the economy, minimizing government imports, and enhancing climate resilience: by harnessing local fields, farmers improve the economy and address environmental challenges. The concept of environmental surveillance, such as monitoring microplastics and endocrine-disrupting chemicals, is consistent with the One Health approach to consider a combination of effects: the effect of pollutants, the effect of diet, and the effect of genetics on the risk of diabetes. Mobile health systems are extended into rural and underserved regions, ensuring that all communities receive tailored interventions (Dai et al., 2025).

Synthetic Biology, Stem Cells, and National Strategy Future-Oriented Integration. NexDi not only manages diseases in the traditional way but also introduces innovations in regenerative medicine and synthetic biology (Reyed et al., 2026). Synthetic biology provides the ability to program a circuit that can control cells precisely, whereas stem cell technologies have the potential to regenerate the pancreas to address both T1D and T5D in terms of their biological causes (Elnaggar et al., 2025; Galvan et al., 2024; Marei, 2025; Yan et al., 2023). Together with AI, multiomic data, and polybiome science, these technologies create a potent ecosystem of dynamic and individualized interventions that span prevention, management, and restoration. Eventually, the NexDi framework will not be a mere scientific model but rather a national innovation pathway that complements Egypt Vision 2030. NexDi offers a scalable, predictive AI-coupled, culturally adapted blueprint that combines predictive AI with personalized nutrigenomics, polybiome-informed strategies, and regenerative interventions to be a sustainable blueprint for adaptation. It follows that advanced scientific discoveries will be directly applied to the health policy of the population, and Egypt is becoming a leader in globalization maturity to achieve health equity and metabolic disease resilience, with molecular innovation, digital simulation, and community-level translation working together as a single national approach (Davies, 2016; Yan et al., 2023).

24.6 Polybiome-Nutrigenomic Dynamics in T1D and T5D Integrative Analysis in NexDi.

The interaction of gastrointestinal microbiome organization and nutrigenomic cues is a key axis for explaining type 1 diabetes (T1D) and type 5 diabetes (T5D) pathogenesis in the NexDi conceptualization. Although these two entities are etiologically involved, they converge mechanistically via pathways that involve immune activation, epithelial permeability, β-cell vulnerability, and environmental exposure. NexDi envisiones T1D and T5D as conditions imposed by dynamic interactions between microbial communities, genomic vulnerability and dietary stimuli as well as between exposomic pressures, dietary stimuli and inflammatory mediators. This is a systems-based paradigm that allows the prediction of metabolic destabilization preceding actual β-cell failure and thus presents prevention as a continuous and adaptive process as opposed to a reactive clinical intervention.

Early immune-regulation microbial colonization defects, including diminished beneficial taxa, including *Akkermansia muciniphila, Fgalaxibacterium prausnitzii*, and inflammatory signaling interactions among genetic factors, contribute to T1D. These microbial changes make the epithelium more permeable, antigenic, and more autoimmune responsive to pancreatic β-cells. The microbiome is a mediator of toxin absorption, inflammatory priming and metabolite stress in T5D, which is modified under the influence of exposure to environmental toxicants, endocrine disruptors, nanoplastics, and chronic dietary exposures. The sensitivity of the AMPK and PPAR pathways, oxidative stress responses, and detoxification genes to environmentally stimulated β cells, which operate on a genetic basis, increases susceptibility to environmentally induced β-cell exhaustion.

NexDi unites these spheres with a synthesis of genomic markers, microbiome signatures, exposomic indicators, inflammatory biomarkers and digital phenotypes into a real-time learning engine that can refine intervention strategies. The key factors that enable NexDi to regulate β-cells and immune resilience include dietary inputs, a microbiome-supportive diet, dietary nutrition during childhood, and environmental toxin reduction. From this perspective, T1D and T5D are such conditions, the development of which can be predetermined by an algorithmic inference on the basis of molecular, microbial, and environmental intelligence.

24.6.1 Strategic Alignment to Egypt Vision 2030.

NexDi serves perfectly with the vision of 2030 in Egypt, as the translation of multiomics and exposomic information into a policy-yielding and operational framework capable of handling early-life immune risk (T1D) and environmentally mediated metabolic frailty (T5D). The framework focuses on the life-course determinants of β-cell susceptibility via a national multiomics portal, exposom surveillance networks, AI-based early-life screening programs, and culturally grounded microbiome-promoting nutrition.

The micronutrient optimization of maternal–child programmes, minimal exposure to endocrine disruptors, increased protection of the breastmilk microbiome, and improved food safety offer a platform on which to increase immune tolerance and reduce T1D. In the case of T5D, mitigating environmental toxicity, especially with the use of packaging, contaminated food systems, and microplastics, is in line with the sustainability priorities of Egypt. A boost in local fermented food systems, including *laban rayebs and kishks*, has led to the development of a low-cost, culturewise, entrenched route for reconstructing microbial dexterity in rural and municipal environments. These actions support the Vision 2030 pillars of prevention, digital transformation, equity, and sustainability. The analysis of similar developing economies responding to the risks of early life and environmental diabetes is a valuable source of an analytical framework. Placing Egypt in a broader South-South policy frame, particularly relative to India, helps highlight common structural weaknesses and strategic opportunities that help justify the warranting of NexDi.

24.6.2 Structural Comparative Advantages between Emerging Economies: The Egypt–India Model.

Analogous vulnerabilities to type 1 diabetes (T1D) and type 5 diabetes (T5D) are shown by the epidemiological profiles of both Egypt and India, which can be explained by similar structural determinants of emerging economies. These factors include high rates of urbanization, increased environmental contamination, westernization of the diet, and unequal access to primary healthcare services. Based on their pediatric populations, both countries seem to have more autoimmune susceptibility attributed to the impaired maturation of microbiomes, exposure to antibiotics at a young age, and reduced microbiome diversity, which is beginning to be recognized as early antecedents of T1D (Ismail et al., 2025; Tamahane et al., 2024). Moreover, environmental toxicants such as endocrine disruptors, pesticides, heavy metals, and microplastics are hypothesized to increase β-cell susceptibility to exposome-mediated T5D. The Ayushman Bharat Digital Mission in India represents the viability of setting up multilayered precision health systems where microbial, genomic, environmental, and behavioral information is amalgamated at the national level (Arjun et al., 2024). The NexDi program builds on this paradigm with more extensive biological integration conformed to the exposomic characteristics, food system configurations, and risk stratification at an early period in Egypt. With such coordinated efforts, NexDi can help Egypt become a regional leader in endorsing multi-omic, environmentally aware, diabetes prevention guidelines to the rest of Africa, the Middle East, and South Asia.

24.6.3 Analysis of Critical Gap

1. Scientific Gaps

- The absence of longitudinal multiomic cohorts that relate microbiome development to autoimmune inducers (T1D) or toxin-induced β-cell dysfunction (T5D).
- Lack of mechanistic understanding of the global effects of nanoplastics and microplastics on β-cell immunity and metabolic robustness.
- Low adoption of microbiome-exposome biomarkers in daily pediatric and adult clinical use.

2. Policy Gaps

- Lack of a national ethics framework on AI necessary to ensure safety in integrating pediatric and exposomic data regarding children.

- Lack of support for the local manufacture of microbiome-supportive and detoxifying nutraceuticals.
- Lack of control over packaging materials and environmental pollutants that increase the risk of T5D.

3. Gaps in Infrastructure and Workforce.

- Data systems that are in different fragments and that cannot integrate datasets based on genomic data, microbial data, and environmental data.
- Low level of computerization at rural maternal–child health centers.
- Clinical training gaps in the interpretation of early-life microbiome disturbances and environmental risk profiles.

NexDi addresses these deficiencies with interoperable data architecture, community-based dietary measures, environmental policy reform, exposure surveillance, and national-scale capacity-building programs. These measures convert structural vulnerabilities into the building blocks of a contemporary, proactive metabolic-health ecosystem.

24.6.4 Forward Outlook

NexDi offers a blueprint of global leadership in tissue engineering prevention of T1D and T5D with biological intelligence, environmental awareness and algorithmic flexibility in Egypt. Multiomic prevention is needed at the national level as exposure to stressors increases and more autoimmune diseases are diagnosed. With early-life interventions, reforms of a sustainable food system, environmental governance, and state-of-the-art AI analytics, Egypt will be the first country to enter the future in which β-cell resilience can be not merely monitored but engineered. NexDi would therefore form one of the pillars of long-term national health independence and scientific supremacy.

24.6.5 Novel Vision and Strategic Gap-Bridging Potential of the NexDi Framework.

The NexDi Framework is a progressive, future-thinking strategic plan designed to overcome the multidimensional barriers that block the contraction of precision nutrition and AI-powered metabolic care in the health systems of low- and middle-income countries (LMICs) directly. As a scalable and adaptive framework, NexDi is a blueprint that consolidates technology, clinical, economic and governance facets into a single operational model that can potentially be fixed to national health priorities in a variety of settings. Its key advantage lies in its alignability with national digital health strategies: by enabling AI-based analysis of locally edited datasets, NexDi consolidates technology localization requirements in Egypt's Vision 2030 and facilitates high-fidelity predictive models of type 5 and type 1 diabetes (T5D and T1D) risk perception. This localized modeling strategy maintains not only the abstract notion of data sovereignty but also ethical governance and adherence to new standards of AI-in-health, which will provide a credible basis for continuing to adopt the system.

In the same way, the shift of NexDi toward a prevention-oriented paradigm, with a focus on nutrition-based interventions, low-cost nutraceuticals, and culturally competent dietary mechanisms based on native resources, significantly transforms the situation. The given approach directly challenges the cost-effectiveness barriers in LMICs and advances the goals of health equity by providing scalable population-level interventions that can help mitigate dependence on pharmacological mechanisms. In this way, the framework fills one of the historical gaps in implementation between scientific evidence and the operational use of public health practices, especially in communities where financial, geographic, and infrastructural constraints limit access to specialized diabetes care.

Moreover, the structural parallelism between NexDi and the Ayushman Bharat Digital Mission (ABDM) in India is the position of NexDi as a globally transferable model of reference. The similarity of the challenges, i.e., fragmented datasets, diverse clinical settings, and the need to have integrated digital policy ecosystems, means that successful Egyptian implementation can be an experimental success that can be replicated more widely across LMICs. This intercontextual relevance increases the global strategic significance of the chapter, providing an idea of how an AI-empowered public health system can be duplicated to speed up the integration of precision nutrition in a variety of sociotechnical contexts. This comprehensive outlook puts Egypt at the center of AI-driven health innovation and provides an approach to bridging existing structural and translational gaps, as well as the foundation of a groundbreaking public health structure for the generations to come. Figure 7 below condenses the strategic value proposal of NexDi: how integrated technology, clinical pathways, governance systems, and culturally informed nutrition interventions align with national priorities and facilitate translation to low - and middle-income countries.

Figure 8. Strategic Value of the NexDi Framework

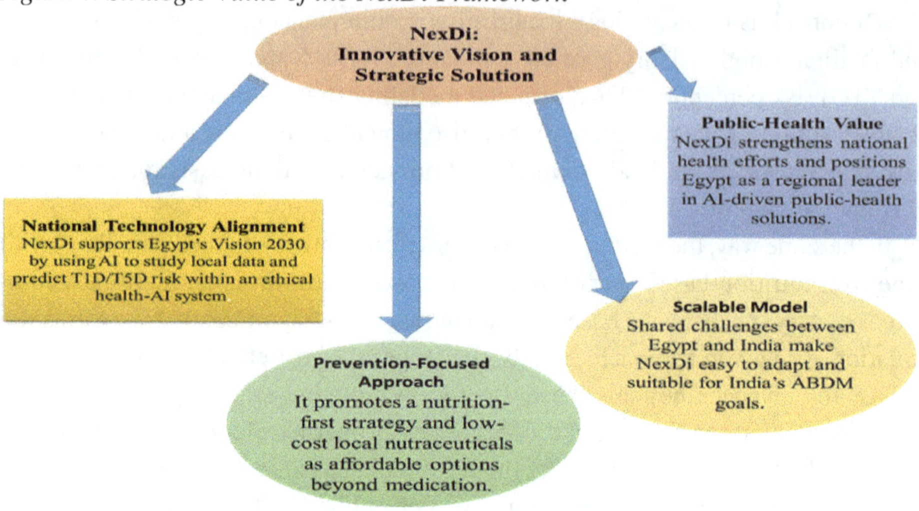

this diagram summarizes the strategic proposal presented by NexDi: implementation of multiomics and exposomic intelligence through national capabilities, such as a national multiomics portal and networks of exposome surveillance, AI-based early-life screening, and culturally sensitive microbiome-promoting nutrition. The figure connects policy levers (prevention, digital transformation, equity, sustainability) with tangible implementation nodes and depicts scaling and international transferability pathways (e.g., alignments with national digital health strategies and analogues like ABDM). The interpretive note explains that NexDi translates both molecular and environmental cues into policy-ready outputs, which can mitigate early immune risk (T1D), environmentally modulated metabolic frailty (T5D), and maintain data sovereignty and ethical governance. This statistic shows how NexDi provides a systems-level strategy by utilizing biological understandings with national health infrastructure. The model depicts how Egypt can embrace a coherent, scalable precision health structure by connecting multi-omic and exposomic data streams with policy processes. The visualization clarifies how both immune vulnerabilities and toxicant-related metabolic frailty in early life can be managed when AI-powered analytics and culturally based interventions are integrated into the national governance frameworks. It further suggests the possibility of Egypt as a regional center for multi-omic, environmentally conscious diabetes prevention (Author-generated figure).

25. CONCLUSION

The integrated examination of type 1 diabetes (T1D) and type 5 diabetes (T5D) and the NexDi model shows that diabetes is no longer a condition characterized by dysregulated glucose or binary diagnostic types. Rather, it is a spectrum of β-cell vulnerability conditioned by interacting biological, nutritional, microbiomic, psychosocial, and environmental forces throughout the life course. The present chapter holds the thesis that autoimmune-mediated β-cell death (T1D) and nutrition-induced developmental insufficiency (T5D) must be seen as variants of a common scale and not deterministic diseases. They are dependent on exposures at the early stages of life, maternal health, genetic inheritance, and socio-environmental realities. A good real-world example of this changing spectrum can be found in Egypt. Increased maternal anemia, fast food shifts, micronutrient deprivations, exposure to plastics, and urban environmental alterations have led to an environment in which hybrid and atypical diabetic phenotypes among children are increasingly hastened. New Egyptian generations are reporting cases that do not fit within classical categories and exhibit a combination of autoimmune, metabolic, nutritional, and microbial traits. These results confirm the necessity of context-specific and integrative models like NexDi. NexDi reformulates the concept of diabetes as a multi- systemic developmental illness that needs predictive, adaptive, and exposure- conscious treatments based on nutrigenomics, multi-omic intelligence, environmental health science, and AI-enabled analytics. In this context, the initial 1000 days of life are redefined as the high-risk period for programming, where micronutrient adequacy, breastfeeding habits, early dieting, and environmental amelioration significantly determine metabolic outcomes . The result of this reorientation is that diabetes is no longer seen as an issue of genetic inheritance but as a preventable course of action that the country needs to prioritize at an early stage. Moreover, the similarities between the alignment of NexDi and the national visions, like Egypt Vision 2030, reveal how interventions grounded in culture, with decreased costs, and with the help of technology, can redesign the outcomes of the relations between people and health. Fermented traditional foods, less exposure to plastics, and AI-powered digital health applications in rural community environments can demonstrate how scientific accuracy can be aligned with local conditions. NexDi is therefore a prototype that can be scaled by nations interested in a middle ground between science and community-based approaches. NexDi is a precision nutrition, microbiome-informed care, and AI-based diabetes governance blueprint that can be tested globally in low- and middle-income countries. It can inform policy design, the clinical decision-making process, and innovation ecosystems by providing a model that incorporates T1D and T5D into a systems architecture and utilizes environmental and developmental determinants. NexDi's envisioned future horizons, such as synthetic biology, microbiome-based regeneration, immune modulation, and β-cell restoration, position NexDi as a paradigm shift in chronic disease survivability.

Key Takeaways - Nutrigenomics and AI integration offer a combined systems perspective of diabetes risk prediction and stratification.

- Long-term prevention and metabolic resilience are better leveraged by early-life intervention and maternal intervention.
- For responsible adoption, it is necessary to address ethical considerations, such as data governance, genomic privacy, and fair AI implementation.
- Clinical translation can be achieved using low-cost diagnostics, polybiome-based nutrition, and scalable digital tools adapted to local environments.

- NexDi becomes a future-ready model that combines the resolution of scientific accuracy with national goals and community realities .

REFERENCES

Abd El Razek, N., Wissa, N., Nassar, H., & Omar, S. (2016). Association of RS 7903146 (C/T) Single-Nucleotide Polymorphism at Transcription Factor 7 Like 2 Gene with Type 2 Diabetes Mellitus in Egyptian Patients. *Suez Canal University Medical Journal, 19*(1), 29–3. DOI: 10.21608/scumj.2016.43911

Abdu, S. M., Youness, S. M., El-Gilany, A.-H., & Sehsah, R. (2025). Knowledge and practice of the public about use and disposal of plastic bags: An online Egyptian descriptive study. *The Egyptian Journal of Community Medicine, 43*(1), 33–44. DOI: 10.21608/ejcm.2024.315491.1323

Adel, M., Khedr, R. A., Sayed, A. A., Shalaby, L., Diab, A. A., Yahia, A., Elanany, M., Lehmann, L. E., Ahmed, S., Aziz, R. K., & Elhaddad, A. (2025). Changes in Gut Microbial Diversity and Correlation with Clinical Outcome in Children with Acute Myeloid Leukemia Receiving Induction Chemotherapy. *Children (Basel, Switzerland), 12*(9), 1176. DOI: 10.3390/children12091176

Agrawal, K., Goktas, P., Kumar, N., & Leung, M. F. (2025). Artificial intelligence in personalized nutrition and food manufacturing: A comprehensive review of methods, applications, and future directions. *Frontiers in Nutrition, 12*, 1636980. DOI: 10.3389/fnut.2025.1636980

Agrawal, P., Kaur, J., Singh, J., Rasane, P., Sharma, K., Bhadariya, V., Kaur, S., & Kumar, V. (2024). Genetics, Nutrition, and Health: A New Frontier in Disease Prevention. *Journal of the American Nutrition Association, 43*(4), 326–338. DOI: 10.1080/27697061.2023.2284997

Ahmadzadeh, A., & Rezaei-Tavirani, M. (2025). Pathogenesis and genetics of celiac disease; a systematic review. *The Egyptian Journal of Medical Human Genetics, 26*, 85. DOI: 10.1186/s43042-025-00713-8

Ahmed, E. A. A., Osman, Y. M. S., & Mohamed, A. F. A. (2024). Artificial intelligence and the future of health care: Is it threatening the existence of nursing? Nurses' perception and attitude. *Egyptian Journal of Health Care, 15*(2), 1101.

Ahmed, N., Saad, N., Khafagy, G., ElSayed, I., & Tawfik, N. M. (2025). Low-carbohydrate diet: A promising strategy for type 2 diabetes management among Egyptian patients. *The Egyptian Journal of Internal Medicine, 37*, 67. DOI: 10.1186/s43162-025-00456-3

Ahn, C., & Jeung, E.-B. (2023). Endocrine-Disrupting Chemicals and Disease Endpoints. *International Journal of Molecular Sciences, 24*(6), 5342. DOI: 10.3390/ijms24065342

Akbar, B., Tayara, H., & Chong, K. T. (2024). Unveiling dominant recombination loss in perovskite solar cells with a XGBoost-based machine learning approach. *iScience*, *27*(3), 109200. DOI: 10.1016/j.isci.2024.109200

Akhlaq, S., Ara, S. A., Itrat, M., Fazil, M., Ahmad, B., Akram, U., Haque, M., Quddusi, N., & Sayeed, A. (2024). An Exploratory Review on the Hypoglycemic Action of Unani Anti-diabetic Drugs via Possible Modulation of Gut Microbiota. *Current Drug Targets*, *25*(1), 1–11. DOI: 10.2174/0113894501275731231215101426

Allakany, A. I., Elbanna, A. A., Rohoma, K. H., Ahmed, S. M., Ibrahim, A. E., Fawzy, M. A., & Header, D. A. (2023). Study of the gut microbiome in Egyptian patients with type 1 diabetes mellitus. *Przeglad Gastroenterologiczny*, *18*(2), 190–197. DOI: 10.5114/pg.2023.126055

Almoghrabi, Y. M., Eldakhakhny, B. M., Bima, A. I., Sakr, H., Ajabnoor, G. M. A., Gad, H. M., Mohammad, F. A. H., Elsamanoudy, S. A., Awan, A. Z., & Elsamanoudy, A. Z. (2025). The interplay between nutrigenomics and low-carbohydrate ketogenic diets in personalized healthcare. *Frontiers in Nutrition*, *12*, 1595316. DOI: 10.3389/fnut.2025.1595316

Aloisi, M., & Poma, A. M. G. (2025). Nanoplastics as Gene and Epigenetic Modulators of Endocrine Functions: A Perspective. *International Journal of Molecular Sciences*, *26*(5), 2071. DOI: 10.3390/ijms26052071

Alum, E. U., Uti, D. E., Ugwu, O. P., Alum, B. N., Edeh, F. O., & Ainebyoona, C. (2025). Unveiling the microbial orchestra: Exploring the role of microbiota in cancer development and treatment. *Discover. Oncology*, *16*(1), 646. DOI: 10.1007/s12672-025-02352-2

Alzahrani, H. A. (2024). Relation between variation in biomarkers of celiac disease and presence of diabetic type-1: A single center experience. *Egyptian Academic Journal of Biological Sciences. C, Physiology and Molecular Biology*, *16*(1), 305–311. DOI: 10.21608/EAJBSC.2024.347063

American Diabetes Association Professional Practice Committee. (2024). 14. Children and Adolescents: Standards of Care in Diabetes-2024. *Diabetes Care*, *47*(Suppl 1), S258–S281. DOI: 10.2337/dc24-S014

An, J., Liu, Y., Wang, Y., Fan, R., Hu, X., Zhang, F., Yang, J., & Chen, J. (2022). The Role of Intestinal Mucosal Barrier in Autoimmune Disease: A Potential Target. *Frontiers in Immunology*, *13*, 871713. DOI: 10.3389/fimmu.2022.871713

Annunziata, G., Arnone, A., Ciampaglia, R., Tenore, G. C., & Novellino, E. (2020). Fermentation of Foods and Beverages as a Tool for Increasing Availability of Bioactive Compounds. Focus on Short-Chain Fatty Acids. *Foods*, *9*(8), 999. DOI: 10.3390/foods9080999

Apostolopoulou, M., Lambadiari, V., Roden, M., & Dimitriadis, G. D. (2025). Insulin Resistance in Type 1 Diabetes: Pathophysiological, Clinical, and Therapeutic Relevance. *Endocrine Reviews*, *46*(3), 317–348. DOI: 10.1210/endrev/bnae032

Aquino-Domínguez, A. S., Gómez-López, M., & Aguilar-Ruiz, S. R. (2025). Harmonious Allies: The Synergy of Antimicrobial Proteins and Microbes in Breast Milk to Protect Neonatal Health. *Hygiene*, *5*(2), 19. DOI: 10.3390/hygiene5020019

Archibald, A., & Joffe, Y. (2021). The role of nutrigenetics and nutrigenomics in clinical nutrition practice. *ADCES in Practice*, *9*(2), 34–40. DOI: 10.1177/2633559X20984137

Arjun, M. C., Poorvikha, S., Kurpad, A. V., & Thomas, T. (2024). Knowledge, Attitude and Practice about Ayushman Bharat Digital Mission and Digital Health among hospital patients. *Journal of Family Medicine and Primary Care*, *13*(10), 4476–4481. DOI: 10.4103/jfmpc.jfmpc_255_24

Awadalla, N. J., Hegazy, A. A., Abd El-Salam, M., & Elhady, M. (2017). Environmental Factors Associated with Type 1 Diabetes Development: A Case Control Study in Egypt. *International Journal of Environmental Research and Public Health*, *14*(6), 615. DOI: 10.3390/ijerph14060615

Bacaloni, S., & Agrawal, D. K. (2025). Nutrition, Gut Microbiota, and Epigenetics in the Modulation of Immune Response and Metabolic Health. *Cardiology and Cardiovascular Medicine*, *9*, 111–124. DOI: 10.26502/fccm.92920430

Baimakhanova, B. B., Sadanov, A. K., Ratnikova, I. A., Baimakhanova, G. B., Orasymbet, S. E., Amitova, A. A., Aitkaliyeva, G. S., & Kakimova, A. B. (2025). In Silico Modeling of Metabolic Pathways in Probiotic Microorganisms for Functional Food Biotechnology. *Fermentation (Basel, Switzerland)*, *11*(8), 458. DOI: 10.3390/fermentation11080458

Bankole, T., & Li, Y. (2025). The early-life gut microbiome in common pediatric diseases: Roles and therapeutic implications. *Frontiers in Nutrition*, *12*, 1597206. DOI: 10.3389/fnut.2025.1597206

Bao, L., Cui, X., Zeng, T., Liu, G., Lai, W., Zhao, H., Gao, F., Wu, J., Leong, K. W., & Chen, C. (2025). Incorporation of polylactic acid microplastics into the carbon cycle as a carbon source to remodel the endogenous metabolism of the gut. *Proceedings of the National Academy of Sciences of the United States of America, 122*(19), e2417104122. DOI: 10.1073/pnas.2417104122

Bell, V., Ferrão, J., & Fernandes, T. (2018). Fermented Food Guidelines for Children. *J Pediatr Pediatr Med, 2*(1), 1–4. https://www.pediatricsresearchjournal.com/articles/fermented-food-guidelines-for-children.pdf

Beverly, E. A., & Gonzalez, J. S. (2025). The Interconnected Complexity of Diabetes and Depression. *Diabetes Spectrum, 38*(1), 23–31. DOI: 10.2337/dsi24-0014

Bezie, M. M., Asebe, H. A., Asnake, A. A., Fente, B. M., Asmare, Z. A., Gebrehana, A. K., Melkam, M., Seifu, B. L., Tsega, S. S., & Negussie, Y. M. (2025). Factors associated with concordance of mother-child (6-23 months) dietary diversity in Sub-Saharan Africa. *PLoS One, 20*(2), e0318493. DOI: 10.1371/journal.pone.0318493

Boege, S., Milne-Ives, M., & Meinert, E. (2024). Impact of Responsible AI on the Occurrence and Resolution of Ethical Issues: Protocol for a Scoping Review. *JMIR Research Protocols, 13*, e52349. DOI: 10.2196/52349

Bora, S. S., Gogoi, R., & Sharma, M. R., Anshu, Borah, M. P., Deka, P., Bora, J., Naorem, R. S., Das, J., & Teli, A. B. (2024). Microplastics and human health: Unveiling the gut microbiome disruption and chronic disease risks. *Frontiers in Cellular and Infection Microbiology, 14*, 1492759. DOI: 10.3389/fcimb.2024.1492759

Britannica. (2024). *Human intelligence*. Retrieved from https://www.britannica.com/science/human-intelligence-psychology

Brügger, V., Kowatsch, T., & Jovanova, M. (2025). Predicting postprandial glucose excursions to personalize dietary interventions for type-2 diabetes management. *Scientific Reports, 15*(1), 25920. DOI: 10.1038/s41598-025-08003-4

Buonocore, G. (2024). Microbiota and gut immunity in infants and young children. *Global Pediatrics, 9*, 100202. DOI: 10.1016/j.gpeds.2024.100202

Calvez, V., Becherucci, G., Covello, C., Piccirilli, G., Mignini, I., Esposto, G., Laterza, L., Ainora, M. E., Scaldaferri, F., Gasbarrini, A., & Zocco, M. A. (2024). Navigating the Intersection: Sarcopenia and Sarcopenic Obesity in Inflammatory Bowel Disease. *Biomedicines, 12*(6), 1218. DOI: 10.3390/biomedicines12061218

Campanale, C., Massarelli, C., Savino, I., Locaputo, V., & Uricchio, V. F. (2020). A Detailed Review Study on Potential Effects of Microplastics and Additives of Concern on Human Health. *International Journal of Environmental Research and Public Health*, *17*(4), 1212. DOI: 10.3390/ijerph17041212

Chaudhary, D., Guleria, D., Aggarwal, H., Mishra, V., Chauhan, A., Dufossé, L., & Joshi, N. C. (2025). Nutrigenomics and personalized diets - Tailoring nutrition for optimal health. *Applied Food Research*, *5*(1), 100980. DOI: 10.1016/j.afres.2025.100980

Chen, T., Lin, Q., Gong, C., Zhao, H., & Peng, R. (2024). Research Progress on Micro (Nano)Plastics Exposure-Induced miRNA-Mediated Biotoxicity. *Toxics*, *12*(7), 475. DOI: 10.3390/toxics12070475

Chen, Y., Li, Q., Zhao, S., Sun, L., Yin, Z., Wang, X., Li, X., Iwakiri, Y., Han, J., & Duan, Y. (2023). Berberine protects mice against type 2 diabetes by promoting PPARγ-FGF21-GLUT2-regulated insulin sensitivity and glucose/lipid homeostasis. *Biochemical Pharmacology*, *218*, 115928. DOI: 10.1016/j.bcp.2023.115928

D'Ambrosio, V., Ferraro, P. M., Lombardi, G., Friso, S., & Gambaro, G. (2022). Unraveling the Complex Relationship between Diet and Nephrolithiasis: The Role of Nutrigenomics and Nutrigenetics. *Nutrients*, *14*(23), 4961. DOI: 10.3390/nu14234961

Dai, M., Liu, J., Hu, M., Zhang, F., Wang, Y., Dai, F., Qu, R., Fang, Z., & Yang, J. (2025). Air Pollution Exposure and Gestational Diabetes Mellitus Risk: A Retrospective Case–Control Study with Multi-Pollutant Analysis in Wuhan, Hubei Province. *Toxics*, *13*(2), 141. DOI: 10.3390/toxics13020141

Daley, S. F., & Haseeb, M. (2025 Jan). Celiac Disease. [Updated 2025 Feb 4] In *StatPearls* [Internet]. StatPearls Publishing., Available from https://www.ncbi.nlm.nih.gov/books/NBK441900/

Davies, J. A. (2016). Synthetic Biology: Rational Pathway Design for Regenerative Medicine. *Gerontology*, *62*(5), 564–570. DOI: 10.1159/000440721

Dima, C., Assadpour, E., Dima, S., & Jafari, S. M. (2020). Bioavailability of nutraceuticals: Role of the food matrix, processing conditions, the gastrointestinal tract, and nanodelivery systems. *Comprehensive Reviews in Food Science and Food Safety*, *19*(3), 954–994. DOI: 10.1111/1541-4337.12547

Ding, F., Yu, Y., Zhang, Y., Wei, S., Han, J. H., Li, Z., Jiang, H. B., Ryu, D., Park, W., Ha, K. T., & Geng, L. (2025). Harnessing nutrients and natural products for sustainable drug development against aging. *Frontiers in Pharmacology*, *16*, 1579266. DOI: 10.3389/fphar.2025.1579266

El-Sayed HH and Saleh RT (2024): Review of Nutritional Genomics and Precision Nutrition. BNNI (64) 204-. doi:DOI: 10.21608/bnni.2024.396356

El-Sayed, R. E. H., Abd El Raaouf, S. E. H., & Laimon, W. (2019). Effect of Vitamin D Supplementation on Glycemic Control in Children with Type 1 Diabetes Mellitus: A Randomized Clinical Trial. *American Journal of Nursing Research, 7*(4), 534–541. DOI: 10.12691/ajnr-7-4-15

El-Ziny, M. A., Salem, N. A., El-Hawary, A. K., Chalaby, N. M., & Elsharkawy, A. A. (2014). Epidemiology of childhood type 1 diabetes mellitus in Nile Delta, northern Egypt - a retrospective study. *Journal of Clinical Research in Pediatric Endocrinology, 6*(1), 9–15. DOI: 10.4274/Jcrpe.1171

Elnaggar, K. S., Gamal, O., Hesham, N., Ayman, S., Mohamed, N., Moataz, A., Elzayat, E. M., & Hassan, N. (2025). A Guide in Synthetic Biology: Designing Genetic Circuits and Their Applications in Stem Cells. *SynBio, 3*(3), 11. DOI: 10.3390/synbio3030011

Enache, R. M., Roşu, O. A., Profir, M., Pavelescu, L. A., Creţoiu, S. M., & Gaspar, B. S. (2025). Correlations Between Gut Microbiota Composition, Medical Nutrition Therapy, and Insulin Resistance in Pregnancy-A Narrative Review. *International Journal of Molecular Sciences, 26*(3), 1372. DOI: 10.3390/ijms26031372

Faja, F., Esteves, S., Pallotti, F., Cicolani, G., Di Chiano, S., Delli Paoli, E., Lenzi, A., Lombardo, F., & Paoli, D. (2022). Environmental disruptors and testicular cancer. *Endocrine, 78*(3), 429–435. DOI: 10.1007/s12020-022-03171-z

Farag, M., Sakr, M., Tayel, N., & Abdelrahman, A. (2024). Association of TCF7L2 polymorphisms rs7903146 and rs4506565 with risk of type 2 diabetes mellitus in the Egyptian patients. *African Journal of Biological Sciences, 20*(2), 109–119. DOI: 10.21608/ajbs.2024.385610

Fatumo, S., Chikowore, T., Choudhury, A., Ayub, M., Martin, A. R., & Kuchenbaecker, K. (2022). A roadmap to increase diversity in genomic studies. *Nature Medicine, 28*(2), 243–250. DOI: 10.1038/s41591-021-01672-4

Galvan, S., Teixeira, A. P., & Fussenegger, M. (2024). Enhancing cell-based therapies with synthetic gene circuits responsive to molecular stimuli. *Biotechnology and Bioengineering, 121*, 2987–3000. DOI: 10.1002/bit.28770

Gambino, I., Bagordo, F., Grassi, T., Panico, A., & De Donno, A. (2022). Occurrence of Microplastics in Tap and Bottled Water: Current Knowledge. *International Journal of Environmental Research and Public Health, 19*(9), 5283. DOI: 10.3390/ijerph19095283

García-García, I., Grisotto, G., Heini, A., Gibertoni, S., Nusslé, S., Gonseth Nusslé, S., & Donica, O. (2024). Examining nutrition strategies to influence DNA methylation and epigenetic clocks: A systematic review of clinical trials. *Frontiers in Aging*, *5*, 1417625. DOI: 10.3389/fragi.2024.1417625

Ge, S., Yang, M., Cui, Y., Wu, J., Xu, L., Dong, J., & Liao, L. (2022). The Clinical Characteristics and Gene Mutations of Maturity-Onset Diabetes of the Young Type 5 in Sixty-One Patients. *Frontiers in Endocrinology*, *13*, 911526. DOI: 10.3389/fendo.2022.911526

Gharanizadeh, K., Ravanbod, H., Poursalehian, M., Medhat, A., Aminian, A., Rajei, M., Hassanzadeh, M., & Abolghasemian, M. (2025). High Risk of Venous Thromboembolism With Aspirin Prophylaxis After THA for High-riding Developmental Dysplasia of the Hip: A Retrospective, Comparative Study. *Clinical Orthopaedics and Related Research*. Advance online publication. DOI: 10.1097/CORR.0000000000003482

Ghazi, H. A. (2025). Diabetes Technology in Egypt: Present Situation, Difficulties, and Prospects. *International Journal of Diabetes and Technology*, *4*(3), 67–69. DOI: 10.4103/ijdt.ijdt_17_25

Gómez-Peralta, F., Pinés-Corrales, P. J., Santos, E., Cuesta, M., González-Albarrán, O., & Azriel, S., & on behalf the AGORA Diabetes Collaborative Group. (2024). Diabetes Management Based on the Phenotype and Stage of the Disease: An Expert Proposal from the AGORA Diabetes Collaborative Group. *Journal of Clinical Medicine*, *13*(16), 4839. DOI: 10.3390/jcm13164839

González, A., Fullaondo, A., & Odriozola, A. (2025). In Search of Healthy Aging: A Microbiome-Based Precision Nutrition Approach for Type 2 Diabetes Prevention. *Nutrients*, *17*(11), 1877. DOI: 10.3390/nu17111877

Grewe-Salfeld, M. (2021). Biohacking, bodies and do-it-yourself: The cultural politics of hacking life itself (American Culture Studies, 36). transcript Verlag. https://doi.org/DOI: 10.14361/9783839460047

Guevara-Ramírez, P., Paz-Cruz, E., Cadena-Ullauri, S., Ruiz-Pozo, V. A., Tamayo-Trujillo, R., Felix, M. L., Simancas-Racines, D., & Zambrano, A. K. (2023). Molecular pathways and nutrigenomic review of insulin resistance development in gestational diabetes mellitus. *Frontiers in Nutrition*, *10*, 1228703. DOI: 10.3389/fnut.2023.1228703

Hafez, M., Hassan, M., Musa, N., Atty, S. A., & Azim, S. A. (2017). Vitamin D status in Egyptian children with type 1 diabetes and the role of vitamin D replacement in glycemic control. *Journal of Pediatric Endocrinology & Metabolism*, *30*(4), 389–394.

Hamadou, M. (2025). Bioactive peptides and metabolic health: A mechanistic review of the impact on insulin sensitivity, lipid profiles, and inflammation. *Applied Food Research*, *5*(2), 101056. DOI: 10.1016/j.afres.2025.101056

Haroen, H., Juniarti, N., Sari, C. W. M., Sari, S. P., Harun, H., Cahyati, P., Arovah, N. I., & Pardosi, J. F. (2025). Integrating Telehealth into Community-Based Palliative Care: A Systematic Review. *Journal of Multidisciplinary Healthcare*, *18*, 7339–7354. DOI: 10.2147/JMDH.S555226

Hasin, Y., Seldin, M., & Lusis, A. (2017). Multiomics approaches to disease. *Genome Biology*, *18*(1), 83. DOI: 10.1186/s13059-017-1215-1

Hassan, F. M., El Desouky, E. D., Salem, M. R., Abdelsabour, M. A., Abdelmoneim, M. A., Elsaieed, M. M., & Ali, M. M. (2023). Plastic use for food and drinks and related knowledge, attitudes, and practices among a sample of Egyptians. *Frontiers in Public Health*, *11*, 1146800. Advance online publication. DOI: 10.3389/fpubh.2023.1146800

Hassan, S. S., & Bayomi, S. S. (2024). Knowledge and Reported Practice about Type 1 Diabetes Mellitus among Preparatory Schools Students at Aswan City, Upper Egypt. *Egyptian Journal of Health Care*, *15*(3), 1463. https://ejhc.journals.ekb.eg/article_399171_ 54a1fd0ad93512fe84dfa50d0a9ae256.pdf

He, K., Chen, R., Xu, S., Ding, Y., Wu, Z., Bao, M., He, B., & Li, S. (2024). Environmental endocrine disruptor-induced mitochondrial dysfunction: A potential mechanism underlying diabetes and its complications. *Frontiers in Endocrinology*, *15*, 1422752. DOI: 10.3389/fendo.2024.1422752

Hsu, C.-N., Lin, Y.-J., Hou, C.-Y., Chen, Y.-W., Chang-Chien, G.-P., Lin, S.-F., & Tain, Y.-L. (2025). Antioxidants, Gut Microbiota, and Cardiovascular Programming: Unraveling a Triad of Early-Life Interactions. *Antioxidants*, *14*(9), 1049. DOI: 10.3390/antiox14091049

Huang, Y., Li, X., Xu, S., Zu, D., Liu, H., He, H., Bao, Q., He, Y., Liang, C., Shi, Y., Cheng, X., Teng, Y., & Ye, Z. (2025). Polyvinyl chloride nanoplastics suppress homology-directed repair and promote oxidative stress to induce esophageal epithelial cellular senescence and cGAS-STING-mediated inflammation. *Free Radical Biology & Medicine*, *226*, 288–301. DOI: 10.1016/j.freeradbiomed.2024.11.012

Hussain, T., Murtaza, G., Kalhoro, D. H., Kalhoro, M. S., Yin, Y., Chughtai, M. I., Tan, B., Yaseen, A., & Rehman, Z. U. (2022). Understanding the Immune System in Fetal Protection and Maternal Infections during Pregnancy. *Journal of Immunology Research*, *7567708*. Advance online publication. DOI: 10.1155/2022/7567708

Hussein, A. M. S., El-Aal, H. A. A., Morsy, N. M., & Hassona, M. M. (2024). Chemical, rheological, and sensorial properties of Baladi bread supplemented with buckwheat flour produced in Egypt. *Scientific Reports*, *14*(1), 3127. DOI: 10.1038/s41598-023-48686-1

Iheagwam, F. N., Joseph, A. J., Adedoyin, E. D., Iheagwam, O. T., & Ejoh, S. A. (2025). Mitochondrial Dysfunction in Diabetes: Shedding Light on a Widespread Oversight. *Pathophysiology : the Official Journal of the International Society for Pathophysiology*, *32*(1), 9. DOI: 10.3390/pathophysiology32010009

Ikegami, H., Babaya, N., & Noso, S. (2021). β-Cell failure in diabetes: Common susceptibility and mechanisms shared between type 1 and type 2 diabetes. *Journal of Diabetes Investigation*, *12*(9), 1526–1539. DOI: 10.1111/jdi.13576

International Diabetes Federation. (2025a). *IDF launches new Type 5 Diabetes working group*. IDF. Retrieved from https://idf.org/news/new-type-5-diabetes-working-group/

International Diabetes Federation. (2025b). IDF Diabetes Atlas, 11th ed. Retrieved from https://diabetesatlas.org/data-by-location/country/egypt/

Ismail, H. M., Perera, D., Mandal, R., DiMeglio, L. A., Evans-Molina, C., Hannon, T., Petrosino, J., Cregeen, S. J., & Schmidt, N. W. (2025). Gut Microbial Changes Associated With Obesity in Youth With Type 1 Diabetes. *The Journal of Clinical Endocrinology and Metabolism*, *110*(2), 364–373. DOI: 10.1210/clinem/dgae529

Jiao, X., Cao, Q., & Deng, Z. (2025). Microplastics and exercise: Impacts on performance and physiological health. *Frontiers in Sports and Active Living*, *7*, 1611255. DOI: 10.3389/fspor.2025.1611255

Johnson-Pitt, A., Catchpole, B., & Davison, L. J. (2024). Exocrine pancreatic inflammation in canine diabetes mellitus – An active offender? *Veterinary Journal (London, England)*, *308*, 106241. Advance online publication. DOI: 10.1016/j.tvjl.2024.106241

Kalam, M. A., McCann, J. K., Shakil, Z., Gambari, A., Ochieng, M., & Jeung, J. (2025). Maternal Mental Health and Child Dietary Diversity in Rural Kenya: Findings from a Pooled Analysis of 2 Baseline Studies. *Current Developments in Nutrition*, *9*(7), 107497. DOI: 10.1016/j.cdnut.2025.107497

Kassem, N. M., Abdelmegid, Y. A., El-Sayed, M. K., Sayed, R. S., Abdel-Aalla, M. H., & Kassem, H. A. (2023). Nutrigenomics and microbiome shaping the future of personalized medicine: A review article. *Journal of Genetic Engineering and Biotechnology*, *21*, 134. DOI: 10.1186/s43141-023-00599-2

Keskesiadou, G.-N., Tsokkou, S., Konstantinidis, I., Georgaki, M.-N., Sioga, A., Papamitsou, T., & Karachrysafi, S. (2024). Endocrine-Disrupting Chemicals and the Development of Diabetes Mellitus Type 1: A 5-Year Systematic Review. *International Journal of Molecular Sciences*, *25*(18), 10111. DOI: 10.3390/ijms251810111

Kiran, M., Xie, Y., Anjum, N., Ball, G., Pierscionek, B., & Russell, D. (2025). Machine learning and artificial intelligence in type 2 diabetes prediction: A comprehensive 33-year bibliometric and literature analysis. *Frontiers in Digital Health*, *7*, 1557467. DOI: 10.3389/fdgth.2025.1557467

Kotsiri, I., Xanthi, M., Domazinaki, C. M., & Magiorkinis, E. (2025). The Role of Viral Infections in the Immunopathogenesis of Type 1 Diabetes Mellitus: A Narrative Review. *Biology (Basel)*, *14*(8), 981. DOI: 10.3390/biology14080981

Lalani, B., Herur, R., Zade, D., Collins, G., Dishong, D. M., Mehta, S., Shim, J., Valdez, Y., & Mathioudakis, N. (2025). Applications of Artificial Intelligence and Machine Learning in Prediabetes: A Scoping Review. *Journal of Diabetes Science and Technology*, *19322968251351995*. Advance online publication. DOI: 10.1177/19322968251351995

Leesang, T., Lyon, P., Pinzone, J., & Cimmino, L. (2024). Micronutrient regulation of the DNA methylome. *Frontiers in Epigenetics and Epigenomics*, *2*, 1409355. DOI: 10.3389/freae.2024.1409355

Leslie, H. A., van Velzen, M. J. M., Brandsma, S. H., Vethaak, A. D., Garcia-Vallejo, J. J., & Lamoree, M. H. (2022). Discovery and quantification of plastic particle pollution in human blood. *Environment International*, *163*, 107199. DOI: 10.1016/j.envint.2022.107199

Li, J., Lin, Y.-C.-D., Zuo, H.-L., Huang, H.-Y., Zhang, T., Bai, J.-W., & Huang, H.-D. (2025). Dietary Omega-3 PUFAs in Metabolic Disease Research: A Decade of Omics-Enabled Insights (2014–2024). *Nutrients*, *17*(11), 1836. DOI: 10.3390/nu17111836

Li, Y., Ma, Y., Yao, L., Li, J., Zhou, X., Wang, M., Gui, M., Li, D., Chen, X., Dong, Y., Lu, B., & Fu, D. (2025). A Review of the Mechanisms of Astragaloside IV and Berberine in Vascular Dysfunction Associated with Obesity and Diabetes. *Drug Design, Development and Therapy*, *19*, 4911–4932. DOI: 10.2147/DDDT.S520323

Lim, M. H., Chae, H., Yoon, J., & Kim, M. (2025). A deep learning framework for virtual continuous glucose monitoring and glucose prediction based on life-log data. *Scientific Reports*, *15*(1), 16290. DOI: 10.1038/s41598-025-01367-7

Lu, S., Wang, C., Ma, J., & Wang, Y. (2024). Metabolic mediators: Microbe-derived metabolites as key regulators of anti-tumor immunity, immunotherapy, and chemotherapy. *Frontiers in Immunology*, *15*, 1456030. DOI: 10.3389/fimmu.2024.1456030

Łukowski, W. (2025). Reframing type 1 diabetes through the endocannabinoidome-microbiota axis: A systems biology perspective. *Frontiers in Endocrinology*, *16*, 1576419. DOI: 10.3389/fendo.2025.1576419

Lundberg, S. M., Erion, G., Chen, H., DeGrave, A., Stapor, J., Ho, T. J., & Lee, S. I. (2020). From local explanations to global understanding with explainable AI for trees. *Nature Machine Intelligence*, *2*(1), 56–67. DOI: 10.1038/s42256-019-0138-9

Macura, B., Kiecka, A., & Szczepanik, M. (2024). Intestinal permeability disturbances: Causes, diseases and therapy. *Clinical and Experimental Medicine*, *24*(1), 232. DOI: 10.1007/s10238-024-01496-9

Maghraby, M. H., Fathy, R., Atta, S., Mostafa, M. H., Sayed, A. A., Ismael, T. A., Abdelraouf, H. M., & Abdelrahman, G. (2022). Association Between Type 2 Diabetes Mellitus and TCF7L2 And FTO Gene Variants Among Upper Egyptian Population. *The Egyptian Journal of Hospital Medicine*, *86*(1), 933–942. DOI: 10.21608/ejhm.2022.219966

Marei, H. E. (2025). Stem cell therapy: A revolutionary cure or a pandora's box. *Stem Cell Research & Therapy*, *16*, 255. DOI: 10.1186/s13287-025-04334-1

Martínez-Augustin, O., Tena-Garitaonaindia, M., Ceacero-Heras, D., Jiménez-Ortas, Á., Enguix-Huete, J. J., Álvarez-Mercado, A. I., Ruiz-Henares, G., Aranda, C. J., Gámez-Belmonte, R., & Sánchez de Medina, F. (2025). Macronutrients as Regulators of Intestinal Epithelial Permeability: Where Do We Stand? *Comprehensive Reviews in Food Science and Food Safety*, *24*(3), e70178. DOI: 10.1111/1541-4337.70178

Martínez-Pinna, J., Sempere-Navarro, R., Medina-Gali, R. M., Fuentes, E., Quesada, I., Sargis, R. M., Trasande, L., & Nadal, A. (2023). Endocrine disruptors in plastics alter β-cell physiology and increase the risk of diabetes mellitus. *American Journal of Physiology. Endocrinology and Metabolism*, *324*(6), E488–E505. DOI: 10.1152/ajpendo.00068.2023

Misra, A., Joshi, S., & Mithal, A. (2025). Malnutrition-related diabetes mellitus: Rushing toward "type 5" amid unresolved questions and limited evidence. *Diabetes & Metabolic Syndrome*, *19*(5), 103250. DOI: 10.1016/j.dsx.2025.103250

Mitra, T, & Anil Kumar, AH Sruthi, Thangaraju R, S. M. SM, Kumari R. S and Janardhanan R. (2025). Impact of phthalate exposure on gestational diabetes mellitus: A systematic review. *Frontiers in Endocrinology*, *16*, 1638655. DOI: 10.3389/fendo.2025.1638655

Mohamed, F., Mohammed, Z., Ali, A., & Bello, K. (2025). Influence of mode of delivery on neonatal microbial colonization and susceptibility to infections. *Microbes and Infectious Diseases*, *6*(2), 565–574. DOI: 10.21608/mid.2025.356332.2491

Monlezun, D. J., Omutoko, L., Oduor, P., Kokonya, D., Rayel, J., Sotomayor, C., Girault, M. I., Los Ríos Uriarte, M. E., Sinyavskiy, O., Aksamit, T., Dugani, S. B., Garcia, A., & Gallagher, C. (2025). Digitalization of health care in low- and middle-income countries. *Bulletin of the World Health Organization*, *103*(2), 148–154. DOI: 10.2471/BLT.24.291643

Muguerza-Rodríguez, L., Mier, A., Ponce-González, J. G., Casals, C., & Corral-Pérez, J. (2025). This is a systematic review on the importance of the gut microbiota in the regulation of type 2 diabetes through physical activity and exercise. *Current Issues in Molecular Biology*, *47*(7), 505. DOI: 10.3390/cimb47070505

Muruganantham, J., & Veerabathiran, R. (2025). Meta-analysis of TCF7L2 and SLC30A8 gene polymorphisms concerning type 2 diabetes risk. *The Egyptian Journal of Internal Medicine*, *37*, 76. DOI: 10.1186/s43162-025-00464-3

Naredi Scherman, M., Melin, J., & Agardh, D. (2025). Celiac disease screening in children: Evaluating the evidence, benefits, and challenges. *Frontiers in Pediatrics*, *13*, 1562073. DOI: 10.3389/fped.2025.1562073

Nartea, A., Kuhalskaya, A., Fanesi, B., Orhotohwo, O. L., Susek, K., Rocchetti, L., Di Vittori, V., Bitocchi, E., Pacetti, D., & Papa, R. (2023). Legume byproducts as ingredients for food applications: Preparation, nutrition, bioactivity, and techno-functional properties. *Comprehensive Reviews in Food Science and Food Safety*, *22*, 1953–1985. DOI: 10.1111/1541-4337.13137

Naylor, R. N., Patel, K. A., & Kettunen, J. L. T.. (2024). Precision treatment of beta-cell monogenic diabetes: A systematic review. *Communications Medicine*, *4*(145). Advance online publication. DOI: 10.1038/s43856-024-00556-1

Nihart, A. J., Garcia, M. A., & El Hayek, E.. (2025). Bioaccumulation of microplastics in decedent human brains. *Nature Medicine*, *31*(11), 1114–1119. DOI: 10.1038/s41591-024-03453-1

Nurkolis, F., Syahputra, R. A., Nugrahanto, A. P., Taslim, N. A., Mustika, A., Tjandrawinata, R. R., Harbuwono, D. S., & Soegondo, S. (2025). Emerging phenotype: Maturity-onset diabetes of the young type 5 (MODY-5) - mechanisms, clinical spectrum, and unmet needs. *Diabetology & Metabolic Syndrome*, *17*(1), 252. DOI: 10.1186/s13098-025-01762-0

Obaied, M. H. R., Ozer, N., Alawneh, H. I. F., & Dalmizrak, O. (2025). Implication of the HLA-DQA1, HLA-DQB1 and CTLA-4 alleles in the susceptibility to type 1 diabetes in Jordanian population. *Molecular Biology Reports*, *52*(1), 330. DOI: 10.1007/s11033-025-10438-x

Ouabbou, S., He, Y., Butler, K., & Tsuang, M. (2020). Inflammation in Mental Disorders: Is the Microbiota the Missing Link? *Neuroscience Bulletin*, *36*(9), 1071–1084. DOI: 10.1007/s12264-020-00535-1

Pammi, M., Aghaeepour, N., & Neu, J. (2023). Multiomics, artificial intelligence, and precision medicine in perinatology. *Pediatric Research*, *93*(2), 308–315. DOI: 10.1038/s41390-022-02181-x

Pammi, M., Shah, P. S., Yang, L. K., Hagan, J., Aghaeepour, N., & Neu, J. (2025). Digital twins, synthetic patient data, and in-silico trials: Can they empower pediatric clinical trials? *The Lancet. Digital Health*, *7*(5), 100851. DOI: 10.1016/j.landig.2025.01.007

Panteli, D., Adib, K., Buttigieg, S., Goiana-da-Silva, F., Ladewig, K., Azzopardi-Muscat, N., Figueras, J., Novillo-Ortiz, D., & McKee, M. (2025). Artificial intelligence in public health: Promises, challenges, and an agenda for policy makers and public health institutions. *The Lancet. Global Health*, *10*(5), e428–e432. DOI: 10.1016/S2468-2667(25)00036-2

Park, I., & Mannaa, M. (2025). Fermented Foods as Functional Systems: Microbial Communities and Metabolites Influencing Gut Health and Systemic Outcomes. *Foods (Basel, Switzerland)*, *14*(13), 2292. DOI: 10.3390/foods14132292

Podobnik, J., & Prentice, K. J. (2025). Metabolic interventions as adjunctive therapies to insulin in type 1 diabetes: Current clinical landscape and perspectives. *Diabetes, Obesity & Metabolism*, *27*(3), 1032–1044. DOI: 10.1111/dom.16154

Povaliaeva, A., Pigarova, E., Zhukov, A., Bogdanov, V., Dzeranova, L., Mel'nikova, O., Pekareva, E., Malysheva, N., Ioutsi, V., Nikankina, L., & Rozhinskaya, L. (2020). Evaluation of Vitamin D Metabolism in Patients with Type 1 Diabetes Mellitus in the Setting of Cholecalciferol Treatment. *Nutrients*, *12*(12), 3873. DOI: 10.3390/nu12123873

Powell, S. (2018). The book of why: The new science of cause and effect [Review of the book The Book of Why: The New Science of Cause and Effect, by J. Pearl & D. Mackenzie]. *Journal of Multidisciplinary Evaluation, 14*(31), 47–54. DOI: 10.56645/jmde.v14i31.507

Prentice, S. (2017). They Are What You Eat: Can Nutritional Factors during Gestation and Early Infancy Modulate the Neonatal Immune Response? *Frontiers in Immunology, 8*, 1641. DOI: 10.3389/fimmu.2017.01641

Purkiewicz, A., Regin, K. J., Mumtaz, W., & Pietrzak-Fiećko, R. (2025). Breastfeeding: The Multifaceted Impact on Child Development and Maternal Well-Being. *Nutrients, 17*(8), 1326. DOI: 10.3390/nu17081326

Qin, B., Wang, Y., & Ding, J. (2025). The role of mitochondrial function in the pathogenesis of diabetes. *Frontiers in Endocrinology, 16*, 1607641. DOI: 10.3389/fendo.2025.1607641

Radwan, S., Gilfillan, D., Eklund, B., Radwan, H. M., El Menofy, N. G., Lee, J., Kapuscinski, M., & Abdo, Z. (2020). A comparative study of the gut microbiome in Egyptian patients with Type I and Type II diabetes. *PLoS One, 15*(9), e0238764. DOI: 10.1371/journal.pone.0238764

Rein, M., Ben-Yacov, O., Godneva, A., Shilo, S., Zmora, N., Kolobkov, D., Cohen-Dolev, N., Wolf, B. C., Kosower, N., Lotan-Pompan, M., Weinberger, A., Halpern, Z., Zelber-Sagi, S., Elinav, E., & Segal, E. (2022). Effects of personalized diets by prediction of glycemic responses on glycemic control and metabolic health in newly diagnosed T2DM: A randomized dietary intervention pilot trial. *BMC Medicine, 20*(1), 56. DOI: 10.1186/s12916-022-02254-y

Reyed, R.M., & Prabhakar, P. K. M. (2025a). Polybiome Derived Therapies and Fermentobiometrics: Novel Strategies for Blocking Cancer Pathways and Progression. In *Cancer Etiology, Diagnosis and Treatments* (pp. 145-212). Nova Medicine and Health/Pathways to Healing: Advancements in Cancer Therapy. DOI: 10.52305/JKMU3682

Reyed, R.M., & Prabhakar, P. K. M. (2025b). An In-Depth Assessment of Dietary Methods and Nutritional Treatments to Improve Female Athletes. In *Examining Physiology, Nutrition, and Body Composition in Sports Science* (pp. 35-66). IGI Global. DOI: 10.4018/979-8-3693-6317-1.ch002

Reyed, R. M., & Prabhakar, P. K. (2025c). Personalized Antiviral Therapy Tailoring Treatments to Genetic Environmental and Lifestyle Factors. In Aslam, M., Yingngam, B., Javed, T., & Riaz, S. (Eds.), *Global Perspectives on Antiviral Drug Development* (pp. 63–138). IGI Global Scientific Publishing. DOI: 10.4018/979-8-3693-9276-8.ch003

Reyed, R. M., & Prabhakar, P. K. (2025d). Impact of Dietary Antioxidants on Redox Homeostasis and Its Effects on Health and Disease. In Prabhakar, P. (Ed.), *The Role of Reactive Oxygen Species in Human Health and Disease* (pp. 81–136). IGI Global Scientific Publishing. DOI: 10.4018/979-8-3693-7919-6.ch003

Reyed, R. M., & Prabhakar, P. K. (2026). Modulating Neuroinflammation and Autophagy in Psychiatric Disorders Through Mechanisms and Therapies: Therapeutic Strategies and Pathways. In Mokhtari, T., & Uludag, K. (Eds.), *Autophagy and Inflammation in Neuropsychological Disorders* (pp. 61–124). IGI Global Scientific Publishing. DOI: 10.4018/979-8-3693-5908-2.ch003

Reyed, R. M. (2007). Probiotics: A new strategies for prevention and therapy of diarrhea disease. *Journal of Applied Sciences Research*, *3*(4), 291–299.

Reyed R.M. (2023a) Simulate the Dynamic Interactions Across Nutritional Supplements, the Polybiome, and the Biochemical Approach Dependent on Patient-Derived Gastrointestinal Knowledge. *Journal of Clinical Epidemiology & Toxicology, 4*, 133. DOI: 10.47363/JCET/2023(4)133

Reyed, R. M. (2023b). Focusing on individualized nutrition within the algorithmic diet: An in-depth look at recent advances in nutritional science, microbial diversity studies, and human health. *Food Health*, *5*(1), 5. DOI: 10.53388/FH2023005

Reyed, R. M., Patel, R., Patel, S., & Prajapati, B. G. (2026). Application of synthetic biology in genetic disorder. In *Synthetic Biology and its Consequences in Present Healthcare* (pp. 201–213). DOI: 10.1016/B978-0-443-33649-2.00001-0

Riddle, M. C., Philipson, L. H., Rich, S. S., Carlsson, A., Franks, P. W., Greeley, S. A. W., Nolan, J. J., Pearson, E. R., Zeitler, P. S., & Hattersley, A. T. (2020). Monogenic Diabetes: From Genetic Insights to Population-Based Precision in Care. Reflections From a Diabetes Care Editors' Expert Forum. *Diabetes Care*, *43*(12), 3117–3128. DOI: 10.2337/dci20-0065

Rizaldi, M. A., Azizah, R., Sulistyorini, L., & Ali, K. (2025). Environmental health risk analysis of microplastics due to consumption of squid and mussels at coastal area. *Environmental Analysis, Health and Toxicology*, *40*(1), e2025009. DOI: 10.5620/eaht.2025009

Rooks, M. G., & Garrett, W. S. (2016). Gut microbiota, metabolites and host immunity. *Nature Reviews. Immunology, 16*(6), 341–352. DOI: 10.1038/nri.2016.42

Roshdy, A. M. (2024). Microbial Functional Foods as a Magic Secret to Healthy Life Style. *Egyptian Journal of Chemistry, 67*(2), 161–173. DOI: 10.21608/ejchem.2023.202051.7926

Ruera, C. N., Guzman, L., Menendez, L., Orellano, L., Girard Bosch, M. C., Catassi, C., & Chirdo, F. G. (2025). Typing of HLA susceptibility alleles as complementary tool in diagnosis of controversial cases of pediatric celiac disease. *Frontiers in Nutrition, 12*, 1500632. DOI: 10.3389/fnut.2025.1500632

Rukavina Mikusic, N. L., Prince, P. D., Choi, M. R., Chuffa, L. G. A., Simão, V. A., Castro, C., Manucha, W., & Quesada, I. (2025). Microbiota, mitochondria, and epigenetics in health and disease: Converging pathways to solve the puzzle. *Pflügers Archiv, 477*(5), 635–655. DOI: 10.1007/s00424-025-03072-w

Salem, M. L., Zeid, W. A., & Ismail, M. A. (2019). Prevalence and predictors of gestational diabetes mellitus among pregnant women attending Fanara Family Center, in Egypt. *Suez Canal University Medical Journal, 22*(1), 64–72.

Sandeep, B., Liu, X., Huang, X., Wang, X., Mao, L., & Xiao, Z. (2024). Feasibility of artificial intelligence its current status, clinical applications, and future direction in cardiovascular disease. *Current Problems in Cardiology, 49*(2), 102349. DOI: 10.1016/j.cpcardiol.2023.102349

Santos, E. W., Khatoon, S., Zheng, Y. M., & Wang, Y. X. (2025). Mitochondrial Reactive Oxygen Species Production in Vascular Dementia Following Experimental Diabetes. *Cells, 14*(16), 1260. DOI: 10.3390/cells14161260

Sarigiannis, D., Karakitsios, S., Anesti, O., Stem, A., Valvi, D., Sumner, S. C. J., Chatzi, L., Snyder, M. P., Thompson, D. C., & Vasiliou, V. (2025). Advancing translational exposomics: Bridging genome, exposome and personalized medicine. *Human Genomics, 19*(1), 48. DOI: 10.1186/s40246-025-00761-6

Selim, S. (2025). Type 5 Diabetes Mellitus: Global Recognition of a Previously Overlooked Subtype. *Bangladesh Journal of Endocrinology and Metabolism, 4*(1), 59–60. DOI: 10.4103/bjem.bjem_13_25

Servida, S., Piontini, A., Gori, F., Tomaino, L., Moroncini, G., De Gennaro Colonna, V., La Vecchia, C., & Vigna, L. (2024). Curcumin and Gut Microbiota: A Narrative Overview with Focus on Glycemic Control. *International Journal of Molecular Sciences, 25*(14), 7710. DOI: 10.3390/ijms25147710

Shaffer, M., Armstrong, A. J. S., Phelan, V. V., Reisdorph, N., & Lozupone, C. A. (2017). Microbiome and metabolome data integration provides insight into health and disease. *Translational Research; the Journal of Laboratory and Clinical Medicine*, *189*, 51–64. DOI: 10.1016/j.trsl.2017.07.001

Shalaby, E., Gaballah, I., & Kadry, A. (2020). Evaluation of phthalate in breast milk and urine of lactating women in Egypt. *Biomedical & Pharmacology Journal*, *13*(1), 429–439. DOI: 10.13005/bpj/1903

Sharma, Y., Kannan, A., & Lee, J. M.. (2024). Frederick G Banting (1891-1941): A Pioneer in Diabetes Treatment. *Cureus*, *16*(11), e73806. DOI: 10.7759/cureus.73806

Shawki, H. A., Abo-Hashem, E. M., Youssef, M. M., Shahin, M., & Elzehery, R. (2022, January 3). PPARγ2, aldose reductase, and TCF7L2 gene polymorphisms: Relation to diabetes mellitus. *Journal of Diabetes and Metabolic Disorders*, *21*(1), 241–250. DOI: 10.1007/s40200-021-00963-4

Sikalidis, A. K. (2019). From Food for Survival to Food for Personalized Optimal Health: A Historical Perspective of How Food and Nutrition Gave Rise to Nutrigenomics. *Journal of the American College of Nutrition*, *38*(1), 84–95. DOI: 10.1080/07315724.2018.1481797

Singh, S., Raucci, A., Glovi, A., Iula, G., Mutti, L., De Laurentiis, M., Giordano, A., & Cinti, S. (2025). Socioeconomic impact of artificial intelligence-driven point-of-care testing devices for liquid biopsy in the OncoCheck system. *Cancer and Metastasis Reviews*, *44*(3), 64. DOI: 10.1007/s10555-025-10281-3

Singh, V. (2023). Current challenges and future implications of exploiting the omics data into nutrigenetics and nutrigenomics for personalized diagnosis and nutrition-based care. *Nutrition (Burbank, Los Angeles County, Calif.)*, *110*, 112002. DOI: 10.1016/j.nut.2023.112002

Sivri, D., & Gezmen-Karadağ, M. (2024). Effects of phytochemicals on type 2 diabetes via microRNAs. *Current Nutrition Reports*, *13*, 444–454. DOI: 10.1007/s13668-024-00549-5

Stein, R. A., & Riber, L. (2023). Epigenetic effects of short-chain fatty acids from the large intestine on host cells. *microLife*, *4*, uqad032. Advance online publication. DOI: 10.1093/femsml/uqad032

Sun, A., & Wang, W.-X. (2023). Human exposure to microplastics and its associated health risks. *Environment & Health*, *1*(3), 139–149. DOI: 10.1021/envhealth.3c00053

Symeonides, C., Aromataris, E., Mulders, Y., Dizon, J., Stern, C., Barker, T. H., Whitehorn, A., Pollock, D., Marin, T., & Dunlop, S. (2024). An Umbrella Review of Meta-Analyses Evaluating Associations between Human Health and Exposure to Major Classes of Plastic-Associated Chemicals. *Annals of Global Health*, *90*(1), 52. DOI: 10.5334/aogh.4459

Tamahane, V., Bhanushali, S., Shah, N., Gupta, A., Khadilkar, V., Gondhalekar, K., Khadilkar, A., & Shouche, Y. (2024). A comparative study of the gut microbiome in Indian children with type 1 diabetes and healthy controls. *Journal of Diabetes*, *16*(5), e13438. DOI: 10.1111/1753-0407.13438

Thakur, A., Sharma, P., Sharma, A. J., Sharma, N., & Singh, S.. (2025). Probiotics: Microbes for human health and beyond. *Food Materials Research*, *5*, e001. DOI: 10.48130/fmr-0024-0029

Tian, W. W., Liu, L., Chen, P., Yu, D. M., Li, Q. M., Hua, H., & Zhao, J. N. (2025). Curcuma Longa (turmeric): From traditional applications to modern plant medicine research hotspots. *Chinese Medicine*, *20*(1), 76. DOI: 10.1186/s13020-025-01115-z

Tilahun, W. M., Simegn, M. B., Abate, A., Golla, E. B., Ali, M. A., Kumbi, H., Wondie, S. G., & Geremew, H. (2025). Cesarean section delivery and its associated factors in Ghana: A multilevel analysis. *PLoS One*, *20*(2), e0318223. DOI: 10.1371/journal.pone.0318223

Topol, E. (2019). *Deep Medicine: How Artificial Intelligence Can Make Healthcare Human Again*. Basic Books.

Trasande, L., & Sargis, R. M. (2024). Endocrine-disrupting chemicals: Mainstream recognition of health effects and implications for the practicing internist. *Journal of Internal Medicine*, *295*(2), 259–274. DOI: 10.1111/joim.13748

Ullah, S., Ahmad, S., Guo, X., Ullah, S., Ullah, S., Nabi, G., & Wanghe, K. (2023). A review of the endocrine disrupting effects of micro and nano plastic and their associated chemicals in mammals. *Frontiers in Endocrinology*, *13*, 1084236. DOI: 10.3389/fendo.2022.1084236

Uta, M., Neamtu, R., Bernad, E., Mocanu, A. G., Gluhovschi, A., Popescu, A., Dahma, G., Dumitru, C., Stelea, L., Citu, C., Bratosin, F., & Craina, M. (2022). The Influence of Nutritional Supplementation for Iron Deficiency Anemia on Pregnancies Associated with SARS-CoV-2 Infection. *Nutrients*, *14*(4), 836. DOI: 10.3390/nu14040836

Valitutti, F., Mennini, M., Monacelli, G., Fagiolari, G., Piccirillo, M., Di Nardo, G., & Di Cara, G. (2025). Intestinal permeability, food antigens and the microbiome: A multifaceted perspective. *Frontiers in Allergy*, *5*, 1505834. DOI: 10.3389/falgy.2024.1505834

Vasilev, G., Kokudeva, M., Siliogka, E., Padilla, N., Shumnalieva, R., Della-Morte, D., Ricordi, C., Mihova, A., Infante, M., & Velikova, T. (2025). T helper 17 cells and interleukin-17 immunity in type 1 diabetes: From pathophysiology to targeted immunotherapies. *World Journal of Diabetes*, *16*(4), 99936. DOI: 10.4239/wjd.v16.i4.99936

Verma, S., Srivastava, N., & Banerjee, M. (2021). Genetic polymorphisms in TCF7L2 and PPARG genes and susceptibility to Type 2 diabetes mellitus. *Meta Gene*, *28*, 100864. DOI: 10.1016/j.mgene.2021.100864

Vincentini, O., Pricci, F., Silano, M., Agrimi, U., Iacoponi, F., & Villa, M.. (2025). Study protocol of D1Ce Screen: A pilot project of the Italian national screening program for type 1 diabetes and coeliac disease in the pediatric population. *PLoS One*, *20*(8), e0328624. DOI: 10.1371/journal.pone.0328624

Vodovotz, Y., Barnard, N., Hu, F. B., Jakicic, J., Lianov, L., Loveland, D., Buysse, D., Szigethy, E., Finkel, T., Sowa, G., Verschure, P., Williams, K., Sanchez, E., Dysinger, W., Maizes, V., Junker, C., Phillips, E., Katz, D., Drant, S., & Parkinson, M. D. (2020). Prioritized Research for the Prevention, Treatment, and Reversal of Chronic Disease: Recommendations From the Lifestyle Medicine Research Summit. *Frontiers in medicine*, *7*, 585744. DOI: 10.3389/fmed.2020.585744

Vonderau, J. S., & Desai, C. S. (2022). Type 3c: Understanding pancreatogenic diabetes. *JAAPA: Official Journal of the American Academy of Physician Assistants*, *35*(11), 20–24. DOI: 10.1097/01.JAA.0000885140.47709.6f

Wachamo, S., & Gaultier, A. (2025). The emerging role of microbiota derived SCFAs in neurodegenerative disorders. *Brain, Behavior, & Immunity - Health*, *46*, 101012. DOI: 10.1016/j.bbih.2025.101012

Weiner, H. L. (2025). Immune mechanisms and shared immune targets in neurodegenerative diseases. *Nature Reviews. Neurology*, *21*(2), 67–85. DOI: 10.1038/s41582-024-01046-7

WHO. (2024a). *Ethics and governance of artificial intelligence for health: Guidance on large multi-modal models*. Retrieved from https://www.who.int/publications/i/item/9789240084759

WHO. (2024b, January 18). *WHO releases AI ethics and governance guidance for large multi-modal models.* Retrieved from https://www.who.int/news/item/18-01-2024-who-releases-ai-ethics-and-governance-guidance-for-large-multi-modal-models

Xu, J., Lao, Y., Zhang, W., Chen, L., Zhang, H., Liu, H., Yan, T., Huang, R., Xu, Y., Ye, L., Zhi, F., & Yang, H. (2025). Berberine alleviates metabolic dysfunction-associated steatohepatitis by enhancing the abundance of *Akkermansia muciniphila. The Journal of Nutritional Biochemistry, 146*, 110069. DOI: 10.1016/j.jnutbio.2025.110069

Yan, F., Wang, X., Du, Y., Zhao, Z., Shi, L., Cao, T., Shen, Y., Sun, L., & Liu, X. (2025). Pumpkin Soluble Dietary Fiber instead of Insoluble One Ameliorates Hyperglycemia via the Gut Microbiota-Gut-Liver Axis in db/db Mice. *Journal of Agricultural and Food Chemistry, 73*(2), 1293–1307. DOI: 10.1021/acs.jafc.4c08986

Yan, X., Liu, X., Zhao, C., & Chen, G.-Q. (2023). Applications of synthetic biology in medical and pharmaceutical fields. *Signal Transduction and Targeted Therapy, 8*, 199. DOI: 10.1038/s41392-023-01440-5

Yang, T., Wu, C., Li, Y., Wang, C., Mao, Z., Huo, W., Li, J., Li, Y., Xing, W., & Li, L. (2024). Association of short-chain fatty acids and the gut microbiome with type 2 diabetes: Evidence from the Henan Rural Cohort. Nutrition, metabolism, and cardiovascular diseases. *Nutrition, Metabolism, and Cardiovascular Diseases, 34*(7), 1619–1630. DOI: 10.1016/j.numecd.2024.03.014

Zhang, D., Jian, Y., & Zhang, Y.. (2023). Short-chain fatty acids in diseases. *Cell Communication and Signaling : CCS, 21*(1), 212. DOI: 10.1186/s12964-023-01219-9

Zhang, J. X., Chen, P. P., & Li, X. Q.. (2024). Deficiency of thiosulfate sulfurtransferase mediates the dysfunction of renal tubular mitochondrial fatty acid oxidation in diabetic kidney disease. *Cell Death and Differentiation, 31*, 1636–1649. DOI: 10.1038/s41418-024-01365-8

Zhong, Y., Xiao, Y., Gao, J., Zheng, Z., Zhang, Z., Yao, L., & Li, D. (2022). Curcumin improves insulin sensitivity in high-fat diet-fed mice through the gut microbiota. *Nutrition & Metabolism, 19*(1), 76. DOI: 10.1186/s12986-022-00712-1

Zhou, M., & Pang, X. (2025). Polyphenols and miRNA interplay: A novel approach to combat apoptosis and inflammation in Alzheimer's disease. *Frontiers in Aging Neuroscience, 17*, 1571563. DOI: 10.3389/fnagi.2025.1571563

Chapter 2

Decoding the Human Polybiome Through AI and Nutrigenomics in Next Generation Personalized Health:

Integrating Multi-Omics for Precision Medicine

Hina Bansal
http://orcid.org/0000-0003-1683-1581
Amity Institute of Biotechnology, India

Arshita
http://orcid.org/0009-0008-2354-4728
Amity Institute of Biotechnology, India

Kalidindi Divya Srija
http://orcid.org/0009-0001-6135-6647
Amity Institute of Biotechnology, India

ABSTRACT

This chapter investigates the integration of artificial intelligence, nutrigenomics, and the human polybiome to enhance precision healthcare. It outlines how deep learning and multiomics approaches decode host–microbe interactions influencing nutrition, gene expression, and disease risk. This chapter introduces advanced analytical frameworks linking microbial metabolites, epigenetic regulation, and

DOI: 10.4018/979-8-3373-3121-8.ch002

metabolic balance. By merging computational biology with clinical insight, AI-driven strategies for predictive nutrition, individualized therapy, and early disease intervention can be developed. Furthermore, this chapter discusses the ethical and interpretability dimensions of AI in biomedical data, emphasizing transparent algorithms and privacy-preserving models that ensure reliability, reproducibility, and responsible innovation in personalized healthcare systems.

1. OVERVIEW OF THE POLYBIOME

Healthcare is a sphere that has ushered in a new era in recent years with the incorporation of artificial intelligence (AI), polybiome science, and nutrigenomics. While most up-to-date research frames these communications through the conventional microbiome lens, this chapter figures on and extends the model to the polybiome, a wide-ranging and more dynamic framework first introduced by Reyed (2023a). This convergence has helped fundamentally redefine our concept of health, disease, and biological individuality and has changed the research paradigms toward more personalized and data-driven strategies (Jenko et al., 2025; Porcari et al., 2025; Reyed, 2023b). Instead of illustrating these domains as separate fields, this chapter frames them as interventions with one another as constituents of a single, dynamic (environment) that forms the foundations of personalized medicine.

This chapter starts by explaining the conceptual underpinnings of AI in health, which explains how highly skilled computational frameworks can examine large biological and clinical examples, identify latent examples, and assist in making sound decisions (Porcari et al., 2025; Kuziel & Rakoff-Nahoum, 2022; Kumar et al., 2024)). Furthermore, the human polybiome is considered a dynamic and adaptive system that affects immunity, metabolism and the risk of disease (Przymus et al., 2025; Ghadermazi & Chan, 2024). The third section addresses the concept of nutrigenomics, one of the interdisciplinary domains connecting genetic diversity and nutritional reactions, thus identifying the nature of those interactions between dietary components and individual genotypes and shaping metabolic and immune phenotypes (Hernández-Calderón et al., 2022; Kok et al., 2025).

Expanding the points stated above, the chapter emphasizes that AI technologies are capable of integrating genomic and microbial data with clinical data to provide more accurate, predictive, and preventive healthcare schemes (Ren et al., 2025; Pati et al., 2024; Reyed, 2023b). This integrative approach enables the shift toward generalized treatment to interventions on the basis of the biological distinctiveness of an individual.

Offering a conceptual framework that is both unified and clearly connected, the introduction aims to trace a reader through the clear and logically related narrative:

starting with the molecular understanding and computational principles and moving to their application to clinical and population health outcome practices. Finally, this chapter does not contain only a list of emerging scientific fields but rather a systematic outlook of the vision through which their convergence is redefining the future of medicine (Jenko et al., 2025; Porcari et al., 2025).

Table 1. Comparative Overview of the Microbiome and Polybiome Concepts and Their Implications for Future Precision Medicine.

Aspect	Microbiome Concept	Polybiome Concept (Reyed & Prabhakar 2025; Reyed, 2023a)
Definition	It chiefly investigates the bacterial composition in a specific location, such as that found in the digestive system.	It consists of a diverse array of microscopic organisms inhabiting the human organism, such as bacteria, fungi, archaea, viruses, protists, and species that are unculturable under laboratory conditions.
Scope	It largely explores the study of bacterial interactions and its impact on it impacts physiological condition.	It illustrates a more comprehensive understanding of how hosts and their microscopic companions interface while addressing numerous biological classifications and biological frameworks.
Functional Focus	It examines the existing microorganisms, irregularities related to microbial abundance, and these organisms' physiological processes.	It explores how metabolic processes, immune responses, and signaling routes in host organisms connect by way of networks of multiomics data.
Technological Integration	The outcome is dependent on strategies for example 16S rRNA gene sequencing and whole genome metagenomics.	It integrates various omics approaches alongside AI modeling with the aim of understanding the dynamics throughout the duration.
Research Perspective	It considers bacterial populations as balanced complex environments.	It understands the microbial community as a complex system which modifies, maintains its own balance, and progresses interdependently.
Clinical Translation	It is employed for diagnostic purposes and when formulating probiotics that target specific bacterial species.	. It finds application in personalized medicine, artificial intelligence-driven therapies, as well as the surveillance of an individual's health via modulation of the polymicrobial community.
Philosophical View	It centers on bacteria as distinct entities.	It interprets microbes as integral factors in the complex human biological system that have an impact on holistic health.

Table 1 outlines the methodological tools of polybiome and nutrigenomics, which therefore have provided a paradigm of comprehension to clarify the mutual regulation between host genotype, nutrition and native microbial networks.

Reyed (2023a) pioneered this concept. Although the 2025 publication includes Prabhakar as a co-author due to editorial collaboration, the intellectual ownership and original development of the concept remain solely attributed to Reyed (Reyed &

Prabhakar, 2025). Within this framework, the microbiome is depicted as a dynamic network that transforms, is regulated through complex interactions, and is highly interrelated. It engages with biochemical elements, immunological components, nutritional factors, and genomic aspects, highlighting its significance in maintaining equilibrium within the host and enabling individualized nutrition. This redefinition goes beyond the conventional understanding of the microbiome as a passive collection of microbial taxa, redefining it as an active multisystem of cross-kingdom interactions, including bacteria, fungi, viruses, archaea, and host metabolic networks. An integrated model of this type emphasizes the mutualistic signaling between host and microbial constituents and its role in the coevolution of physiological balance and health resilience. On the basis of this framework, modern artificial intelligence-enabled analysis systems now possess the ability to solve the complex interactions of this ecosystem, enabling polybiome-informed precision medicine. By making predictions on the basis of multiomic, dietary, and clinical data, these systems can be used to forecast individualized nutrition, therapy, and environmental response and thus represent a groundbreaking step toward data-driven, personalized health care.

This holistic view goes beyond a naive list of microbial groups, an in-depth definition of interactions between microbial constituents and their adaptive interaction with the host. As a result, metabolic balance, immune activity, vulnerability to diseases, and therapeutic responses are defined (Kuziel et al., 2022; Porcari et al., 2025; Przymus et al., 2025). The gut polybiome is conceptualized as a biological interface that converts nutritional inputs into molecular signals that are able to regulate gene expression, metabolic control, and immune resilience in this paradigm. Nutrigenomics increases this model by explaining how nutrient consumption can affect the expression of genes and vice versa (Reyed, 2021a; Nielsen and El-Sohemy, 2012; Lagoumintzis and Patrinos, 2023). For example, SNP variation associated with metabolic pathways may modulate the response of an individual to macronutrient composition and predispose him or her to chronic diseases such as obesity, diabetes, or cancer when nutritional exposure does not match SNP variation.

Importantly, the incorporation of AI-driven multiomics analytics has turned this interaction into a more descriptive model of a predictive and adaptive precision health system. Transformer-based models and machine learning have the ability to map polybiomic relationships between high-dimensional relationships among dietary inputs, gene expression patterns, and polybiomics to provide customized nutritional and therapeutic advice (Madan et al., 2024; Linares-Blanco et al., 2022; Pati et al., 2024; Mienye et al., 2024). These technologies make it possible to decipher microbial–host–diet interactions in real time to predict disease progression and act in advance with personalized nutrition and specific treatments.

Synergizing polybiome profiling, nutrigenomic data and AI-powered decision models, contemporary precision medicine goes beyond traditional population-wide prescriptions and advances to personalized, evidence-based health and wellness approaches. This approach is especially relevant to metabolic and immune-mediated disease, cancer, and cardiovascular pathologies, where the interactions of genes with diets and microbes significantly affect disease onset, disease progression, and response to treatment (Paul et al., 2025; Rouskas et al., 2025; Mousa & Al Ali, 2024). Moreover, the multiscale view of the polybiome allows the consideration of viral and fungal elements, which have remained largely overlooked in previous studies on the microbiome, thereby reflecting the complexity of the ecological environment in the gut.

Such a theoretical development is a paradigm shift between reactive care and predictive and preventive medicine. With these areas combined, AI-enabled nutrigenomics and polybiome systems will aid customized dietary planning, optimized drug–nutrient interactions, and patient-specific therapeutic strategies on the basis of the biological signature unique to the patient. The combination of these technologies, when looked at as a whole, is an indicator of a future where polybiomic intelligence will be a pillar of sustainable precision healthcare.

1.2 Personalized Medicine

Personalized medicine is another name for precision medicine. Genetics, lifestyle, the environment, and personal habits are all factors considered by precision medicine when choosing a treatment plan for a patient. As a result, patients are less likely to use therapies that can cause side effects. It is fair to call Archibald E. Garrod the originator of precision medicine. He proposed the idea of individuality among chemicals. As a result, every individual's biochemistry is unique to them and connects to their genes. The idea that every person is different is at the heart of developing precision medicine, which treats each patient on the basis of their traits. Rather than depending on a single solution for all, personalized medicine uses wireless health monitoring, proteomics, DNA sequencing, and modern imaging to determine how individuals respond to illnesses, respond to treatments, and develop various diseases differently. As an example, because everyone has different genetic and metabolic changes, two patients with the same disease may react in diverse ways to the same medicine. Personalized medicine uses customized risk models and each person's health baseline, rather than average statistics for a group, to manage both treating and preventing diseases. To assess how these treatments work, researchers are performing N-of-1 studies, matching interventions, and carrying out adaptive clinical trials that consider a patient's personal results. Using advanced methods for reproduction, technologies that track health in real time, digital treatments tailored to individuals, and patient-based cells for testing new drugs are all emerging in this field. Personalized medicine presents several issues, including that it is costly, the privacy of data is uncertain, there are many regulations, and personnel must have specific training (Goetz & Schork, 2018). These innovations, however, illustrate only part of the broader evolution of personalized medicine. In this chapter, the conceptual underpinnings of personalized medicine are examined. It also introduces emergent technologies, including artificial intelligence, polybiome profiling, and nutrigenomics. These technologies now enable rapid translation of personalized medicine into real-time, predictive clinical applications. Together, they support a more integrated and cross-sectoral vision for the future of medicine. Recent publications (Shao et al., 2025) have shown that AI-based multiomics and polybiome research can revolutionize personalized medicine through enhanced disease forecasting and personalized nutrition. Global efforts such as *All of Us and Genomics England* demonstrate an increase in the trend of fair, data-driven healthcare.

1.3 Role of Artificial Intelligence in Personalized Health

Artificial intelligence in precision medicine discusses how AI and ML can be used in healthcare to improve patient prognosis, diagnosis, treatment, and illness detection. AI can quickly address appropriately chosen medical information on a large scale, helping to manage and analyze big datasets for each patient's exceptional care needs: their genomic sequences, laboratory results, heterogeneous imaging, and electronic health records. AI supports various medical disciplines, including oncology for cancer therapy, as well as identifying, screening, identifying molecular patterns and assisting with drug discovery; cardiology for prediagnosis evaluations; endocrinology for diagnosing thyroid, pituitary and similar conditions and performing personal health assessments; and reproductive medicine for embryo checking. In addition, AI has been useful in digital pathology by creating algorithms for analyzing whole-slide images to help analyze tumors, AI and digital pathology with prognostics concerning cancer diagnosis. AI integrated with radiomics makes it possible for radiation therapy to evaluate how patients react to therapy. While trying to use these modern technology tools, we experience problems with data quality and integrity, the clarity of algorithms, and related social or ethical matters in AI. Regardless of the issues with AI, using AI in healthcare will put medicine on a course for even better, more individualized, and smoother treatments in the future (Mukherjee et al., 2025). Even though AI-powered innovations have already changed several clinical fields, their greatest potential will be revealed when they are used to analyze multiomics data. The combination of polybiome and nutrigenomic data with AI is the next important step in moving toward very precise individualized interventions rather than generic predictive analytics. As part of a strategy aimed at achieving the greatest transformative potential, artificial intelligence (AI) should increasingly be used in the process of integrating advanced multi-omics strata, namely, polybiome and nutrigenomic data. It should do so into dynamic and clinically relevant models. These models can help refine therapeutic decision-making. They can also form the basis of nutritional interventions in the precision-medicine paradigm (Shao et al., 2024). However, the full potential of these computational frameworks emerges when they are integrated with biological layers — particularly polybiome profiling and nutrigenomics — which together offer unprecedented precision in understanding individual variability. This integrated approach reflects a paradigm shift from reactive treatment to predictive, preventive and adaptive health care — a vision in which polybiomic intelligence and AI-driven nutrigenomics form the backbone of next-generation precision medicine.

1.4 Chapter Objectives

This chapter explores the integration of artificial intelligence (AI) within polybiome research and nutrigenomics to advance personalized medicine. The foundation of personalized medicine can be traced back to early scientific milestones—beginning with Mendel's work on heredity; the discovery of DNA structure by Watson and Crick in 1953; and the mid-20th century emergence of pharmacogenetics, which revealed how genetic variability influences drug responses Feero & Guttmacher (2021).. The completion of the Human Genome Project further revolutionized this field by mapping the entire human genetic code, enabling associations between genes, diseases, and therapeutic outcomes. These developments, coupled with advances in genomics, proteomics, and metabolomics, have laid the groundwork for today's precision health strategies.

Building upon this historical evolution, this chapter aims to examine how AI-driven analytical tools and computational models enhance the interpretation of complex biological data, optimize diagnostic precision, and inform individualized therapeutic strategies. It emphasizes empirical studies and practical applications that illustrate the synergy between AI, microbial science, and nutritional genomics in improving health outcomes. Additionally, the discussion addresses key challenges—including ethical considerations, regulatory constraints, and interoperability issues—that must be overcome to ensure the responsible and sustainable implementation of AI-enabled tailored medicine (Jørgensen, 2019).

2. ADVANCING PERSONALIZED MEDICINE

2.1 Key Technologies in Personalized Medicine

The area of personalized medicine is developing quickly by incorporating high-impact technologies, including genomics, bioinformatics, artificial intelligence (AI), nanotechnology, and multiomics platforms. These tools make interventions individualized, real-time, and add to the diagnostic accuracy and help to make adaptive therapeutic decisions. Specifically, generative AI models have become powerful medicinolinguistic accelerators (drug discovery, predictive modeling, and clinical diagnostics) with the ability to simulate drug–target interactions, stratify patients, and interpret high-resolution medical imaging (MRI and CT) (Ho et al., 2020).

The heart of individual care is genomics, which, through next-generation sequencing (NGS), can allow large-scale mapping of genetic differences as well as disease-related mutations quickly. These genomic findings form the basis of guided and precision therapy, such as HER2-targeted therapy for breast cancer. Bioinformatics provides an analytical foundation for the organization and analysis of large quantities of biological data to facilitate the discovery of pathogenic variants, drug response markers and therapeutic susceptibilities. This data-centric approach converts therapeutic paradigms into predictive paradigms, which can allow a transition to precision-based interventions.

No less radical is nanotechnology, which enables the delivery of smarter drug delivery systems based on targeted nanoparticles that can release therapeutic agents selectively at disease sites to optimize effects and reduce systemic toxicity (Ghanem et al., 2024). Immunotherapy and gene therapy continue to broaden personalized treatment by relying on immune pathways or solving genetic defects, which were previously deemed incurable.

New tools such as single-cell RNA sequencing (scRNA-seq) and state-of-the-art biosensors allow the detection of molecular heterogeneity earlier, real-time tracking of therapeutic response, and the creation of new possibilities for adaptive intervention strategies. Likewise, lab-on-a-chip diagnostic methods can be used to detect tumor DNA or exosomes in circulating fluids or exosomes rapidly with minimal invasiveness at a lower cost and greater accessibility. Multiomics integration, including genomics, transcriptomics, proteomics, and metabolomics, is a novel systems-level perspective of disease, providing a molecular signal-to-phenotype linkage with greater precision than ever before (Reyed, 2023b).

Additionally, clinical decision-making can be enhanced by the introduction of real-world evidence via electronic health records and patient registries, particularly in cases where conventional clinical trials are not possible. All of these developments, as summarized in Table 2, point to the cross-sectoral innovation that is the future of precision healthcare. The table describes the fundamental contributions, critical issues, innovative solutions, and future perspectives of these technologies, which presents a strategic framework for translating these technologies into clinical applications (Ho et al., 2020; Reyed, 2023b; Ghanem et al., 2024).

Table 2. Strategized Scientific Frontiers Shape the Future for Personalized Medicine: Key Gaps, Innovative Resolutions, and Future Visions

Technical Alignment	Fundamental Supports to Precision Medicine	Emerging Gaps and Research Bottlenecks	Strategic and Innovative Responses	Projected Clinical Courses	Selected References
Genomics & Multi-Omics Integration	It proposes a molecular map to customized diagnosis and treatment incorporating next-generation sequencing and high-resolution omics data.	Unequal datasets, limited interpretability, as well as unequal access to genomic substructures.	Artificial intelligence-enabled annotation systems, federated data-sharing models, and pipeline adaptive variant understanding.	An interconnected genomic ecosystem that enables real-time, clinically actionable decision-making worldwide.	Lagoumintzis & Patrinos, 2023; Ma et al., 2022
Artificial Intelligence	Acts as an accelerator for drug development, predictive modeling, and clinical decision support.	The lack of explainability, the underdevelopment of regulatory systems, and the barriers to the harmonization of data all hinder further progress.	Clear explainable AI architectures, strong validation requirements, and embedded decision-support layers are also important elements.	AI plays the role of a transparent and intrinsic copilot in precision care delivery settings.	Madan et al., 2024; Mohapatra et al., 2025
Polybiome Science	Illuminates systemic microbial interactions (gut-oral-skin-lung axes) shaping immune, metabolic, and neuroendocrine networks.	Significant interindividual heterogeneity, limitations to causal inference, and a lack of longitudinal data make it difficult to draw conclusive findings.	Multilayer network analysis, poly-omics modeling, and microbial signatures tailored to individuals.	Precision interventions with predictive and preventive abilities were bestowed by polybiome-driven precision..	Kuziel & Rakoff-Nahoum, 2022; Reyed, 2023a; Porcari et al., 2025
Nutrigenomics & Precision Nutrition	Aligns genetic and metabolic profiles with tailored dietary strategies to reduce disease risk and enhance treatment response.	A lack of population-specific data, cross-border discrepancies, and uncertainty in regulations hinder equal application.	Adaptive dietary algorithms, AI-assisted feedback loops, and personalized nutrition frameworks.	Precision nutrition embedded into standard care pathways.	Pati et al., 2024; Nielsen & El-Sohemy, 2012

continued on following page

Table 2. Continued

Technical Alignment	Fundamental Supports to Precision Medicine	Emerging Gaps and Research Bottlenecks	Strategic and Innovative Responses	Projected Clinical Courses	Selected References
Nanotechnology & Smart Drug Delivery	Enhances therapeutic targeting, reduces systemic toxicity, and enables controlled release.	Scalability, thorough safety profile, and complex regulatory landscapes are among the key challenges in manufacturing complex chemical entities.	Biocompatible nanoparticles, hybrid nanobio delivery systems and adaptive release systems.	Personalized medicine with nanotechnology-enabled medicine.	Ghanem et al., 2024; Phillips, 2013
Real-World Data & Digital Health	This model converts quantitative data created by patients and clinical data into individualized and dynamic care paths.	There are data interoperability issues, privacy issues, and inconsistent evidence translation.	It suggests implementing secure multilayer data ecosystems, monitoring that is wearable-integrated, and adaptive clinical analytics.	Cybernetic health systems that can develop by constant and precisely targeted evolution are imagined.	National Academies, 2022; Rouskas et al., 2025

As shown by the table, the role of core technologies, including genomics, bioinformatics, AI, nanotechnology, and multiomics, in creating a paradigm shift in care, which makes it more precise, predictive, and individualized. This review summarizes their principal clinical uses, their primary implementation obstacles, and their innovative polybiome interventions that maximize their translational potential in clinical practice. This systematic synthesis stresses the importance of cross-sectoral interaction and technology integration to promote personalized medicine.

On this basis, genomics and other omics-based technologies, including proteomics and metabolomics, are the next necessary steps in unraveling the complex pathophysiology of diseases and delivering therapy tailored to the individual.

2.2 Genomics, Proteomics, and Metabolomics

The integration of genomics into personalized medicine has significantly advanced the understanding of disease susceptibility and drug response. This has greatly promoted disease susceptibility and drug response through the incorporation of genomics in personalized medicine. With next-generation sequencing, it has become possible to sequence genomes rapidly to analyze disease-linked mutations and single nucleotide polymorphisms (SNPs). BRCA1/2 mutations can be used in oncology; for example, they are used to predict the risk of hereditary breast and ovarian cancers and to inform the selection of targeted therapies, such as PARP inhibitors. Pharmacogenomics, which is based on tailoring drug treatment according to a personal genetic reading, is also rooted in genomics (Ho et al., 2020).

In the context of the genomic approach, proteomics offers complementary functional information through the examination of the whole body of proteins. Disease phenotypes are more closely linked with proteins than with genes; thus, proteomic profiling is a strong way to identify disease biomarkers and therapeutic targets. The posttranslational modifications of protein and protein interactions may be revealed via methods such as mass spectrometry or protein microarrays. Proteomics has already been used to develop targeted cancer therapy in a clinical setting, including HER2-based therapy using trastuzumab in breast cancer (Karczewski & Snyder, 2018).

Metabolomics is the measurement of the profile of small-molecule metabolites and provides a clear picture of the biochemical activity of a person. Metabolomics is especially valuable in monitoring the course of disease and therapeutic response because metabolites are the result of not only the expression of genetic material but also the environment. It has appeared promising in fields such as pharmacometabolomics, whereby metabolic signatures may be utilized to determine patient reactions to drugs, including glucose-lowering medicines, in diabetes management (Hasin et al., 2017). Personalized medicine can only be very effective when it is supplemented with multiomics data with a combination of genomics, proteomics, and metabolomics. This stratified approach provides an overview of disease pathologies and personal heterogeneity. Multiomics platforms complemented with artificial intelligence and bioinformatics have the power to detect intricate molecular interactions and increase diagnostic accuracy as well as precision with therapy plans on the basis of individual molecular profiles (Karczewski & Snyder, 2018).

2.3 EHR Integration (Electronic Health Records)

The integration of electronic health records (EHRs) with artificial intelligence (AI) applications is important in actualizing the potential of personalized medicine. Integrating EHRs with powerful machine learning algorithms allows clinicians to analyze a patient in a granulated, multilevel format by generating a well-organized, scalable record including their overall health data, such as medical history, diagnostic details, lifestyle, and genomics. This integration will help with predictive analytics, which can allow physicians to foresee the risks of developing various diseases, create a personal treatment plan and monitor results in real time. The ability of the EHR to include AI-driven decision support tools that aid in the accuracy of the diagnosis also contributes to the reduction of the trial-and-error approach to choosing the therapy. For example, a machine learning algorithm might be used to predict patient reactions to a certain drug on the basis of the correlations among clinical and genetic data obtained through EHR, decreasing adverse results and increasing the effectiveness of treatment. Such trends encourage the transition of relatively universal healthcare approaches toward dynamism in patient-centered care models on the basis of responsiveness to individual health profiles. As illustrated in the case studies noted in the article, such an integrated platform has already been implemented in organizations, such as AIIMS or Biocon, to enable better chronic disease management and streamline clinical trials, providing evidence of the revolutionary potential of EHR-AI integration in precision medicine practices (Paramasivan, 2023).

2.4 Limitations of Traditional Approaches

Conventional medical practices can be considered one-size-fits-all, as diagnosis and treatment plans are developed according to the average of the population instead of individual differences. Although this model provides a good background for both public health and clinical practice, it does not consider the biological diversity of individuals, especially at the genetic, molecular and metabolic stages. This may lead to many patients not responding well to regular treatments, and some patients can experience adverse reactions. An example would be two patients with identical diagnoses who react much differently when subjected to the same medication because there are many variations in genes, which may influence drug breakdown or the development of the disease. In addition, traditional diagnostics are insensitive enough to recognize early or subtle molecular modifications, depriving them of the possibility of early treatment or personalized medicine. Such weaknesses delay the possible implementation of genuinely personalized medicine, and they emphasize the importance of such integrative and data-based disciplines as genomics, proteomics, and AI-enabled decision systems in current personalized medicine (Schork, 2015).

2.5 Future Integration and Practical Vision

New AI-based multiomics data spaces will sweep precision medicine by incorporating data generated by genomics, proteomics, metabolomics, and polybiome mapping. The increased resiliency of biomarker discovery, improved understanding of biological systems, and better forecasting of disease risk and response to therapies made possible by this synthesis open new areas of research in the developing science of personalized medicine. Furthermore, the emergence of single-cell multiomics analysis presents an entirely new level of resolution of cellular heterogeneity and can serve as a tool to promote our understanding of disease biology, as well as enable highly personalized treatment. In the future, explainable AI (xAI) frameworks, federated learning, and quantum-powered analytics tools are expected to handle the complexities of multiomics datasets without sacrificing explainability and without breaching the privacy of the patient (Ren et al., 2025).

2.6 Ethical, Regulatory, and Practical Deployment Challenges

In the chapter in question, the authors highlight the significant setbacks that come with artificial intelligence (AI) when combined with multiomics data analysis. These challenges are dominated by metadata heterogeneity, discrimination in algorithm results, high computational demands, and the inherent challenge in ensuring the generalizability of findings to an entire population. The legal considerations are also high, requiring strict protection of the confidentiality of genomic and polybiome data and strict adherence to data protection directives such as the General Data Protection Regulation (GDPR), the Health Insurance Portability and Accountability Act (HIPAA), and other data protection policies that are specific to polybiome-based therapeutic applications that are on the rise (Konnoth, 2024). The chapter is therefore directed to conclude that there will be long-term interdisciplinary cooperation, comprising clinicians, bioinformaticians, AI experts, and policymakers, to build interoperable, scalable systems, as well as support open-access multiomics repositories, to transform innovation into medical practice. These research directions define an ecosystem that places its direction and focus in line with the overall objective of this volume, specifically, the pathway of patient-centric precision health.

3. POLYBIOME AND ITS ROLE IN HEALTH

3.1 Gut Polybiome and Its Function

A healthy gut microbiome protects us by affecting our digestion, immune system, and general health. Findings from animal research have shown that the absence of microbiota makes the immune system less efficient, slows metabolic processes and changes gut characteristics, highlighting the key role of colonization by microbes in the proper progress of a host (Kuziel & Rakoff-Nahoum, 2022 ; Reyed & Prabhakar, 2025).

Bile acids, short-chain fatty acids (SCFAs), and tryptophan derivatives are metabolites that help the polybiome and the host communicate. They manage immune reactions in the gut, improve gut barriers, and send messages that control activity in distant organs. In particular, SCFAs support the health of the intestines and their ability to fight inflammation by inhibiting histone deacetylase activity and promoting T-regulatory cells (Kuziel & Rakoff-Nahoum, 2022 ; Reyed & Prabhakar, 2026). The gut microbiota is known to have a major impact on epigenetic regulation. Butyrate and trimethylamine N-oxide metabolites can affect gene activity by affecting DNA methylation and histone changes. As a result, these molecules can influence both cell differentiation and immune function (Liu et al., 2023). The signals from the microbiota are sensed by the host and affect both the body's nervous system and immune system (Reyed & Prabhakar, 2026).

Additionally, changes in microbial composition and the substances they make can occur in response to what we eat, antibiotic use, and different health conditions. Evidence indicates that microbial synthesis of branched-chain and aromatic amino acids can shift how host cells function and grow. More changes, such as these, are being seen as indications that host–microbe interactions are starting to go wrong. Using multiomics techniques such as metagenomics and metabolomics, researchers have shown that the gut microbiota responds to the body's conditions. Owing to these technologies, profiling microbial and metabolic activity has become more precise. This helps reveal how diverse microbes and their products can affect the host body under various physical and environmental circumstances (Kuziel & Rakoff-Nahoum, 2022).

Although mechanistic findings commonly guide the focus of polybiome research, new evidence also suggests that it has promise as a diagnostic tool. The study of both microbial species and metabolites in feces has shown promise for identifying biomarkers, revealing early differences related to important health changes even in those without symptoms (Coker et al., 2022). These results suggest that polybiome-based detection could assist clinicians in addition to standard diagnostic methods.

3.2 Functions of the Polybiome in Immunity, Metabolism, and Health

The gut polybiome, which is composed of bacteria, archaea, fungi, and viruses, is essential for many metabolic and immunological functions and represents a complex and dynamic ecosystem that closely affects host physiology (Reyed & Prabhakar, 2025; Reyed, 2023a; Cox et al., 2022; Fujisaka et al., 2023). One of the primary roles of this microbial network is fermentation of otherwise indigestible dietary fibers into short-chain fatty acids (SCFAs), which include acetate, propionate and butyrate. These metabolites are involved in cell communication, maintaining intestinal barrier integrity, modulating immune responses and lipid and carbohydrate processes, and maintaining whole-body energy in a steady state. (Reyed & Prabhakar, 2026; Cox et al., 2022). Importantly, primary bile acids produced by microbial biochemical activity are transformed into secondary metabolites that interact with FXR and TGR5 receptors, with consequences for cholesterol homeostasis, thermogenesis and lipid absorption (Wu et al., 2021). There is also evidence that the microbial production of branched-chain amino acids (BCAAs) and indoles derived from tryptophan has indirect consequences on insulin sensitivity and the balance of systemic inflammation. Germ-free animal studies provide compelling evidence that the presence and composition of the polybiome directly regulate obesity, total energy exchange and susceptibility to metabolic disorders, revealing the causative rather than correlative role of microbial communities (Cox et al., 2022; Reyed & Prabhakar, 2025). In addition to metabolism, the polybiome critically affects the development and function of the immune system. Early colonization by good microbes causes the differentiation of regulatory T cells and calibrates innate immune responses, laying the foundation for immune competence (Reyed & Prabhakar, 2026; Sanidad & Zeng, 2020).

SCFAs exhibit anti-inflammatory properties, whereas metabolites derived from bile acids further suppress proinflammatory T helper 17 activities to promote and sustain immunological homeostasis (Fujisaka et al., 2023). The depletion of microbial diversity, which promotes dysbiosis, incites systemic inflammation that increases the tendency toward type 2 diabetes, inflammatory bowel disease, and cardiovascular disorders. Age-related reductions in microbial diversity enhance low-grade chronic inflammation, which reduces the adaptability of the immune system and metabolic control (Bosco & Noti, 2021).

The polybiome is therefore an ideal target for therapeutic intervention given its primary role in host physiology. Dietary changes along with prebiotics, probiotics and fecal polybiota transplantation (FPT) are under investigation to restore microbial balance and improve outcomes in terms of human health (Reyed & Prabhakar, 2025; Aggarwal et al., 2022;). New progress in synthetic biology and polybiome engineering has made it possible to rationally design microbial communities for specific metabolic or immunologic goals, thus expanding the delivery methods possible in precision medicine.

The understanding of the mechanisms of polybiome activity has been accelerated by the combination of high-throughput sequencing, multiomic, and metabolomics data with computational technologies. The size and complexity of these datasets require the use of AI and machine-learning tools to recognize indicative patterns that predict host responses while allowing the identification of dysbiosis phenotypes associated with obesity, allergic disease and dermatoses. Moving from descriptive associations to mechanisms of action of the agents via graph analysis and network-based computations will increasingly yield better results in understanding host–microbiome interactions. By integrating genomic analysis, transcriptomics, metabolomics and AI modeling, it will become possible to plot host-microbiome signaling pathways, including interactions with FXR and TGR5 receptors, thereby determining the causal mechanisms behind immune and metabolic regulation (Shawky et al., 2025, Reyed, 2023b). To translate these mechanistic insights into clinical benefits, computational tools have become indispensable. Algorithms allow the use of polybiome science in customized treatments, such as specific diets or microbial therapies (Pati, et al., 2024; Reyed, 2023b). This aspect emphasizes the high significance of the polybiome for prospective patient-specific healthcare (Reyed & Prabhakar, 2025).

3.3 Dysbiosis: From Pathological Trajectory to AI-Driven Personalized Therapy

The current literature identifies dysbiosis as a significant disturbance and disproportional balance in the gut microbiota that alters fundamental physiological functions and is strongly linked to the onset of gastrointestinal, metabolic, allergic, and inflammatory diseases (Singh et al., 2021). The human gastrointestinal tract supports trillions of microorganisms, most of which belong to the phyla *Firmicutes* and *Bacteroidetes*. In the eubiosis state, these organisms promote the absorption of nutrients, regulate the immune system and provide protection against colonizing pathogens. Dysbiosis can be triggered by exposure to excessive antibiotics, poor dietary habits, infection, and chronic stress, leading to reduced microbial variance and a change in the metabolite profile (Singh et al., 2021).

An increase in the abundance of the *Firmicutes* phylum, which increases the ability of the host to obtain caloric energy in the diet, is likely one of the most common dysbiotic signature responses to obesity. This disequilibrium is usually coupled with decreased production of short-chain fatty acids (SCFAs), the metabolites necessary to maintain the integrity of the intestinal barrier and control inflammation. Empirical research has shown that probiotic treatments can maximize the fermentation of inedible carbohydrates into SCFAs to prevent and treat obesity (Reyed, 2021b). By implication, increased intestinal permeability allows the translocation of endotoxins into the systemic circulation, which is referred to as metabolic endotoxemia. This low-grade inflammatory environment is an important contributor to insulin resistance and fat tissue proliferation (Amabebe et al., 2020). Small intestinal bacterial overgrowth (SIBO) is a dysbiosis-related motility disorder that impairs epithelial integrity and facilitates the development of digestive symptoms and malabsorption. These bacterial changes are closely linked to health issues such as irritable bowel syndrome (IBS) and diabetic gastroparesis (Singh et al., 2021).

In addition to local gastrointestinal effects, dysbiosis has systemic effects, as it upsets immune homeostasis. The gut microbiota is in contact with remote organ systems through integrated neuroendocrine and immune signaling pathways. For example, microbial changes have been linked to skin disorders such as psoriasis and atopic dermatitis, where low levels of SCFA production and the growth of proinflammatory bacteria interact to weaken the skin barrier and trigger systemic immune dysregulation (De Pessemier et al., 2021) (Figure 1).

Figure 1. Association of Gut Polybiome Imbalance with Common Health Conditions

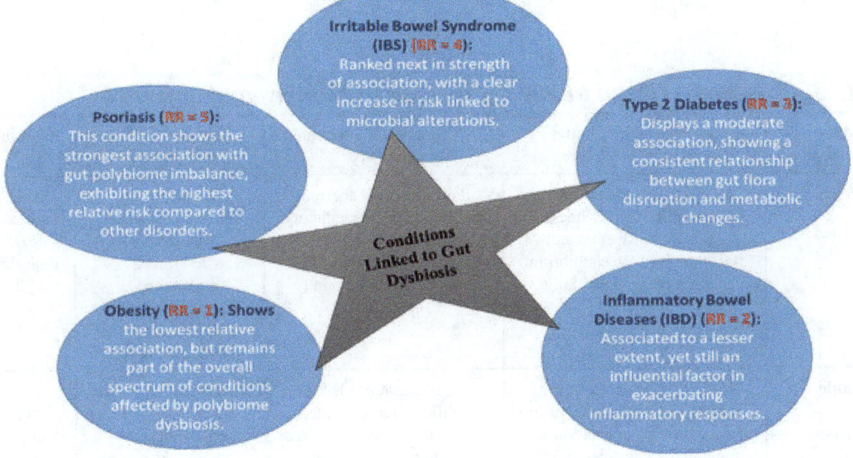

Relative Risk (RR): Indicates how much more likely a condition is associated with gut Polybiome imbalance compared to a reference. Higher RR values mean a stronger association.

Furthermore, a growing body of evidence indicates that dysbiosis influences adaptive immunity, thus playing a role in allergy disorders. Microbiota changes may reduce the activity of regulatory T2 cells and promote Th2-driven inflammation, increasing their vulnerability to food allergies, asthma, and atopic dermatitis (Reyed & Prabhakar, 2025; Augustine et al., 2023). In this context, the polybiome is thus a perfect target for therapeutic intervention because of its central position with respect to host physiology. Table 2 provides a brief summary of dysbiosis-related disorders, with emphasis on fundamental changes in the core microbes and functional pathways. Population-wide interventions involving dietary changes, probiotics, prebiotics, and fecal polybiota transplantation (FPT) have been used as traditional methods to recondition microbial equilibrium, which is needed to prevent or treat these associated conditions (Reyed & Prabhakar, 2025).

Table 3 provides an overview of significant ailments resulting from dysbiosis, indicating typical polybiome changes, the mechanisms involved, and their consequences for the host's body. The table demonstrates how dysfunction in SCFA-producing groups, shifts in the composition of bile acids, and signals between the immune system and the microbiome add to inflammation processes, metabolic issues, and compromised barrier integrity. These ideas support specific interventions, such as the use of engineered microbial communities, prebiotic protocols informed by AI, and personalized fecal polybiota transplantation, to reestablish microbial homeostasis and optimize the control of immune and metabolic processes in line with the principles of nutrigenomics and precision medicine.

Table 3. Conditions linked to gut dysbiosis, key changes and impact.

Condition	Key Polybiome Changes	Mechanistic/Functional Insights	Impact on Host Physiology	References
Psoriasis	Psoriasis is correlated with ↓diminished the richness of microbial species and fluctuations in the equilibrium between bacteria and fungi.	It also involves aberrant regulation of the regulatory T-cell/T helper 17 cell balance and changes in the level of short-chain fatty acids.	Such elements lead to prolonged inflammatory responses, immunological complications, and deterioration of the epidermal barrier.	Reyed & Prabhakar, 2025; Fujisaka et al., 2023
Irritable Bowel Syndrome (IBS)	Irritable bowel syndrome is characterized by variations in the proportions of bacteria and fungi in the intestines, which influences the intestinal-brain communication system.	These alterations lead to modified short-chain fatty acid and bile acid-mediated communication, coupled with diverse activity of the intrinsic nervous system of the digestive tract.	This often causes can manifest as altered gastrointestinal function, increased sensitivity to intestinal sensations, along with abdominal swelling.	Reyed, 2023a; Sanidad & Zeng, 2020
Type 2 Diabetes	↓ Reduced amounts of short-chain fatty acid (SCFA)-producing bacteria is linked to dysregulation in branched-chain amino acid (BCAA) and tryptophan metabolites.	This is associated with reduced glucagon-like peptide-1 (GLP-1) signaling, insulin resistance, and widespread inflammation.	Such determinants are implicated in dysfunctional glucose homeostasis and increased endotoxins in the bloodstream.	Cox et al., 2022; Reyed & Prabhakar, 2025
Inflammatory Bowel Disease (IBD)	↓ The species *Faecalibacterium prausnitzii* has shifted virome-fungal dialogs.	The wall of the intestinal tract is compromised, resulting in increased immune-mediated reactions.	Recurrent gut inflammation and mucosal damage.	Fujisaka et al., 2023; Reyed, 2023a
Obesity	An elevated Firmicutes/ Bacteroidetes ratio is often seen alongside changes in short-chain fatty acid and bile acid processes.	This effect can lead to increased calorie retrieval from ingested nutrients, subtle inflammation, and different communication patterns involving the gut-brain axis.	These shifts are linked to gaining weight, an accumulation of body fat, and metabolic dysfunctionalities.	Cox et al., 2022; Reyed & Prabhakar, 2025
Age-related Immune Decline	↓ As individuals age, the body's defense mechanisms usually diminishes, in connection with a decline in the complexity of their gut microbiome, consequently causing the production of short-chain fatty acids is reduced.	This phenomenon might produce a reduced number of immunomodulatory T cells being created and a sustained, low-grade inflammation.	Due to this, elderly individuals experience diminished resistance and are more susceptible to infections.	Bosco & Noti, 2021; Aggarwal et al., 2022
Potential Therapeutic Interventions	Restoration of SCFA-producing taxa; modulation of bile acids; engineered microbial consortia	Targeted prebiotics, probiotics, FMT, polybiome engineering	Personalized metabolic and immunological regulation	Shawky et al., 2025;

However, a critical scientific and clinical gap still exists: the fact that these correlations in belief have not yet been converted into individualized and predictive therapies. This weakness lies in the natural interindividual variability and the enormous complexity of host–microbe interaction networks.

An understanding of these complex processes has been significantly accelerated through computational strategies that assimilate high-throughput sequencing and multiomic and metabolomic data into complex analytical systems. The very magnitude and intricacy of such datasets require the use of artificial intelligence (AI) and machine-learning systems to identify the patterns predictive of host responses, which would help identify particular types of dysbiosis depending on metabolic disorders, allergic diseases, and dermatoses. To increase the understanding of descriptive associations to a mechanism-based understanding, it is critical to consider the strength of graph analysis and network-based computations to outline the interactions between hosts and microbiomes. Using genomic, transcriptomic, metabolomic, and AI modeling, it is possible to map host microbiome signaling pathways, such as those associated with interactions with the FXR and TGR5 receptors, and, by means of such maps, explain causal mechanisms that control immune and metabolic regulation (Shawky et al., 2025; Reyed, 2023b).

Building on this mechanistic insight, Table 4 outlines a practical and analytically constructive vision for next-generation personalized medicine. It goes beyond a mechanism description to outline the fundamental therapeutic goals (e.g., the Heal phase or Halt phase) and justify AI and nutrigenomics as the facilitators of these objectives.

Table 4. Integration of the Holobiome Restoration Strategy (HRS) with AI-Driven Nutrigenomics for Polybiome-Based Cancer and Metabolic Therapy.

Condition	HRS Phase & Core Therapeutic Objective	Nutrigenomic Pathways and Dietary Targets	AI-Augmented Microbiome Applications	Expected Clinical/ Physiological Outcomes	References
Psoriasis	**Halt Phase:** To prevent inflammation due to dysbiosis and restore SCFA balance	controlling the metabolism of PUFAs, influencing Treg/ Th17 genes like IL10 and RORC, and optimizing omega-3 lipid signals better.	Intelligent algorithms can establish associations between immune responses and metabolic substances, generating individualized inflammation-reducing nutrition plans.	This approach alleviates inflammation throughout the body and re-establishes the immune system of the skin to its proper function.	Reyed & Prabhakar, 2025; Reyed, 2023b)
Irritable Bowel Syndrome (IBS)	**Harbor Phase:** Re-establish gut-friendly bacteria and modulate neural pathways between gut and brain.	Variations in target genes in the transport of serotonin (also known as SLC6A4) and the processing of bile acids (NR1H4).	Use virtual models to establish a connection between microbial metabolites concerning the way the neurological system interacts with the intestines.	Anticipated outcomes are healthier digestive system, improved tolerance, and standard quantities concerning short-chain fatty acids and bile acids.	Reyed & Prabhakar, 2025; Reyed, 2023b; Sanidad & Zeng, 2020
Type 2 Diabetes	**Help Phase:** Contribute to promoting fermentobiometric balance and cellular responsiveness to insulin.	Regulate the variation within TCF7L2 and PPARG genes by utilizing particular interactions between fiber and amino acids.	Apply algorithms to forecast glucose level changes to personalized blend of prebiotics and polyphenols formulations.	Improve GLP-1 pathway activation, minimize circulating endotoxins, and restore insulin sensitivity.	Cox et al., 2022; Reyed & Prabhakar, 2025; Reyed, 2023b)
Inflammatory Bowel Disease (IBD)	**Heal Phase:** Rebuild the barrier function of the mucosa and minimize overabundant the generation of cytokines.	Genes related to intercellular connections (OCLN, ZO-1) and cytokines with anti-inflammatory properties exhibit enhanced expression by nutrigenomics.	Probiotics are developed through the use of deep learning to stimulate gut-associated immunological acceptance.	The epithelial integrity is regenerated, and the Th17-mediated inflammatory cascade is mitigated.	Fujisaka et al., 2023; Reyed & Prabhakar, 2025; Reyed & Prabhakar, 2026)
Obesity	**Hold Phase:** Ensure metabolic equilibrium and prevent recurrence.	SCFA and the return to normal bile acid levels cause epigenetic changes in fat mass and obesity-associated and leptin genes.	An AI system assesses fermentobiometric feedback in order to boost calorie utilization.	Preserve a healthy weight, recover physiological harmony, and support an inflammation-reducing physiological condition	Wu et al., 2021; Shawky et al., 2025; Reyed & Prabhakar, 2025
Age-related Immune Decline	**Hold/Heal Phase:** Reinforce resistance by optimizing the gut ecosystem.	Re-energize FOXO3 and SIRT1, genetic factors influencing lifespan with the help of polyphenolic compounds and short-chain fatty acids utilizing the principles of nutrigenomic approaches	Implement future-oriented computational intelligence modeling to age-related resident microorganisms for approaches to immune rejuvenation.	Mitigate inflammation associated with aging, boost the regeneration of T cells, and strengthen immune system function.	(Aggarwal et al., 2022; Bosco & Noti, 2021; Reyed & Prabhakar, 2025)

The 5H Holobiome Restoration Strategy (HRS) phases—Halt, Harbor, Help, Heal, and Hold—are relevant to significant ailments related to dysbiosis, as detailed in the table. Every stage integrates key interventions aimed at nutrigenetic aims and artificial intelligence-led microbe-based treatments. This combined strategy promotes targeted adjustments in fermentobiometric profiles, the regulation of immunity and metabolism, and long-lasting well-being of the host–microbiome interaction within precision medicine. Fermentobiometric science, as defined by Reyed & Prabhakar (2025), is the quantitative investigation of biotransformation through microbial action in biological and alimentary environments. It integrates multiomic, metabolic, and AI data to quantify how microbial fermentation modulates the host's well-being, the processing of nutrients, and the equilibrium of the gut's ecological balance.

Table 4 highlights how the Holobiome Restoration Strategy (HRS), which employs nutrigenomics and artificial intelligence, translates the notion of tailored polybiome therapeutics into a readily applicable method in clinics. To ensure that the efficacy of these interventions is achieved, it is necessary to develop refined approaches to elucidate the configuration of microorganisms, their performance, and the manner in which they interface with the host. The next section focuses on the key molecular techniques and computational strategies used in investigations of microbiomes. These approaches, such as 16S rRNA gene sequencing and metagenomic and functional multiomics, provide the foundation for training, validation, and personalization of artificial intelligence and nutrigenomic models for application in clinical settings (Reyed & Prabhakar 2025).

3.4 Techniques Related to Studies of Microbiome Approaches

Although the comprehensive conceptual context of this study adopts a polybiome perspective, in progress, molecular techniques for microbial community outlining are still principally established on the basis of classical microbiome methods.

3.4.1 Marker Gene Sequencing

The bacterial taxa in a community are identified by examining the 16S rRNA gene via marker gene analysis. Conserved parts on both sides of unstable areas in the 16S rRNA gene were tested with this technique, allowing organisms to be sorted to the genus or sometimes to the species level. Noise caused by sequencing errors is removed by using OTUs or ASVs (Reyed, 2023b; Galloway-Peña & Hanson, 2020). Using references from SILVA and Greengenes, tools such as QIIME, DADA2 and Mothur can both process sequences and identify taxa (Reyed, 2023b; Peterson et al., 2021).

3.4.2 Shotgun Metagenomics

The method of shotgun metagenomics allows researchers to capture data from all parts of a sample's DNA. As a result, scientists can find bacteria, archaea, fungi, and viruses and examine their genetic activity with annotation (Reyed, 2023b; Bharti & Grimm, 2021). Amplicon methods provide less information on both taxonomy and function, whereas shotgun sequencing provides more information but needs higher levels of sequencing and computing power (Reyed, 2023b; Peterson et al., 2021).

3.4.3 Predictive and Direct Functional Analysis

Predicting what genes do in an environment can be performed via PICRUSt and Tax4Fun, which rely on 16S data; metagenomic sequencing proves this directly. For more detailed information, metatranscriptomic, metaproteomic and metabolomic approaches were used to access RNA, protein and metabolite profiles. Omics technologies make it possible to study microbial action and relationships with the host more clearly (Galloway-Peña & Hanson, 2020).

4. NUTRIGENOMICS: THE GENETICS OF NUTRITION

4.1 Definition: Interaction Between Diet and Genes

Nutrigenomics is an emerging discipline at the crossroads of nutrition, genetics and medicine that investigates the effects that food has on how our genes are expressed themselves. Nutrigenomics seeks to determine the effects of various nutrients in the human body, as all people have different genetic makeups instead of using a diet blanket approach. Food is not only a fuel but also a source of bioactive compounds that can affect genes that may provoke metabolic changes, as well as a potential risk of illness (Arshad, et al., 2025; Reyed, 2021a). This scientific discipline has gained its second weight since even slight genetic differences, such as in the case of single nucleotide polymorphisms (SNPs), which were identified by the efforts of the Human Genome Project, have been found to have a large effect on healthcare. For example, some vitamins or plant factors can activate or deactivate some of the genes that mediate inflammatory processes, cell growth, or detoxification. With such knowledge of interactions, nutrigenomics is intended to develop specific dietary plans that prevent or control chronic diseases such as diabetes, obesity, and heart diseases. In short, nutrigenomics can provide a customized health solution--whereby nutrition is crafted to operate in concert with the individual genetic make-up of the individual Sales et al. (2014).

4.2 Genetic Variations (e.g., SNPs) and Nutrient Metabolism

Individual variation at the genetic level, e.g., single-nucleotide polymorphisms (SNPs), is important in their processing and response to nutrients. To understand how to manage these distinctions and respond to them better, nutrigenomics, or the science of the association between genes and nutrition, can be used in the study of individual dietary requirements and risks. For example, individuals affected by some of these changes in the enzyme MTHFR might have difficulty transforming into an active form, which makes them more prone to cardiovascular diseases or neural tube malformations (Nielsen & El-Sohemy, 2012). Nevertheless, the risks can be alleviated through targeted folate supplementation, which indicates how genetic specifics of nutrition can be helpful in enhancing health outcomes.

Genetics also determines our capacity to metabolize fats, carbohydrates and vitamins. SNPs associated with cholesterol levels, obesity risk, or blood sugar responses are examples of some SNPs. It enables them to formulate customized diets that enhance chronic conditions such as diabetes and heart disease prevention or management (Pati et al., 2024; Paul et al., 2025). Nutrigenomics takes the idea of one-size-fits-all nutritional recommendations and shifts to a more individual approach to nutrition depending on a person's genetic makeup.

The process of gene–nutrient interaction is two-way and complex. Nutrients not only power the body but also regulate gene expression. For example, omega-3 fatty acids have the capacity to control the genes involved in inflammation, whereas polyphenols found in plants control oxidative stress. However, these effects differ across individuals. What works out effectively on one person may not be effective on the other or even dangerous, depending on their own genetic composition (National Academies of Sciences, Engineering, and Medicine, 2022). These differences are the key focus of precision nutrition, which attempts to provide individualized nutritional advice, depending on how the body of an individual actually responds to nutrients. Nutrigenomics will lead to the eventual realization of individualized nutrition plans that will help people better metabolize, avoid becoming sick, and live longer (German et al., 2011).

4.3 Nutrigenetics vs Nutrigenomics

Nutrigenetics and nutrigenomics are two related, although different fields institute personalized nutrition. Although they both delve into the issue of nutrition and genes, the area of concern varies dramatically. Nutrigenetics refers to research in which the individual genetic makeup influences the effects of nutrients. This approach aims to determine why individuals vary in their ability to absorb, metabolize or use certain nutrients (Lagoumintzis & Patrinos, 2023). An example can be provided through genetic polymorphisms in enzymes such as methylenetetrahydrofolate reductase (MTHFR), which are capable of shifting folate metabolism and increasing the folate requirement in some people. This variance is the basis on which individualized dietary plans are tailored to one based on one's genetic makeup to prevent or treat disease (Paul et al., 2025).

Conversely, nutrigenomics studies the effects of nutrients on the expression of genes (Rouskas et al., 2025). Research has focused on the effects of bioactive food compounds that are considered signaling molecules that regulate the transcription and translation of genes (Lagoumintzis & Patrinos, 2023). Nutrients, which act through this mechanism, can affect metabolic pathways, inflammation, detox pathways, and disease susceptibility. Nutrigenomics focuses on the responsive and dynamic interaction between the genome and diet to develop precision dietary guidelines to ensure clinical progress beyond avoiding illness to lifelong health (Rouskas et al., 2025; Reyed 2021a). Combined with nutrigenomics, nutrigenetics serves as a suitable personal health alternative that makes use of both genetic and dietary data, thereby shaping personalized nutrition plans that favor disease prevention and the effectiveness of treatment approaches Kassem et al. (2023).

4.4 Applications in Obesity, Diabetes, Cancer, and Cardiovascular Diseases

Nutrigenomics redefines the chronic disease management process by revealing how genes regulate our response to food. In the case of obesity, genes such as FTO and MC4R increase the likelihood of gaining weight on high-fat diets and weight gain, but people who possess such genes tend to fare well on personalized diets that are composed of unsaturated fats (Phillips, 2013).

In type 2 diabetes, insulin sensitivity and glucose control are influenced by genetic variations such as the Pro12Ala variant in PPAR and risk variants in TCF7L2. Individualized diets, such as those that are high in healthy fats or low glycemic load diets, can be very effective in the treatment of these individuals (Corella & Ordovas, 2014).

Nutrigenomics can also be used in cancer prevention. Individuals carrying mutations in folate metabolism might also require increased folate intake to promote DNA repair. Other nutrients, such as selenium and lycopene, may turn on protective genes and curb oxidative stress and inflammation (GarcÃ-Bailo et al., 2012).

With respect to cardiac health, variations in the genes encoding APOA5, CETP, and MC4R affect fat and salt metabolism. Low sodium is more advantageous in people with specific profiles, particularly in a Mediterranean-style intervention. Overall, gene-based nutrition allows more effective individual approaches to the prevention and management of chronic conditions and provides general dietary recommendations for precision health.

4.5 Critical Synthesis and Future Outlook

Newly developed multiomics technologies, a combination of metagenomics, metabolomics, and transcriptomics, are revolutionizing our mechanistic understanding of host–polybiome interactions via integrated analysis. Nevertheless, challenges of data inconsistency, methodological variations, and scalability are still evident in limiting the translation of such findings into clinical practice (Acharya and Mukhopadhyay, 2024). Algorithms that learn via artificial intelligence offer an effective approach for solving high-dimensional data and causal inference but can potentially reflect biases implicitly and are unvalidated in diverse populations. The capabilities of therapeutic methods are also enhanced by synthetic biology-based methods that design engineered consortia to suit the needs of individuals; however, developing regulatory frameworks is necessary to test safety and efficacy (Lamsaf, et al., 2025). Explainable AI is particularly important in the clinical environment, where clinicians need to understand and have confidence in model reasoning in the context of taking algorithmic recommendations to ensure the safety of patients and receive regulatory clearance. Moreover, to safeguard genomic and polybiome data privacy, as well as ensure equal access to personalized interventions, there must be strong support systems and policy frameworks. The gap between basic science and patient-focused health care requires the combination of data standards to be interoperable to distribute data across different jurisdictions, federated learning to ensure that data remain privacy-bound, and machine learning to be trusted and explainable (Alkhanbouli, et al., 2025). Tight clinical trials are needed to verify polybiome-based diagnostics and therapeutics, expanding following the demonstrated efficiency of polybiome interventions for *Clostridioides difficile* and the potential of their use in Alzheimer's disease. Moreover, there needs to be coherent and lengthy ethical and regulatory directions for consent, ownership, and equitable access to achieve a sustainable predictive, preventive, and genuinely personalized healthcare environment (Porcari et al., 2025).

5. AI/ML TECHNIQUES APPLIED

5.1 Data Preprocessing and Integration

Preprocessing is an essential initial step in personalized medicine in general and when analyzing complex polybiome and nutrigenomic data in particular.

Noisy, inconsistent, and heterogeneously formatted raw data presented by genomic sequences, diet logs, polybiome profiles, and even clinical records typically mirror different, even opposing, treatment effects. It is important to clean, normalize, and transform these data so that machine learning (ML) models can be confidently and accurately built (Reyed, 2023b).

The imputation of missing values (e.g., identified in gene expression or abundance of microbial profiles) is normally performed with statistical imputation (e.g., mean, median, or model-based approaches). Through structural fixes and standard encoding, duplicate and inconsistent labels and symbolic errors are handled. Then, normalization is used to apply different characteristics, such as gene expression and microbial counts, to the same scale to analyze them evenly.

After cleaning, the next step is to integrate these multisource databases. Since biological data are often both multilevel (e.g., genetic, metabolic, and taxonomic levels) and heterogeneous (e.g., structured and unstructured data), sophisticated techniques such as schema mapping or data warehousing are employed to homogenize data. Because of the complexity of these methods, dimensionality reduction strategies, such as PCA or t-SNE, are used to simplify such datasets while retaining biologically meaningful patterns, reducing the challenge involved in visualizing the data and performing clustering.

Such well-processed and combined data are subsequently utilized in ML pipelines through applications such as disease forecasts, nutritional advice, or patient classification. Otherwise, models can be biased or overfit or fail to capture important insights, but this can be done by ignoring or performing preprocessing inadequately. Nevertheless, preprocessing and integration are not merely technical procedures but rather essential steps toward the optimal possible value offered by AI-assisted personalized healthcare (Kassem et al., 2023).

5.2 Machine Learning Approaches

5.2.1 Supervised Learning

The process of learning by using labeled data to anticipate future data is known as supervised learning, and it has become increasingly important in microbiome-based classification. The methodology allows the identification of microbial trends that are related to health and diseases, although some challenges, such as sparseness and compositional bottlenecks, are involved (Reyed, 2023b; Huang et al., 2023).

Some popular algorithms are random forests, support vector machines, and XGBoost algorithms, which have shown efficiency in predicting the occurrence of diseases such as colorectal cancer according to the profile of the gut microbiota (Topcuoglu et al., 2020). However, a model's interpretability may be poor. Because algebraic structures are understandable and the accuracy is good, symbolic regression has been able to deliver a solution (Haldar et al., 2024).

More kernel-based models, such as KernelBiome, consider the special geometry of compositional rdata to ensure a path to better model fit (Huang et al., 2023). These developments allow for the pursuit of precision and interpretability while also enabling the interpretation of machine learning applications in clinical microbiome diagnostics with more clarity and confidence (Figure 2). Approaches to learning on the basis of labeled data are presently geared toward the microbiome, as demonstrated by the empirical evidence we possess and our investigative approach, although the polybiome concept integrates the complex relationships among the host species, the microbial population, and the ambient surroundings.

Figure 2.

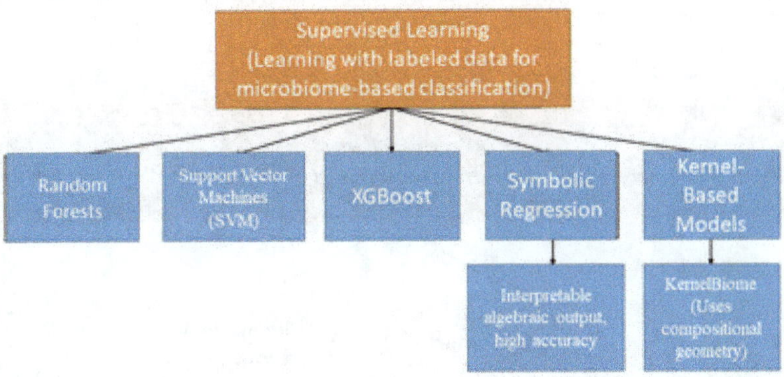

5.2.2 Unsupervised Learning

Unsupervised learning plays a key role in detecting hidden patterns and population structures in microbiome data, where no labeling or result variable is known. The common methods of doing this include clustering and reducing dimensionality to identify subtypes of a microbial community or gradients that could indicate disease severity or responsiveness to treatment (Reyed, 2023b).

As an example, the MMUPHin statistical framework can be used to conduct large-scale meta-analyses on microbes, which makes it possible to identify both discrete cluster structures and continuous structure corrections in batch effects over studies (Ma et al., 2022). These methods revealed that a continuum, rather than distinct subgroups, would be a more realistic way to characterize the microbiome structure in inflammatory bowel disease (IBD).

Bayesian sparse tree-based clustering was used in studies of the cancer microbiome to separate patient populations on the basis of microbiome composition via both phylogeny and selection of features with biological relevance (Shi et al., 2023). In addition, other methods, such as differential network analysis and microbial hub detection, facilitate interest-driven investigations of microbial interactions but are vulnerable to overoptimism when unguarded methodological decisions are present (Ullmann et al., 2023).

Unsupervised methods can inform us about microbial heterogeneity, although successful sequencing without validation and with limited methodological transparency makes interpretation of results difficult and beyond the scope of regulation (Figure 3). Approaches focused on the microbiome, interpreted within a systems biology framework, reinforce the more expansive polybiome notion. This framework integrates host, diet, and microbial traits to develop algorithmically supported diet–gene interaction models with the goal of elucidating pathology and intervention.

Figure 3.

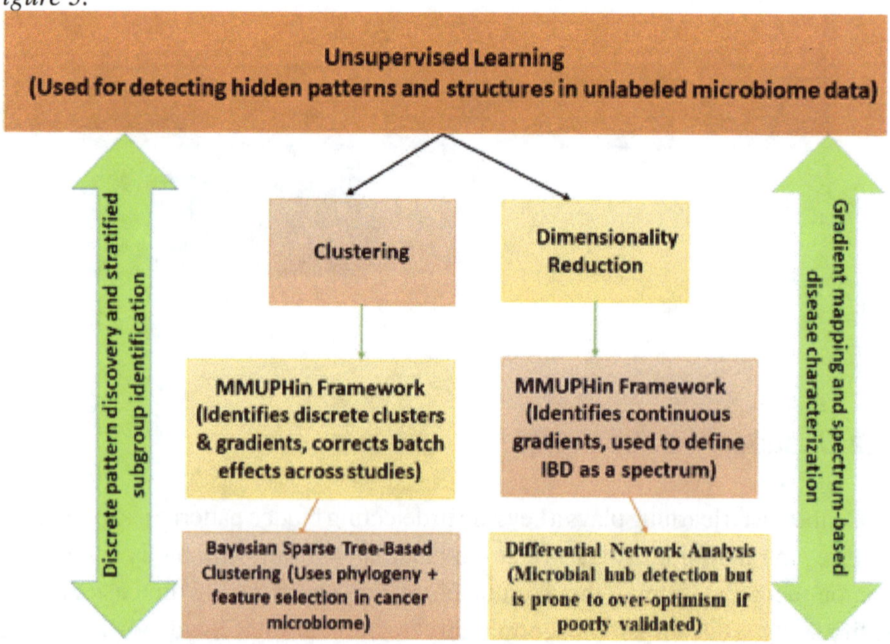

5.2.3 Reinforcement Learning

RL, or reinforcement learning, is a powerful machine learning algorithm capable of recursive, real-time decision-making via trial-and-error learning to identify the best strategies through the dynamics of the environment. In a personalized medicine setting, RL will be used to aid treatment regimens that are customized for the feedback of individual patients and their health and long-term effects. Its application has been employed in chemotherapy scheduling, insulin dosing, and management in the ICU (Banumathi et al., 2025).

RL agents for microbiome modeling are simulations of microbial behavior in variable environments. An example is the SPAM-DFBA algorithm, which combines RL with dynamic flux balance analysis to obtain a prediction of the eventual metabolic strategy to promote microbial fitness in the long term without assuming prior knowledge of the metabolic patterns of microbes (Ghadermazi & Chan, 2024). This enables microbial agents to develop behaviors that maximize the use of nutrients and make them attempt to achieve ecosystem stability.

The reinforcement learning that was collaboratively used with collaborative filtering in the field of nutritional planning improved the personal salient meal recommendations with the clusters of the individual users and their interests, user feedback, and objectives associated with health-related outcomes. The CFRL algorithm is trained to learn dynamic patterns of meals and to adjust and provide better recommendations regarding user satisfaction and adherence (Amiri et al., 2024). Overall, RL offers a framework of individual, adaptive interventions in the clinical and microbial realms and does so by maximizing the outcomes via incessant learning and source-specific adjustments (Figure 4). Within the framework of systems nutrigenomics, microbiome modeling that leverages reinforcement learning plays a key role in the polybiome framework. The analysis establishes a relationship between individualized nutritional plans and the genetic predispositions of the host, along with microbiome composition.

Figure 4.

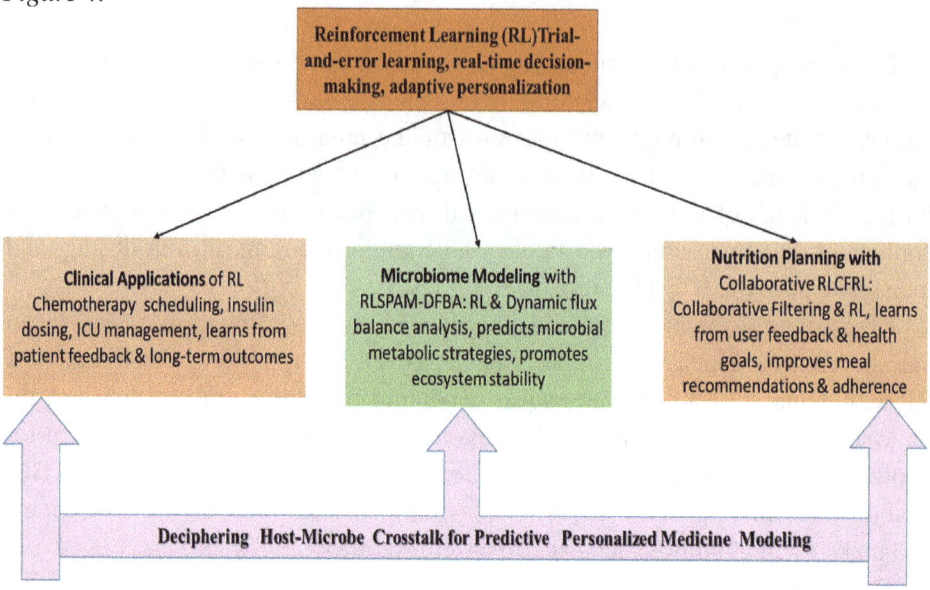

5.3 Deep Learning and Neural Networks

5.3.1 Convolutional Neural Networks (CNNs)

One type of deep learning framework that is commonly used to handle data in spatial and grid-like patterns is convolutional neural networks (CNNs). CNNs have already been adapted to work with complex matrices of microbial abundance data used in microbiome research in terms of phylogenetic trees. These models incorporate taxonomic connections within two-dimensional matrices that enable CNNs to trap spatial associations among the microbial species to increase the ability to guess phenotypes (Reiman et al., 2017).

Other advanced systems, such as iMic and gMic, continue polishing CNN applications by converting microbe data into a graph or image, making effective use of taxonomic depth as well as microbes being present. The approaches prove to have better performance in disease prediction tasks, including IBD and colorectal cancer, outperforming conventional machine learning methods .

An advantage of this approach is the PM-CNN model, which combines multipath CNNs and can be organized according to phylogenetic clustering to provide multistatus classification with high precision when combining a broad range of health conditions (Wang et al., 2024). They are more interpretable because they better utilize gradient-based visualizations than CNN-based frameworks are, and they are compatible with temporal data and therefore have much potential to be applied in longitudinal nutrigenomics. CNNs are the strength in developing AI-based personalized health diagnostics and healing interventions, as they work well with sparse and high-dimensional data (Figure 5). In the realm of polybiome studies, convolutional neural network architectures supply mechanisms that combine the identification of microbial patterns with an individualized understanding of genomes and metabolism.

Figure 5.

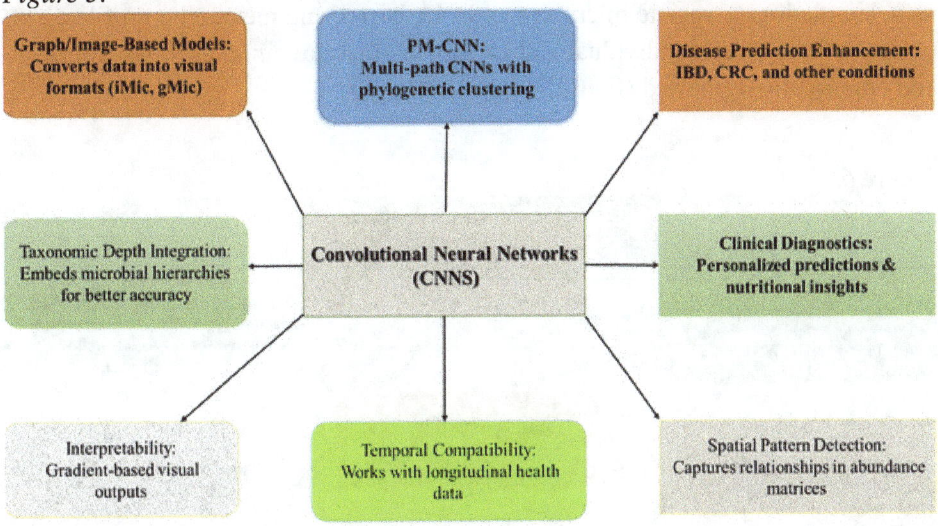

5.3.2 Recurrent Neural Networks (RNNs)

Recurrent neural networks (RNNs) and their variants, long short-term memory (LSTM) networks, are specifically required for microbiome dynamics and longitudinal nutrigenomics to replicate sequential biological data. These models also incorporate temporal relationships into data by using time series data; thus, they are closely suited to the modeling of microbial interactions and downstream phenomena such as the production of metabolites and the individual risk of diseases.

Furthermore, with respect to forecasting the abundance of species and pertinent metabolites such as butyrate in the artificial gut microbiome, LSTM models have performed better than traditional ecology models such as the generalized Lotka–Volterra model. (Baranwal et al., 2022). With the objective of forecasting the probability of food allergies on the basis of longitudinal microbiome history in early childhood, RNNs can likewise perform better than traditional machine learning algorithms (Metwally et al., 2019).

To improve the biological realism of RNNs, bounded RNNs such as the microbiome RNN (MiRNN) incorporate physical constraints to prevent the output of physically unrealistic values, e.g., negative species abundance but also allow one to specify a Bayesian experimental design to optimize the role of the microbiome in solving a problem (Thompson et al., 2023). Customized networks offer scalable possibilities for personalized nutritional and treatment planning on the fly, reduce the number of tests, and improve model accuracy. With respect to the polybiome area, simulations associate microbiome, DNA-based, and metabolite profiles. The objective is to design individualized treatments that are simultaneously dynamic and empirically supported (Figure 6).

Figure 6.

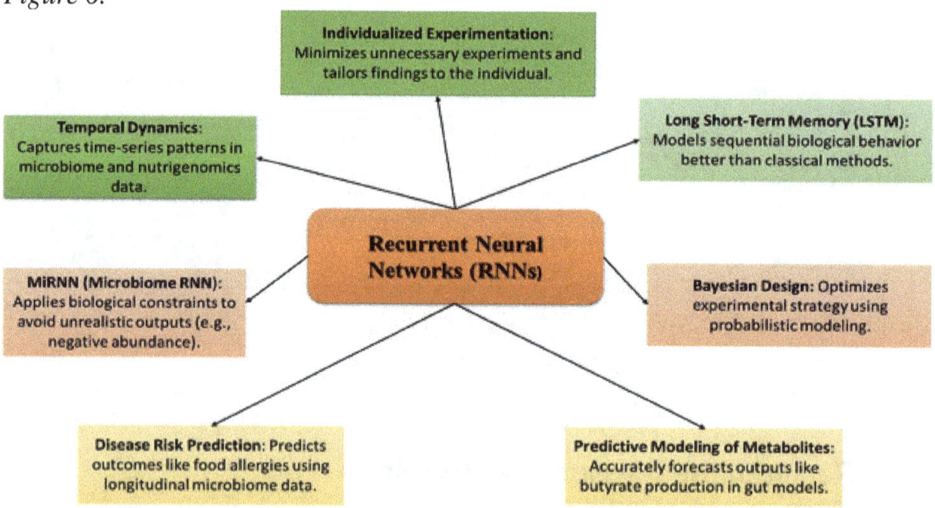

5.3.3 Transformers

Transformers were first developed for natural language processing, but because of their self-attention mechanisms and capacity to extract long-range dependencies from data, they have found widespread use in biological applications. In contrast to conventional sequential models, transformers allow parallelizing the work with the input sequence, so they are efficient and scalable (Madan et al., 2024). Transformer-based models such as DNABERT and MetaTransformer have been found to perform better in microbiome and nutrigenomics tasks involving the diagnostic of taxonomic signals and the prediction of functional genomic sequences via the embedding of k-mers (Wichmann et al., 2023).

There has also been success in the use of pretrained domain-specific transformers (such as BioBERT and ClinicalBERT), which are fine-tuned using biomedical corpora to reach state-of-the-art on biomedically focused entity recognition, relation extraction, and disease classification. The models allow the handling of biological sequences, structured electronic health records, and metagenomic data well, and integration analyses across modalities are possible (Madan et al., 2024).

Attention visualization can also aid the explainability of transformers and can be helpful in clinical acceptance and a regulatory context. Other models, such as the Enformer model, are suitable for stretching transformer applications to predict gene expression given the DNA, applying long-range genomic interactions, and representing a step toward more accessible precision medicine applications. The ability of transformers to digest a variety of biomedical information puts them at the forefront of improving AI-Augmented Medicine and the field of personalized nutrition.

5.3.4 Data Integration Methods for Multiomics

Multiomics data can offer a comprehensive picture of biological systems by integrating information from transcriptomics, proteomics, metabolomics, microbiomics, and genomes. The hypothesis links the predictive value of phenotype outcome with genotype, environmental contribution, or intermediate endophenotypes in terms of genome expression and metabolism (Wu & Xie, 2024). Multiomics models have become possible with personalized medicine and can elucidate the molecular etiology of complex diseases as well as define biomarkers and personalized therapeutic targets.

Unsupervised learning in medicine assumes common latent representations with multimodal omics entries such as variational autoencoders (VAEs). The most recent incorporating frameworks of high fidelity, e.g., MultiVI and scMVP, consider modality-specific noise and polarity to augment the polarity of integration between single-cell RNA and chromatin. The attention mechanism is applied in transformer-based models, including scGPT and GeneCompass, which scale cross-modality and cross-species representation learning (Wu & Xie, 2024). In polybiome studies, integrative pipelines that evaluate the unit taxonomic, functional, and metabolomic signatures of microbes are utilized to explain host polybiome interactions (Arıkan & Muth, 2023).

Even though progress has been made, challenges remain, such as data heterogeneity, missing values, and interpretability. The innovations found in pipelines, AI models, and spatial single-cell technologies have been acquired, but they require a process of standardization to allow the implementation of the full potential of translating multiomics into a personalized approach to health.

5.3.5 NLP and Literature Mining

Before using NLP as a revolutionary software to solve the larger but exponentially expanding biomedical literature, it became necessary to find a significant correlation between the polybiome and disease, as well as an association between diet and the polybiome. To identify and classify complex relationships between bacteria, nutrients, and illness states with a high rate of accuracy, researchers have implemented automated pipelines through refined domain-specific language models, such as BioLinkBERT, BioGPT, and PubMedBERT (Karkera et al., 2023). The models are superior to traditional manual curation because of their scalability and reduced labor intensity, which enables near-real-time updating of knowledge bases that are essential to tailored medicine.

Manual annotation projects such as DiMB-RE have also opened the field, building annotated corpora with a wide variety of entity types and relationships of interest to personalized nutrition. The identification of subtle and context-related relationships between food components, microbial taxa, and physical outcomes is made possible by the combination of named entity recognition, connection extraction, and factuality detection (Hong et al., 2025). Although generative models are promising, discriminative models are more accurate now, especially when cross-sentence relations and uncertainty come into the game. They provide clients with an important intermediary between unstructured literature and structured knowledge graphs to promote integrative AI initiatives in polybiome and nutrigenomics research. Figure 7 shows a microbiome-related network constructed by means of NLP-based models. The resulting network demonstrates the manner in which artificial intelligence-driven text analysis is capable of identifying interdependencies connecting host organisms, the microbiome, and nutrition.

Figure 7.

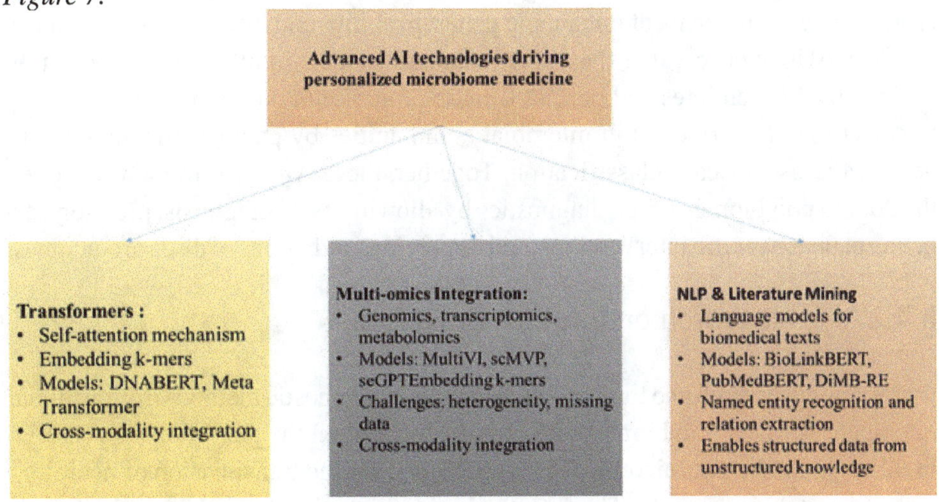

5.4 AI-Driven Approaches to Polybiome Research

5.4.1 Polybiome Profiling and Classification

The profiling of polybiomes has become important in the identification of microbial signatures related to diseases as well as the promotion of precision medicine. Understanding microbial diversity in large populations has been made easier by population-scale programs such as the Human Microbiome Project and Lifelines, which have been made possible by 16S rRNA sequencing (Wu et al., 2024). These research projects, even though they focus on the human microbiome, provide a springboard to support the polybiome concept. This concept focuses on exploring the relationships among microorganisms between distinct body locations (gut, mouth, and skin) and their surroundings collectively. The goal is to advise personalized healthcare strategies.

Taxa4Meta avoids technical and demographic biases because it allows variable sequence lengths and controls taxonomic classification. It accurately differentiates gastrointestinal conditions, including inflammatory bowel disease (IBD), irritable bowel syndrome (IBS), and *Clostridium difficile* infection (CDI), and this model greatly improves the classification of these conditions via a two-step binary model. (Wu et al., 2024)

Deep learning approaches, including algorithms such as MV-CVIB, incorporate data regarding quantity and interactions between neighboring cells to distinguish colon and rectal cancer, which has metastasized from its nonmetastatic counterpart

(Cui et al., 2023). The method under study is a multiview learning technique, which is advantageous in terms of enhancing generalizability and interpretability and has a higher AUC metric value than other standard models in all datasets above 0.9. MV-CVIB differentiates and defines disease progression on the basis of its ability to detect minute variations in microbial communities by paying little attention to healthy/disease–disease classification. Together, these AI-based frameworks open the door to polybiome-driven diagnostics by allowing early diagnosis, precisely targeting patient-specific interventions, and improving predictions of disease outcomes.

5.4.2 AI in Research on the Gut–Brain Axis

Emerging research and further developments in diagnostic methods from artificial intelligence-driven polybiome analysis have focused on the gut–brain axis (GBA); in this system, the gut microbiota is fundamental for the transmission of neurological and metabolic signals. The intestinal microbiota regulates the gut–brain axis (GBA), a complex communication system that includes neurological, endocrine, immunological, and metabolic networks. Disorders such as multiple sclerosis, anxiety, depression, and autism are thought to be associated with disturbances in the axis. AI, which creates interpretable, predictive models using multimodal data, including genetic, metabolomic, behavioral, and neurophysiological data, helps to change these complex relationships (Arif et al., 2024). Currently, microbiological indicators, mental health hazards, and individually customized therapy strategies are all determined by machine learning and deep learning algorithms. The RNNs are used to model time series cases of microbiome data, whereas the CNNs are those that extract the patterns in space that are connected to neuropsychiatric outcomes.

Fine-tuning our understanding of GBA signaling involves models that use molecular communication (MC). For example, SCFAs can be used as messengers to activate vagal pathways, control neurotransmitter discharge, and nourish the brain (Ortlek & Akan, 2024). Such biological channels present a measure of signal fidelity quantified under information-theoretic measures. To compensate for this difference, AI-driven digital twins, i.e., virtualized GBA models enabled with real-time biosensors, have been proposed as instruments for dynamic diagnostics and personalized feedback cycles (Meystre et al., 2023). These advances are transforming individualized mental health approaches via polybiome modeling coupled with real-time neural feedback specific to the individual.

5.4.3 AI for Microbial Signature Discovery in Disease

Artificial intelligence is transforming the detection of microbial signatures of diseases through accurate and fast processing of high-dimensional metagenomic data.

Machine learning algorithms trained on taxonomic abundance profiles enable the accurate classification of diseases such as colorectal cancer (CRC), type 2 diabetes (T2D), and inflammatory bowel disease (IBD) (Davis, 2023). As such, microbiome GSM allows one to identify taxonomic biomarkers with a near-perfect rate of success, with an AUC of 98% in IBD focused on order-level taxa (Bakir-Gungor et al., 2023).

Another study involved training machine learning models on American Gut Project data to predict a genus-level signature of microbial presence that predicts subtypes of IBD. Other methods, such as the Kruskal–Wallis test and fast correlation-based filtering, allow the identification of taxa with high discriminatory power, and validated models have shown AUCs of 0.74–0.76 in external cohorts, both in Crohn's disease and ulcerative colitis (Liñares-Blanco et al., 2022). These AI-related discoveries point to taxa, such as Streptococcus and Clostridium, as controlling forces in the stratification of diseases. They are scalable ways to identify biomarkers of the microbiome that will spur early-stage diagnoses and interventions in personalized medicine (Figure 8).

Figure 8.

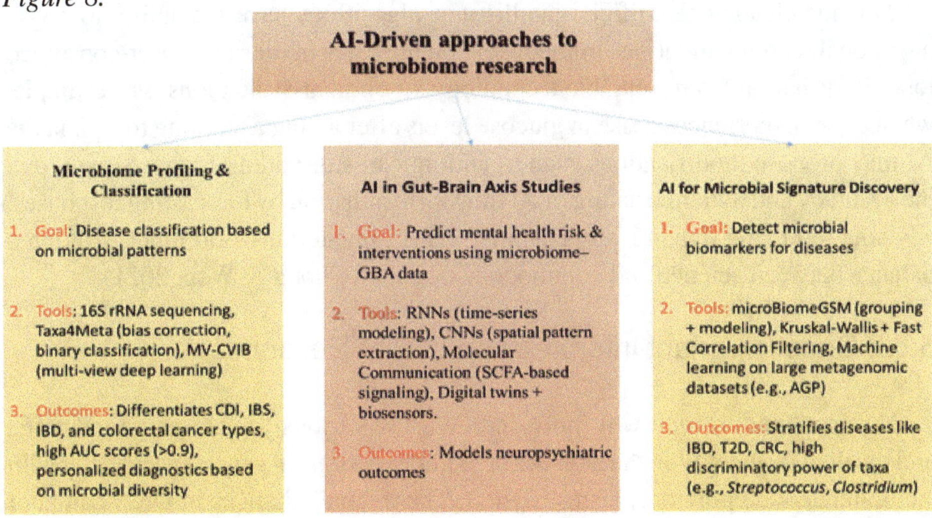

5.5 AI-Enabled Nutrigenomic Analysis

5.5.1 Personalized Diet Plans Using AI Models

Artificial intelligence (AI) is changing nutrition by making nutrition advice increasingly personal and customized to the genetics, biology, and lifestyle of an individual. The root–to-this transition involves nutrigenomics, the study of gene–nutrient interactions. AI is used to analyze massive amounts of data, such as genomic variations, microbiome profiles, and biomarkers, to provide customized nutrition plans that are much more useful than generic rules (Paul, et al.,2025).

Machine learning, particularly supervised algorithms such as random forests, SVMs, and the gradient boosting approach, can be used to predict the response of people with given gene variants (e.g., FTO or MTHFR) to various specific nutrients. These tools contribute to the development of individual diet schemes on the basis of specific needs and risks.

Machine learning algorithms have the potential to enhance personalized platforms through the proposal of foods drawn from user data, preferences, and the outcomes of similar users, as seen in examples such as Netflix and Amazon. For example, assuming that a dietary approach emphasizing fiber regulates the body's sugar processing appropriately for persons with a particular genetic predisposition, the algorithm could propose this approach for recently diagnosed individuals possessing that characteristic. These algorithms develop with increasing experience and a greater volume of data. A major positive aspect that AI offers for dietary applications is its dynamic character. Artificial intelligence algorithms have the ability to revise nutritional recommendations immediately on the basis of diaries of user-consumed food, information from wearable technology, or clinical evaluations. For example, when a user experiences a rise in glucose levels after acting according to a plan, the AI may propose modifications such as cutting carbs or adding fibers. Simply put, the accuracy and dynamic nature of AI transform nutrition by forecasting responses, advising food, and dynamically adapting plans to material world input, striking a balance between science and commonplace health (Chong & Woo, 2021).

5.5.2 Genetic Variant Interpretation in Diet Response

Nutrigenomics reveals why individuals react so variously to ordinary foodstuffs, and much of it depends on the nuances of slight variations in genes, specifically with single-nucleotide polymorphisms (SNPs). These small changes in DNA can alter the processes of nutrient processing within our bodies. For example, individuals who have specific variations in the FADS1 or FADS2 genes might have difficulty effectively converting important fatty acids, limiting the dietary necessities of omega-3 s and omega-6 s. Notably, gene polymorphisms in MTHFR can decrease body utilization of folate and increase the risk of cardiovascular complications or prenatal abnormalities unless they are counteracted by a diet (Corella & Ordovás, 2020).

Decoding these genetic signals allows specialists to identify the nutritional susceptibilities of a person, such as how the person handles caffeine or processes vitamins, and design special diets to fit their biology. In another example, an individual whose genetics slow down the breakdown of caffeine by taking longer to metabolize caffeine because of a unique CYP1A2 variant would require individuals to take the substance in small doses that do not increase their blood pressure, but another patient may not be affected by caffeine in this regard.

Such a personalized level is especially beneficial for preventing and treating chronic illnesses such as diabetes, obesity, and heart disease. Instead of using generic guidelines, through nutrigenomics, it will be possible to construct diets on the basis of the exact genetic makeup of an individual to produce better outcomes and minimize risks. Finally, the method is a combination of the scientific accuracy of molecular science and practical nutrition that allows individuals to be ahead of disease instead of being behind it later (Kok et al., 2025).

5.5.3 Integrating Dietary Intake, Genomics, and Lifestyle

The ability to unite genomics, dietary patterns, and lifestyle behavior is vital for providing actionable insights into personalized nutrition, especially in the age of big data. Artificial intelligence (AI) offers game-changing capability in this way in the form of interpretations of multidimensional, high-dimensional datasets that arise through the use of omics platforms, wearable health monitors, and self-reported lifestyle data.

The current AI algorithms can interpret the genomics of an individual with his or her regular diet and physical exercise to understand the ways in which it can interact to affect health. Software such as MiBiOmics and Omics Integrator is used to combine transcriptomic, proteomic, and metabolomic data to identify disease-associated biomarkers and nutritional targets. However, these models can do much more than merely unravel the complex biological pathways impacted by nutrients because they could also predict how a person would react to a particular diet on the grounds of his/her genetic blueprint and environmental exposures. (Kok et al., 2025)

The predictive capability of AI can lead to tailored, nutritional plans that are not only gene-specific but also lifestyle adaptive. For example, the Chat Diet is a chatbot based on AI that augments personal genomic data and population trends to deliver real-time diet recommendations by offering the most personalized dietary suggestions. This change will enable constant surveillance and modification of nutritional advice to enhance compliance and outcomes.

In addition, AI can be used to create decision-support systems to help clinicians and dietitians use less time and expert knowledge to analyze complex omics data manually. It detects vulnerable people, advises measures, and even predicts the far-reaching outcome of certain dietary decisions. Nevertheless, there are also downsides of AI when it is used in nutrigenomics, such as data privacy, algorithmic transparency, and access at parity. These issues must be addressed so that there is responsible implementation. Nevertheless, through the convergence of genomics, diet, and lifestyle into one common package, AI-driven nutrigenomics can be an effective method for promoting precision medicine and nutrition (Hosseiniara & Hosseini 2024).

5.6 Analytical Conclusion and Future Directions

There have been numerous breakthroughs in precision nutrition and microbiome-based dietetics through the establishment of AI-enhanced multiomics techniques that allow the unification of stringent data preprocessing procedures with complex machine learning, deep learning, transformer-based architectures, and NLP (Przymus et al., 2025). These approaches enable localized fingerprinting of host–microbe interactions, disease foot printing, and personalized dietary phenomenology. There are still, however, some significant weaknesses. Despite explainability being a crucial determinant of clinical acceptance, a wide range of systems (or deep-learning models and transformers) are often defined as black boxes, thus limiting the degree of trust in such systems on the part of clinicians and regulatory approval (Mohapatra et al., 2025). There is also a lack of interoperable data standards and consistent validation datasets, which reduces the possibility of cross-study benchmarking and generalizability.

Furthermore, algorithmic bias and unequal access impede fair results; the majority of models are taught using homogeneous samples, which can lead to bias, excluding underrepresented groups. With respect to the potential of precision-driven healthcare, future studies must thus ensure the following:

- Explanatory methods that allow interpretable feature attribution (e.g., SHAP, LIME), thus including clinician reliability and confidentiality (Mienye et al., 2024);
- Federated learning and data governance to protect the privacy of patients and allow training in a diverse and inclusive model across institutions. Given that tax rates (or taxes), held taxes (or held taxes), and the weighted average of tax rates (or taxes) on labor productivity are affected by the field of study, the latter question was answered by a model that was already established in the literature (Ferreira et al., 2025):
- Normalized benchmarking data and validation trials, particularly in diverse populations and clinical environments, to maintain the robustness and repeatability of AI models (Nasef et al., 2025);
- Ethical principles and regulation principles covering informed consent, data ownership, algorithmic transparency, and fairness in the allocation of AI-enabled nutrigenomic and polybiome treatments (Detopoulou, 2023).

6. AI APPLICATIONS IN MEDICINE AND PERSONALIZED NUTRITION

6.1 AI-Based Dietary Recommendation Systems

By employing deep learning algorithms to create individualized nutritional regimens, artificial intelligence has transformed personalized nutrition. They use data about a person, such as age, BMI, exercise level, and medical records, to provide dietary suggestions that follow the set nutritional standards (Papastratis et al., 2024). This approach applies a variational autoencoder (VAE) to organize personal data in accordance with a compressed representation, which in turn captures unique diet preferences. A recurrent neural network (RNN) designs regular eating plans. Thereafter, a system adjusts the size of servings to maintain total calorie consumption within the recommendations of the EFSA and the WHO. This yields food options that are balanced in nutrients and specifically designed for each individual.

By using ChatGPT, the system finds nutritionally similar meals from many regional cuisines. The use of structured prompts ensures that the added meals still meet the original nutritional standards, helping the model maintain accuracy (Papastratis et al., 2024). Applications that follow this hybrid approach become more culturally relevant and can be applied to many users. The accuracy of macronutrients was approximately 87%, and calories were aligned with target values after working on both synthetic and real-world examples. By comparison, using ChatGPT alone led to up to 17% mistakes in energy estimates, indicating that careful planning and proper loss functions are important (Raj, 2025). The latent space made it easy to see discrete clusters of users with similar health needs, which backed up the model's transparency and ease of explanation.

Additional work has shown that connecting feedback loops, adding wearable devices, and using machine learning improve the management and prevention of chronic illnesses (Raj, 2025). Moreover, ensuring that data are handled ethically, biases are reduced, and users trust that the system is key to safe deployment (Figure 9).

Figure 9.

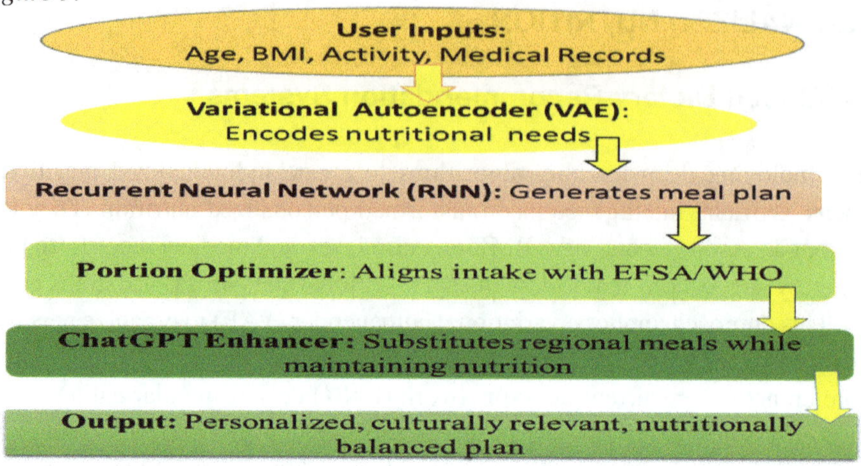

6.2 Predictive Modeling for Disease Risk

Healthcare has improved because AI is now used in predictive modeling, so consultants can identify diseases early by studying many types of data. They rely on advanced approaches to identify relationships between aspects of health, genes, and lifestyle (Behera et al., 2025).

Deep learning (DL) and machine learning (ML) are the primary drivers of this shift. While neural networks examine data with medical labels, unsupervised techniques identify patient groupings that may be at risk for cardiac or chronic conditions (Singh et al., 2024), and AI and imaging advancements both contribute to the prediction of heart failure and the identification of renal disease and stroke risk (Behera et al., 2025); (Peter, 2025). Care can now concentrate on stopping illnesses that would normally require help. With AI, an individual's illness can be predicted ahead of symptoms, allowing for fast intervention. They assist in personalized medicine by recommending personalized advice from their genetic and health information (Singh et al., 2024).

Political challenges continue to exist. A model may not function well if its data are unfairly selected or missing information. Being clear, simple to explain, and ethical is vital for the use of AI in health care settings. Taking steps such as supporting fairness, being clear about how it works, and obeying regulations support trust and dependability in medical AI (Peter, 2025). Overall, AI-assisted models help doctors evaluate disease risk more accurately and begin targeted treatments quickly. Space architecture improves health care because it helps patients recover faster and safely (Figure 10).

Figure 10.

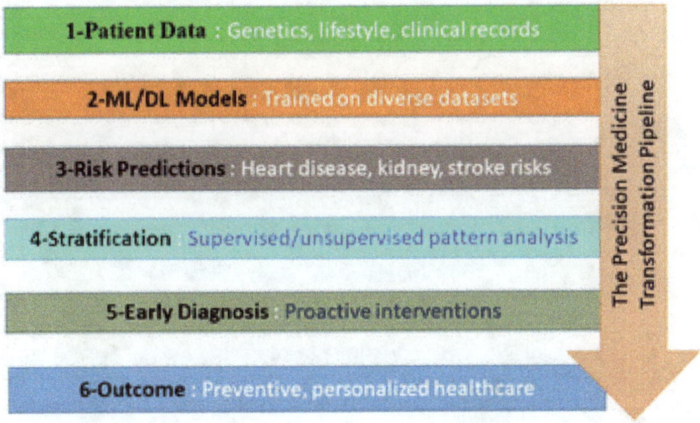

6.3 Gut Microbiota-Based Diagnostics

Microbiome-based diagnostics provide a promising, safe, and noninvasive method for the diagnosis of gastrointestinal and systemic diseases via microbial signatures. The gut microbiota is altered in several disorders, such as inflammatory bowel disease (IBD) and colorectal cancer (CRC), which makes it easier to create biomarker panels and prediction models. Metagenomic sequencing and droplet digital PCR are novel technologies that have revealed disease-specific microbial signatures. In the case of IBD, bacterial species-based diagnostic models have been able to reach accuracy levels of more than 90% in the differentiation of patients and healthy individuals compared with traditional fecal calprotectin metrics (Zheng et al., 2024). Tumors are associated with the presence of species such as *Fusobacterium nucleatum*, *Peptostreptococcus anaerobius*, and *Parvimonas micra* in colorectal cancer. Fecal immunochemical testing in conjunction with microbiological testing seems to improve the sensitivity and specificity of early detection (Zhou et al., 2020). Enhanced biomarker validation is currently performed through a curated knowledge base across-study comparison that assists in microbial signatures.

With the aim of increasing the accuracy of diagnosis, the integration of microbiome profiles, metabolomics, and host genetics allows for individual differences. Because dysbiosis causes a shift in microbial byproducts and a decrease in helpful microbial occupants, it is linked to the severity of IBD and treatment responsiveness (Mousa & Al Ali, 2024). The application of microbiome-based diagnostics is progressing despite challenges such as sample variability and geographic differences, and it has become one of the primary instruments of personalized medicine (Figure 11).

Figure 11.

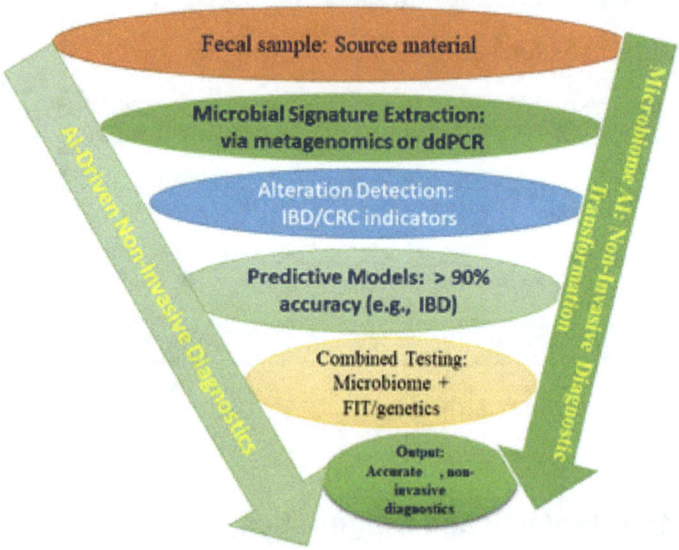

6.4 Personalized Supplementation and Nutrient Planning

Profiling of the gut microbiome is now making nutrition a precision health strategy, making custom supplementation and meal planning possible. The mainstream systems recommend that all people consume the same types of diets regardless of their metabolic rates, evolutionary gut microbiome profiles, and dietary preferences. Although there is currently little information on this technique, new research suggests that modifying nutrition to account for microbial content can improve metabolic protein results and can lower the chance of developing chronic illnesses (Hernandez-Calderon et al., 2022).

A disturbed intestinal microbiome strongly impacts nutrient absorption, energy metabolism, and inflammation. Microbiota-tailored diets are an interesting treatment approach for obesity and type 2 diabetes. These interventions in clinical trials resulted in HbA1c, blood pressure and C-reactive protein reductions, as well as an increase in microbial diversity and improvement in the bacterial profile associated with metabolic health (Kallapura et al., 2024). Artificial intelligence-enabled platforms go a step further and incorporate microbiome, dietary, lifestyle and biometric data to personalize further. An AI-mediated scheme of six weeks positively affects microbial stratification, microbial putrefaction, waist circumference and macronutrient consumption (Rouskas et al., 2025).

Diet responsiveness could also be made accurate by using baseline microbial signatures as predictors of improved compliance and effectiveness. The science of the polybiome is also becoming used to evolve individual nutrients that are optimal for health, microbe control, and the prevention of metabolic disorders. The convergence between machine learning, genomic science, and intestinal microbiome mapping (Figure 12) is facilitating the development of individualized healthcare. In addition to the gut alone, the burgeoning discipline of polybiome research has attempted to merge diverse microbial environments with the genetics of the host organism, lifestyle practices, and diet. This strategy enables a more holistic pathway for understanding tailored nutritional plans and metabolic well-being.

Figure 12.

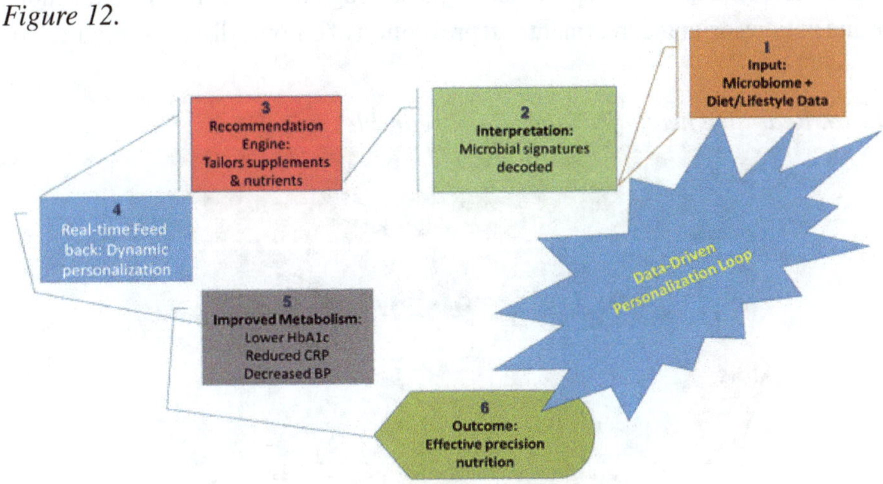

6.5 Real-Time Feedback via Wearables and Mobile Apps

By providing a continuous display of physiological indicators, including heart rate, heart rate variation (HRV), oxygen status, and sleep, the use of gadget space and personal health apps is transforming personalized healthcare. This continuous monitoring provides a dynamic picture of health status more than what one would get on a visit to clinics.

In inflammatory bowel disease (IBD), as well as other gastrointestinal diseases, wearables can identify slight physiological shifts five--seven weeks in advance of a flare, including decreased HRV, elevated resting heart rate, step count and circadian rhythm disturbances, and intervene early (Hirten et al., 2025). They are helpful in disease management and prompt therapeutic decisions.

In addition to diagnostics, mobile applications encourage behavior adjustment through monitoring diet, exercise, stress, and bowel habits and providing immediate feedback to users. They support collaborative decision making and relay information to health professionals to encourage compliance with treatment (Chong & Woo, 2021).

Patient self-management is also enabled by innovations in noninvasive biosensors, including those that use sweat to provide early warning of inflammatory markers such as CRP and IL-1 (Lagoumintzis & Patrinos, 2023). These technologies are changing as they adapt artificial intelligence to multisensor platforms to become adaptive digital health coaches, which can provide personalized, real-time interventions. The increased clinical importance of these drugs indicates the need for more active and patient-oriented treatment and prevention of chronic diseases (Figure 13).

Figure 13. Real-time health feedback from wearables and apps

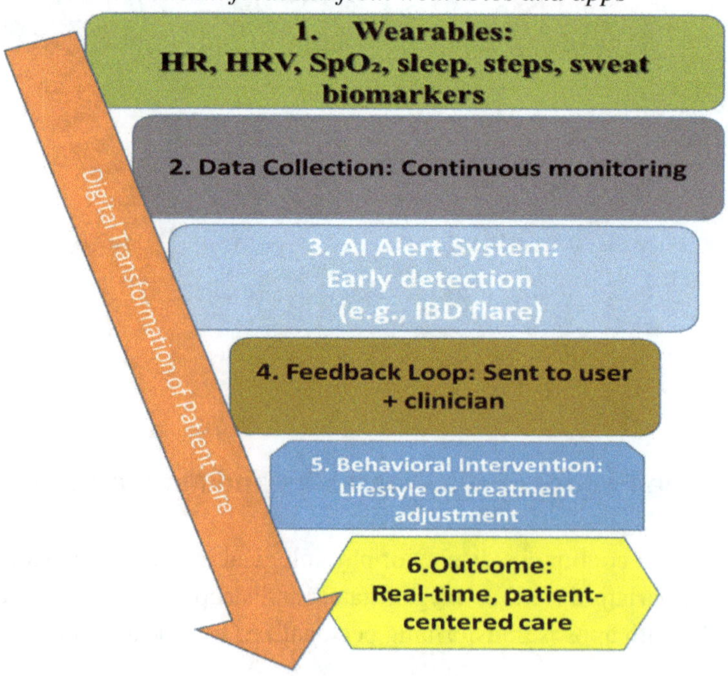

6.6 Analytical Conclusion and Research Imperatives

The next generation of healthcare is catalyzed by precision medicine in Section 6, which includes AI-portrayed dietary recommendation systems, predictive disease modeling, polybiome-based diagnostics, and personalized supplementation.

Nonetheless, some essential areas of gaps and implementation issues need to be resolved to realize these innovations into equitable, patient-centered practices:

1. Algorithm transparency latent-space clustering and attention visualization are frequently employed tools in emerging models, although these tools remain black boxes in many cases. Explainable algorithms such as SHAP, LIME, or saliency-map visualizations have become necessary to enable clinicians to feel more comfortable with AI-generated decisions, especially when they are concerned with nutrition or diagnosis. Research has shown that explanations that are clear, concise, and contextually meaningful encourage clinician trust, whereas confusing or complex explanations decrease confidence levels (Vimbi et al., 2024).
2. Clinical Validation The majority of existing evidence is based on pilot or observational short-term studies. It is time to conduct long-term randomized controlled trials (RCTs) to assess health improvements, adherence to diets, and metabolic changes in practical and diverse conditions. RCTs that are nutrition-specific are not without their challenges but are possible and needed to establish statements of AI efficacy (Wang et al., 2025).
3. Equity and Access Large proportions of AI nutrition and polybiome platforms are developed using homogeneous data, where training datasets pose biases and reduce generalizability. The inclusion of an inclusive design or federated learning frameworks would be necessary to guarantee the inclusion of individuals with insufficient resources and a large variety of people, which would ensure fairness and reduce bias in the algorithm (Agrawal et al., 2025).

Various policies are needed to promote the responsible use of AI-based health technologies.

1. Consistent protocols are defined to explain abilities that are computable across recommendation and diagnostic pipelines and therefore enhance interpretation by clinicians and strengthen regulatory confidence.
2. Multicenter intervention studies that could make valid AI models cross-sectional and longitudinal in practice settings should be conducted. 3. Introduce federated learning and other privacy-conscious and architecturally inclusive systems to ensure fair access and to suppress bias (Reyed 2023b).

7. REAL-WORLD CASE STUDIES AND PLATFORMS

7.1 Viome: AI-Driven Metatranscriptomics

The viome has led to the use of meta-transcriptomics to study gut health. In contrast to conventional assays of the microbiome based on DNA, which reveal the presence of microbes, Viome is based on RNA to reveal which microbial genes are being utilized. With this real-time vision, it is easier to better comprehend the importance of microbiomes in the processes of digestion, immunity, and inflammation. Through a combination of this information with AI algorithms, Viome provides personal recommendations about health, including diet and supplements, on the basis of the activity of the unique microbes that make each unique. They also work on early disease detection (i.e., detecting oral cancers) as an example of the wider potential of metatranscriptomic diagnostics in preventive care.

7.2 Nutrigenomix: Genetic-Based Nutrition

The business offers tailored nutrition counseling via input provided by the individual genetic makeup. Their tests analyze genetic factors that influence the metabolism of nutrients, dietary intolerance, and reactions to food substances such as caffeine or fat. Nutrigenomix also hopes to improve the effectiveness and success of nutritional transformation by focusing strategies on personal genetic makeup. Research indicates that gene-informed nutrition patients are more prone to proceed with the intended behavioral transformations, implying that such an approach has an operational advantage over blanket nutrition advice. A rapidly rising overlap of nutrigenomics and lifestyle medicine exists in this platform in terms of prevention and management of chronic diseases (Wheatley, 2024; Nielsen & El-Sohemy, 2012).

7.3 PREDICT Studies (ZOE Project)

The studies known as PREDICT produced by the researchers of King College London and the company ZOE have shaken our approach to thinking about metabolic reactions to food. The project, which has been conducted by studying more than 1,000 participants, has gathered detailed information such as the level of glucose, fat level in the blood, composition of the polybiome in the gut, genetics, and sleeping mode. The difference in the individual reactions of people to the same food is one of the most significant discoveries that undermines official recommendations about what is good to eat and what is not. On the basis of these data and machine learning models, PREDICT provides a more detailed, individualized picture of nutrition. Such a practice opens the door to an individual approach to nutrition that will enhance long-term metabolic health. (Berry et al., 2020; Asnicar et al., 2021).

7.4 Day Two and Microbiome-Driven Meal Planning

DayTwo is a blood glucose management system based on the knowledge of a microbiome. The platform uses personal health data and the sequencing of a gut microbiome to determine how an individual responds to various foods. It is based on such predictions, creating individual dietary recommendations that may support the management of conditions such as type 2 diabetes and prediabetes. In contrast with generic low-carb diets, DayTwo uses information about the personal microbiome-driven glucose response of every user. According to clinical trials conducted by companies, this approach may enhance glycemic control above the current nutritional guidelines. (Zeevi et al., 2015).. Prevalent approaches analyze a specific microbiome area, but next-generation methods have the potential to exploit data from various microbial locations in the human system. These findings can contribute to advancing individualized dietary intake choices and estimating how the body metabolizes food.

7.5 Regional Implementation and Healthcare Integration

The application of microbiome- and genomics-based nutrition tools is diverse across regions and regions and is dependent mostly on available infrastructure, research funding, and clinical preparedness. Additionally, in Europe, these tools are becoming more widely available, and regulatory, standardization, and data-sharing issues continue to hamper efforts to integrate these tools into healthcare economics. In addition, in countries such as Sweden, government-driven health plans have started to incorporate activities such as whole-genome and microbiome sequencing within overall national health strategies. In Africa, the activity targets research capacity development and fair access to the tools of personalized medicine. Such regional variations support the necessity of flexible, ethical, and culturally relevant models, as microbiome-informed healthcare becomes more conventional. (D'Amato et al., 2020; Forslund et al., 2023). Present-day techniques commonly focus on individual microbial ecosystems, specifically those found in the gastrointestinal tract or mouth. The next step consists of bringing data from a number of these communities together. Upon integrating distinct microbial ecosystems—such as the microbiota found in the intestines, oral cavity, skin, and urogenital region—in conjunction with details concerning a subject's genetics, metabolic functions, and daily life, mathematical models are capable of producing a complete overview of their microbial system. This comprehensive technique provides possibilities for more effective dietary recommendations, enhanced assessments of the body's metabolism and inflammatory process, and a deeper understanding of how microbes interact between different organs. This broadens individualized treatment in addition to considering specific microbial ecosystems.

7.6 Future Implications and Analytical Conclusions

The empirical cases presented (Viome, Nutrigenomix, ZOE/PREDICT, and DayTwo) indicate that the implementation of AI-powered microbiome and nutrigenomic services has already started to transform personalized nutrition and personalized diagnostics (National Academies of Sciences, Engineering, and Medicine, 2022).

To achieve the full potential of these innovations in the healthcare field, however, a strong response to some general weaknesses, as well as three critical challenges, is needed:

1-Scientific validation and clinical validation: A snapshot of cellular activity at a given moment, metatranscriptomic RNA profiling offered by Viome, characterizes the prevalence of active microbial genes. It is a molecular database that can be used to personalize dietary advice and underpin screening activities, a point proven with oral cancer, among others. In a pilot study (n = 27, 12 intervention arms, 15 controls), the Viome Precision Nutrition Program (VPNP) resulted in a clinically meaningful reduction in HbA1c of approximately 0.42% following an ~90-day dietary intervention (p = 0.028) (Vuyisich et al., 2024). The small sample size is a limitation because the results cannot be generalized. A larger multicenter randomized controlled trial (V-197; ClinicalTrials.gov NCT05465616) of an estimated 150 participants is currently ongoing, with projected enrollment completion scheduled in late 2026. The results will require comprehensive logistical planning, long-term funding, and collaboration with institutions to confirm both efficacy and scalability.

2-Explainable AI clinician trust: Viome and ZOE/PREDICT use deep learning algorithms to make highly accurate but opaque predictions on complex omics data. To reduce the obstacles to clinician trust, it will be necessary to integrate explainable AI techniques, including explainable AI techniques such as attention visualization, SHAP, and LIME, into the system and to present results in clinically friendly formats (National Academies of Sciences, Engineering, and Medicine, 2022).

3-Scalability and accessibility: The success of these services is limited to only a small percentage of the population because each product is still costly and access is unequal. Wider use will be possible with the creation of scalable and cost-effective platforms that promote geographic and socioeconomic equity

(Suárez et al., 2025). The balance and accessibility of real-world evidence (RWE) populations have been the major focus of present-day personalized medicine investigations. Most platforms of precision medicine that are based on molecules rely on homogeneous or direct-to-consumer data, such as in the case of the multimodal monitoring trial PREDICT 1. The study was conducted on more than 1,000 participants in the United Kingdom and the United States through the ZOE (Berry et al., 2020). Despite its impressive scale, the demographics of the cohort highlighted ongoing disparities in representation: approximately 95% of the UK participants were of European origin, and the preliminary racial analysis revealed that the majority of respondents referred to themselves as White, with significantly smaller proportions of Asian, Black, and other ethnic backgrounds (Bermingham, 2024).

These findings help highlight the strong necessity of becoming more inclusive along socioeconomic- and ethnic-based divisions. Future directions must be considered:

1. It is possible to implement standardized explainability techniques to integrate latent data clustering and active gene expression with meaningful clinical text that can be understood by healthcare providers.
2. Conducting prospective controlled, longitudinal clinical studies in various settings to prove the efficacy of algorithmic prognostication and treatment
3. The incorporation of antibias design principles ensures that data and privacy are preserved, i.e., representative of federated learning.
4. Tracking has changed to the regulatory landscape, such as the Food and Drug Administration (FDA) artificial intelligence in healthcare guidance and the future EU Artificial Intelligence Act (FDA, 2023; Jenko et al., 2025).

Conformance with such structures will be key to ensuring algorithmic visibility, data security, and clinical performance factors that form the antecedents of popular adoption and social trust in AI-facilitated personalized medical services.

Scientific validation, explainability, and fairness are the three necessary components that must be met simultaneously to realize the successful implementation of precision healthcare services powered by artificial intelligence. Given such proportionality to each of these factors, section 7 can graduate from being simply a descriptive catalog of relevant activities into a rational overall strategic framework of the field. This development enhances the moral soundness of the approach and facilitates both better clinical performance and positive perceptions of patients simultaneously. As a result, the current discussion directly fits into the general aim expressed in the text comprehensively, i.e., the promotion of a transformative and equitable vision of AI-enabled personalized medicine. Integrating findings originating from numerous microbiomes is the next step. Rather than just concentrating on sole environments such as the intestinal tract, innovative AI approaches have the ability to merge data from microbial ecosystems of the gut, oral cavity, skin, and urinary system with host information such as genomics, metabolomics, and lifestyle data. This creates a complete view of a person's microbial environment. With this framework, it is possible to establish more effective dietary strategies with a better chance of predicting metabolic and inflammatory responses, allowing for a more comprehensive grasp of how microbes interact between organs. This expands personalized medicine beyond looking at just one microbiome.

8. CHALLENGES AND LIMITATIONS

8.1 Data Heterogeneity and Quality Issues

One of the key setbacks to the development of AI-based microbiome and nutrigenomics research is data heterogeneity and limited data quality. Highly heterogeneous microbiome datasets, particularly those produced by high-throughput sequencing, are the result of diversity in experimental design, sample processing, sequencing protocols, and data processing (Kool et al., 2023). Owing to the sensitivity of microbial profiles to even minor modifications (such as storage temperature, PCR cycles, or stabilizing compounds), the results may be biased.

Even microbiome data are messy, often high-dimensional, sparse, and compositional. These features present a problem with conventional statistical approaches and contribute to interstudy discrepancies, such as contradictory associations among microbial taxa and diseases (Kleine Bardenhorst et al., 2021).

The other more pertinent question is the nonexistence of standardized metadata. Gaps and inconsistencies in metadata restrict the ability of the procedure to stratify and incorporate datasets, a necessary phase in training powerful machine learning algorithms (Kumar et al., 2024). Additionally, the variation in bioinformatics pipelines, including quality control mechanisms and taxonomic classifiers, may generate incongruent results for the same raw data, which implies a decrease in the quality of AI predictions (Bharti & Grimm, 2021).

Standardized protocols, better metadata practices, and community-approved workflows are needed to address such challenges. These issues are essential for overcoming the potential of AI related to personalized nutrition and microbiome studies (Figure 14).

The investigation of unique microbiomes faces difficulties due to differences, datasets that are sparse and contain various variables, and poorly standardized annotations. This demonstrates the requirements for agreed-upon protocols, processes supported by community consensus, and quality assurance measures. As this area assimilates polybiomes, resolving these difficulties is paramount for developing dependable AI algorithms that are equipped to address many microbial niches at once, broadening the scope of customized nutrition and medical care breadth and precision.

Figure 14.

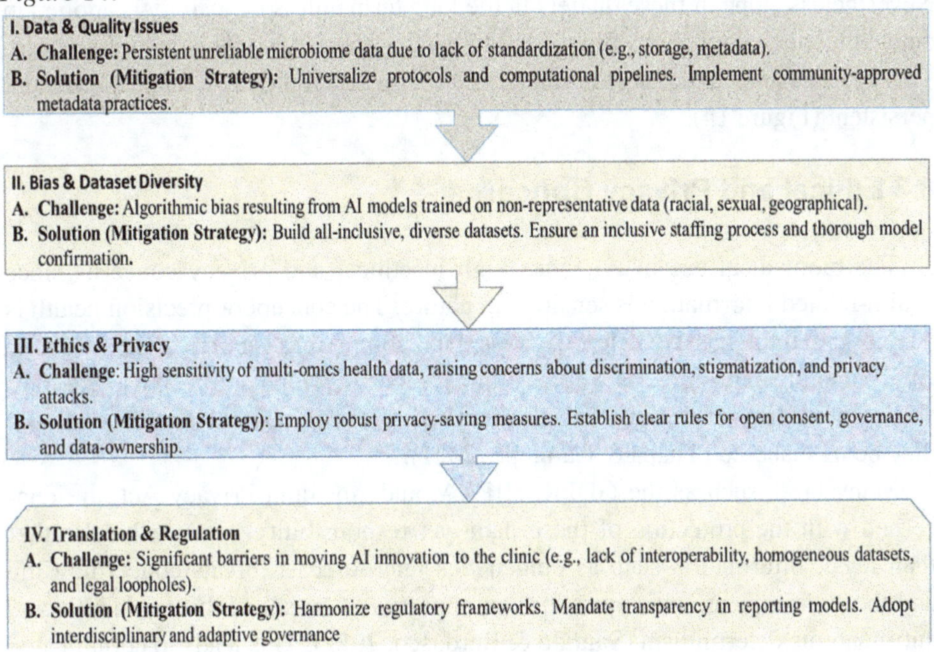

8.2 Algorithmic Bias and Limited Diversity in Datasets

Although AI has unrealized potential in biomedicine, the trend remains prominent problems of bias in algorithms and unrepresentative datasets. Models that have been trained on biased data (excluding some races, genders, or geographical areas) may have poor results or results that lead to discrimination, which is more dangerous to those who have a disadvantaged social space (Yang et al., 2024).

An obvious example is the racial discrimination of pulse oximeters, which tends to lead to misinterpretation of the oxygen values of people with dark skin. This may lead to defective clinical decisions and poor predictions by the AI (Zou & Schiebinger, 2021). Likewise, the representativeness of disease risk models to an ethnically diverse population varies because genomic databases such as TCGA are Eurocentric in nature and thus do not index well-represented ethnicities (Sarkar & Maniatakos, 2021).

These problems require structural and technical intervention measures. The exposure to model bias introduced in the process of development can be mitigated through adversarial learning, data reweighting, and federated learning (Yang et al., 2024). Methods such as TRAPDOOR exist to identify unconscious biases in genetic data without diminishing the performance of the models (Sarkar & Maniatakos, 2021). Nevertheless, none of these matters in the long term unless inclusivity is applied in data collection, algorithm validation, and heterogeneous people in research teams. This is because AI in healthcare can reduce disparities rather than make them more persistent (Figure 14).

8.3 Ethical and Privacy Concerns

The biomedical use of AI ranks high in ethical and privacy concerns since health-related information is sensitive in nature. The concept of precision health is heavily based on massive heterogeneous data, comprising genomics, lifestyle, and environmental exposures, as well as clinical records, which could be used to their detriment in terms of exposure, discrimination, or loss of jobs and insurance, should they be mishandled (Thapa & Camtepe, 2021).

Many laws, such as the GDPR, HIPAA, and Australia Privacy Act, are concerned with the protection of these data, yet responsibility is also ethical rather than legal. Information such as epigenetics may forecast threats concerning life and the environment, which provides an opportunity for social discrimination and anti-genomic determinism (Santaló & Berdasco, 2022). This leads to complicated issues regarding autonomy, the right not to know, and equitable accountability of health outcomes. Privacy in other examples of AI being vulnerable to privacy attacks includes membership inference, whereby attackers find individuals in training data. Techniques that enable models to be trained collaboratively without raw data, such as federated learning, homomorphic encryption, and differential privacy, can combat this (Torkzadehmahani et al., 2022).

Altered microbial strains and next-generation probiotics (NGPs) are feasible methods for promoting nutrition but are associated with ethical and safety issues. The engineered microbes may also harbor undesirable mutations, circuits may be lost, or the transfer of genes may occur to native gut bacteria, posing a biosafety risk. The error rates of gene-editing tools are off-target, and they need to be validated. Regulatory uncertainty remains due to the lack of harmonized standards for implementing NGPs as live biotherapeutic products. Dual-use risks are associated with the acceleration of the creation of agents in synthetic biology, which requires monitoring. Combining metabolomic and microbiome data presents a challenge and requires informed consent with effects (Carolak et al., 2025).

In addition to technology-based protection, ethical AI development requires openness, informed consent, clarity of ownership, and trust among stakeholders. These multidimensional issues should be addressed to ensure that AI achieves the interests of the population in the field of health without violating personal rights. Because research on microbiome integration has evolved, ethical and safety concerns will address not only standalone microbial communities but also numerous microbial locations within the body. Guaranteeing public understanding and agreement, ensuring data privacy, and complying with biosafety regulations will be essential in the process of integrating data sourced from the gastrointestinal, oral, epidermal, and other microbiome environments in conjunction with genomic and metabolomic information, as artificial intelligence applications become more sophisticated and common.

8.4 Regulatory and Clinical Translational Barriers

Although the progress of AI-driven personalized medicine is currently accelerating, regulatory and translational research constitute a major challenge. One of its greatest obstacles is ineffective data interoperability in healthcare systems, which prevents data on the microbiome, nutrigenomics, and clinical datasets from being integrated into machine-readable formats (Carini & Seyhan, 2024). This restricts the extendibility of multiomics-based AI for personalized health interventions.

Clinical AI tools tend not to be generalizable; in most cases, they are developed on the basis of homogeneous data or single centers of trials. This is particularly problematic in the case of studies on the microbiome and nutrigenomics because exposure to the environment and differences in populations have a significant impact (El Arab et al., 2025). These restrictions, when combined with algorithmic bias, run the danger of worsening health disparities and render the models unreliable. Adoption is also penalized by ethical and regulatory concerns. Sensitive genetic and polybiome data further highlight the issues of data ownership, permission, and privacy (Shaman, 2024). In addition, there are many AI-based diagnostic methods, particularly those based on dynamic polybiome or diet–gene relationships, with no obvious regulatory systems. Thus, clinicians are deprived of any guidelines.

To overcome these challenges, AI-focused applications in the microbiome and nutrigenomics must be employed strategically. This begins with the conscious incorporation of diverse, high-quality datasets. The modeling process itself should be subject to transparent reporting and clear documentation. Finally, cross-disciplinary governance systems must align with emerging regulatory requirements. The development of these applications into more practical tools for healthcare delivery in person could be accomplished only through coordinated and inclusive initiatives. These surviving constraints also indicate the presence of an imbalance in the structure, which needs severe criticism. One of the key issues is the overall fragmentation of data pipelines, including variability in sequencing and ethical regulations, which leads to poor and often nonreproducible data and follows the epistemological gap between translational preparedness and technological progress. From this perspective, applying AI cannot necessarily be defined as transformative unless it is contextualized in inclusive, ethically regulated, and medically relevant models. As a result, the domain should transition away from siloed algorithm creation and toward an ecosystem perspective, where data science, clinical validation, bioethics, and social equity can be coordinated. In the future, the path of precision nutrition will be determined by three innovations: (1) flexibility to dynamic models across populations; (2) decentralized, real-world clinical validation; and (3) transparent, participatory governance norms with integration between patients, clinicians, and regulators. Instead of strictly trying to reduce the current bias or enhance technical performance, the domain should aim at achieving so-called interpretability equity: AI tools should work explicitly well with diverse populations, as well as have them understand and trust them. This kind of paradigm shift would not only free the field of technological determinism but also shift it toward participatory precision health, which more closely reflects the ethical and practical potential of personalized medicine. This shift toward ecosystem-level integration and interpretability equity is also complemented by new guidance on the governance of AI and ethical nutrition (Reyed, 2023b) and explanation frameworks in precision medical practice (Mohapatra et al., 2025). The way of integrating the considerations important both to the implementation of the practice and proper ethical control cannot be avoided in the development of research in the fields of personalized nutrition and microbiome science. With the development of polybiome integration studies, ethical governance, data sharing, and clinical application considerations will expand from individual microbiomes to multiple microbial areas in the body. Machine learning algorithms integrating data from the gut, oral cavity, skin, and other microbiomes in conjunction with information on genetics, metabolic processes, and lifestyle choices should incorporate increased accountability, informed permission, and stakeholder involvement and supervision to promote unbiased and health-promoting individualized nutrition strategies.

9. FUTURE DIRECTIONS AND INNOVATIONS

9.1 Federated Learning and Privacy-Preserving Models

The movement toward AI in healthcare has made the privacy of patients a more valuable concern than ever before. Conventional data-sharing approaches are risky, yet federated learning (FL) could present a solution, which involves training AI models collaboratively without sharing raw data between hospitals and labs. Local training at each institution protects sensitive information because only encrypted updates are shared.

Although federative learning decreases the risk of invasions of privacy, it is not perfect. To pursue this goal, they employ techniques such as secure multiparty computation, homomorphic encryption, and differential privacy to ensure that the data remain secure throughout the process, even when processing is performed. Confidential computing provides one step toward further protection by enhancing hardware environments.

Despite not being able to eliminate challenges entirely, including data format differences and computing capacity, FL allows a responsible application of AI at a large scale in personalized medicine. It encourages innovation but makes it possible to keep patient data confidential and safe (Pati et al., 2024).

9.2 Digital Twins and Predictive Modeling

Digital twins (DTs) are innovative beginning in the future of personalized medicine. TTs were originally applied in aerospace and manufacturing, and DTs are also coming to healthcare as dynamic, data-driven virtual twins of bodies, organs, or body systems. Owing to the inclusion of real-time information, including biometrics, genomics, and lifestyle patterns, DTs provide an extremely personalized solution to care, simulate and forecast how the body of a person under consideration reacts to prescribed treatments or behavioral variations.

With AI capabilities, digital twins can be used to conduct robust predictive modeling, allowing them to predict disease dynamics, patient response, and even symptoms of the preclinical disease phase, e.g., mild dysfunction of the heart. For example, real-time data could be monitored with a digital twin of the cardiovascular system to help predict a cardiac event and provide preventive actions prior to the onset of symptoms. These models improve each time new patient data are added to them, so a real-time feedback loop similar to that used in engineering to perform predictive maintenance is created. In healthcare, this implies that health professionals will be able to practice a therapy plan on a digital model of the patient before transplanting it to the real one, minimizing the number of risks and improving accuracy.

Nevertheless, regardless of the potential promised, there are still limitations: simulations of human biology are technically complicated, data collected on wearables and clinical devices need to be aligned, and synchronization of digital models with patient-based data presents ethical and security issues. Nevertheless, digital twins promise to transform healthcare into proactive care and deliver personalized predictive care on the basis of an individual and his or her biology and lifestyle (van Dinter et al., 2022).

9.3 Advances in AI for Multiomics Integration

A flood of biological data at various omics levels, including genomics, transcriptomics, proteomics, and metabolomics, has provided not only existing opportunities but also analytical challenges. These datasets can be complex and large, and conventional approaches to data analysis usually fail with them. Here, artificial intelligence (AI) comes in handy. Methods for simplifying and inferring complex biological signals, such as deep learning, are increasingly being utilized. For example, a special type of neural network, such as an autoencoder, can compress a large amount of data (omics) to retain significant patterns. They allow scientists to reveal obscured biological relationships, e.g., disease signatures or targets of treatments that cannot be discovered via standard methods (Acharjee et al., 2023).

In addition to pattern spotting, AI also provides researchers with the ability to examine the interactions between various biological systems. The graph neural network (GNN) is one of the most influential tools that causes this change since genes, proteins, and other biomolecules are put into the network as nodes that are connected to others. The models assist in visualizing the effects of a perturbation in a biological layer, such as gene expression, on the behavior of proteins or the level of metabolites. Moreover, supervised learning algorithms, which use models of support vector machines and ensemble algorithms, have also demonstrated their usefulness in both classifying patient subtypes and in the future prediction of clinical outcomes. In contrast to conventional methods, such AI approaches are able to identify subtle, nonlinear relationships among biological features that can be central in identifying complex diseases (Thapa & Camtepe, 2021).

9.4 Nutritional Epigenomics and AI

Nutritional epigenomics studies the effects of food and nutrients on gene expression by regulating events such as DNA methylation and histone modification without changing the underlying DNA molecules. These epigenetic changes may be significant in either causing or averting diseases, including cancer, diabetes, and autoimmune diseases (Thapa & Camtepe, 2021).

In the future, AI will radically change the landscape, as there will be assistance in deciphering the seemingly complicated interplay between diet, genes, and health. Machine learning models are able to examine enormous sets of data that contain genetic, dietary, and lifestyle data and predict how the body of a specific individual will react to a specific set of nutrients. This lays the foundation for truly personalized nutrition, in which diet regimes are customized by both genetic profiles and change dynamically depending on shifts in health, the environment and lifestyle (Reyed, 2021a).

AI is also streamlining the process of determining which nutrients or bioactive compounds can have a relevant effect on gene expression and, as a consequence, new diet-based therapies. With the onset of wearable technology and home tests, AI take over and provide real-time nutrition recommendations on the basis of genes to ensure lifelong health (Reyed, 2023b; Santalo & Berdasco, 2022).

The development of next-generation probiotic formulations and polybiome-based therapies is opening new possibilities because of the generation of artificial intelligence. It also allows the simulation of complex microbial communities, which makes it possible to generate probiotic mixtures that lead to a particular therapeutic response to a given clinical condition, bypassing the traditional trial–and-error approach. These efforts can involve the engineering of novel bacterial strains with increased resistance, inventive metabolic potential, or disease-modulatory properties (Carolak et al., 2025).

An individualized approach is one of the most conspicuous characteristics of this style. Combining polybiome measurements, host genomic data, and lifestyle data, generative AI can be used to design personalized probiotic regimens, thus turning treatment paradigms into condition-specific regimens, especially in cases of gastrointestinal and systemic diseases. Through artificial gut models, such as ABIOME, generative AI is capable of modeling bacterial interactions, metabolism, and environmental interactions.

Such *in silico* models would assist in supporting virtual clinical trials where thousands of combinations of probiotics are tested before laboratory or clinical validation (Reyed, 2021a). As a result, the speed of new probiotic discovery has significantly increased, the overall price of development has decreased, and real-time, patient-specific probiotic treatment can be personalized on the basis of the unique biology and clinical presentation of each person.

9.5 Bridging Gaps for a Unified Future in AI-Driven Personalized Medicine

The process of artificial intelligence (AI) integration into the healthcare sector still faces several major difficulties related to long-term interoperability, strict validation, and ethical issues. Despite the potential transformative power of innovative approaches such as federated learning, digital twins, or generative AI to develop probiotic and polypeptide-based interventions, these methods are inherently limited by data compatibility concerns, the distribution and inequality of computational resources, and regulatory uncertainty (van Dinter et al., 2022; Carolak et al., 2025). Digital twins are highly capable of simulation but cannot achieve biological complexity and the combination of different microbiome and polybiome data, thus limiting their applicability to real-world situations purely to demonstrations. Similarly, multiomics AI is capable of discovering latent patterns but requires mechanistic validation to convert results into practical clinical information (Acharjee et al., 2023). Nutritional epigenomics and microbiome studies add another point of individualization, and this scalability depends on high-quality longitudinal results that include the gut microbiome, polybiome interactions, and genetic variability. Gerative AI has accelerated the development of probiotics and polybiome-based therapeutics, but it is also prone to overfitting *in silico*; thus, it should undergo robust wet laboratory and clinical validation. To overcome these gaps and achieve the full potential of AI, an ecosystem-level approach in which microbiome and polybiome data are incorporated into hybrid validation pipelines involving predictive analytics, laboratory experiments, and longitudinal trials is necessary to achieve efficacy, safety and fairness.

Considerable improvements in the transition to precision healthcare require an available form of modular hardware and universal open data schema and holistically accessible polybiome datasets, next allowing federated learning in heterogeneous institutions and the creation of a friendly AI-empowered information ecosystem (National Academies of Sciences, Engineering, and Medicine, 2022). The lack of transdisciplinary consortia, including bioengineering, data science, clinical practice, and ethics, is necessary to generalize the understanding acquired from digital twins, the microbiome, the polybiome, and epigenomic analyses into individual interventions. The active integration of explainable AI models should also foster trust in clinicians, patient understanding and compliance with regulations (Mohapatra et al., 2025).

The ideas of personalized medicine point toward an AI-based microbiome and polybiont profile analysis, using genomics, metabolomics, lifestyle, and other environmental data to design unique health strategies. Examples of the practical implementation of such integrative approaches include platforms such as Viome, Nutrigenomix, ZOE, and DayTwo, which provide individualized dietary and probiotic advice on the basis of the microbiome and polybiome profile. Ethical, regulatory, and data-representativeness challenges are also significant, and careful governance is needed to ensure fair, safe, and clinically viable AI implementation.

In the future, AI models that preserve privacy, digital twins, generative AI of polybiome-informed probiotics and diets are the future of predictive, preventive, and precise healthcare. Microbiome and polybiome analytics coupled with AI-infused interpretation enable individuals to take charge of their health, beyond generalized prescriptions, toward genuinely personalized, analytically rigorous, and socially equitable precision medicine. The integration of machine learning and polybiome data has the potential to significantly enhance individualized healthcare. Nevertheless, the suitable implementation of these resources is contingent upon a combination of technological validation and ethical rules. The following section addresses the ethical aspects essential to the role of AI within the context of nutribiome investigations.

9.6 Ethical AIs in Nutribiome Studies

Ethical Foundations: Microbiome Ethics with Polybiome Ethics. The ethics of traditional microbiomes have mostly revolved around the responsible gathering, documentation, and visualization of bacterial community information, prioritizing patient privacy, informed consent, and the governance of biospecimens. Although this framework is still necessary, it is increasingly unable to represent the enormous biological and ethical complexity of microbial ecosystems associated with humans. This is extended by the new paradigm of polybiome ethics to more integrative and systems-level approaches that recognize the human holobiont as an active network of bacterial, fungal, viral, archaeal, and symbiotic components that interact on environmental, nutritional, and genomic planes. This shift from microbiome to polybiome ethics fundamentally reconfigures the field of ethical responsibility in AI-based nutribiome studies. Instead of managing individual microbiomes, researchers increasingly maintain a multidimensional ecological and digital lifeform, a construct that involves biological diversity, information sovereignty, cultural forms of diet, and algorithmic control. As a result, any integration of polybiome information into AI pipelines requires new standards of consent, equity, and disclosure so that an informational model that has been trained on the basis of complex microbial interactions not only observes data integrity but also acknowledges the interdependence between humans and their environment. This transformation can be strongly associated with the theories of One Health and sustainable well-being, which combine human nutrition, microbial ecology, and artificial intelligence into a single moral framework. In this approach, polybiome ethics encourages internationally rational and locally focused policies, including the Ethical AI Guidelines (ICMR, 2023) and the Charter for Responsible AI (MCIT, 2023) in Egypt and India, respectively, to empower technological innovation in nutribiomics to be distributed for the common good of humanity and society, as opposed to serving the selfish goals of institutional and commercial stakeholders. Nutribiome research has been transformed at its core by the adoption of artificial intelligence (AI) as a means of entering the news of nutritional science and microbiome analytics. However, this technological advancement also brings with it a very important ethical aspect upon which the believability, openness, and inclusivity of scientific results depend. With the processing and analysis of complicated biological, dietary, and microbial data by AI algorithms, a variety of ethical issues lie ahead in the definitions of data confidentiality, informed consent, bias in algorithms, transparency, and fairness (Kassem et al., 2025). Such issues can be considered particularly timely, especially given that the goal is to personalize nutrition according to individual microbiome profiles—a procedure that is inherently associated with delicate health and genomic information. Research projects on the nutribiome are often based on multiomics datasets that include genomic and metabolomic data as well as dietary intake, all of which are quite personalized and cultural. Increased stewardship of such data is essential in an ethical manner, which entails strong governance of these data by ensuring that it is securely stored and shared through encrypted platforms and that consent to these data is clear. The Egyptian Charter of Responsible AI (Ministry of Communications and Information Technology, Egypt, 2023) emphasizes these principles by promoting data minimization, anonymization, and a right to algorithmic disclosure, which ensures that AI technologies are responsive to the principles of majority values and individual autonomy. Similarly, the ICMR Ethical Guidelines for AI in Biomedical Research (2023) in India provide a governance model to ensure that the ethical considerations of fairness, traceability, and accountability in AI-based health research are upheld, making them the best example to be considered for countries with emerging digital systems. Another ethical issue introduced by algorithmic bias is that it can be based on excessive dependence on Western-centric data, which insufficiently capture the variations in microbiome compositions and diets of other parts of the world, such as Africa, the Middle East, and South Asia. Such biases may lead to misclassification, incorrect nutritional suggestions, or even systematic deprivation of localized groups from the advantages of AI-inspired health innovations. To resolve this problem, it is necessary to develop locally trained, culturally contextualized AI systems, which are the priorities of not only the National AI Strategy of Egypt (Digital Watch Observatory, 2025) but also the National Programme on AI in Health and Nutrition in India (Sharma & Gupta, 2023). These frameworks support inclusivity, as they entrench local data ecosystems and therefore increase the accuracy, dependability, and social importance of nutribiome analytics. Figure 15 shows the conceptual interdependence between artificial intelligence ethical management, AI-motivated nutribiome innovation, and national policy regimes in Egypt and India. The figure outlines a conceptual map that explains the relationships among ethical AI governance, national policy frameworks, and the nutribiome research ecosystems in use in both countries. The diagram explains the intersection of algorithmic transparency, data power, and unbiased access to develop nutribiome innovations that are well founded on ethical principles.

Figure 15.

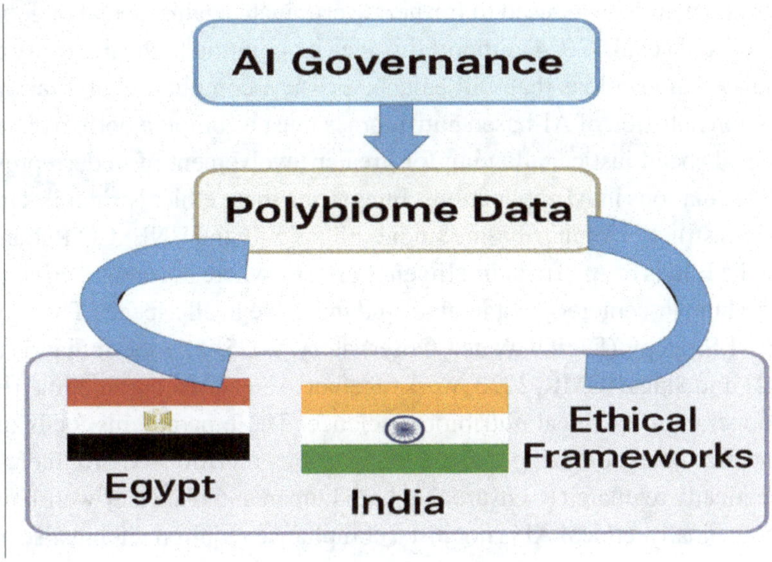

Moreover, explainable artificial intelligence (xAI) is one of the foundations of creating ethical and trustworthy nutribiome models, whereby interpretation allows clinicians, nutritionists, and policymakers to comprehend the derivation of predictions and suggestions (Naik & Reddy, 2025). This is important in healthcare and nutritional decision-making settings because opaque, black-box models may undermine professional confidence and trust in the community. The OECD AI Review (2024) of Egypt emphasized the need to be more transparent and to introduce national oversight committees that control the fairness of AI, data security, and clinical impact. Similarly, the UNESCO program Ethics in Action (2025) in India recommends the use of participatory forms of governance, which entail the introduction of ethical audits, cultural diversity, and algorithmic transparency systems in national AI ecosystems. Both India and Egypt can be considered emerging examples of the Global South, which promote what scholars now call ethical AI sovereignty. This is the idea that AI development must also be consistent with local socioeconomic and dietary cultural trends, as well as health priorities, and not merely with the importation of Western frameworks. The Indian Responsible AI Vision (NITI Aayog, 2021) and the Egyptian Charter for Responsible AI (MCIT, 2023) share common values, including the development of open-source solutions, the inclusivity of stakeholders, transparency of algorithms, and fair, inclusive access to information. These projects illustrate how two countries, operating within a context of strong cultural diversity and scientific potential, can reshape global narratives on AI ethics in a decolonial and people-focused manner. There is an urgent need to enact ethical frameworks that prevent the manifestation of digital inequality so that AI-driven solutions to nutrition and health are not used to further discriminate against social or economic disparities. Voola et al. (2024) support this idea by pointing to the necessity of creating balanced innovations that will balance privacy, compliance, and fair access. The precision nutrition of AI-based nutribiomes must be tailored both to diagnostic accuracy and social justice, allowing for greater involvement of underrepresented and rural populations in AI applications. Integrating these ethical principles into the scientific construction of nutribiome studies aligns with the UNESCO Recommendation on the Ethics of Artificial Intelligence (2021), where the ethical principles of AI must be human-centered, sustainable, and inclusive in all aspects. The Egyptian National AI Strategy (Digital Watch Observatory, 2025) and the Indian National Health AI Guidelines (ICMR, 2023) work together to create a robust roadmap toward global cooperation in ethical nutribiome science. These paradigms focus on One Health integration, recognizing that the microbiome–nutrition–environment nexus must be ethically regulated to ensure that both human and planetary well-being are maintained. Finally, ethical AI is not just a compliance requirement but also a game changer that guarantees that technological advancements in nutribiome science are in line with human dignity, sustainability, and equity worldwide. By following ethical AI principles, nutribiome research may transform from a data-rich project into an ethical AI science that leads to food security, health equity, and sustainable well-being, all of which are central to the Egyptian Vision 2030 and the Indian ethical AI roadmap. This principled context can lead to national nutrition inventiveness and sustenance the institution of an African–Asian consortium for ethical AI-driven nutrition exploration.

10. CONCLUSION

Although artificial intelligence (AI) has made tremendous advancements and computational biology has been applied in numerous health institutions, many potential initiatives for AI-driven system implementation in healthcare are still limited by outstanding concerns about system interoperability, algorithm validation, and ethical leadership. New technologies, including federated learning, digital twins, and generative models to create probiotics, have the potential to be transformative, with critical barriers to translation persisting. Federated learning still faces scalability challenges due to heterogeneous institutional systems, unsynchronized data formats, and differences in computing power. Digital twins are still unable to simulate complex human biology in a clinically actionable manner, especially in acute or high-risk cases, despite their potential to simulate physiology in detail (van Dinter et al., 2022). Although artificial intelligence multiomics models are competent at revealing hidden molecular trends, they cannot be incorporated into practical clinical decisions until the insights provided by these models are validated mechanistically via rigorous methods (Acharjee et al., 2023). Likewise, the cost, lack of population diversity in study samples, and lack of long-term cohort studies are barriers to the development of nutritional epigenomics. It is also possible that generative models that hypothesize new probiotics will overfit in vivo, requiring the use of wide wet labs and controlled clinical trials before use (Carolak et al., 2025; Reyed, 2021a).

To overcome these obstacles, a hybrid translational model that combines computationally quick prediction and an effort to target laboratory experimental work with a solid clinical examination is essential. This would provide an integrative pipeline; the output of AI-based interventions is not only innovative but also biologically plausible and safe for the patient. At the core of this vision is developing the existence of a universally open-access data schema, modular computational infrastructures and uniform hardware ecosystems that work together to support scalable federated learning and just data representation (Reyed, 2023b). Such frameworks also allow safe, multi-institutional analyses and encourage global scientific cooperation and reproducibility (National Academies of Sciences, Engineering, and Medicine, 2022).

The intersection of AI, multiomics, and microbial science has resulted in the creation of the polybiome framework, an enhanced conceptual framework that extends the gut microbiome across all the other microbial contexts of the oral, dermal, respiratory and urogenital environments. In contrast to conventional compartmentalized microbiome models, the polybiome is complex in that it involves cross-kingdom interactions and ecological interdependencies and influences overall human health. The framework allows AI systems to model complex cross-talk between host genetics, microbial communities, metabolomics, and environmental exposures. Combined with digital-twin platforms, polybiome-informed models have the potential to predict systemic metabolic or immune diversion, whereas generative AI can compile next-generation probiotics to address many microbial niches concomitantly.

The translational aspects of AI–microbiome integration can already be seen through commercially existing products such as Viome, Nutrigenomix, ZOE, and DayTwo, which are multimodal platforms that integrate microbiome sequencing with metabolomics and lifestyle measurements to provide individualized dietary advice. These systems are already focused on gut microbial profiling, but the addition of whole-body polybiome analysis will significantly increase their predictive power and biological applicability. However, the progress of AI-powered microbiome systems should be backed with strict care regarding data quality, the reduction of algorithmic bias, and the creation of effective privacy-sensitive designs, especially those supported by federated learning.

With the advent of AI systems where microbial ecosystems are beginning to be affected and potentially re-engineered, the ethics and regulatory aspects of microbiome science will need adaptation as well. Nontrivial ecological risk factors are introduced by polybiome manipulation, which includes unintended changes in metabolic responses and horizontal gene transfer; hence, next-generation probiotics and microbial therapeutics will have to be carefully subjected to stringent mechanistic, biosafety and regulatory evaluation before implementation in clinical settings (Carolak et al., 2025). Explainable AI (XAI) must be incorporated to guarantee clinical transparency and promote clinician trust and accountability in processes driven by algorithms. Moreover, privacy-safe computing methods still need to ensure the integrity of genomics and microbial data, thus ensuring the preservation of patient autonomy and ensuring the continued trust of society in these technologies (Mohapatra et al., 2025).

In the future, AI, polybiome science, and nutrigenomics convergence will lead to an initial transformation of the precision medicine landscape. Healthcare will take a sharp turn in favor of predictive, preventative and highly personalized care models based on ecosystem-level analyses that define the dynamic interaction of communities of microbes, host genetics, nutritional exposures and lifestyle influences. The more powerful version, known as generative AI, will create individualized probiotic formulas and nutrition plans, and digital twins will help to test therapeutic interventions in complex biological systems risk-free. Collectively, these approaches constitute a cohesive and morally sound paradigm of precision health in which AI provides individuals with the tools they need to proactively control their health, averts the prevalent practice of clinicians offering precisely targeted therapy, and provides researchers with an unparalleled understanding of the multilayered structure of human biology. This composite framework, finally, creates the groundwork for the future, with technologically advanced, bioinformed and socially responsible healthcare being truly universal.

REFERENCES

Acharjee, A., Prasad, T. S. K., & Gowda, H. (2023). Integrated multiomics and machine learning approaches in nutrigenomics. *Briefings in Bioinformatics*, *24*(1), bbac582. Advance online publication. DOI: 10.1093/bib/bbac582 PMID: 36545800

Acharya, D., & Mukhopadhyay, A. (2024). A comprehensive review of machine learning techniques for multiomics data integration: Challenges and applications in precision oncology. *Briefings in Functional Genomics*, *23*(5), 549–560. DOI: 10.1093/bfgp/elae013 PMID: 38600757

Aggarwal, N., Kitano, S., Puah, G. R. Y., Kittelmann, S., Hwang, I. Y., & Chang, M. W. (2022). Microbiome and human health: Current understanding, engineering, and enabling technologies. *Chemical Reviews*, *123*(1), 31–72. DOI: 10.1021/acs.chemrev.2c00431 PMID: 36317983

Agrawal, K., Goktas, P., Kumar, N., & Leung, M. F. (2025). Artificial intelligence in personalized nutrition and food manufacturing: A comprehensive review of methods, applications, and future directions. *Frontiers in Nutrition*, *12*, 1636980. DOI: 10.3389/fnut.2025.1636980 PMID: 40771216

Alkhanbouli, R., Matar Abdulla Almadhaani, H., Alhosani, F., & Simsekler, M. C. E. (2025). The role of explainable artificial intelligence in disease prediction: A systematic literature review and future research directions. *BMC Medical Informatics and Decision Making*, *25*(1), 110. DOI: 10.1186/s12911-025-02944-6 PMID: 40038704

Amabebe, E., Robert, F. O., Agbalalah, T., & Orubu, E. S. (2020). Microbial dysbiosis-induced obesity: Role of gut microbiota in homoeostasis of energy metabolism. *British Journal of Nutrition*, *123*(10), 1127–1137. DOI: 10.1017/s0007114520000380 PMID: 32008579

Amiri, M., Sarani Rad, F., & Li, J. (2024). Delighting Palates with AI: Reinforcement Learning's Triumph in Crafting Personalized Meal Plans with High User Acceptance. *Nutrients*, *16*(3), 346. DOI: 10.3390/nu16030346 PMID: 38337630

Arif, A., Garg, P., & Srivastava, P. (2024). Microbiome-Gut-Brain Axis: AI Insights. *Insights in Biology and Medicine, 8*(1), 001-010. https://dx.doi.org/DOI: 10.29328/journal.ibm.1001027

Arıkan, M., & Muth, T. (2023). Integrated multiomics analyses of microbial communities: A review of the current state and future directions. *Molecular Omics*, *19*(8), 607–623. DOI: 10.1039/D3MO00089C PMID: 37417894

Arshad, M. T., Ali, M. K. M., Maqsood, S., Ikram, A., Ahmed, F., Aljameel, A. I., Al-Farga, A., & Hossain, M. S. (2025). Personalized Nutrition in the Era of Digital Health: A New Frontier for Managing Diabetes and Obesity. *Food Science & Nutrition, 13*(10), e71006. DOI: 10.1002/fsn3.71006 PMID: 41019177

Asnicar, F., Berry, S. E., Valdes, A. M., Nguyen, L. H., Piccinno, G., Drew, D. A., & Segata, N. (2021). Microbiome connections with host metabolism and habitual diet from 1,098 deeply phenotyped individuals. *Nature Medicine, 27*(2), 321–332. DOI: 10.1038/s41591-020-01183-8 PMID: 33432175

Augustine, T., Kumar, M., Al Khodor, S., & van Panhuys, N. (2023). Microbial dysbiosis tunes the immune response toward allergic disease outcomes. *Clinical Reviews in Allergy & Immunology, 65*(1), 43–71. DOI: 10.1007/s12016-022-08939-9 PMID: 35648372

Bakir-Gungor, B., Temiz, M., Jabeer, A., Wu, D., & Yousef, M. (2023). microBiomeGSM: The identification of taxonomic biomarkers from metagenomic data using grouping, scoring and modeling (GSM) approach. *Frontiers in Microbiology, 14*, 1264941. DOI: 10.3389/fmicb.2023.1264941 PMID: 38075911

Banumathi, K., Latha, V., Sonia, B. L., Vijayalakshmi, K., & Sathya, S. N. (2025). Reinforcement Learning in Personalized Medicine: A Comprehensive Review of Treatment Optimization Strategies. *Cureus, 17*(4). Advance online publication. DOI: 10.7759/cureus.82756 PMID: 40406780

Baranwal, M., Clark, R. L., Thompson, J., Sun, Z., Hero, A. O., & Venturelli, O. S. (2022). Recurrent neural networks enable design of multifunctional synthetic human gut microbiome dynamics. *eLife, 11*, e73870. DOI: 10.7554/eLife.73870 PMID: 35736613

Behera, B., Irshad, A., Rida, I., & Shabaz, M. (2025). AI-driven predictive modeling for disease prevention and early detection. *SLAS Technology, 31*. Advance online publication. DOI: 10.1016/j.slast.2025.100263 PMID: 40032202

Bermingham, K. M., May, A., Asnicar, F., Capdevila, J., Leeming, E. R., Franks, P. W., & Berry, S. E. (2024). Snack quality and snack timing are associated with cardiometabolic blood markers: The ZOE PREDICT study. *European Journal of Nutrition, 63*(1), 121–133. DOI: 10.1007/s00394-023-03241-6 PMID: 37709944

Berry, S., Mazidi, M., Franks, P., Valdes, A., Sattar, N., Wolf, J., Hadjigeorgiou, G., Drew, D., Chan, A., Segata, N., Asnicar, F., Spector, T., & Hall, W. (2020). *Impact of postprandial lipemia and glycemia on inflammatory factors in over 1000 individuals in the US and UK: Insights from the PREDICT 1 and InterCardio studies.* Current Developments in Nutrition, 4(Suppl 2), nzaa068_003. https://doi.org/DOI: 10.1093/cdn/nzaa068_003

Berry, S. E., Valdes, A. M., Drew, D. A., Asnicar, F., Mazidi, M., Wolf, J., & Spector, T. D. (2020). Human postprandial responses to food and potential for precision nutrition. *Nature Medicine*, *26*(6), 964–973. DOI: 10.1038/s41591-020-0934-0 PMID: 32528151

Bharti, R., & Grimm, D. G. (2021). Current challenges and best-practice protocols for microbiome analysis. *Briefings in Bioinformatics*, *22*(1), 178–193. DOI: 10.1093/bib/bbz155 PMID: 31848574

Bosco, N., & Noti, M. (2021). The aging gut microbiome and its impact on host immunity. *Genes and Immunity*, *22*(5), 289–303. DOI: 10.1038/s41435-021-00126-8 PMID: 33875817

Carolak, E., Czajkowska, J., Stypułkowska, A., Waszczuk, W., Dutkiewicz, A., & Grzymajlo, K. (2025). Being a better version of yourself: Genetically engineered probiotic bacteria as host defense enhancers in the control of intestinal pathogens. *Gut Microbes*, *17*(1), 2519696. DOI: 10.1080/19490976.2025.2519696 PMID: 40530826

Chong, K. P., & Woo, B. K. (2021). Emerging wearable technology applications in gastroenterology: A review of the literature. *World Journal of Gastroenterology*, *27*(12), 1149. DOI: 10.3748/wjg.v27.i12.1149 PMID: 33828391

Coker, O. O., Liu, C., Wu, W. K. K., Wong, S. H., Jia, W., Sung, J. J., & Yu, J. (2022). Altered gut metabolites and microbiota interactions are implicated in colorectal carcinogenesis and can be noninvasive diagnostic biomarkers. *Microbiome*, *10*(1), 35. DOI: 10.1186/s40168-021-01208-5 PMID: 35189961

Corella, D., & Ordovas, J. M. (2014). Nutrigenomics in cardiovascular medicine. *Circulation: Cardiovascular Genetics*, *7*(2), 159–173. DOI: 10.1161/CIRCGENETICS.113.000129 PMID: 20031645

Corella, D., & Ordovás, J. M. (2020). Can genotype be used to tailor treatment of obesity? *Nutrition Reviews*, *78*(2), 145–158. DOI: 10.1093/nutrit/nuz082 PMID: 31532485

Cox, T. O., Lundgren, P., Nath, K., & Thaiss, C. A. (2022). Metabolic control by the microbiome. *Genome Medicine*, *14*(1), 80. DOI: 10.1186/s13073-022-01092-0 PMID: 35906678

Cui, Z., Wu, Y., Zhang, Q. H., Wang, S. G., He, Y., & Huang, D. S. (2023). MV-CVIB: A microbiome-based multiview convolutional variational information bottleneck for predicting metastatic colorectal cancer. *Frontiers in Microbiology*, *14*, 1238199. DOI: 10.3389/fmicb.2023.1238199 PMID: 37675425

D'Amato, A., Di Cesare Mannelli, L., Lucarini, E., Ghelardini, C., & Pacini, A. (2020). Personalized nutrition and gut microbiota: Where are we now? *Nutrients*, *12*(10), 3080. DOI: 10.3390/nu12103080 PMID: 33050315

Davis, K. (2023, April 25). *Microbiome tests: What to know. Medical News Today*. https://www.medicalnewstoday.com/articles/microbiome-testing

De Pessemier, B., Grine, L., Debaere, M., Maes, A., Paetzold, B., & Callewaert, C. (2021). Gut–skin axis: Current knowledge of the interrelationship between microbial dysbiosis and skin conditions. *Microorganisms*, *9*(2), 353. DOI: 10.3390/microorganisms9020353 PMID: 33670115

Detopoulou, P., Voulgaridou, G., Moschos, P., Levidi, D., Anastasiou, T., Dedes, V., Diplari, E.-M., Fourfouri, N., Giaginis, C., Panoutsopoulos, G. I., & Papadopoulou, S. K. (2023). Artificial intelligence, nutrition, and ethical issues: A mini-review. *Clinical Nutrition Open Science*, *50*, 46–56. DOI: 10.1016/j.nutos.2023.07.001

Digital Watch Observatory. (2025). *Egypt's national AI strategy – Second edition 2025–2030.* MCIT/Egypt. https://dig.watch/resource/egypts-national-ai-strategy-second-edition-2025-2030

El Arab, R. A., Abu-Mahfouz, M. S., Abuadas, F. H., Alzghoul, H., Almari, M., Ghannam, A., & Seweid, M. M. (2025, March). Bridging the Gap: From AI Success in Clinical Trials to Real-World Healthcare Implementation—A Narrative Review. In *Healthcare* (Vol. 13, No. 7, p. 701). MDPI. https://doi.org/DOI: 10.3390/healthcare13070701

Feero, W. G., & Guttmacher, A. E. (2021). Genomic medicine—An updated primer. *The New England Journal of Medicine*, *384*(10), 966–976. DOI: 10.1056/NEJMra1802873 PMID: 20505179

Ferreira, D. D., Ferreira, L. G., Amorim, K. A., Delfino, D. C. T., Ferreira, A. C. B. H., & Souza, L. P. C. E. (2025). Assessing the Links Between Artificial Intelligence and Precision Nutrition. *Current Nutrition Reports*, *14*(1), 47. DOI: 10.1007/s13668-025-00635-2 PMID: 40087237

Forslund, S. K., Chakaroun, R., Zimmermann-Kogadeeva, M., Markó, L., Aron-Wisnewsky, J., Nielsen, T., & Bäckhed, F. (2023). Microbiome-host interactions in personalized nutrition and metabolic health. *Nature Reviews. Endocrinology*, *19*, 27–44. DOI: 10.1038/s41574-022-00672-5

Fujisaka, S., Watanabe, Y., & Tobe, K. (2023). The gut microbiome: A core regulator of metabolism. *The Journal of Endocrinology*, *256*(3). Advance online publication. DOI: 10.1530/joe-22-0111 PMID: 36458804

Galloway-Peña, J., & Hanson, B. (2020). Tools for Analysis of the Microbiome. *Digestive Diseases and Sciences*, *65*, 674–685. DOI: 10.1007/s10620-020-06091-y PMID: 32002757

German, J. B., Zivkovic, A. M., & Barile, D. (2011). Nutrigenomics and personalized diets: What will they mean for food? *Annual Review of Food Science and Technology*, *2*, 97–123. DOI: 10.1146/annurev.food.102308.124155 PMID: 22129377

Ghadermazi, P., & Chan, S. H. J. (2024). Microbial interactions from a new perspective: Reinforcement learning reveals new insights into microbiome evolution. *Bioinformatics (Oxford, England)*, *40*(1), btae003. Advance online publication. DOI: 10.1093/bioinformatics/btae003 PMID: 38212999

Ghanem, M., Ghaith, A. K., & Bydon, M. (2024). Artificial intelligence and personalized medicine: transforming patient care. In *The New Era of Precision Medicine* (pp. 131–142). Academic Press., DOI: 10.1016/b978-0-443-13963-5.00012-1

Goetz, L. H., & Schork, N. J. (2018). Personalized medicine: motivation, challenges, and progress. *Fertility and sterility, 109*(), 952-963. https://doi.org/DOI: 10.1016/j.fertnstert.2018.05.006

Haldar, S., Stein-Thoeringer, C., & Borisov, V. (2024). Interpreting Microbiome Relative Abundance Data Using Symbolic Regression. *arXiv preprint arXiv:2410.16109*. https://doi.org//arXiv.2410.16109DOI: 10.48550

Hasin, Y., Seldin, M., & Lusis, A. (2017). Multiomics approaches to disease. *Genome Biology*, *18*, 83. DOI: 10.1186/s13059-017-1215-1 PMID: 28476144

Hernández-Calderón, P., Wiedemann, L., & Benítez-Páez, A. (2022). The microbiota composition drives personalized nutrition: Gut microbes as predictive biomarkers for the success of weight loss diets. *Frontiers in Nutrition*, *9*, 1006747. DOI: 10.3389/fnut.2022.1006747 PMID: 36211501

Hirten, R. P., Danieletto, M., Sanchez-Mayor, M., Whang, J. K., Lee, K. W., Landell, K., & Sands, B. E. (2025). Physiological Data Collected From Wearable Devices Identify and Predict Inflammatory Bowel Disease Flares. *Gastroenterology*. Advance online publication. DOI: 10.1053/j.gastro.2024.12.024 PMID: 39826619

Ho, D., Quake, S. R., McCabe, E. R. B., Chng, W. J., Chow, E. K.-H., Ding, X., & Lee, Y. Y. (2020). Enabling technologies for personalized and precision medicine. *Trends in Biotechnology*, *38*(5), 497–518. DOI: 10.1016/j.tibtech.2019.12.021 PMID: 31980301

Hong, G., Hindle, V., Veasley, N. M., Holscher, H. D., & Kilicoglu, H. (2025). DiMB-RE: Mining the scientific literature for diet-microbiome associations. *Journal of the American Medical Informatics Association : JAMIA*, *32*(6), 998–1006. DOI: 10.1093/jamia/ocaf054 PMID: 40152137

Hosseiniara, S. M., & Hosseini Zijoud, S. S. (2024). Nutrigenomics: A promising frontier in chronic disease prevention. *Journal of Preventive and Complementary Medicine*, *3*(4), 195–200. DOI: 10.22034/jpcm.2024.496108.1196

Huang, S., Ailer, E., Kilbertus, N., & Pfister, N. (2023). Supervised learning and model analysis with compositional data. *PLoS Computational Biology*, *19*(6), e1011240. DOI: 10.1371/journal.pcbi.1011240 PMID: 37390111

Indian Council of Medical Research (ICMR). (2023). *Ethical guidelines for artificial intelligence in biomedical research and healthcare*. Indian Council of Medical Research. https://www.icmr.gov.in/icmrobject/custom_data/pdf/Ethical-guidelines/Ethical_Guidelines_AI_Healthcare_2023.pdf

Jenko, S., Papadopoulou, E., Kumar, V., Overman, S. S., Krepelkova, K., Wilson, J., Dunbar, E. L., Spice, C., & Exarchos, T. (2025). Artificial Intelligence in Healthcare: How to Develop and Implement Safe, Ethical and Trustworthy AI Systems. *AI*, *6*(6), 116. DOI: 10.3390/ai6060116

Jørgensen, J. T. (2019). Companion and complementary diagnostics: From biomarker discovery to clinical implementation. *Expert Review of Molecular Diagnostics*, *19*(6), 491–492. DOI: 10.1080/14737159.2019.1620152

Kallapura, G., Prakash, A. S., Sankaran, K., Manjappa, P., Chaudhary, P., Ambhore, S., & Dhar, D. (2024). Microbiota based personalized nutrition improves hyperglycemia and hypertension parameters and reduces inflammation: A prospective, open label, controlled, randomized, comparative, proof of concept study. *PeerJ*, *12*, e17583. DOI: 10.7717/peerj.17583 PMID: 38948211

Karczewski, K. J., & Snyder, M. P. (2018). Integrative omics for health and disease. *Nature Reviews. Genetics*, *19*(5), 299–310. DOI: 10.1038/nrg.2018.4 PMID: 29479082

Karkera, N., Acharya, S., & Palaniappan, S. K. (2023). Leveraging pretrained language models for mining microbiome-disease relationships. *BMC Bioinformatics*, *24*(1), 290. DOI: 10.1186/s12859-023-05411-z PMID: 37468830

Kassem, H., Beevi, A. A., Basheer, S., Lutfi, G., Cheikh Ismail, L., & Papandreou, D. (2025). Investigation and assessment of AI's role in nutrition—An updated narrative review of the evidence. *Nutrients*, *17*(1), 190. DOI: 10.3390/nu17010190 PMID: 39796624

Kassem, N. M., Abdelmegid, Y. A., El-Sayed, M. K., Sayed, R. S., Abdel-Aalla, M. H., & Kassem, H. A. (2023). Nutrigenomics and microbiome shaping the future of personalized medicine: A review article. *Journal of Genetic Engineering and Biotechnology*, *21*(1), 134. DOI: 10.1186/s43141-023-00599-2 PMID: 37993702

Kleine Bardenhorst, S., Berger, T., Klawonn, F., Vital, M., Karch, A., & Rübsamen, N. (2021). Data analysis strategies for microbiome studies in human populations—A systematic review of current practice. *mSystems*, *6*(1), 10–1128. DOI: 10.1128/msystems.01154-20 PMID: 33622856

Kok, D. E. G.. (2025). Future directions in precision nutrition: Integrating omics and microbiome data. *Nutrients*, *17*(1), 65. DOI: 10.3390/nu17010065 PMID: 41515183

Konnoth, C. (2024). AI and data protection law in health. In Solaiman, B., & Cohen, I. G. (Eds.), *Research Handbook on Health, AI and the Law*. Edward Elgar Publishing., DOI: 10.4337/9781802205657.ch07

Kool, J., Tymchenko, L., Shetty, S. A., & Fuentes, S. (2023). Reducing bias in microbiome research: Comparing methods from sample collection to sequencing. *Frontiers in Microbiology*, *14*, 1094800. DOI: 10.3389/fmicb.2023.1094800 PMID: 37065158

Kumar, B., Lorusso, E., Fosso, B., & Pesole, G. (2024). A comprehensive overview of microbiome data in the light of machine learning applications: Categorization, accessibility, and future directions. *Frontiers in Microbiology*, *15*, 1343572. 10.3389v/fmicb.2024.1343572 PMID: 38419630

Kuziel, G. A., & Rakoff-Nahoum, S. (2022). The gut microbiome. *Current Biology*, *32*(6), R257–R264. DOI: 10.1016/j.cub.2022.02.023 PMID: 35349808

Lagoumintzis, G., & Patrinos, G. P. (2023). Triangulating nutrigenomics, metabolomics and microbiomics toward personalized nutrition and healthy living. *Human Genomics*, *17*(1), 109. DOI: 10.1186/s40246-023-00561-w PMID: 38062537

Lamsaf, A., Carrilho, R., Neves, J. C., & Proença, H. (2025). Causality, Machine Learning, and Feature Selection: A Survey. *Sensors (Basel)*, *25*(8), 2373. DOI: 10.3390/s25082373 PMID: 40285063

Liñares-Blanco, J., Fernandez-Lozano, C., Seoane, J. A., & López-Campos, G. (2022). Machine learning based microbiome signature to predict inflammatory bowel disease subtypes. *Frontiers in Microbiology*, *13*, 872671. DOI: 10.3389/fmicb.2022.872671 PMID: 35663898

Liu, Y., Lau, H. C. H., & Yu, J. (2023). Microbial metabolites in colorectal tumorigenesis and cancer therapy. *Gut Microbes*, *15*(1), 2203968. DOI: 10.1080/19490976.2023.2203968 PMID: 37095682

Ma, S., Shungin, D., Mallick, H., Schirmer, M., Nguyen, L. H., Kolde, R., & Huttenhower, C. (2022). Population structure discovery in meta-analyzed microbial communities and inflammatory bowel disease using MMUPHin. *Genome Biology*, *23*(1), 208. DOI: 10.1186/s13059-022-02753-4 PMID: 36192803

Madan, S., Lentzen, M., Brandt, J., Rueckert, D., Hofmann-Apitius, M., & Fröhlich, H. (2024). Transformer models in biomedicine. *BMC Medical Informatics and Decision Making*, *24*(1), 214. DOI: 10.1186/s12911-024-02600-5 PMID: 39075407

Metwally, A. A., Yu, P. S., Reiman, D., Dai, Y., Finn, P. W., & Perkins, D. L. (2019). Utilizing longitudinal microbiome taxonomic profiles to predict food allergy via long short-term memory networks. *PLoS Computational Biology*, *15*(2), e1006693. DOI: 10.1371/journal.pcbi.1006693 PMID: 30716085

Meystre, S. van STIPHOUT, R., Goris, A., & Gaitan, S. (2023). AI-Based Gut-Brain Axis Digital Twins. In *Caring is Sharing–Exploiting the Value in Data for Health and Innovation* (pp. 1007-1008). IOS Press. https://doi.org/DOI: 10.3233/SHTI230327

Mienye, I. D., Obaido, G., Jere, N., Ebikella Mienye, I. D., Aruleba, K., Emmanuel, I. D., & Ogbuokiri, B. (2024). A survey of explainable artificial intelligence in healthcare: Concepts, applications, and challenges. *Informatics in Medicine Unlocked*, *51*, 101587. Advance online publication. DOI: 10.1016/j.imu.2024.101587

Ministry of Communications and Information Technology (MCIT). (2023). *Egyptian charter for responsible AI, v1.0*. MCIT. https://ai.gov.eg/SynchedFiles/en/Resources/EgyptianCharterForResponsibleAIEnglish-v1.0.pdf

Mohapatra, R. K., Jolly, L., & Dakua, S. P. (2025). Advancing explainable AI in healthcare: Necessity, progress, and future directions. *Computational Biology and Chemistry*. Advance online publication. DOI: 10.1016/j.compbiolchem.2025.108599 PMID: 40743677

Mousa, W. K., & Al Ali, A. (2024). The Gut Microbiome Advances Precision Medicine and Diagnostics for Inflammatory Bowel Diseases. *International Journal of Molecular Sciences*, *25*(20), 11259. DOI: 10.3390/ijms252011259 PMID: 39457040

Mukherjee, S., Khatun, R., Sengupta, S., & Bhattacharya, M. (2025). ARTIFICIAL INTELLIGENCE: IN PRECISION MEDICINE. [IJARET]. *Technology*, *16*(2). Advance online publication. DOI: 10.34218/IJARET_16_02_022

Naik, S., & Reddy, B. (2025). A framework for the governance of AI in healthcare for India: An exploratory model. *Indian Journal of Medical Ethics*, *6*(1). Advance online publication. DOI: 10.20529/IJME.2025.037

Nasef, D., Nasef, D., Sher, M., & Toma, M. (2025). A Standardized Validation Framework for Clinically Actionable Healthcare Machine Learning with Knee Osteoarthritis Grading as a Case Study. *Algorithms*, *18*(6), 343. DOI: 10.3390/a18060343

National Academies of Sciences, Engineering, and Medicine; Health and Medicine Division; Food and Nutrition Board; Food Forum; Callahan, E. A. (Ed.). (2022, April 15). *Challenges and opportunities for precision and personalized nutrition: Proceedings of a workshop (Chapter 3: Innovative Methodologies and Technologies)*. National Academies Press (US). https://doi.org/ stockton.primo.exlibrisgroup.com+9DOI: 10.17226/26299

Nielsen, D. E., & El-Sohemy, A. (2012). A randomized trial of genetic information for personalized nutrition. *Genes & Nutrition*, *7*(4), 559–566. DOI: 10.1007/s12263-012-0292-7 PMID: 22407352

Organization for Economic Co-operation and Development (OECD). (2024). *OECD artificial intelligence review of Egypt*. OECD Publishing., DOI: 10.1787/2a282726-

Ortlek, B. E., & Akan, O. B. (2024). Modeling and Analysis of SCFA-Driven Vagus Nerve Signaling in the Gut-Brain Axis via Molecular Communication. *arXiv preprint arXiv:2412.19945*. https://doi.org//arXiv.2412.19945DOI: 10.48550

Papastratis, I., Konstantinidis, D., Daras, P., & Dimitropoulos, K. (2024). AI nutrition recommendation using a deep generative model and ChatGPT. *Scientific Reports*, *14*(1), 14620. DOI: 10.1038/s41598-024-65438-x PMID: 38918477

Paramasivan, A. (2023). AI-Powered Population Health Management: Addressing Public Health Challenges with Predictive Insights. *International Journal on Science and Technology*, *14*(1), 1–18. DOI: 10.5281/zenodo.14551607

Pati, R., Mondal, S., & Ranjan, S. (2024). AI-integrated nutrigenomics: A step toward disease prevention and health optimization. *OMICS: A Journal of Integrative Biology*, *28*(2), 85–97. DOI: 10.1089/omi.2024.0012

Paul, S., Barman, R. K., & Das, S. (2025). Decoding the role of nutrigenomics and gut microbiome in metabolic diseases: Current insights and future directions. *Genes & Diseases*. Advance online publication. DOI: 10.1016/j.gendis.2025.01.001

Peter, H. (2025). Artificial Intelligence and Predictive Modeling for Early Detection of Chronic Diseases in Healthcare Networks. https://www.researchgate.net/publication/391428110

Peterson, D., Bonham, K. S., Rowland, S., & Pattanayak, C. W.Resonance Consortium, & Klepac-Ceraj, V. (2021). Comparative analysis of 16S rRNA gene and metagenome sequencing in pediatric gut microbiomes. *Frontiers in Microbiology*, *12*, 670336. DOI: 10.3389/fmicb.2021.670336 PMID: 34335499

Phillips, C. M. (2013). Nutrigenetics and metabolic disease: Current status and implications for personalized nutrition. *Nutrients*, *5*(1), 32–57. DOI: 10.3390/nu5010032 PMID: 23306188

Porcari, S., Ng, S. C., Zitvogel, L., Sokol, H., Weersma, R. K., Elinav, E., Gasbarrini, A., Cammarota, G., Tilg, H., & Ianiro, G. (2025). The microbiome for clinicians. *Cell*, *188*(11), 2836–2844. DOI: 10.1016/j.cell.2025.04.016 PMID: 40446358

Przymus, P., Rykaczewski, K., Martín-Segura, A., Truu, J., Carrillo De Santa Pau, E., Kolev, M., Naskinova, I., Gruca, A., Sampri, A., Frohme, M., & Nechyporenko, A. (2025). Deep learning in microbiome analysis: A comprehensive review of neural network models. *Frontiers in Microbiology*, *15*, 1516667. DOI: 10.3389/fmicb.2024.1516667 PMID: 39911715

Raj, P. (2025). AI-Based Personalized Dietary Planning Tools. *International Journal of Forensic Medicine and Research*, *7*(2). Advance online publication. DOI: 10.36948/ijfmr.2025.v07i02.41432

Reiman, D., Metwally, A., & Dai, Y. (2017, July). Using convolutional neural networks to explore the microbiome. In *2017 39th annual international conference of the IEEE engineering in medicine and biology society (EMBC)* (pp. 4269-4272). IEEE. https://doi.org/DOI: 10.1109/EMBC.2017.8037799

Ren, F., Wei, J., Chen, Q., Hu, M., Yu, L., Mi, J., Zhou, X., Qin, D., Wu, J., & Wu, A. (2025). Artificial intelligence-driven multiomics approaches in Alzheimer's disease: Progress, challenges, and future directions. *Acta Pharmaceutica Sinica. B*. Advance online publication. DOI: 10.1016/j.apsb.2025.07.030 PMID: 41049729

Reyed, R.M. (2021a). Live biotherapeutic probiotics human microbiome: Machine learning *in silico* clinical trials. Novel practical approach to the era of personalized probiotics medicine. In Proceedings of the 4th European Summer School on Nutrigenomics (ESSN 2021), June 21–25, 2021 (*Lifestyle Genomics, 14*(3), 91-116). https://karger.com/lfg/article-pdf/14/3/91/3699066/000517609.pdf. DOI: 10.1159/000517609

Reyed, R. M. (2021b). The short-term impact of probiotic intervention on the human gut microbiome facilitates fermentation potential of indigestible carbohydrates to short-chain fatty acids for preventing and treatment of obesity. *Journal of Food and Clinical Nutrition, 4*(1), 1. https://www.pulsus.com/abstract/the-shortterm-impact-of-probiotic-intervention-on-the-human-gut-microbiome-facilitate-fermentation-potential-of-indigest-7559.html

Reyed, R.M. (2023a). Simulate the dynamic interactions across nutritional supplements, the polybiome, and the biochemical approach dependent on patient derived gastrointestinal knowledge. *Journal of Clinical Epidemiology & Toxicology, 4*, 133. DOI: 10.47363/JCET/2023(4)133

Reyed, R. M. (2023b). Focusing on individualized nutrition within the algorithmic diet: An in depth look at recent advances in nutritional science, microbial diversity studies, and human health. *Food & Health, 5*(1), 5. DOI: 10.53388/FH2023005

Reyed, R.M., & Prabhakar, P. K. M. (2025a). Polybiome Derived Therapies and Fermentobiometrics: Novel Strategies for Blocking Cancer Pathways and Progression. In Cancer Etiology, Diagnosis and Treatments (pp. 145-212). Nova Medicine and Health/Pathways to Healing: Advancements in Cancer Therapy. /DOI: 10.52305/JKMU3682

Reyed, R. M., & Prabhakar, P. K. (2026). Modulating Neuroinflammation and Autophagy in Psychiatric Disorders Through Mechanisms and Therapies: Therapeutic Strategies and Pathways. In Mokhtari, T., & Uludag, K. (Eds.), *Autophagy and Inflammation in Neuropsychological Disorders* (pp. 61–124). IGI Global Scientific Publishing. DOI: 10.4018/979-8-3693-5908-2.ch003

Rouskas, K., Guela, M., Pantoura, M., Pagkalos, I., Hassapidou, M., Lalama, E., & Argiriou, A. (2025). The Influence of an AI-Driven Personalized Nutrition Program on the Human Gut Microbiome and Its Health Implications. *Nutrients, 17*(7), 1260. DOI: 10.3390/nu17071260 PMID: 40219016

Sales, N. M. R., Pelegrini, P. B., & Goersch, M. C. (2014). Nutrigenomics: Definitions and advances of this new science. *Journal of Nutrition and Metabolism, 202759*. Advance online publication. DOI: 10.1155/2014/202759 PMID: 24795820

Sanidad, K. Z., & Zeng, M. Y. (2020). Neonatal gut microbiome and immunity. *Current Opinion in Microbiology, 56*, 30–37. DOI: 10.1016/j.mib.2020.05.011 PMID: 32634598

Santaló, J., & Berdasco, M. (2022). Ethical implications of epigenetics in the era of personalized medicine. *Clinical Epigenetics, 14*(1), 44. DOI: 10.1186/s13148-022-01263-1 PMID: 35337378

Sarkar, E., & Maniatakos, M. (2021). Trapdoor: Repurposing backdoors to detect dataset bias in machine learning-based genomic analysis. *arXiv preprint arXiv:2108.10132*. https://doi.org//arXiv.2108.10132DOI: 10.48550

Schork, N. J. (2015). Personalized medicine: Time for one-person trials. *Nature, 520*(7549), 609–611. DOI: 10.1038/520609a PMID: 25925459

Shaman, J. A. (2024). The Future of Pharmacogenomics: Integrating Epigenetics, Nutrigenomics, and Beyond. *Journal of Personalized Medicine, 14*(12), 1121. DOI: 10.3390/jpm14121121 PMID: 39728034

Shao, Y., Lv, X., Ying, S., & Guo, Q. (2024). Artificial Intelligence-Driven Precision Medicine: Multi-Omics and Spatial Multi-Omics Approaches in Diffuse Large B-Cell Lymphoma (DLBCL). *Frontiers in Bioscience (Landmark Edition), 29*(12), 404. DOI: 10.31083/j.fbl2912404 PMID: 39735973

Sharma, N., & Gupta, D. (2023). Artificial intelligence in medicine: Navigating ethical challenges. *Journal of the Epidemiology Foundation of India, 1*(1), 36–37. https://efi.org.in/journal/index.php/JEFI/article/view/10

Shawky, H., Fayed, D. B., Abd El-Karim, S. S., Rezk, H., Esawy, M. A., & Farrag, E. K. (2025). Immunotherapeutic effects of de novo benzimidazole derivative and prebiotic bacterial levan against triple-negative breast tumors by harnessing the immune landscape to intercept the oncogenic transcriptome. *International Journal of Biological Macromolecules, 289*, 138844. DOI: 10.1016/j.ijbiomac.2024.138844 PMID: 39701264

Shi, Y., Zhang, L., Do, K. A., Jenq, R., & Peterson, C. B. (2023). Sparse tree-based clustering of microbiome data to characterize microbiome heterogeneity in pancreatic cancer. *Journal of the Royal Statistical Society. Series C, Applied Statistics*, *72*(1), 20–36. DOI: 10.1093/jrsssc/qlac002 PMID: 37034187

Singh, M., Kumar, A., Khanna, N. N., Laird, J. R., Nicolaides, A., Faa, G., & Suri, J. S. (2024). Artificial intelligence for cardiovascular disease risk assessment in personalized framework: A scoping review. *EClinicalMedicine*, *73*. Advance online publication. DOI: 10.1016/j.eclinm.2024.102660 PMID: 38846068

Singh, R., Zogg, H., Wei, L., Bartlett, A., Ghoshal, U. C., Rajender, S., & Ro, S. (2021). Gut microbial dysbiosis in the pathogenesis of gastrointestinal dysmotility and metabolic disorders. *Journal of Neurogastroenterology and Motility*, *27*(1), 19. DOI: 10.5056/jnm20149 PMID: 33166939

Suárez, R., Bautista-Valarezo, E., Matos, A., Calderón, P., Fascì-Spurio, F., Castano-Jimenez, J., Zambrano-Villacres, R., Sarno, G., & Frias-Toral, E. (2025). Obesity and nutritional strategies: Advancing prevention and management through evidence-based approaches. *AIDS Care*. Advance online publication. DOI: 10.1080/09540105.2025.2491597

Thapa, C., & Camtepe, S. (2021). Precision health data: Requirements, challenges and existing techniques for data security and privacy. *Computers in Biology and Medicine*, *129*, 104130. DOI: 10.1016/j.compbiomed.2020.104130 PMID: 33271399

Thompson, J. C., Zavala, V. M., & Venturelli, O. S. (2023). Integrating a tailored recurrent neural network with Bayesian experimental design to optimize microbial community functions. *PLoS Computational Biology*, *19*(9), e1011436. DOI: 10.1371/journal.pcbi.1011436 PMID: 37773951

Topçuoğlu, B. D., Lesniak, N. A., Ruffin, M. T.IV, Wiens, J., & Schloss, P. D. (2020). A framework for effective application of machine learning to microbiome-based classification problems. *mBio*, *11*(3), 10–1128. DOI: 10.1128/mbio.00434-20 PMID: 32518182

Torkzadehmahani, R., Nasirigerdeh, R., Blumenthal, D. B., Kacprowski, T., List, M., Matschinske, J., ... & Baumbach, J. (2022). Privacy-preserving artificial intelligence techniques in biomedicine. *Methods of information in medicine, 61*(S 01), e12-e27. https://doi.org/DOI: 10.1055/s-0041-1740630

Ullmann, T., Peschel, S., Finger, P., Müller, C. L., & Boulesteix, A. L. (2023). Over-optimism in unsupervised microbiome analysis: Insights from network learning and clustering. *PLoS Computational Biology*, *19*(1), e1010820. DOI: 10.1371/journal.pcbi.1010820 PMID: 36608142

UNESCO. (2025, June 11). *Ethics in action: India continues its journey toward ethical AI with the final RAM consultation in New Delhi.* UNESCO. https://www.unesco.org/en/articles/ethics-action-india-continues-its-journey-towards-ethical-ai-final-ram-consultation-new-delhi

U.S. Food and Drug Administration (FDA). **(2023).** Proposed Regulatory Framework for Modifications to AI/ML-Based Software as a Medical Device (SaMD). https://www.fda.gov/media/122535/download

van Dinter, J., Sinha, R., & Weinstock, G. M. (2022). Multiomics for personalized nutrition and health. *Trends in Genetics*, *38*(8), 787–800. DOI: 10.1016/j.tig.2022.05.002 PMID: 35490031

Vimbi, V., Shaffi, N., & Mahmud, M. (2024). Interpreting artificial intelligence models: A systematic review on the application of LIME and SHAP in Alzheimer's disease detection. *Brain Informatics*, *11*(1), 10. DOI: 10.1186/s40708-024-00222-1 PMID: 38578524

Voola, P. K., Pakanati, D., Cherukuri, H., & Goel, P. (2024). Ethical AI in healthcare: Balancing innovation with privacy and compliance. *Shodh Sagar*, 12(3), 97–108. https://dira.shodhsagar.com/index.php/j/article/download/97/108/202

Vuyisich, M., Antoine, G., Mehrtash, M., Keiser, H., Connell, J., Ogundijo, O., Shen, N., Julian, C., Garimella, M., Banavar, G., & Tanton, D. (2024). Data-Driven Personalized Nutrition Reduces HbA1c in a Prediabetic and Diabetic Population – a Pilot, Randomized, Placebo-Controlled Clinical Trial. *Preprints*. https://doi.org/DOI: 10.20944/preprints202408.1740.v1

Wang, Q., Fan, X., Wu, S., & Su, X. (2024). PM-CNN: Microbiome status recognition and disease detection model based on phylogeny and multipath neural network. *Bioinformatics Advances*, *4*(1), vbae013. Advance online publication. DOI: 10.1093/bioadv/vbae013 PMID: 38371919

Wang, X., Sun, Z., Xue, H., & An, R. (2025, June). Artificial Intelligence Applications to Personalized Dietary Recommendations: A Systematic Review. In *Healthcare* (Vol. 13, No. 12, p. 1417). MDPI. https://doi.org/DOI: 10.3390/healthcare13121417

Wheatley, M. C. (2024). Nutrigenomics: Tailoring Nutrition Based on Genes. *Genetics*, *1*, 100002. DOI: 10.70389/PJG.100002

Wichmann, A., Buschong, E., Müller, A., Jünger, D., Hildebrandt, A., Hankeln, T., & Schmidt, B. (2023). MetaTransformer: Deep metagenomic sequencing read classification using self-attention models. *NAR Genomics and Bioinformatics*, *5*(3), lqad082. Advance online publication. DOI: 10.1093/nargab/lqad082 PMID: 37705831

Chapter 3
Biointelligent Reframing of Colorectal Cancer:
From Somatic Mutation to Systemic Coherence

Mona A. Esawy
http://orcid.org/0000-0001-5306-4946
Pharmaceutical and Drug Industries Research Institute, Egypt

Reyed M. Reyed
http://orcid.org/0000-0001-5027-3478
Genetic Engineering and Biotechnology Research Institute, Egypt & City of Scientific Research and Technological Applications, Egypt

ABSTRACT

This chapter presents a critical analysis of the existing textual paradigm of linear somatic mutation in colorectal cancer (CRC) and suggests reinterpretation as a failure of distributed biological intelligence in the holobiome. It presents the idea of the cognitive oncology system (COS) as an ontological requirement, conceptualizing cancer as an adaptive process that is caused by the loss of systemic coherence along microbial, nutrigenomic, and metabolic axes. The framework integrates polybiome–nutrigenome analytics, biointelligent modeling and neuroecological organoid interfaces. Together, these factors form the basis of circular medicine, which assumes the systematic recycling of biological intelligence to attain therapeutic efficacy and secure national autonomy. This view is in line with Egypt Vision 2030, which promotes the development of scientific infrastructure at the forefront of a biointelligent civilization. The chapter ends with a translational blueprint connecting disease reframing to a sustainable national health strategy.

DOI: 10.4018/979-8-3373-3121-8.ch003

INTRODUCTION: COGNITIVE ONCOLOGY AS AN ONTOLOGICAL SHIFT.

The conceptual framework of colorectal cancer (CRC) has long been established on the basis of the principles of biomedical paradigms that recently shifted to include molecular genetics and pathological morphology (Zitvogel et al., 2025). Classical theories describe carcinogenesis as a slow process of accumulation of somatic mutations leading to cells being inappropriately adapted, which results in therapeutic interventions focused on oncogenic drivers, signaling cascades, and pharmacological vulnerabilities (Yang et al., 2025). However, the emergent complexity of CRC as a multiscalar ecological dysfunction, a disturbance that appears in microbial, metabolic, immunological, and informational planes and coevolves the human host, does not sufficiently explain such reductionist explanations. The failure of therapeutic interventions across the board, the plateauing effect of response rates by drugs, and the ineffectiveness of a single-size-fits-all mode of treatment all point to the necessity of an ontological change in the approach in which cancer is no longer viewed as a failed system but as an adaptive, self-organizing process working on a platform of a dysregulated biological network (Wang et al., 2025).

This is why cognitive oncology should be integrated: it redefines cancer not as a result of a linear journey of somatic mutations but as a failure of distributed cognitive intelligence, a breakdown of the ability of the biological system to perceive, make decisions and organize collective behavior. This broadened ecologically aware view identifies CRC as a consequence of systemic incoherence — a breakdown in communication between the elements of the holobiome, microbial communities, epithelial barriers, immunometabolic circuits and neuroendocrine signaling. These interactive failures require redefinition of CRC in a biointelligent paradigm where the pathogenesis of a disease is hypothesized to be affected by perturbation of the distributed cognitive functions of the biological system. This is consistent with the strategic national direction of Vision Egypt 2030, which specifically concerns health, sustainability, and the digital-biological interface as a single developmental ecosystem of scientific sovereignty in the long term.

The conceptual basis of this paradigm shift is biointelligence, which is represented in this chapter. CRC is a complex process involving biological, computational, and ecological systems, and the overall interpretation of CRC as a learning, dynamic, polybiomic ecosystem instead of a pathological one (Reyed & Prabhakar, 2025a). The model locates cancer on a continuum of microbial-metabolic-immune relationships and focuses on adaptive feedback loops, metabolic plasticity, and ecological communication failures as contributors to malignant transformation. Figure 1 integrates the biointelligence systems framework, which summarizes the key elements of this cohesive view within the context of cognitive systems biology, polybiomic ecology and the dynamics of tumor evolution. The graph is used to outline a multilevel feedback system of CRC and visualizes the relationships and interplay between the cognitive, bioinformatic and therapeutic layers and how they are destabilized and disrupted under the influence of pathological pressures (Cuisiniere et al., 2025; Karam et al., 2025; Wu et al., 2025).

Figure 1 serves as the intellect map of the chapter, providing a visual display of how the interplay between disrupted microbial consortia, disrupted fermento-biometric signaling, disrupted immune sensing and inability to control metabolic processes together results in the systemic degradation that is observed in CRC. With this applied systems-style framework at the beginning, the chapter brings an integrated prism through which the rest of the discourse may be perceived, including Cognitive Oncology Systems (COS), Polybiome–Nutrigenome interactions, organ-oid modeling, fermentabiometrics, and national-scale applications in the context of Vision 2030 in Egypt.

Figure 1.

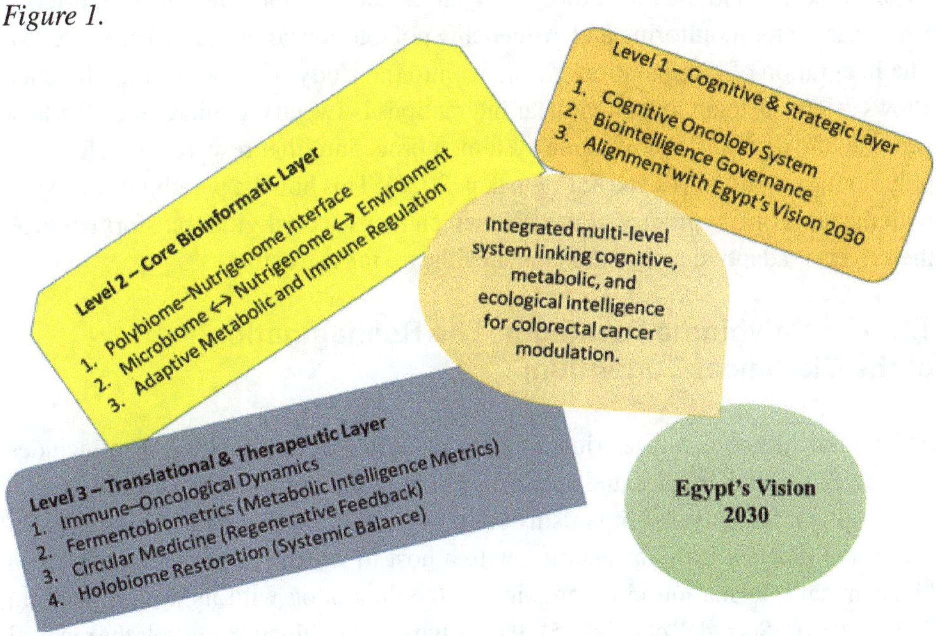

1.1. Genetic Determinism to Thinking of Biointelligent Systems.

Classical oncology has established drug-based therapy, which has historically contextualized causation via gene-based causal accounts and has made DNA mutations central to the causation of malignancy. These models are good because they have been able to explain what should happen but lack epistemological depth. The emergence of systems medicine and the development of multiomics have given new light to the fact that tumorigenesis is the result of a combination of genomic, epigenomic, metabolic, and microbial involvement, which constitute a network of self-sustaining feedbacks (Kim, et al., 2025). In this modern context, biointelligence is seen not only as data fusion but also as the natural ability of living things to sense, manipulate and adjust to information, suggesting that the biological system has an inherent cognitive architecture (Reyed & Prabhakar, 2026). In this context, cancer can be deemed the breakdown of such cognitive coherence whereby the cellular network becomes dissociated from its ecological and informational environment, thus turning a polybiome-mediated ecosystem into a pathophysiological condition. The polybiome concept supports the view that the microbiome, host genome, nutrigenome, and metabolome interact intelligently as opposed to separate compartments (Reyed & Prabhakar, 2025a; Silva, et al., 2025). Moreover, the fermentobiometrics methodological toolkit, which involves the quantitative mapping of microbial fermentation processes and the metabolomic signatures of microbial intelligence, offers a new measure for monitoring and influencing polybiomic cognition (Li et al., 2025). The integration of computational systems into the study of biological intelligence allows scientists to uncover nonlinear interactions between the microbiota, the host genome, the diet, and the immune system, a procedure that researchers refer to as polybiomic cognition (Reyed & Prabhakar, 2026). This holistic paradigm correlates with the aims of Egyptian Vision 2030, which focuses on digital transformation as the driver of adaptive and sustainable healthcare infrastructure.

1.2. The Polybiome Paradigm: The Reimagination of the Biological Continuum.

The polybiome is not restricted to the bacterial microbiome but also includes viruses, fungi, and archaea and networks of their metabolism with each other. It is not simply a collective of organisms but an extended intelligence, a biomatrix of connected biochemical conversation with a host in which it encodes and decodes biochemical information in an ongoing metabolic dialog with its host (Liu et al., 2022; Carrillo-Serradell et al., 2025). This polybiotic intelligence controls the renewal of the epithelia, immune activity, and metabolic homeostasis in the colon. When this communication fails as a result of nutritional imbalance, excessive antibiotic usage, or pollutant use, the dysregulation of polybiotics leads to systemic inflammation and genomic instability, which are antecedents of colorectal carcinogenesis (Takamaru et al., 2025).

This paradigm disputes the traditional ontology of cancer. Instead of using tumors as an isolated rogue population of cells, it places them in a more global context of systemic dissonance. The colon becomes a location of failed ecological negotiation, where microbial metabolites and host immune signaling generate feedback mechanisms of stress, healing, and discommunication (Reyed & Prabhakar, 2025a). The conceptualization of CRC as a biointelligent disorder helps us realize that the disorder can be cured only by reinstating the informational balance, as opposed to just eliminating malignant cells. This shift from pathogenic to eco-systemic oncology is the intellectual center of the Egyptian school of thought in biomedical systems, an epistemology that combines medical knowledge with environmental, nutritional and cognitive sustainability (Reyed, 2023b).

1.3 Methodological Framework: A Transdisciplinary Systems Approach.

The conceptual systems methodology used to analyze the chapter was a transdisciplinary approach based on synthesizing conceptual frameworks, systems modeling and bioinformatic inference. The epistemological goal is not to validate one causal process but rather to build an integrative interpretative framework in which the biological, computational, and ecological components of colorectal cancer (CRC) are discussed in the context of biointelligence. This method contrasts with reductive empiricism, since life processes are viewed as dynamic, cognitive and self-organizing information exchange systems and not mechanistic biochemical phenomena.

Recent multiomics studies have reiterated the importance of integrated molecular cartography in redefining our understanding of tumor ecosystems and patterns of adaptive signaling (Xu et al., 2025). These results support the shift from linear genetic determinism to network-based system cognition in oncological biology. Similarly, comprehensive microbial–genomic correlation analyses revealed that tumorigenesis arises as a result of the coadaptation of microbial, metabolic, and genetic subsystems (Zou et al., 2024). Simultaneously, computational modeling and data-driven integration have provided the epistemic basis of such biointelligent inference and, therefore, the ability to simulate complex host–microbiota–nutrigenome interactions (Karam et al 2025).

This methodology is, fundamentally, described in terms of four mutually supporting analytical levels that can be defined in relation to the different facets of the biointelligent paradigm:

1. Conceptual-Epistemological Analysis:

This tier is devoted to the reinvention of main constructs such as Polybiome, Fermentobiometrics, and Holobiome Restoration. Its methodological orientation is based on ontological synthesis and cognitive systems theory, which seek to reconceptualize the ontological synthesis of biological causation into an informational and adaptive process.

2. Systems and Network Analysis:

This level is used to map the interactions of microbial, metabolic and genetic systems. It incorporates systems thinking and multiomics integration to uncover the dynamic interactions and feedback loops that drive the evolution of colorectal cancer, as per the integrative genomic–microbial frameworks detailed by Zou et al. (2024)

3. Biointelligent Modeling:

This element focuses on the modeling of adaptive behaviors on the host-microbe-nutrigenome interfaces. The relationships between biological intelligence and metabolic and ecological adaptability are examined through computational reasoning and cognitive modeling to further detail the oncology strategies provided by Karam et al 2025.

4. Contextual-Strategic Analysis:

This tier coordinates the biointelligent model with the vision of sustainable health innovation by Egypt. It focuses on policy integration and socioecological analysis, ensuring that scientific frameworks are linked to national development, ethics, and ecological sustainability.

This framework acknowledges that the CRC cannot be fully described in terms of either gene-centric or cell-autonomous paradigms but is instead a result of complex adaptative interactions of microbial communities, metabolic feedback, and informational entropy. As such, the methodological position is conceptual and systemic but not strictly empirical. The peer-reviewed literature (2021--2025) was synthesized in this chapter to find theoretical coherence and no statistical generalization, and its validity relies on the depth of explanations, logicality, and transdisciplinaryism.

Using such analytical devices as comparative systems mapping at the domains of the microbiome, metabolome, and nutrigenome, the application of fermentobiometric indices (e.g., butyrate/acetate ratio and secondary bile acid ratio) as dynamic biomarkers of ecological communication and environmental simulations of cognitive oncology that model tumor behavior as a bioinformational process are used. Insight into organoid biology, synthetic bioengineering, and nutrigenomic profiling also supports these multiomics interpretations (Xu et al., 2025).

Finally, this methodological orientation represents the spirit of the Egyptian School of Biointelligent Medicine, where scientific research and discovery merge a blend of ecological knowledge, cultural sense, and technological accuracy. In this philosophy, method is not the instrument of measurement but the gesture of cognitive synthesis—a way of thought that is consistent with the spirit of the integration and regeneration of Vision 2030 in Egypt.

1.4. Fermentobiometrics The Metabolism of Hidden Language.

Fermentobiometrics is a young field that offers an appendage in terms of quantification between metabolic activities and biointelligent perception. Short-chain fatty acids (SCFAs), bile acids, and polyamines are a dynamic set of biomarkers of interactions between the host and microbe that are measured by this emerging science to coregulate. These molecules are informational intermediates of the gut that pass through biochemical messages that regulate immune tone, cell proliferation and epigenetic control (Liu et al., 2024; Cao et al., 2025). An elevated butyrate-to-acetate ratio, for example, is associated with improved mucosal structural maintenance and regulatory T cell-cells and with the dominance of secondary bile acids associated with oxidative stress and tumor-promoting signaling (Herlo et al., 2024; Saylam et al., 2025).

They do not represent inert byproducts in the context of a biointelligent model but rather a repertoire of metabolites by which the polybiome communicates with the host genome. On the basis of fermentobiometric data, artificial intelligence can be used to determine metabolism, which can provide an early indication of precancerous conditions; preventing oncology has become descriptive in biochemistry and predictive of cognition (Reyed & Prabhakar, 2025a; Reyed & Prabhakar, 2026). This strategy echoes the values of circular medicine, in which the waste of biology (metabolic byproducts) becomes a diagnostic tool, which is what sustainability in the ecological and medical dimensions is all about. The integrative relationships among key fermentobiometric markers, biointelligent signaling, and circular medicine sustainability are illustrated in Figure 2.

Figure 2.

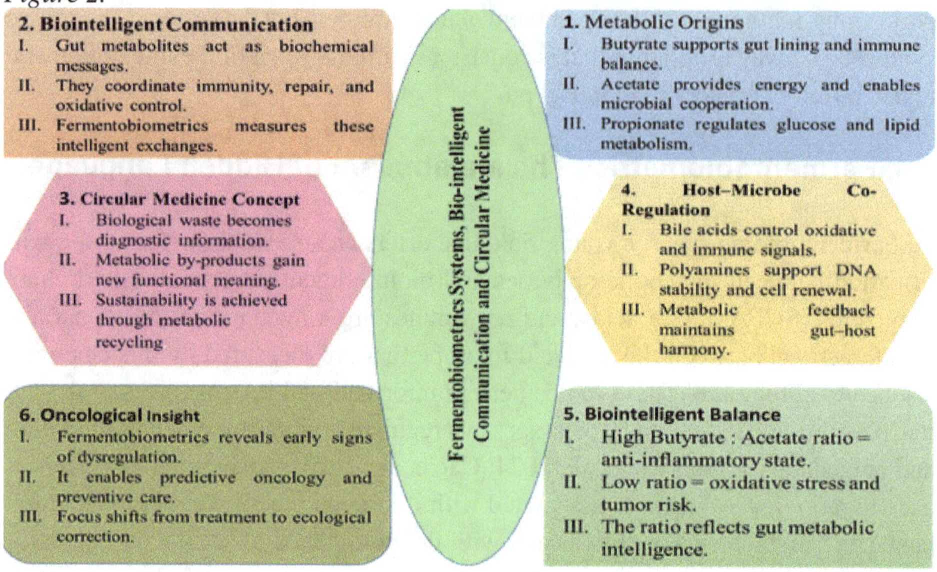

1.5 Case studies and clinical evidence: A comparative and vision-driven analysis

New multidisciplinary clinical reports support the empirical basis of the fermentobiometric paradigm, demonstrating the modulation of colorectal carcinogenesis by microbial–metabolite feedback systems on the basis of biochemical signaling networks.

Patients with colorectal cancer (CRC) in a Polish presurgical cohort presented significantly lower levels of fecal short-chain fatty acids (SCFAs), especially butyrate and propionate, than healthy controls did, which points to a breakdown in the fermentation balance of the microbial population (Kaźmierczak-Siedlecka et al., 2023). Such depletion represents early disruption of the immune homeostatic conversation that determines mucosal stability and regulatory T-cell activity, which is the core belief of the fermentobiometric model. Similarly, a longitudinal prospective study comparing serum bile acids and SCFAs with CRC incidence revealed that high levels of secondary bile acids and low levels of circulating SCFAs predict tumor progression in the future, supporting the notion that metabolic dysregulation is the precursor to morphological pathology (Loftfield, et al., 2022).

The Egyptian clinical setting offers initial indications within the local context, bringing global results into a localized metabolic ecology and dietary framework.

When Bacteroides and *Clostridium difficile* became dominant, a cross-sectional study involving 40 Egyptian patients with CRC and 40 healthy controls revealed that microbial diversity (Shannon index) was significantly decreased, indicating a high prevalence of polybiotic dysbiosis (Fawzy et al., 2025). Khattab et al. (2023) conducted an integrative multiomics analysis (metagenomics, metabolomics, and proteomics) of fecal samples from Egyptian CRC cohorts. They reported that the expression of conjugated bile acids (taurocholic and cholic acid) and increased polyamine/tryptophan metabolism, whereas a decrease in butyrate-producing taxa was noted (their study). The partial loss of SCFAs in fecal matrices, explained by rapid absorption in the colon, highlights the methodological gaps in existing metabolomic quantification pipelines. Together, these findings outline a practical axis between bile acids, polyamines, and microbial fermentation, defining the biochemical substrate of fermentobiometric intelligence in the Egyptian population.

Similarly, Abo-Hammam et al. (2024) employed shotgun metagenomic sequencing in postcolectomy Egyptian CRC patients and discovered dynamic changes in the microbiological network. Pathogenic taxa such as *Fusobacterium nucleatum* and *Bacteroides fragilis* remained detectable, whereas beneficial genera (Akkermansia and Bifidobacterium) were more abundant after surgical re-excision, suggesting partial ecological rebalancing. Analysis of functional pathways also revealed a disturbed route of bile acid transformation and energy metabolism markers indicative of postoperative reorganization of fermentative metabolism. Taken together, these results suggest that the state of host-microbiome metabolic crosstalk in Egyptian CRC patients is dynamic and can be influenced by nutritional, microbial, or biotechnological measures.

1.6 Critical Comparative Appraisal

The combination of Polish, international, and Egyptian studies reveals a common biological insight: fermentative metabolites are not merely passive byproducts of metabolism but also active biological agents that maintain cellular and immune harmony by functioning as informational intermediates. However, comparative analysis reveals structural research gaps, such as the lack of standardized fermentobiometric reference panels and population-specific baseline datasets. The small sample sizes and cross-sectional designs used in Egyptian pilot studies (Khattab et al., 2023; Abo-Hammam et al., 2024) limit statistical generalization and study harmonization. Moreover, fecal SCFA measurements are statistically insufficient because of differences in absorption kinetics; combining plasma and mucosal SCFA analyses, alongside dynamic bile acid assessments, is essential for establishing reliable systemic biomarkers. These methodological discrepancies represent a short-term research challenge in translating fermentobiometrics to precision oncology.

1.7 Integrative Therapeutic and Future Research Vision

To address these gaps, future Egyptian and regional research efforts should adopt a transomics longitudinal paradigm that integrates metagenomic, metabolomic, and clinical measurements across varying ecological and dietary environments. By establishing a national fermentobiometric registry, it would be possible to calculate population-specific cutoff values for critical ratios, including the butyrate/acetate index and the secondary bile acid ratio, which could then be used to develop early detection algorithms and posttreatment surveillance. A combination with cognitive oncology systems (Reyed & Prabhakar, 2025a; Reyed & Prabhakar, 2026) would additionally support the analytical modeling of microbial–metabolite–host connections across oncologic and neurobiological systems, advancing the translational vision of biointelligent medicine in line with Egypt's Vision 2030.

This approach aligns with other national objectives of sustainable health innovation, digital transformation, and bioeconomic development as part of the broader framework of Egypt's Vision 2030. By merging fermentobiometrics with the principles of circular medicine (where metabolic waste is transformed into diagnostic data), Egypt can become the first to establish a regenerative biomedical framework in which the polybiome is regarded not merely as a diagnostic object but also as a therapeutic partner.

1.8. Cognitive Oncology Systems: to a Living Disease.

The young field of cognitive oncology systems (COSs), as a discipline, views the tumor ecosystem as a dynamic computer system. COS involves the modeling of bioinformatic and organoid-simulated models as well as pattern recognition by AI to replicate the adaptive behavior of living tissues. With a longitudinal patient dataset, including the genomic sequences, microbial profiles and metabolic fluxes of data, researchers can build virtual models of the disease that are capable of self-learning (Calvino et al., 2025). The cognitive systems not only predict tumor progression but also have the ability to predict the therapeutic interventions that could be used, and the most adaptive solutions of the intervention are identified in silico before they can be applied in the clinic. This overlap of computation and biology represents the emergence of a different epistemology, biocybernetic medicine, where rather than studying biological systems, the biological systems are cognitively interactive (Tiwari, et al., 2025). The Vision 2030 of Egypt prioritizes such digital infrastructures and places the country in the position of becoming the first country to introduce AI through the ACCA, which molds the union of laboratory research, clinical practice, and national data ecosystems (Reyed & Prabhakar, 2026, Senthil et al., 2023).

1.9. Cure to Coherence: The Strategy of Restoring the Holobiome.

The drawbacks of curative models determine the need to shift the paradigm of the Holobiome Restoration approach, in which healing is viewed as the balance of ecological and informational support. The Holobiome refers to all symbiotic organisms, including human, microbial, and environmental ones, involved in the maintenance of systemic homeostasis. This restoration can be described in colorectal cancer and can incorporate three convergent axes: microbial modulation, nutrigenomic reprogramming, and biointelligent feedback monitoring:

1. Microbial modulation complements targeted probiotics, prebiotics and microbiota transplantation on the basis of patient-specific polybiomic signatures.
2. Nutrigenomic remodeling combines the diet as a bioalgorithm, retuning the expression of genes by resetting them via nutritionally induced epigenetic modifications.
3. Biological intelligence monitoring uses machine learning to extract meaning from real-time fermentobiometric and immunometabolic signals to make dynamic adjustments to therapeutic decisions.

This three-tiered model heralds the shift in reactive intervention to adaptive eco-therapy, a philosophy that is engraved in the sustainable health of Egypt, where medical innovation is balanced by environmental ethics and community needs (Reyed & Prabhakar, 2025a; Liu et al., 2024; Cao et al., 2025).

Figure 3.

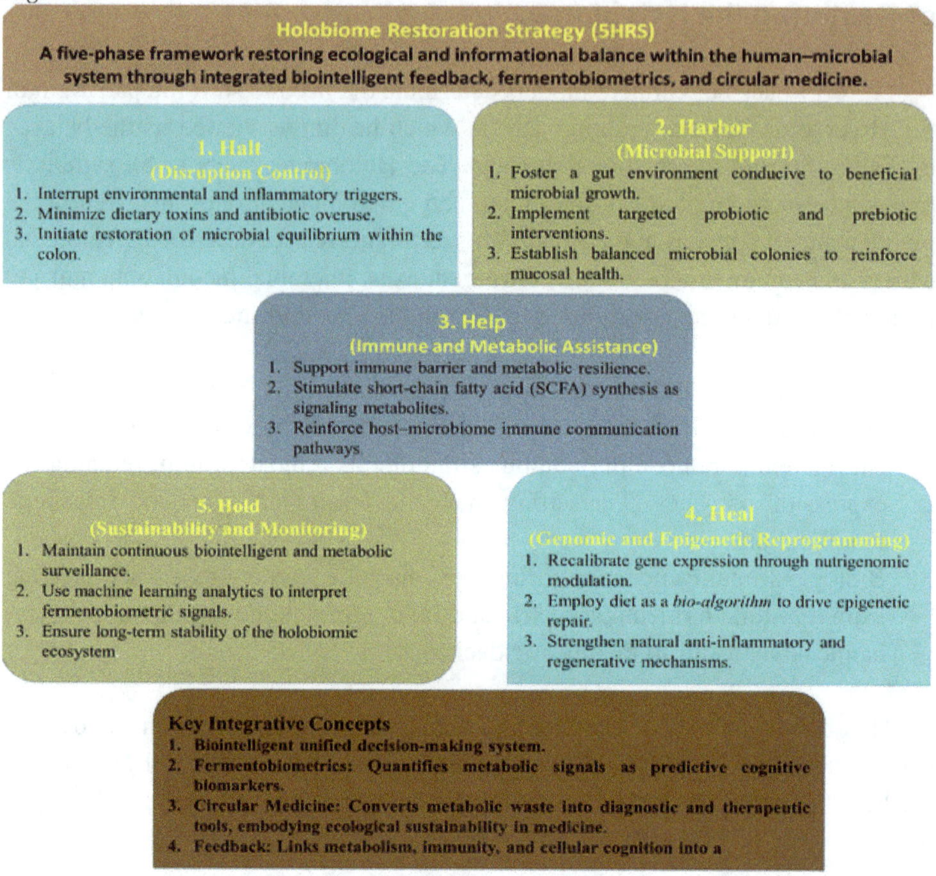

1.10 Biointelligent Egypt: Philosophical and strategic implications

When one puts colorectal cancer into the context of biointelligence, the definition of the disease becomes not merely a molecular anomaly but also a manifestation of a disequilibrium in civilization with regard to the environment and information. Such philosophical change requires a correspondingly revolutionary approach: the creation of National Biointelligence Centers, which would combine research in the fields of medicine, agriculture, data science, and environmental engineering. These hubs become the drivers of the shift in Egypt toward a knowledge-based health economy, where local innovations are born on the basis of the specifics of the microbial, nutritional, and genetic legacy of the country.

This vision must be accompanied by education reform, the introduction of systems thinking and biointelligent reasoning into medical education, and the training of scientists who will navigate complexity and not reduce it to reductionist bits. The ecological aspect of Vision 2030 in Egypt will be inseparable oncology because the combination of pollution, diet, and microbial biodiversity defines the course of health. The circular medicine model therefore reflects the Egyptian spirit of moral and scientific ethos: a medicine that preferentially rejuvenates the ecosystem when it preferentially rejuvenates the person and transforms the entropy of disease into the energy of renewal.

1.11 Integrative Synthesis

The Biointelligent Reframing of Colorectal Cancer provides the philosophical background for the following parts of the chapter. It reconceptualizes CRC not as a tissue-predominant pathology but as an emergent characteristic of impaired polybiomic intelligence. The presented concepts, Polybiome, Fermentobiometrics, Cognitive Oncology Systems, Holobiome Restoration, and Circular Medicine, can be viewed as an intertwined system in which a vision of medicine can think, sense, and develop in the future. In harmony with Vision 2030 of Egypt, this model is a welcome new form of scientific humanism that includes unitarizing data with wisdom, ecology with technology, and biology with consciousness (Reyed & Prabhakar, 2025a; Reyed & Prabhakar, 2025b; Chen, et al., 2025).

2 MICROBIOME TO POLYBIOME: GUT SYSTEM EVOLUTIONARY ECOLOGY.

The transgressional transformation of the microbiome into the polybiome is not an extension of terminology but rather a paradigmatic metamorphosis of our insight into biological organization, ecologically symbiotic structuring, and oncogenic formation in the human gut. However, some research has focused exclusively on compositional diversity, such as quantifying the richness and abundance of bacterial species, and has ignored all cross-domain interactions that occur between bacteria, archaea, viruses, and fungi as well as eukaryotic cells. The polybiome, in turn, redefines the gut ecosystem as a coevolutionary and holistic network in which microbial taxa, host cells, and environmental inputs build a coordinated metabolic and functional continuum (Reyed, 2023a; Wong and Yu, 2023) (Figure 4). This systems-level vision can be seen as going beyond reductionist biology, positioning colorectal cancer (CRC) in the context of ecological mismatch and irreversible metabolic entropy and forming perturbed symbiogenesis.

Figure 4.

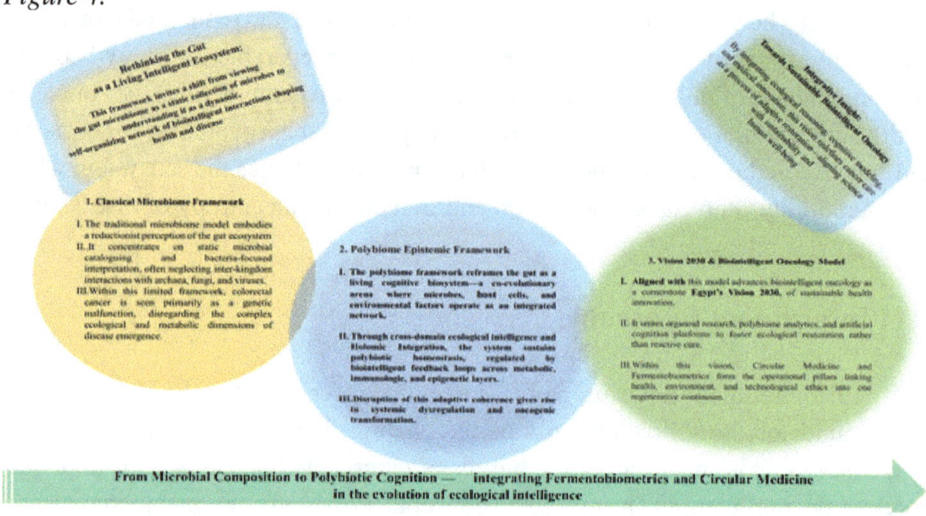

The key to this change is the realization that the gut is not a single digestive organ but an adaptive, cognitive, and evolutionary biosystem, i.e., a biointelligent organ. The interactions between the gut and the brain, the immune system, and metabolic pathways represent a multilayered feedback network that involves the active involvement of microbial communities in the process of making regulatory decisions (Park et al., 2025). This action encompasses what can be called holomic integration, i.e., the smooth interconnection of tiny biosignaling via microbial, immunological and neuroendocrine mechanisms with a spontaneous homeostatic framework. In this context, CRC is not a chance mutation but rather a failure of polybiotic intelligence, where microbial dysbiosis is interrupted in coordinating metabolic signaling, immune tolerance, and genomic integrity all at once (Karam et al., 2025).

In this context, the polybiome is the basic substrate on which an emergent ecological process, such as tumorigenesis, can be understood. Alterations in microbial–metabolic symbioses, which are usually provoked by dietary disturbances, environmental toxins, or excessive use of antibiotics, are what can be termed polybiotic dysregulation. It is a chronic, low-grade condition characterized by inflammation, nonregulated fermentative metabolic pathways, and intensive microbial loss, leading to pathogenic overgrowth. The central metabolic products in this ecosystem, including the short-chain fatty acids butyrate, propionate, and acetate, act as metabolic intermediates of cell differentiation, apoptosis and immune regulation (Xu, et al.,

2025). The depleted status, which is observed in people at high risk of CRC, is an indicator not only of metabolic deficiency but also of the breakdown of microbial control over epithelial homeostasis.

Traditional genomics and microbiome-focused methods are frequently unable to capture the dynamic interdependence between microbial activity and host genomics. In the polybomics model, the interplay between dietary substrates, microbial metabolism, and host epigenetics constitutes a very long running feedback system, a living bioalgorithm, by which the information is processed by biochemical transactions. This recursive mechanism can be referred to as metaecogenomic dynamics; it combines the plasticity of microbial genomes and host adaptive reactions to create emergent behaviors, which are beyond the linear causal relationship (Pandey & Bhattacharyya 2024). An example of this includes the reprogramming of the expression of host genes by microbial metabolites via histone acetylation and dynamic host immune signaling related to microbial gene regulation. This bilateral informational economy reiterates the fact that the disease cannot be reduced only to the host or the microbiome—it appears at the border of the two.

Recent research on organoid biology and synthetic biology offers strong experimental support for this point of view. Polybiome–host interactions can now be tested using human colonic organoids that are three-dimensional minimal-size replicas of the gut. Coculture of organoids with microbial consortia: By culturing microbial consortia in the presence of organoids, researchers have started recreating the multi-kingdom gut ecology in vitro, where changes in the microbial composition can influence epithelial morphogenesis, immune responses, and carcinogenic transformation (Liu et al., 2024). Synthetic biology, in turn, proposes an additional dimension of control, namely, the generation of engineered bacterial strains that can sense and regulate tumor-related metabolites, potentially becoming biointelligent sensors in the gut ecosystem. These developments represent convergence between biological and computational reasoning, striking a chord between evolutionary microbiology and accuracy oncology (Siguenza et al., 2024).

However, the possibility of applying these insights to clinical oncology is crippled by gaps in conceptualization and interpretation. The prevailing biomedical model still views CRC as a cell-autonomous pathology that develops on the basis of genetic errors, disregarding the environmental and microbial co-determinants of tumor development. Furthermore, microbiome studies in clinical settings are commonly based on methods of static sampling in which the temporal variation or nonlinear behavior of microsystems is not accurately modeled. To fill these gaps, an epistemological transformation of taxonomy to teleology is necessary: the end of the cataloging of the species into order, toward the decoding of the purposeful dynamics of microbial behavior in the framework of public health.

Vision 2030 provides Egyptian scientific vision with a special platform to operationalize this transformation (Ministry of Planning and Economic Development, 2023). Owing to its high leaf intake of fiber as well as fermented foods, the nutritional environment of Egypt offers an experimental platform for exploring polybiotic adaptation and nutritional resilience (Ministry of Health and Population, 2024). The introduction of polybiome analytics into the national health policy might allow the recognition of the risk of CRC disease in the initial stages, the recognition of metabolic and microbial patterns, instead of invasive or exclusively genetic diagnostics. Moreover, having Egyptian polybiome centers built that are multiomics, organoid, and AI infused would make Egypt a leader in biointelligent oncology studies in the region and turn its paradigm of healthcare delivery not into a reactive treatment model but rather into a model of proactive ecological restoration (Figure 5).

Figure 5.

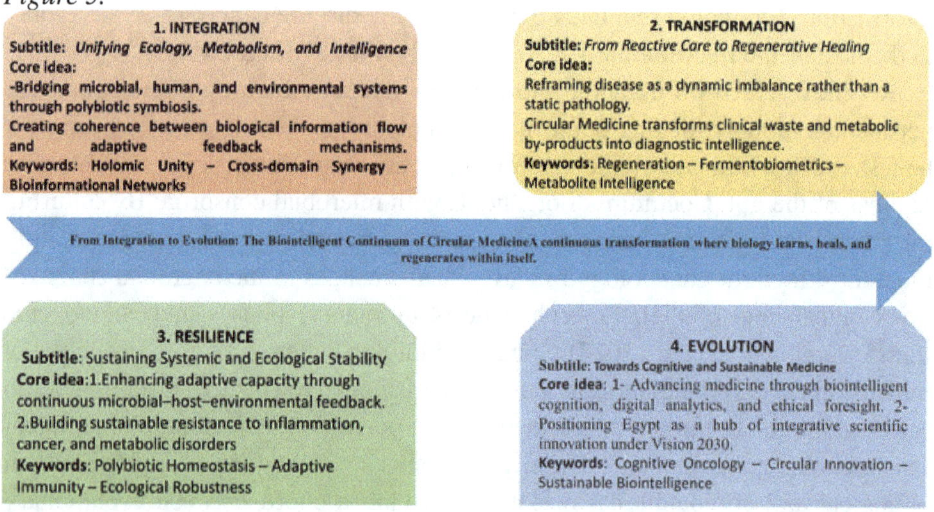

One can conclude that the shift toward the polybiome is not a simple scientific update but rather a cognitive revolution in medical knowledge. It considers colorectal cancer as an ecological intelligence disorder that requires integrative schemas that bring together molecular biology and systems ecology, synthetic bioengineering and computational cognition. Under the guise of science, by integrating this philosophy into biomedical innovation, Egypt can be on the vanguard of a new era of sustainable precision medicine: scientifically advanced and meaningful and aligned with the concept of Vision 2030.

3 FERMENTOBIOMETRICS: THE QUANTITATIVE LANGUAGE OF TUMOR ECOSYSTEM.

The latest branch of fermentobiometrics is a conceptual and methodological shift in the study of the gut–gut-tumor interface, re-codifying how metabolic communication at the polybiome interface controls the development of colorectal cancer (CRC). Fermentobiometrics avoids looking at microbial fermentation as a biochemical byproduct, but as a qualitative language of ecological knowledge, the microbial communities and even the dietary inputs and host tissues attract and respond to regulatory signals. In this context, the tumor microenvironment represents an information field that is mediated by metabolic codes such as short chain fatty acids (SCFAs), secondary bile acid, lactate, and polyamines that regulate the expression of genes, immune regulation, and cell differentiation (Reyed & Prabhakar 2025a).

3.1 Fermentation as an information process beyond metabolism

Fermentation is considered an energy-generating process when seen through traditional biochemistry, but it is actually a communicative action, as seen through biointelligent oncology. Individual microbial products are ecological syntax semantic units in a semantic memory containing biological languages. SCFAs such as butyrate and propionate can not only feed colonocytes but also act as epigenetic switches, affecting histone acetylation and DNA methylation, as well as noncoding RNA expression (Yang et al., 2025). Depletion of butyrate-producing taxa, including *Faecalibacterium prausnitzii,* in CRC results in the silencing of tumor suppressor genes, which indicates that metabolic silencing may evolve into genomic dysregulation.

This explanatory finding represents an existing gap in research that is critical to traditional oncology because of the inability to translate metabolic variability into functional diagnostics. Fermentobiometrics bridges this gap by measuring not only the concentrations of metabolites but also the relative ratios of metabolites to the dynamic fingerprints of system-wide equilibrium. Diagnostic indicators, such as the butyrate-to-acetate index (BAI) and secondary bile acid ratio (SBAR), can be used as diagnostic indicators of the inflammatory load, mucosal resilience, and tumorigenic potential. These indices therefore become a biointelligent diagnostic grammar, which converts the invisibility of microbial talks to therapeutic measures.

3.2 Organoid Frontier: Fat Sphering 3D Systems.

The biology of organoids has become the biology of the living laboratory of fermentobiometric exploration: a dynamic interface between in vitro experimentation and in vivo reality. Colonic organoids are created by cloning patient feces and growing the cells further into structures that are comparable in architecture and genetics to those found on the human gut epithelium (Zhang et al., 2024). In combination with microbial consortia, these miniaturized systems enable real-time monitoring of microbial fermentation, metabolite flux, and immune signaling. More recent research has demonstrated that metabolic imbalance can be reversed through epithelial polarity restoration and inhibition of the expression of inflammatory cytokines in the presence of butyrate supplementation in organoid–microbiome cocultures, confirming therapeutic reversibility (Woo et al. 2025).

Organoids therefore represent biointelligent reflections, indicating the consequences of diet and microbial interactions that are reflected in the form of their molecular structure. They are used to connect synthetic biology and systems oncology to enable scientists to trial engineered microbial strains in physiologically relevant human tissue models. This provides a platform for precision fermentobiomics whereby synthetic bacterial systems are able to detect pH, SCFA gradients or oxidative stress and respond with specific anti-inflammatory actions. However, such inventions represent the conversion of the microbiota into active therapeutic agents through synthetic biology, which is a component of the biointelligence paradigm, and the restoration of the previously missing symphony of the polybiotic ecosystem (Hamamah et al., 2023).

3.3 Fermentobiometric Dysregulation: Mapping the Breakdown of Ecological Intelligence

CRC progression can possibly be considered a slow rupture of fermentobiometric coherence. Homeostasis of microbial metabolites in healthy individuals is achieved by feedback mechanisms that match the efficiency of fermentation and immune tolerance. The harmony crumbles in the case of dysbiosis. This is due to the build-up of toxic fermentation byproducts (hydrogen sulfide, ammonia, and deoxycholic acid), which supersede anti-inflammatory metabolites over a phase that progresses to genomic instability and chronic inflammation. This metabolic blindness presents as polybiotic entropy or an energetic and informational disorder that drives carcinogenesis.

Traditional metabolic profiling captures only a series of snapshots of this process. However, fermentobiometrics has led to some changes through the use of temporal analytics, which allows the monitoring of rhythmic changes in fermentation products in response to circadian and dietary differences (Chen et al., 2025). These oscillations may also be observed in terms of biointelligent signals through machine-directed modeling, enabling early warning of metabolic meltdown without the incidence of histopathological alterations.

3.4 Biointelligent Application of Fermentobiometrics: Design of Ferment-based Diagnosis to Design.

The actual transformative power of fermentobiometrics is not necessarily in the form of a diagnosis but rather in the design forms of therapy. Clinicians can reprogram tumor ecology to health by measuring microbial communication. By incorporating fermentobiometric information into the Cognitive Oncology System (COS), this information can be translated into real-time modifications in diet, probiotics or medication on the basis of feedback algorithms that can simulate biological self-regulation. The profiles of SCFAs and markers of inflammation and data on the microbial population continuously feed predictive models in such a system, forming a learning health network, which improves together with the biology of the patient (Dong et al., 2023).

This strategy has special promise in Egypt. Fermentobiometric research is naturally grounded in the Egyptian nutritional paradigm with high fiber intake and fermented food contents. National Fermentobiometric Labs would provide an integrative approach between traditional knowledge of diet and the development of sophisticated omics analytics, creating the foundation of a circular medicine ecosystem in which food, microbiota, and health become regenerative connections. This is well in line with the objectives of Egyptian sustainable health innovation, biotechnology localization, and equitable digital transformation in Egypt Vision 2030.

3.5 Future of Preventive and Predictive Oncology: A New Research Frontier

The fermentobiometric paradigm redefines preventive oncology and creates new criteria for early disease detection at the same time. Instead of being confined to detecting a late-stage molecular mutation, clinicians are able to track a metabolic shift with early signs of polybiotic disequilibrium. The next step involves individual interventions to recover fermentative balance via the diet, engineered probiotics, or synthetic biotherapies. Artificial intelligence platforms can be continuously trained with population-level data; therefore, regional metabolic patterns can be traced on the basis of environmental exposure or dietary changes.

This biointelligent solution redefines colorectal cancer (CRC) as a fixable system malfunction and not an unchangeable genetic destiny (Jacob, et al., 2024). It places prevention not only within the framework of molecular surveillance but also within ecological restoration, thus reinstating the lost host–microbiota mutualism. Fermentobiometrics, therefore, is an advocate of diagnostics as well as a method of remedy, providing a scientific, philosophical, and cultural grid of ferments of imagining oncology in Egypt and beyond (Reyed & Prabhakar, 2025a; Das 2022).

3.6 Future Projections: Fermentobiometrics into National Innovation Ecosystems

Fermentobiometric introduction within the biomedical innovation system of Egypt requires a transdisciplinary partnership between data science, nutrition, and clinical oncology (Mir. 2025). By integrating the use of fermentobiometric indices into electronic health records, it would become possible to identify vulnerable populations at an early stage. In parallel, collaboration with the agricultural and food industry would create nutritionally smart food systems that are designed to enhance polybiotic resilience (Elsayed & Saleh, 2024).

In this context, biointelligence is not only a tool of modeling but also an epistemic philosophy that orients scientific questioning to sustainable social change. This requires the formation of fermentobiometric knowledge networks whereby clinicians, data scientists, and biotechnologists can liaise and convert ecological measurements into practical health measures (Singh et al., 2024).

Fermentation biometrics is not just an analytical model but also a new epistemology of health, in which the practice of metabolism is understood as cognition, fermentation as communication, and medicine as the art of healing dialogs in the living ecosystem. Through the framework of this strategy built into Vision 2030 of Egypt, it is possible to make the country a leader in translating a model of sustainable precision oncology that strikes a balance between biology, computation, and culture in the quest for holistic well-being (EFESO Management Consultants & Giza Systems. The sequential integration of these six stages is summarized in Figure 6.

Figure 6.

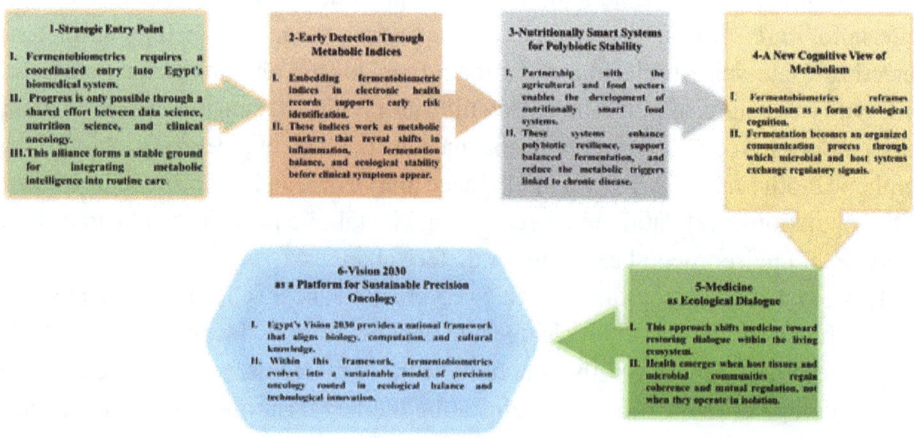

4 NUTRIGENOMIC CONVERGENCE: FOOD AS A BIO-ALGORITHM.

Nutrition and genomics can now be conceived not as a static concept of diet–gene interaction but more as a dynamic, biointelligent process where food across diverse biointelligent forms no longer takes a passive biochemical signal but rather as a living codebook that rewrites genomic and microbial codes (Salunkhe et al., 2023). In this paradigm, food is information, decoded by the polybiome and epigenetically encoded by host cells as physiological results. Consequently, colorectal cancer (CRC) is no longer depicted as an imminent outcome of genetic destiny but rather as a systemic problem in this nutritional biocode, disruption of the ongoing dialogic relationship between diets plus microbial metabolites and host genome expression (Brandolino et al., 2025; Reyed & Prabhakar 2025a; Reyed, 2023a).

4.1 Nourishing our Story: The Epigenetic Language of Food.

Conventional nutritional science measures macro- and micronutrients without elucidating the semantic role of food, which has the potential to express regulatory meaning. Food molecules, as part of the concept of biointelligence, are biosemantic messages conveying detailed messages to cellular receptors, transcription factors, and enzymes of microbes. The nutrigenomic modem includes product categories such as polyphenols, fibers and omega-3 fatty acids, which decode dietary patterns into rhythms of gene expression in a manner facilitated by histone acetylation, methylation, and noncoding RNA activation (Leziak et al., 2025).

Diet, in this sense, is a genomic programming language, a living dialog between the environment and the genome abruptly abacused by ultra-processed foods, antibiotic remnants, and xenobiotics and whose syntax breaks apart the nutrigenomic coherence to initiate pro-inflammatory cascades and metabolic entropy and biases evolution to tumorigenesis. The gap in the research is the reductionist disaggregation of diet into single molecules; nutrigenomic convergence requires a system grammar unifying molecular constitution, microbial translation, and genomic interpretation in one cognitive–ecological framework (Britten & Tosi, 2024).

4.2 Polybiome Mediation: The gut as a nutritional processor

The polybiome is the computing unit of the nutrigenomic system—an evolutionary superorganism that computes dietary data via the metabolism of microbes. Every microbial guild performs as a meta-transcribing machine, converting fibers into a fatty acid backbone and amino acids into signaling peptides and lipids into immunomodulatory molecules. Microbial computation generates metabolic semiotics encoding dietary meaning into chemical messages that tune host cell behavior (Mamun & Rahman 2025).

Dysbiosis interferes with this interpretive circuitry in the pathogenesis of CRC. When species that produce butyrates are reduced and proteolytic or bile-resistant species predominate, the gut transforms into a biointelligent translator into a toxic amplifier, producing carcinogenic metabolites, including hydrogen sulfide and secondary bile acids. This impairment of microbial translation precision, which can be called nutrigenomic noise, is an informational pathology that reveals severe failure in existing oncological nutrition frameworks that view diet as a linear feature as opposed to a polybiotic feedback-based calculation (Jiang et al., 2025; Reyed & Prabhakar 2025c).

4.3 Frontiers of Organoid and Synthetic Biology: Experimental Nutritional Intelligence.

Recent developments in organoid biology have offered unprecedented hardware to learn more about nutrigenomic processes in the context of a controlled yet realistic environment. By visually depicting the effects of a particular diet on gene networks, epithelial integrity, and tumor suppressor activation in real time, colonic organoids cultured using tailored nutrient matrices and microbial consortia allow researchers to examine the effects of a given diet (Liu et al., 2024). In the context of multiomics profiling, it is now feasible to establish nutrigenomic map dynamic models that can be used to relate nutrient patterns to transcriptional and metabolic products (Lemme-Dumit 2025).

Conversely, synthetic biology enables the design of biosensors of nutritional biocircuits: purpose-built bacteria that breakdown toxic substances, release anti-inflammatory metabolites, or ferry therapeutic RNAs to the site of action. Nutrigenomics is no longer a descriptive science but a constructive space with this blending of organoid systems and artificial systems that can be used to design nutritional therapies on the basis of biointelligence. In this sense, the human diet translates a fixed form of consumption into a programmable architecture of biological communication under the guidance of data-driven ecological intelligence (Kang et al., 2020).

4.4 Fermentobiometrics and Nutrigenomic Feedback Loops

The combination of fermentobiometrics with the nutrigenomic modeling framework introduces a quantitative language for decoding nutritional–microbial interactions. The butyrate-to-propionate ratio or lactate flux index fermentative indices are paired quotient ratios that provide quantitative impressions of how dietary preferences reorganize metabolic networks. With this integration of these indices into the Cognitive Oncology Systems (COS), clinicians can model the impact of particular nutritional interventions on tumor behavior, immune tone and microbiome diversity (Zhang et al., 2024; Reyed, 2023a).

This technology adds a novel category of adaptive nutrition algorithms to the list of smart feedback systems: feed forward systems that learn continuously to tailor a diet to prevent and remit CRC on the basis of the fermentobiometric and nutrigenomic data of a patient.

4.5 The Egyptian Context: Food Culture as a Biointelligent Heritage

Egypt is a unique environment for the development of nutrigenomic integration. The naturally polybiotic architecture of the Mediterranean–Nile diet, which is based on legumes, fermented dairy, whole grains, and phytonutrient-rich vegetables, has supported millennia-long metabolic resilience. The combination of conventional Egyptian food culture with the modern use of omics and organoid technologies has the ability to explain the ancestral reasons behind the nutritional intelligence inscribed in local diets (Suarez & Adibi, 2025).

In Vision 2030, the creation of the National Nutrigenomic Centre might consolidate societal health, biotechnology, and AI-enhanced analytics to track the effects of local dietary changes on polybiotic health (CIRAD, 2025). These centers not only increase the pace of preventing colorectal cancer but also transform the agricultural and food sectors of Egypt into nutritionally intelligent ecosystems, which offer sustainable health through culturally based innovation. This paradigm is an example of circular medicine, a regenerative approach that is based on food systems, microbial ecology, and human health, which are symbiotically coevolved in a sustainable societal structure (Agrawal, 2025).

4.6 Philosophical prospects: Cognitive nutrition and circular health

Nutrigenomic convergence alters alimentation as a form of cognitive discourse with the biosphere at the philosophical level. Food is no longer a passive substrate but is intended to encode intelligence, which coordinates cellular behaviors. This understanding drives a new epistemology of nutrition—cognitive nutrition—in which metabolism is conceptualized as thought, digestion is conceptualized as calculations, and disease is conceptualized as a failure of ecological responses.

The idea found in such a shift is in line with the Egyptian tradition of thinking scientifically, in which health is always viewed as a systemic state of harmony and not an independent organ. Through the integration of this philosophy into research, education and innovation, all of Egypt can be at the forefront of a biointelligent civilization where knowledge, ecology and medicine are all part of a continuous culture story.

Taken together, Nutrigenomic Convergence redefines food as the hub of the polybiotic web of life—a living algorithm and the capacity to restore genomic balance, microbial diversity and systemic intelligence. From this perspective, oncology has become remedial, focused on restoring the lost balance between food, biology, and meaning, which is reflected in Vision Egypt 2030 in the spheres of health, education, and sustainable innovation (Ferlisi et al., 2023; Szopa et al, 2005; Kaput & Rodriguez 2004).

5 COGNITIVE ONCOLOGY SYSTEMS: THE QUEST FOR PREDICTIVE BIOINTELLIGENT THERAPIES.

The emergence of cognitive oncology systems (COSs) is creating a conceptual turning point in cancer, as mutations are replaced by cancer being seen as a failure of the information architecture that controls multicellular life. From this perspective, tumors are dissonant cognitive units whose signaling networks are no longer linked to the greater coherence of biological intelligence (Reyed, 2023a). This shift is justified by recent findings that genetic mutations cannot be considered the sole cause of the observed tumor phenotypes and their capacity to enhance metabolic and immunological adaptation to environmental pressures (Fernandez et al., 2025; Zhang et al., 2025). In this biointelligent architecture, communication between genes, proteins, immune cells, and the polybiome contributes to higher-level order, and cancer occurs when the holomic network degenerates into informational entropy (Zitvogel et al., 2025).

The polybiome, comprising bacteria, viruses, and fungi, is a distributed cognitive layer that may influence tumor behavior by signaling metabolites, modulating immune responses, and reprogramming metabolism. Apoptosis, autophagy, and immune surveillance are regulated by biochemical messengers that include microbial metabolites, such as butyrate, propionate, and indole derivatives (Reyed& Prabhakar, 2025a; Pircalabioru et al., 2025). The disruption of this polybiotic signaling intensifies carcinogenic pathways by promoting effects via inflammatory drift and disrupted epithelial communication (Jin, et al., 2025). Within the COS paradigm, the tumor is regarded as an adaptive learning system within its ecological and microbial context that can develop survival strategies and metabolic rewiring, much like synthetic cognition in machine systems. Tumor evolution modeling via biointelligent algorithms capable of capturing microbial transitions, ecological pressures and metabolic timing provides a means of decoding cancer as more of an information disorder than a mere genetic malfunction (Zhang et al., 2025).

5.1 Organoid and Synthetic Biology: Experimental Instantiations of Cognitive Oncology

The synthetic biology of organoids converges with the biology of organoids, forming the experimental basis of COS-based medicine. Colonies of organoids Patients Three-dimensional cultures that maintain genetic and epigenetic markers and that are derived as living avatars of single tumors are called patient-derived colonic organoids. When grown with microbial consortia, immune cells and synthetic biosensors, these organoids recreate the tumor microecology in vitro, which allows real-time mapping of tumor cognition, i.e., the responsiveness, adoption and communication of cancer cells within their biointelligent environment (Reyed & Prabhakar, 2026; Papamichail et al., 2025; Reyed, 2023a; Drost & Clevers, 2018).

Synthetic biology introduces an interventional aspect to this strategy. Biosensing and therapeutic circuits can be engineered into engineered bacterial strains, which monitor oncogenic metabolites and react to these signals by expressing antitumour molecules, restoring microbial homeostasis or by inducing apoptosis in tumor cells. These therapeutic symbionts turn the gut into a living biocomputer, which can constantly monitor and repair itself—an embodiment of circular medicine that we heal and monitor during the same ecological cycle (Jin et al., 2025).

5.2 Predictive Oncology Modeling and Fermentbiometrics.

Fermentobiometrics is a computational-level grammar of decoding the metabolic intelligence of the tumor ecosystem. COS platforms can infer the health of miRNA microbial–epithelial communication by assessing the fluxes of microbial fermentation products, redox status, and metabolite ratios (e.g., the butyrate–acetate index). Machine-augmented systems are based on biointelligent modeling but not traditional AI but use these data streams in layers of multiomics (genomic, metabolomic, and immunomic) to deliver predictions on the basis of tumor behavior in real time (Baumgartner 2025; Cao et al., 2024).

The COS model therefore has far-reaching ethical consequences, which compels practitioners not only to consider the tumor itself but also the cognitive milieu in which the tumor grows. The process of healing, then, should be imagined as communication being restored between cells and microbes and between the mind and society. This integrative philosophy is in agreement with the Egyptian school of scientific thought, where health is viewed as a harmony of various types of intelligence, including biological, ecological, and civilizational intelligence. Medicine has more than a curative role in the re-establishment of such harmony and has emerged as a type of cognitive ecology (Hanahan 2022; Lyon 2015).

5.3 Egyptian Vision 2030: The application of Cognitive Oncology

The Vision 2030 of Egypt provides a strategic framework according to which the COS will be operationalized as a national innovation model. Oncology, data science, and microbiome research may be merged into one synergistic platform with the creation of National Cognitive Oncology Centers. These centers would have multiomics integrated databases, organoid testing facilities, and synthetic biosystem laboratories to produce biointelligent therapeutic guidelines—personalized, adaptive and culturally based (Hassan et al., 2025; Eldridge et al., 2025)).

COS platforms might be integrated into the hospital network in conjunction with the developing bioinformatics infrastructure in Egypt, which can allow clinicians to perform real-time fermentobiometric profiling and nutrigenomic mapping of colorectal cancer patients. This would allow active therapeutic changes under the guidance of ongoing adjustments between the polybiome and tumor behavior of the patient. These models can not only improve survival rates but also save healthcare money because medicine is no longer focused on reacting to an issue but rather on restoring it.

In addition, Egypt has exceptional biodiversity and traditional pharmacognosy that provides special biotechnological sources of natural-synthetic hybrid therapies such as bioengineered probiotics (Woo et al., 2025), polyphenolic nanocarriers, and adaptive nutraceuticals, which reflects the principles of Holobiome Restoration. The reinvention of biomedical innovations with ecological sustainability enabled Egypt to become the first mover of circular precision medicine, spearheading the shift to the regenerative health paradigms of the Global South (Reyed & Prabhakar, 2026; Reyed & Prabhakar, 2025a, 2025d; Reyed, 2022).

5.4 The Cognitive Turn in Oncology: Toward a Biointelligent Future.

The cognitive oncology system is the brain and soul of twenty-first-century medicine. They oppose the epistemic stagnation of conventional biomedicine and map out the way to biointelligent curing, whereby diagnosis, treatment and evolution coalesce into a single self-organizing system. COS represents a biology-cognition hybrid, an integrative model of synthetic bioengineering, organoid experimentation, polybiome analytics, and fermentobiometric modeling, a medical system to heal and learn.

This strategy is within the large-scale context of Vision Egypt 2030, which is the arrangement of knowledge-driven and innovation-driven civilization, the one that aims to strike a balance between technology and life. The oncological future as re-envisioned is no only about eliminating illness but rather about regaining coherence, thus reestablishing the dialog that has been lost between human biology and the biosphere. Egypt can be a first mover in the paradigm shift in which Cognitive Oncology Systems are used to transform cancer care into a paradigm of biointelligent evolution, where health is a manifestation of adaptive, ecological, and cognitive resilience and not the absence of disease. A consolidated visual roadmap for implementing Cognitive Oncology Systems within Egypt Vision 2030 is illustrated in Figure 7.

Figure 7.

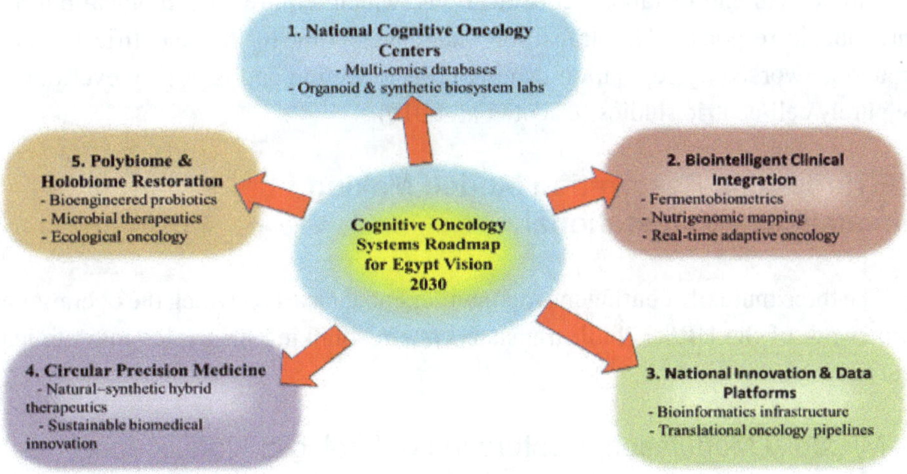

6 THE HOLOBIOME RESTORATION STRATEGY (HRS): A PARADIGM SHIFT TOWARD BIOINTELLIGENT ONCOLOGY

The Holobiome Restoration Strategy (HRS) represents a paradigmatic breakthrough in the oncological field, going beyond reductionist and pharmacologically/suppressive conceptualizations that dominated conventional cancer therapy. HRS reimagines colorectal cancer (CRC) as not a localized cellular alteration to be eliminated but rather a systemic malfunction of ecological intelligence resulting in a loss of dynamical coherence between the host, bodily microbiota, metabolome, and environment, which is collectively referred to as the Holobiome (Tudorache et al., 2025). This approach substitutes the cytotoxic regime-based biological warfare view with bioecological reprogramming, making oncological therapeutic approaches for healing a matter of repairing communication and integrative functionality into the living system (Silver & Nelson, 2018; Pienta et al., 2008).

6.1 Critical Gap: Eradication and Restoration

Classic oncology is supported by an eradicationist logic focusing on the destruction of tumors in the absence of consideration of the long-term ecological and informational impacts of this type of intervention. HRS is a direct accusation of this epistemic stasis and introduces a restoring epistemology where therapy is a kind of biological diplomacy to tune the system in which the tumor has emerged back into harmony. From this perspective, cancer is an outcome of disturbed microbial, immunological and metabolic conversations, which require an ecological but not antagonistic response. This transition challenges point to the scientific vacuum around networked dysregulation driving CRC evolution and is largely overlooked by purely cell-centric studies (Cadet et al. 2016).

6.2 Three-Axis Restoration Model: Making Biointelligence Operational.

The three mutually contingent and interdependent axes on which the operational framework of the HRS is built translate systems-thinking principles into clinical practice

A. Microbial Modulation: Restoration of Ecological Intelligence.

This axis re-establishes the therapeutic object to be the tumor as opposed to the ecological disruption that has made the emergence of the tumor in the first place possible. HRS views the polybiome (bacteria, fungi, viruses, and archaea) as an adaptive cognitive system, which is why the selective reconstitution of microbes is performed via next-generation probiotics, prebiotics and fecal polypeptide transplantation (FPT) (Reyed & Prabhakar, 2025d; Ma et al.,2024). Most importantly, the HRS goes further, suggesting region-specific microbial consortia, which is, as per local microbial ecology, an innovation, as is the case of the Egyptian National Polybiome Initiative. In addition, this axis combines advances in synthetic biology and the design of symbiotic bacterial strains that sense the presence of oncogenic metabolites in the gut (e.g., deoxycholic acid) and secrete counteractive molecules, thus converting the gut into a biointelligible biosphere (Kim et al., 2024).

B. Nutrigenomic Optimization: Nutrition as an Epigenetic Interface

HRS does not consider food simply as sustenance but rather as a biocomputational algorithm that can rewrite gene expression and epigenetic states. This axis exploits fermentobiometric analytics (such as the proportion of butyrate and acetate) to plan accurate nutrition plans on the basis of the metabolic and genomic profile of the patient (Souza et al., 2025). In this intervention, antitumour resilience is enhanced by addressing the diet-metabolite-microbe triad. Its integration with synthetic nutrigenomics, food-based biotechnology, produces functional foods supplemented with microbial metabolites, which practically transforms food into therapy and technology itself and bridges the gap between nutrition and molecular therapy (Kaur & Kumar 2020).

C. Biointelligent Monitoring: Dynamic Adaptation and Clinical adaptation.

The last axis makes clinical oncology a cyberbiological field with adaptive feedback control. Biointelligent systems, which are superior to standard artificial intelligence, involve multimodal, real-time measurements based on fermentobiometrics, metabolomic sensors, and organoid simulations (Farley et al., 2013). This generates a Cognitive Oncology Dashboard, which forecasts inflammatory flare-ups or metabolic breakdowns and automatically optimizes dietary, microbial, or pharmacological therapy. It is an adaptive and self-corrective procedure that breaks the constraints of fixed biomarker regimens, suggesting a dynamic algorithmic approach to treatment (Gagalova et al., 2020).

7. TRANSLATIONAL PILLARS AND OVERLOOKING FUTURE VIEW

7.1 Organoid and Synthetic Biology as Platforms of Regeneration

More recent developments in organoid technology have placed these platforms at the center of the stage in terms of understanding diet–microbiome–immune interactions in colorectal cancer (El-Derby et al., 2024). Although, to date, a completely functional platform of fully integrated clinical organoids utilizing this triad in Egypt has not been reported, the theoretical underpinnings of a given system are now well established through conceptual analyses and translational systems that are emerging. These articles describe how patient-derived colonic organoids can be used as living holobiome avatars capable of replicating microbial fermentation dynamics, immunometabolic signaling, and nutrigenomic modulation in vitro, thus facilitating preclinical optimization of personalized therapeutic protocols (Reyed, 2023a).

Internationally, models incorporating microbial consortia and immune factors into organoid cultures have been shown to provide high-fidelity descriptions of host–microbe–nutrient interactions, supporting the translational potential of organoid-based ecological oncology (Wu et al., 2025; Ekanger, et al., 2025). This international development provides firm empirical ground upon which Egypt may follow in the future in the use of these platforms as part of the national precision-medicine infrastructure. In line with this, although experimental implementation has yet to be achieved, the Egyptian translational research pipeline aligns conceptually with global organoid–ecology paradigms, allowing a swift shift to applied research when national resources and institutional networks converge.

7.2 Global Relevance: Vision 2030 Roadmap

The HRS framework is a strategic goal with the Egypt National Artificial Intelligence Strategy (2025-2030) and Vision 2030 (National AI Council, 2025). Its national rollout will involve developing specialized Holobiome Restoration Clinics and a biointelligence innovation Hub that combines multiomics, organoid diagnostics, and cognitive therapeutic modeling.

This national implementation is a pilot project around the world, the Global South, to promote a circular health economy. In this future vision, medical outcomes are reinvested in environmental and nutritional planning; ecological biodiversity supports the development of probiotics; and health is reflexed as adaptive, ecological, and cognitive resilience, reinforcing the ancient Egyptian philosophical concept of Maat (Systemic Harmony).

The HRS offers a blueprint of sustainable precision health by emphasizing restoration and coherence over just eradication and turning Egypt into a producer of next-generation biointelligent paradigms instead of a consumer of medical technology.

8. COGNITIVE ONCOLOGY SYSTEMS AS A FOUNDATION FOR CIRCULAR MEDICINE AND EGYPT VISION 2030:

The conceptual framework of the cognitive oncology system (COS) provides the framework around which a nationwide shift is effected to sustainable precision medicine and redefines colorectal cancer (CRC) as a failure of neuroecological intelligence instead of a genetic or cellular malady. Oncogenesis within this paradigm can be thought of as a failure in distributed cognition that coordinates microbial symbiosis, epithelial signaling, metabolic homeostasis, and psychoneuroimmunology (Yazdanian et al.,2020). This view aligns with the strategic philosophy of circular medicine and the developmental pillars of Egypt Vision 2030, where health is viewed as an ecological, digital, and cognitive ecosystem and not a small biomedical sector. COS is based on the thesis that biological life is inherently cognitive: the holobiome can be understood as a distributed intelligence network where microbial, neural, immune, and metabolite components are connected by biochemical exchanges via synapses (Han et al., 2024). The loss of such coherence causes the corruption of signaling throughout the microbial-immune-neural axes, thus allowing malignant evolution. This model is operationalized in the neuroecological triad, which incorporates the Mind Axis, which controls the inflammatory tone via psychoneuroimmunology (Willmann & Moita,. 2025); the microbiome axis, which has a second cognitive system synthesizing neuroactive metabolites (Miri et al., 2023); and the metabolic axis, which regulates nutrient flux, redox homeostasis, and cellular fate. COS, through biointelligent modeling, integrates genomic, transcriptomic, metabolomic and immunomic data into cognitive networks that predict how changes in diet, microbial composition or metabolic flux diffuse through communication circuits to modulate tumoral behavior. These strategies make oncology a predictive cognitive science rather than a descriptive pathology (Catalano et al., 2025).

Experimentally, COS take material shape via neuroecological organoid models, where patient-derived colonic organoid models cocultured with microbial and neural elements provide a representation of host-microbe-neural communication in vitro, indicating that cancer is a process of progressive withdrawal (Che et al., 2023). Synthetic cognitive systems—engineered consortia of microbes with the ability to sense oncogenic metabolites and release repair compounds—broaden this paradigm for therapy and act as biointelligent sentinels, restoring signaling fidelity and ecological collaboration. Through replacing the notion of biomedical ethos of control with a notion of coherence, the COS reinvents the nature of healing as underpinning restorative communication between the holobiome (Guo et al 2025). The change in technology and philosophy directly contributes to Egypt's 2030 vision, which aims to ensure that biomedical innovation is compatible with environmental care, digital health nationalism, and sustainable economic development. The three primary components of an independent precision-medicine network within this national framework are polybiome science, nutrigenomics, and biointelligent computation.

To operationalize this transformation, Egypt can develop National Polybiome and Nutrigenome Centers to map biocultural ecosystems through metagenomic, metabolomic, transcriptomic and nutrigenomic profiling, creating population-flexible microbial biomarkers and adaptive dietary signatures. Parallel National Biointelligent Labs incorporating organoid biology, synthetic biology, and COS modeling will produce translational pipelines capable of cyclical exchange in vitro experimentation and clinical innovation (Sinha et al., 2025). The anchoring of data sovereignty, ethical analytics, and SDG-appropriate health governance will be anchored in multiomics databases embedded within the national digital health architecture (Global Health Systems Symposium, 2024). On the philosophical level, such developments bring together an Egyptian School of Evolutionary and Intelligent Medicine, integrating ancient ecological wisdom with modern systems biology. Circular medicine clinically operationalizes this synthesis by means of closed loop biological recycling, regenerative nutrition, and microbial restoration and converts hospitals, food systems, and research centers into biocircular ecosystems. Egypt can produce new biointelligent practitioners and researchers through educational reform, policy integration and eco-biotechnology entrepreneurhip (Shaharudin et al., 2024).

Finally, the COS, Circular Medicine, and Vision Egypt 2030 intersect to form a coherent national initiative, the development of a biointelligent civilization, where ecological intelligence, biological cognition, and digital systems mutually enhance each other. This model assumes that disease prevention, environmental stewardship, food systems, and medical innovation work as a unified adaptive spectrum. By combining health, knowledge and ecological resiliency, Egypt can be a leader in culturally based, scientifically autonomous, precise medicine worldwide (Miehe et al., 2021).

9. CONCLUSION: CIRCULAR MEDICINE: THE AGE OF BIOINTELLIGENT CIVILIZATION.

The new frontier of oncology demands a clear epistemological switch with reductionist paradigms to a unitary perspective of cancer as a systemic failure of distributed biological cognition. Instead of discussing malignancy as a single genetic defect, the view reframes cancer as a failure of the neuroecological intelligence of the Holobiome, the failure of the neuroecological networks of communication that organize microbial symbiosis, immunometabolic regulation, and host cognitive–physiological coherence (Barakat et al. 2024; Rosenthal et al., 2011). This reformulation poses a challenge to the ancient biomedical principle of control and eradication, which historically placed the clinician in a reactive war against malignant cells. What is needed to sustain the future of medicine rather is a transition to restoring systemic coherence.

Within this paradigm, cognitive oncology systems (COSs) operationalize this philosophical change by substituting force-based suppression with approaches that restore resonant biological communication. As a kind of biological diplomacy, therapeutic intervention steers the living system toward self-regulation and does not impose any destructive pressure on it (Menigoz et al., 2020). This is the conceptual paradigm shift of circular medicine, in which biological information in the form of microbial metabolites, immune mediators and metabolic byproducts is constantly recycled and reinvested in adaptive healing processes (Ratto et al., 2025; Wang et al., 2025). The concept of circular medicine redefines the body as a battlefield; rather, the body is an ecological engine that can be renewed recursively.

The technological infrastructure that will facilitate this shift will be the result of polybiome analytics, biointelligent computational modeling, and neuroecological organoid convergence. These systems implement multiomics data as part of predictive, cognition-enabled frameworks with the ability to simulate the propagation of microbial, metabolic or nutrigenomic perturbations through the tumor-host ecosystem (Li, et al., 2025). They are directly aligned with the pillars of innovation in Egypt Vision 2030, which include digital transformation, sustainable health systems, biotechnology sovereignty, and ecological stewardship as national priorities that are closely linked to one another (National AI Council, 2025).

This national strategy will enable Egypt to move beyond the status of an importer of foreign biomedical paradigms to an exporter of locally based, biointelligent systems of innovation. This change forms the conceptual basis of biointelligent civilization—a societal architecture where ecological intelligence, biological cognition, and digital technologies interactively develop into long-term resilience (Rabie, 2021). In this type of model, disease prevention, food systems, environmental policy and medical practice are combined elements within a single adaptive continuum. The process of reestablishing coherence is not only the goal of therapy but also the structuring principle of national health governance (Jiang et al., 2021).

This synthesis will make Egypt stand as an exceptional leader in worldwide exemplar of sustainable precision medicine on the basis of local ecological expertise, cultural identity, and sophisticated scientific modeling. The overlap of polybiome science, nutrigenomics, and cognitive oncology creates a path-shifting roadmap by delivering the strategic aspirations of Vision 2030 and proving that the future of health depends on the recognition and restoration of the deep communicative architecture of life itself.

REFERENCES

Abo-Hammam, R. H., Salah, M., Shabayek, S., Hanora, A., Zakeer, S., & Khattab, R. H. (2024). Metagenomic analysis of fecal samples in colorectal cancer Egyptians patients post colectomy: A pilot study. *AIMS Microbiology*, *10*(1), 148–160. DOI: 10.3934/microbiol.2024008 PMID: 38525041

Agrawal, K., Goktas, P., Kumar, N., & Leung, M.-F. (2025). Artificial intelligence in personalized nutrition and food manufacturing: A comprehensive review of methods, applications, and future directions. *Frontiers in Nutrition*, *12*, 1636980. Advance online publication. DOI: 10.3389/fnut.2025.1636980 PMID: 40771216

Al Mamun, A., & Rahman, S. T. (2025). Microbial Signaling Networks: Bridging Cellular Communication with Applications in Health, Agriculture, and Biotechnology. *Preprints*. https://doi.org/DOI: 10.20944/preprints202504.1833.v1

Alves Costa Silva, C., Almonte, A. A., & Zitvogel, L. (2025). Oncobiomics: Leveraging Microbiome Translational Research in Immuno-Oncology for Clinical-Practice Changes. *Biomolecules*, *15*(4), 504. DOI: 10.3390/biom15040504 PMID: 40305219

Barakat, G. M., Ramadan, W., Assi, G., & Khoury, N. B. E. (2024). Satiety: A gut-brain-relationship. *The Journal of Physiological Sciences; JPS*, *74*(1), 11. DOI: 10.1186/s12576-024-00904-9 PMID: 38368346

Baumgartner, C. (2025). Computational modeling and simulation in oncology. *Clinical and Translational Medicine*, *15*(9), e70456. DOI: 10.1002/ctm2.70456 PMID: 40910400

Brandolino, S., Franzago, M., Murmura, G., Ricci, F., Gatta, V., Stuppia, L., & Vitacolonna, E. (2025). Nutrition, Environment, and Genetics in Colorectal Cancer. Epigenetics and Possible Future Perspective. *Current Nutrition Reports*, *14*(1), 111. DOI: 10.1007/s13668-025-00701-9 PMID: 41108343

Britten, O., & Tosi, S. (2024). The role of diet in cancer: The potential of shaping public policy and clinical outcomes in the UK. *Genes & Nutrition*, *19*, 15. Advance online publication. DOI: 10.1186/s12263-024-00750-9 PMID: 39097687

Cadet, T., Davis, C., Elks, J., & Wilson, P. (2016). A Holistic Model of Care to Support Those Living with and beyond Cancer. *Healthcare (Basel, Switzerland)*, *4*(4), 88. DOI: 10.3390/healthcare4040088 PMID: 27869728

Calvino, G., Farro, J., Zampatti, S., Peconi, C., Megalizzi, D., Trastulli, G., Andreucci, S., Cascella, R., Strafella, C., Caltagirone, C., Grifalchi, F., & Giardina, E. (2025). From Genomics to AI: Revolutionizing Precision Medicine in Oncology. *Applied Sciences (Basel, Switzerland)*, *15*(12), 6578. DOI: 10.3390/app15126578

Cao, C., Yue, S., Lu, A., & Liang, C. (2024). Host-Gut Microbiota Metabolic Interactions and Their Role in Precision Diagnosis and Treatment of Gastrointestinal Cancers. *Pharmacological Research*, *207*, 107321. DOI: 10.1016/j.phrs.2024.107321 PMID: 39038631

Cao, Q.. (2025). How gut microbial metabolites influence colorectal cancer. *Frontiers in Microbiology*, *16*. Advance online publication. DOI: 10.3389/fmicb.2025.1611698 PMID: 40771692

Carrillo-Serradell, L., Liu-Tindall, J., Planells-Romeo, V., Aragón-Serrano, L., Isamat, M., Gabaldón, T., Lozano, F., & Velasco-de Andrés, M. (2025). The Human Mycobiome: Composition, Immune Interactions, and Impact on Disease. *International Journal of Molecular Sciences*, *26*(15), 7281. DOI: 10.3390/ijms26157281 PMID: 40806413

Catalano, M., D'Angelo, A., De Logu, F., Nassini, R., Generali, D., & Roviello, G. (2025). Navigating Cancer Complexity: Integrative Multi-Omics Methodologies for Clinical Insights. *Clinical Medicine Insights. Oncology*, *19*, 11795549251384582. DOI: 10.1177/11795549251384582 PMID: 41147019

Centre de coopération internationale en recherche agronomique pour le développement (CIRAD). (2025, March 26). Improving nutrition: Transforming food systems for better nutrition. https://www.cirad.fr/en/cirad-news/news/2025/improving-nutrition

Chen, C., Su, Q., Zi, M., Liu, Y., Wang, H., & Zhao, J. (2025). *Harnessing gut microbiota for colorectal cancer therapy: From clinical insights to therapeutic innovations. npj Biofilms and Microbiomes, 11,* 190. https://doi.org/DOI: 10.1038/s41522-025-00818-3

Cuisiniere, T., Hajjar, R., Oliero, M., Calvé, A., Fragoso, G., Vennin Rendos, H., Gerkins, C., Taleb, N., Gagnon-Konamna, M., Dagbert, F., Loungnarath, R., Sebajang, H., Schwenter, F., Wassef, R., Ratelle, R., De Broux, É., Richard, C., & Santos, M. M. (2025). Initial gut microbiota composition is a determining factor in the promotion of colorectal cancer by oral iron supplementation: Evidence from a murine model. *Microbiome*, *13*, 100. DOI: 10.1186/s40168-025-02101-1 PMID: 40259408

Das, B. K. (2022). Altered gut microbiota in hepatocellular carcinoma: Insights into the pathogenic mechanism and preclinical to clinical findings. *APMIS*, *130*, 719–740. DOI: 10.1111/apm.13282 PMID: 36321381

Dong, Y., Zhang, K., Wei, J., Ding, Y., Wang, X., Hou, H., Wu, J., Liu, T., Wang, B., & Cao, H. (2023). Gut microbiota-derived short-chain fatty acids regulate gastrointestinal tumor immunity: A novel therapeutic strategy? *Frontiers in Immunology*, *14*, 1158200. DOI: 10.3389/fimmu.2023.1158200 PMID: 37122756

Drost, J., & Clevers, H. (2018). Organoids in cancer research. *Nature Reviews. Cancer*, *18*(7), 407–418. DOI: 10.1038/s41568-018-0007-6 PMID: 29692415

EFESO Management Consultants & Giza Systems. (2024). Digital Health: The Case of Egypt. https://www.gizasystems.com/storage/publications/Digital%20Health%20Report-V3.pdf-1735646045.pdf gizasystems.com

Ekanger, C. T., Ramnefjell, M. P., Guttormsen, M. S. F., Hekland, J., Dahl-Michelsen, K., Lotsberg, M. L., Lu, N., Stuhr, L. E. B., Hoareau, L., Salminen, P. R., Gärtner, F., Aanerud, M., Akslen, L. A., Lorens, J. B., & Engelsen, A. S. T. (2025). An Organoid Model for Translational Cancer Research Recapitulates Histoarchitecture and Molecular Hallmarks of Non-Small Cell Lung Cancer. *Cancers (Basel)*, *17*(11), 1873. DOI: 10.3390/cancers17111873 PMID: 40507353

El-Derby, A. M., Schaaf, C. R., Shelkey, E., Cook, K. L., Votanopoulos, K. I., & Soker, S. (2024). Immune-reactive tumor organoids system to determine the effects of microbial metabolites on cancer immunity and immunotherapies. *Frontiers in Microbiomes*, *3*, 1411322. DOI: 10.3389/frmbi.2024.1411322

Eldridge, L., Goodman, N. R., Chtourou, A., Galassi, A., Monge, C., Cira, M. K., Pearlman, P. C., Loehrer, P. J.Sr, Gopal, S., & Ginsburg, O. (2025). Barriers and Opportunities for Cancer Clinical Trials in Low- and Middle-Income Countries. *JAMA Network Open*, *8*(4), e257733. DOI: 10.1001/jamanetworkopen.2025.7733 PMID: 40293747

Elsayed, H., & Saleh, R. (2024). Review of: Nutritional Genomics and Precision Nutrition. Bulletin of the National Nutrition Institute of the Arab Republic of Egypt, 64(2), 204-234. doi:DOI: 10.21608/bnni.2024.396356

Farley, T., Kiefer, J., Lee, P., Von Hoff, D., Trent, J. M., Colbourn, C., & Mousses, S. (2013). The BioIntelligence Framework: A new computational platform for biomedical knowledge computing. *Journal of the American Medical Informatics Association : JAMIA*, *20*(1), 128–133. DOI: 10.1136/amiajnl-2011-000646 PMID: 22859646

Fawzy, M. A., Afifi, M. T., Ahmed, S. M., & Shamseya, A. (2025). Study of the 16 s rRNA gene sequences of Gut Microbiome in some Egyptian Patients with Colorectal Cancer. *SVU-International Journal of Medical Sciences*, *8*(1), 351–364. DOI: 10.21608/svuijm.2023.220485.1609

Ferlisi, F., Tang, J., Cappelli, K., & Trabalza-Marinucci, M. (2023). Dietary supplementation with olive oil coproducts rich in polyphenols: A novel nutraceutical approach in monogastric animal nutrition. *Frontiers in Veterinary Science*, *10*, 1272274. DOI: 10.3389/fvets.2023.1272274 PMID: 37901105

Fernandez, E., Wargo, J. A., & Helmink, B. A. (2025). The microbiome and cancer: A translational science review. *Journal of the American Medical Association*, *333*(24), 2188–2196. DOI: 10.1001/jama.2025.2191 PMID: 40354071

Gagalova, K. K., Leon Elizalde, M. A., Portales-Casamar, E., & Görges, M. (2020). What You Need to Know Before Implementing a Clinical Research Data Warehouse: Comparative Review of Integrated Data Repositories in Health Care Institutions. *JMIR Formative Research*, *4*(8), e17687. DOI: 10.2196/17687 PMID: 32852280

Global Health Systems Symposium. (2024, November 11). Building Climate-Resilient Health Systems: The Focus of the 2024 Global Health Systems Symposium. Global Health Systems Symposium. https://healthsystemsglobal.org/news/building-climate-resilient-health-systems-the-focus-of-the-2024-global-health-systems-symposium/

Guo, Y., Gao, M., & Jiang, L.. (2025). Synthetic microbial consortia based on quorum-sensing for disease therapy. *Bioresources and Bioprocessing*, *12*(1), 125. DOI: 10.1186/s40643-025-00962-w PMID: 41182556

Hamamah, S., Amin, A., Al-Kassir, A. L., Chuang, J., & Covasa, M. (2023). Dietary Fat Modulation of Gut Microbiota and Impact on Regulatory Pathways Controlling Food Intake. *Nutrients*, *15*(15), 3365. DOI: 10.3390/nu15153365 PMID: 37571301

Han, G. Y. Q., Alexander, M., Gattozzi, J., Day, M., Kirsch, E., Tafreshi, N., Chalar, R., Rahni, S., Gossner, G., Burke, W., & Damaghi, M. (2024). Ecological and evolutionary dynamics to design and improve ovarian cancer treatment. *Clinical and Translational Medicine*, *14*(9), e70012. DOI: 10.1002/ctm2.70012 PMID: 39210542

Hanahan, D. (2022). Hallmarks of Cancer: New Dimensions. *Cancer Discovery*, *12*(1), 31–46. DOI: 10.1158/2159-8290.CD-21-1059 PMID: 35022204

Hassan, A. M., Naeem, S. M., Eldosoky, M. A. A., & Badr, A. W. (2025). Multiomics-based Machine Learning for the Subtype Classification of Breast Cancer. *Arabian Journal for Science and Engineering*, *50*(2), 1339–1352. DOI: 10.1007/s13369-024-09341-7

Herlo, L. F.. (2024). Gut microbiota signatures in colorectal cancer as a biomarker framework. *International Journal of Molecular Sciences*, *25*(14), 7937. DOI: 10.3390/ijms25147937 PMID: 39063179

Jacob, M., Reddy, R. P., Garcia, R. I., Reddy, A. P., Khemka, S., Roghani, A. K., Pattoor, V., Sehar, U., & Reddy, P. H. (2024). Harnessing Artificial Intelligence for the Detection and Management of Colorectal Cancer Treatment. *Cancer Prevention Research (Philadelphia, Pa.)*, *17*(11), 499–515. DOI: 10.1158/1940-6207.CAPR-24-0178 PMID: 39077801

Jiang, L., Lau, H. C., Zeng, R., & Yu, J. (2025). Diet, Gastric Microbiota, and Metabolites in Gastric Tumorigenesis. Research (Washington, D.C.), 8, 0693. https://doi.org/DOI: 10.34133/research.0693

Jiang, P., Klemeš, J. J., Fan, Y. V., Fu, X., & Bee, Y. M. (2021). More Is Not Enough: A Deeper Understanding of the COVID-19 Impacts on Healthcare, Energy and Environment Is Crucial. *International Journal of Environmental Research and Public Health*, *18*(2), 684. DOI: 10.3390/ijerph18020684 PMID: 33466940

Jin, K., Huang, Y., Che, H., & Wu, Y. (2025). Engineered Bacteria for Disease Diagnosis and Treatment Using Synthetic Biology. *Microbial Biotechnology*, *18*(1), e70080. DOI: 10.1111/1751-7915.70080 PMID: 39801378

Kang, M., Choe, D., Kim, K., Cho, B. K., & Cho, S. (2020). Synthetic Biology Approaches in The Development of Engineered Therapeutic Microbes. *International Journal of Molecular Sciences*, *21*(22), 8744. DOI: 10.3390/ijms21228744 PMID: 33228099

Kaput, J., & Rodriguez, R. L. (2004). Nutritional genomics: The next frontier in the postgenomic era. *Physiological Genomics*, *16*(2), 166–177. DOI: 10.1152/physiolgenomics.00107.2003 PMID: 14726599

Karam, F., El Deghel, Y., & Iratni, R. (2025). The Gut microbiome and colorectal cancer: An integrative review of the underlying mechanisms. *Cell Biochemistry and Biophysics*, *83*, 2637–2650. DOI: 10.1007/s12013-025-01683-9 PMID: 39946032

Karam, F., El Deghel, Y., Iratni, R., Dakroub, A. H., & Eid, A. H. (2025). The Gut Microbiome and Colorectal Cancer: An Integrative Review of the Underlying Mechanisms. *Cell Biochemistry and Biophysics*, *83*, 2637–2650. DOI: 10.1007/s12013-025-01683-9 PMID: 39946032

Kaur, S., & Kumar, S. (2020). Crosstalk between food components and microRNAs: Role in metabolism, nutrition, health and diseases. *Integrative Food, Nutrition and Metabolism*, *7*. Advance online publication. DOI: 10.15761/IFNM.1000293

Kaźmierczak-Siedlecka, K., Skonieczna-Żydecka, K., Palma, J., Sobocki, B. K., Świerblewski, M., Siedlecka-Kroplewska, K., Kalinowski, L., & Połom, K. (2023). Microbiota-derived metabolites in colorectal cancer patients in preoperative period. *European Review for Medical and Pharmacological Sciences*, *27*(4), 1443–1449. DOI: 10.26355/eurrev_202302_31384 PMID: 36876684

Khattab, R. H., Abo-Hammam, R. H., Salah, M., Hanora, A. M., Shabayek, S., & Zakeer, S. (2023). Multiomics analysis of fecal samples in colorectal cancer Egyptians patients: A pilot study. *BMC Microbiology*, *23*(1), 238. DOI: 10.1186/s12866-023-02991-x PMID: 37644393

Kim, K., Lee, M., Shin, Y., Lee, Y., & Kim, T.-J. (2025). Optimizing Cancer Treatment Through Gut Microbiome Modulation. *Cancers (Basel)*, *17*(7), 1252. DOI: 10.3390/cancers17071252 PMID: 40227841

Kim, T. H., Cho, B. K., & Lee, D. H. (2024). Synthetic Biology-Driven Microbial Therapeutics for Disease Treatment. *Journal of Microbiology and Biotechnology*, *34*(10), 1947–1958. DOI: 10.4014/jmb.2407.07004 PMID: 39233526

Lemme-Dumit, J. M. (2025). Intestinal organoid models as tools to interrogate the physiology of human mucosal tissues and host-microbe interactions. *MSphere*, *10*(8), e0082024. DOI: 10.1128/msphere.00820-24 PMID: 40772719

Leziak, A., Lipina, J., Reclik, M., & Kocelak, P. (2025). Dietary Modulation of Metabolic Health: From Bioactive Compounds to Personalized Nutrition. *Metabolites*, *15*(9), 624. DOI: 10.3390/metabo15090624 PMID: 41003008

Li, Y., Fu, L., Chen, Y., Yang, H., Xiao, Y., Ren, Y., & Wang, C. (2025). Integrated Metabolome and Microbiome Analysis Reveals the Regulatory Effects of Fermented Soybean Meal on the Gut Microbiota of Late Gestation. *Fermentation (Basel, Switzerland)*, *11*(6), 315. DOI: 10.3390/fermentation11060315

Li, Y., Wang, Y., Liang, T., Li, Y., & Du, W. (2025). A multiomics integration framework using multilabel guided learning and multiscale fusion. *Briefings in Bioinformatics*, *26*(5), bbaf493. Advance online publication. DOI: 10.1093/bib/bbaf493 PMID: 40996147

Liu, N., Jiao, N., Tan, J., Wang, Z., Wu, D., Wang, A., Chen, J., Tao, L., Zhou, C., Fang, W., Cheong, I. H., Pan, W., Liao, W., Kozlakidis, Z., Heeschen, C., Moore, G. G., Zhu, L., Chen, X., Zhang, G., & Wang, H. (2022). Multikingdom microbiota analyses identify bacterial–fungal interactions and biomarkers of colorectal cancer across cohorts. *Nature Microbiology*, *7*, 238–250. https://www.google.com/search?q=https://doi.org/10.1038/s41564-021-01030-7 PMID: 35087227

Liu, Y., Wang, D., Luan, Y., Tao, B., Li, Q., Feng, Q., Zhou, H., Mu, J., & Yu, J. (2024). The application of organoids in colorectal diseases. *Frontiers in Pharmacology, 15*, 1412489. DOI: 10.3389/fphar.2024.1412489 PMID: 38983913

Loftfield, E., Falk, R. T., Sampson, J. N., Huang, W. Y., Hullings, A., Murphy, G., Weinstein, S. J., Albanes, D., Freedman, N. D., & Sinha, R. (2022). Prospective Associations of Circulating Bile Acids and Short-Chain Fatty Acids With Incident Colorectal Cancer. *JNCI Cancer Spectrum*, *6*(3), pkac027. Advance online publication. DOI: 10.1093/jncics/pkac027 PMID: 35583137

Lyon, P. (2015). The cognitive cell: Bacterial behavior reconsidered. *Frontiers in Microbiology*, *6*, 264. DOI: 10.3389/fmicb.2015.00264 PMID: 25926819

Ma, Z., Zuo, T., Frey, N., & Rangrez, A. Y. (2024). A systematic framework for understanding the microbiome in human health and disease: From basic principles to clinical translation. *Signal Transduction and Targeted Therapy*, *9*(1), 237. DOI: 10.1038/s41392-024-01946-6 PMID: 39307902

Menigoz, W., Latz, T. T., Ely, R. A., Kamei, C., Melvin, G., & Sinatra, D. (2020). Integrative and lifestyle medicine strategies should include Earthing (grounding): Review of research evidence and clinical observations. *Explore (New York, N.Y.)*, *16*(3), 152–160. DOI: 10.1016/j.explore.2019.10.005 PMID: 31831261

Miehe, R., Buckreus, L., Kiemel, S., Sauer, A., & Bauernhansl, T. (2021). A Conceptual Framework for Biointelligent Production—Calling for Systemic Life Cycle Thinking in Cellular Units. *Cleanroom Technology*, *3*(4), 844–857. DOI: 10.3390/cleantechnol3040049

Ministry of Health and Population. (2024). *Egypt National Health Strategy 2024–2030* [PDF]. Arab Republic of Egypt. Retrieved from https://www.mohp.gov.eg/Publications.aspx?cat_id=1041. Ministry of Health and Population+1

Ministry of Planning and Economic Development. (2023). *Sustainable Development Strategy: Egypt's Vision 2030* [Updated version]. Arab Republic of Egypt. Retrieved from https://mped.gov.eg/Files/Egypt_Vision_2030_EnglishDigitalUse.pdf. Ministry of Planning

Mir, S. S. (2025). Innovations in medical microbiology for advancing precision medicine: Integrating microbiome data with clinical and genetic information. East African Journal of Biological Sciences, 2(1), 55-68. Retrieved from https://eajbsc.journals.ekb.eg/article_411912_1804af7cb774a9c024771e8d8470b690.pdf eajbsc.journals.ekb.e.g.

Miri, S., Yeo, J., Abubaker, S., & Hammami, R. (2023). Neuromicrobiology, an emerging neurometabolic facet of the gut microbiome? *Frontiers in Microbiology*, *14*, 1098412. DOI: 10.3389/fmicb.2023.1098412 PMID: 36733917

National, A. I. Council. (2025, January 16). Egypt National Artificial Intelligence Strategy, Second Edition (2025–2030). https://ai.gov.eg/SynchedFiles/en/Resources/AIstrategy%20English%2016-1-2025-1.pdf

Pandey, M., & Bhattacharyya, J. (2024). Gut microbiota and epigenetics in colorectal cancer: Implications for carcinogenesis and therapeutic intervention. *Epigenomics*, *16*(6), 403–418. DOI: 10.2217/epi-2023-0382 PMID: 38410915

Papamichail, L., Koch, L. S., Veerman, D., Broersen, K., & van der Meer, A. D. (2025). Organoids-on-a-chip: Microfluidic technology enables culture of organoids with enhanced tissue function and potential for disease modeling. *Frontiers in Bioengineering and Biotechnology*, *13*, 1515340. DOI: 10.3389/fbioe.2025.1515340 PMID: 40134772

Park, J. C., Chang, L., Kwon, H., & Im, S. (2025). Beyond the gut: Decoding the gut–immune–brain axis in health and disease. *Cellular & Molecular Immunology*, *22*, 1287–1312. DOI: 10.1038/s41423-025-01333-3 PMID: 40804450

Pienta, K. J., McGregor, N., Axelrod, R., & Axelrod, D. E. (2008). Ecological therapy for cancer: Defining tumors using an ecosystem paradigm suggests new opportunities for novel cancer treatments. *Translational Oncology*, *1*(4), 158–164. DOI: 10.1593/tlo.08178 PMID: 19043526

Rabie, M. (2021). A roadmap to innovation management and national innovation system in Egypt. *Arab States Journal for Sustainable Science and Technology*, *2*. Advance online publication. DOI: 10.21608/ARABSTI.2021.209210

Ratto, D., Cavalloro, V., Tumminelli, E., Soffientini, I., Martino, E., Rossi, D., Collina, S., & Rossi, P. (2025). From waste to worth: Transforming winemaking residues into high-value ingredients. *ACS Food Science & Technology*. Advance online publication. DOI: 10.1021/acsfoodscitech.5c00519 PMID: 41306649

Reyed, R. M. (2022). Practical implementation of nutritional services, gut microbiome, combination therapy, and nutraceutical supplements. *Microenvironments and Microecology Research*, *4*(2), 8. DOI: 10.53388/MMR2022008

Reyed, R.M. (2023a). Simulate the dynamic interactions across nutritional supplements, the polybiome, and the biochemical approach dependent on patient derived gastrointestinal knowledge. *Journal of Clinical Epidemiology & Toxicology*, *4*, 133. DOI: 10.47363/JCET/2023(4)133

Reyed, R. M. (2023b). Focusing on individualized nutrition within the algorithmic diet: An in-depth look at recent advances in nutritional science, microbial diversity studies, and human health. *Food & Health*, *5*(1), 5. DOI: 10.53388/FH2023005

Reyed, R.M., & Prabhakar, P. K. M. (2025a). Polybiome Derived Therapies and Fermentobiometrics: Novel Strategies for Blocking Cancer Pathways and Progression. In Cancer Etiology, Diagnosis and Treatments (pp. 145-212). Nova Medicine and Health/Pathways to Healing: Advancements in Cancer Therapy. DOI: 10.52305/JKMU3682

Reyed, R. M., & Prabhakar, P. K. (2025b). Mechanisms, Impact, and Therapeutic Strategies for Oxidative Stress in Metabolic Diseases: Understanding and Mitigating Oxidative Stress. In Prabhakar, P. (Ed.), *The Role of Reactive Oxygen Species in Human Health and Disease* (pp. 203–256). IGI Global Scientific Publishing. DOI: 10.4018/979-8-3693-7919-6.ch006

Reyed, R. M., & Prabhakar, P. K. (2025c). Impact of Dietary Antioxidants on Redox Homeostasis and Its Effects on Health and Disease. In Prabhakar, P. (Ed.), *The Role of Reactive Oxygen Species in Human Health and Disease* (pp. 81–136). IGI Global Scientific Publishing. DOI: 10.4018/979-8-3693-7919-6.ch003

Reyed, R. M., & Prabhakar, P. K. (2025d). Personalized Antiviral Therapy Tailoring Treatments to Genetic Environmental and Lifestyle Factors. In Aslam, M., Yingngam, B., Javed, T., & Riaz, S. (Eds.), *Global Perspectives on Antiviral Drug Development* (pp. 63–138). IGI Global Scientific Publishing. DOI: 10.4018/979-8-3693-9276-8.ch003

Reyed, R. M., & Prabhakar, P. K. (2026). Modulating Neuroinflammation and Autophagy in Psychiatric Disorders Through Mechanisms and Therapies: Therapeutic Strategies and Pathways. In Mokhtari, T., & Uludag, K. (Eds.), *Autophagy and Inflammation in Neuropsychological Disorders* (pp. 61–124). IGI Global Scientific Publishing. DOI: 10.4018/979-8-3693-5908-2.ch003

Rosenthal, M., Goldberg, D., Aiello, A., Larson, E., & Foxman, B. (2011). Skin microbiota: microbial community structure and its potential association with health and disease. Infection, genetics and evolution: journal of molecular epidemiology and evolutionary genetics in infectious diseases, 11(5), 839–848. DOI: 10.1016/j.meegid.2011.03.022

Salunkhe, A. A., Tamboli, F. A., Zade, M. S., Kore, M. D., & Mhakle, R. A. (2023). Nutrigenomics: A new approach to the diagnosis and prevention of diseases. *J Nutr Metab Health Sci*, *6*(4), 148–156. DOI: 10.18231/j.ijnmhs.2023.025

Saylam, E., Özden, Ö., Yerlikaya, F. H., Sivrikaya, A., Yormaz, S., Arslan, U., Topkafa, M., & Maçin, S. (2025). Investigation of Intestinal Microbiota and Short-Chain Fatty Acids in Colorectal Cancer and Detection of Biomarkers. *Pathogens (Basel, Switzerland)*, *14*(9), 953. DOI: 10.3390/pathogens14090953 PMID: 41011853

Senthil Kumar, K., Miskovic, V., Blasiak, A., Sundar, R., Pedrocchi, A. L. G., Pearson, A. T., Prelaj, A., & Ho, D. (2023). Artificial Intelligence in Clinical Oncology: From Data to Digital Pathology and Treatment. American Society of Clinical Oncology educational book. American Society of Clinical Oncology. Annual Meeting, 43, e390084. DOI: 10.1200/EDBK_390084

Shaharudin, M. R., Abdullah, D., Zainoddin, A. I., Legino, R., & Wararatchai, P. (2024). The evolution of circular economy: A literature review on sustainability transitions and challenges. *International Journal of Research and Innovation in Social Science*, *8*(10), 102–115. DOI: 10.47772/IJRISS.2024.8100009

Siguenza, N., Brevi, A., Zhang, J. T., Pabani, A., Bhushan, A., Das, M., Ding, Y., Hasty, J., Ghosh, P., & Zarrinpar, A. (2024). Engineered bacterial therapeutics for detecting and treating CRC. *Trends in Cancer*, *10*(7), 588–597. DOI: 10.1016/j.trecan.2024.04.001 PMID: 38693003

Silver, B. B., & Nelson, C. M. (2018). The Bioelectric Code: Reprogramming Cancer and Aging From the Interface of Mechanical and Chemical Microenvironments. *Frontiers in Cell and Developmental Biology*, *6*, 21. DOI: 10.3389/fcell.2018.00021 PMID: 29560350

Singh, V. K., Hu, X.-H., Singh, A. K., Solanki, M. K., Vijayaraghavan, P., Srivastav, R., Joshi, N. K., Kumari, M., Singh, S. K., Wang, Z., & Kumar, A. (2024). Precision nutrition-based strategy for management of human diseases and healthy aging: Current progress and challenges forward. *Frontiers in Nutrition*, *11*, 1427608. DOI: 10.3389/fnut.2024.1427608 PMID: 39183982

Sinha, A., Maibam, M., Jain, R., Aggarwal, K., Sahu, A. K., Gupta, P., Paul, S., Bisht, B., & Paul, M. K. (2025). Organoid: Biomedical application, biobanking, and pathways to translation. *Heliyon*, *11*(10), e43028. DOI: 10.1016/j.heliyon.2025.e43028

Souza, A. P., Marinho, V., & Marques, M. R. (2025). The Fundamental Role of Nutrients for Metabolic Balance and Epigenome Integrity Maintenance. *Epigenomes*, *9*(3), 23. DOI: 10.3390/epigenomes9030023 PMID: 40700039

Suarez, C., & Adibi, S. (2025). Integrating Traditional Nutritional Wisdom into Digital Nutrition Platforms: Toward Culturally Adaptive and Inclusive Health Technologies. *Nutrients*, *17*(12), 1978. DOI: 10.3390/nu17121978 PMID: 40573089

Szopa, M., Wybrańska, I., Malczewska-Malec, M., & Dembińska-Kieć, A. (2005). Nutrigenomika—czyli o genetyce i zywieniu [Nutrigenomics--about genetics and nutrition]. *Przeglad Lekarski, 62*(4), 245–252. PMID: 16229245

Takamaru, H., Tsay, C., Shiba, S., Yachida, S., & Saito, Y. (2025). Microbiome and Colorectal Cancer in Humans: A Review of Recent Studies. *Journal of the Anus, Rectum and Colon, 9*(1), 20–24. DOI: 10.23922/jarc.2024-080 PMID: 39882224

Tiwari, A., Mishra, S., & Kuo, T. (2025). Current AI technologies in cancer diagnostics and treatment. *Molecular Cancer, 24*, 159. DOI: 10.1186/s12943-025-02369-9 PMID: 40457408

Tudorache, M., Treteanu, A. R., Gradisteanu Pircalabioru, G., Lixandru-Petre, I. O., Bolocan, A., & Andronic, O. (2025). Gut Microbiome Alterations in Colorectal Cancer: Mechanisms, Therapeutic Strategies, and Precision Oncology Perspectives. *Cancers (Basel), 17*(14), 2294. DOI: 10.3390/cancers17142294 PMID: 40723178

Wang, Y., Ge, M., Wang, J., Xu, Y., Wang, N., & Xu, S. (2025). Metabolic reprogramming in ischemic stroke: When glycolytic overdrive meets lipid storm. *Cell Death & Disease, 16*(1), 788. DOI: 10.1038/s41419-025-08114-w PMID: 41184254

Willmann, K., & Moita, L. F. (2025). Biological and clinical implications of a model of surveillance immunity. *The Journal of Clinical Investigation, 135*(15), e191645. DOI: 10.1172/JCI191645 PMID: 40759580

Woo, S. G., Kim, S. K., Lee, S. G., Kim, S. H., Park, J. H., & Kim, M. K. (2025). Engineering probiotic *Escherichia coli* for inflammation-responsive indoleacetic acid production using RiboJ-enhanced genetic circuits. *Journal of Biological Engineering, 19*, 10. DOI: 10.1186/s13036-025-00479-y PMID: 39838372

Wu, G., Wu, Y., Wu, X., Shen, H., Chu, J., Qu, Z., & Han, S. (2025). Fecal occult blood affects intestinal microbial community structure in colorectal cancer. *BMC Microbiology, 25*, 34. DOI: 10.1186/s12866-024-03721-7 PMID: 39833681

Wu, Y., Sha, Y., Guo, X., Gao, L., Huang, J., & Liu, S.-B. (2025). Organoid models: Applications and research advances in colorectal cancer. *Frontiers in Oncology, 15*, 1432506. DOI: 10.3389/fonc.2025.1432506 PMID: 39990692

Xu, K., Motiwala, Z., Corona-Avila, I., Makhanasa, D., Alkahalifeh, L., & Khan, M. (2025). The gut microbiome and its multifaceted role in cancer metabolism, initiation, and progression: Insights and therapeutic implications. *Technology in Cancer Research & Treatment, 24*. Advance online publication. DOI: 10.1177/15330338251331960 PMID: 40208053

Yang, Y. C., Chang, S. C., Hung, C. S., Shen, M. H., Lai, C. L., & Huang, C. J. (2025). Gut-Microbiota-Derived Metabolites and Probiotic Strategies in Colorectal Cancer: Implications for Disease Modulation and Precision Therapy. *Nutrients*, *17*(15), 2501. DOI: 10.3390/nu17152501 PMID: 40806086

Yazdanian, A., Ayatollahi, H., & Nahvijou, A. (2020). A Conceptual Model of an Oncology Information System. *Cancer Management and Research*, *12*, 6341–6352. DOI: 10.2147/CMAR.S259013 PMID: 32821154

Zhang, W., Xiang, Y., Ren, H., Yang, H., & Zhou, Y. (2025). The tumor microbiome in cancer progression: Mechanisms and therapeutic potential. *Molecular Cancer*, *24*, 195. DOI: 10.1186/s12943-025-02403-w PMID: 40665305

Zhang, Y., Meng, R., Sha, D., Gao, H., Wang, S., Zhou, J., Wang, X., Li, F., Li, X., & Song, W. (2024). Advances in the application of colorectal cancer organoids in precision medicine. *Frontiers in Oncology*, *14*, 1506606. DOI: 10.3389/fonc.2024.1506606 PMID: 39697234

Zitvogel, L., Derosa, L., Routy, B., Loibl, S., Heinzerling, L., de Vries, I. J. M., & Engstrand, L., ONCOBIOME Network, Segata, N., & Kroemer, G. (2025). Impact of the ONCOBIOME network in cancer microbiome research. *Nature Medicine*, *31*(4), 1085–1098. DOI: 10.1038/s41591-025-03608-8 PMID: 40217075

Zou, D., Yin, Z., Yi, S. G., Wang, G., Guo, Y., Xiao, X., & Chen, W. (2024). CD4+ T cell immunity is dependent on an intrinsic stem-like program. *Nature Immunology*, *25*(1), 66–76. PMID: 38168955

Chapter 4
AI-Guided Monoclonal Antibody Design Integrating Polybiome and Nutrigenomic Insights:
Integrating AI, Polybiome, and Nutrigenomics in Therapy

Ashraf Tabll
http://orcid.org/0000-0001-7041-5445
Egypt Centre for Research and Regenerative Medicine, Egypt

ABSTRACT

This chapter explores how the microbiome, the broader concept of the polybiome, and nutrigenomic factors converge within precision immunotherapy. Monoclonal antibodies (mAbs), once confined to infectious and oncological diseases, are now being designed to correct dysbiosis and nutrigenomic disturbances, restoring immune and metabolic balance. Artificial intelligence (AI)-driven discovery refines antibody specificity and predicts therapeutic outcomes. Case studies demonstrate how mAbs interact with microbial metabolites and individual nutrigenomic profiles. By integrating AI, polybiome profiling, and nutritional genetics, this chapter outlines a new personalized framework for chronic and autoimmune disease management.

DOI: 10.4018/979-8-3373-3121-8.ch004

1. INTRODUCTION

1.1. Background and Rationale:

The cross-linking of the polybiome (defined as the cooperative microbial bio-networks of the human body encompassing bacteria, fungi, viruses, and archaea with host metabolism and immunity) (Reyed & Prabhakar, 2025), immunotherapy, nutrigenomics, and artificial intelligence has provided some of the most reformist information in precision medicine over the past decade (Chang et al., 2025). The human polybiome, which comprises trillions of microorganisms that populate the human body, has become a core determinant of immune modulation, host disease susceptibility, and responsiveness to treatment in a continuum of cancers, autoimmune diseases and metabolic diseases (Theis et al., 2025). Moreover, the results of nutrigenomic studies highlight the complexity of the links between diet, microbial composition, and immune control, which support the necessity of promoting health and preventing disease through sustainable and individualized approaches (Do Rosario et al., 2025).

The development of monoclonal antibody (mAb) technology has shifted the focus of traditional neutralization of pathogens to the now equalized manipulation of polybiome-derived antigens, which has relieved diseases that are related to dysbiosis and thus was previously mediated by broad-spectrum antibiotics, probiotics, or dietary supplements (Alam et al. 2023).

Moreover, machine-learning-based discovery engines make it possible to analyze large-scale multiomics data, build predictive models to understand host–polybiome and nutrigenomic interactions, and design mAbs with improved affinity and specificity. These algorithms have the potential to predict antibody–antigen complexes, isolate individualized microbial and metabolic targets, and optimize therapeutic effects, which significantly decreases the development time and expenditure. Nevertheless, the use of mAb therapeutics in the broadened polybiome-nutrigenomic paradigm is still experimental (de la Cuesta-Zuluaga et al., 2024).

The future obstacles include complex problems: identification of the most appropriate microbial and host targets, elucidation of the complexity of immunological cross-talk of the gastrointestinal polybiome and the host immune system, and incorporation of dietary genomic parameters and maintenance of therapeutic specificity and safety in different patient groups. These barriers can be overcome with precision medicine systems that include polybiome profiles, genetic predispositions, and nutrigenomic maps that provide strategic opportunities to create highly customized treatment options (Chaudhary et al., 2025).

The existing chapter considers the potential to utilize AI-enhanced monoclonal antibody adherents to refine pale and nutrigenomic patterns, potentially facilitating next-generation and stratified advancements in therapeutics. The strategy is based on the idea of combining the knowledge of immunology, computational biology, polybiome science, and nutritional genomics to present a multidisciplinary roadmap to implementing mAbs as a precision option in individual healthcare paradigms. This interdisciplinary integration highlights the scientific and translational opportunities for chronic inflammatory, metabolic and autoimmune disease treatment.

1.2. Scope and Objectives:

This chapter serves as a critical review to address how monoclonal antibody (mAb) technology is converging rapidly with the fields of polybiome science, artificial intelligence (AI), immunotherapy, and precision medicine. With improved knowledge of the human polybiome, therapeutic approaches to manipulate microbial communities to achieve improved health benefits are being developed. Monoclonal antibodies that have historically been used in treating infectious diseases and malignancies are currently under investigation because of their ability to either directly or indirectly tune the polybiome by acting on host–microbe interactions (Gleeson, et al., 2024). This chapter attempts to form a multidisciplinary paradigm that combines immunological engineering, AI-driven biomarker discovery, and personalized therapeutic design.

Building on the concept of integrative thinking, this chapter describes the application of nutrigenomics, which investigates nutrient–gene interactions on the basis of AI-enabled mAb use to enable genuinely personalized therapeutic responses (Lagoumintzis and Patrinos, 2023). Individual nutrigenomic and polybiome profiles, when analyzed and connected to each other, allow the personalization of mAbs to target the context of dietary-genomic specifics, with precision medicine being used at different levels (Singar et al., 2024). This integrative approach is likely not only to significantly expand our immune and microbial regulation capacity but also to reinvent personalized nutrition as a coregulator of mAb efficacy, which is part of future and patient-specific treatment designs (Phugat & Goel, 2025).

By incorporating mAbs in an integrative framework of knowledge that converges immunology, computational science, and polybiome research in a framework supported by AI and precision medicine, this chapter intends to explain the positive potential of mAbs to scholars, clinicians, and biotechnologists alike. These findings suggest that a systems-level solution to therapeutic design is dynamically interactive, individualized, and selectively sensitive to complex interactions between the host immune system and adherent microbes.

1.3. Interdisciplinary Nexus: Polybiome, AI, and Immunotherapy

The extensive field development in the study of polybiomes has transformed our view of human health and disease and illustrated that the microbiota is a living and indispensable component of host physiology, particularly concerning immune activity. This growing appreciation has provoked studies attempting to evaluate polybiome control as a therapeutic approach for diverse disorders, such as chronic inflammation, metabolic disorders, cancer, and infectious diseases. The complexity of microbial communities, their individual features, and the multidimensionality of interactions with the human immune system require solutions in balance with the high levels of complexity of treatment and analysis (O'Riordan et al., 2025). Immunotherapy has revealed a new tool for facilitating specific immune responses by targeting checkpoints or specific receptors with monoclonal antibodies as specific therapeutic agents because of its ability to selectively modulate certain immune circuits, neutralize pathogens and change immune responses. Monoclonal antibodies are traditionally used in the fields of oncology, autoimmune disorders, and infectious diseases, but currently, they are considered effective with respect to antibodies against antigens produced by the microbiota, bacterial metabolites, and host receptors under conditions of polypeptide activity. This conceptual shift suggests a substitutive shift wherein scientists started to focus only on human-derived antigens and prioritize the concept of the polybiome as an available and therapeutically useful area (Aboul-Ella et al., 2024).

Artificial intelligence has become an effective point of integration in this context, providing the computational basis needed to examine high-dimensional polybiome and immunological data. Machine learning algorithms have the potential to discover latent behaviors in datasets of metagenomic and transcriptomic data, infer disease-related microbial predicted signatures and model host microbe interplay with immunologic processes. Such functionalities will be needed to find new antibody therapy targets and to develop improved individual patient treatment protocols on the basis of real-time patient data (Dakal et al., 2025). At the heart of this interdisciplinary convergence is precision medicine, which involves the personalization of medical treatment according to the individual characteristics of patients, such as their genetic composition, immunological phenotype, and microbial flora. Researchers and clinicians can overcome their reliance on generic pharmacological interventions and can achieve more specific, adaptive, and effective therapies by using AI to stratify patients on the basis of their polybiome and immunological phenotypes and by using monoclonal antibodies with unprecedented precision (Dewaker et al., 2025).

Nutrigenomics can be described as the study of how personal genetic variations control reactions to nutrients in food and thus affect the makeup of the polybiome and immune responses. Recent innovations have criticized the old paradigm of interdisciplinarity (Lam et al., 2025). Nutrigenomic datasets assimilated with the assistance of artificial intelligence analytics and acquired on monoclonal platforms further allow the development of individualized therapeutic interventions on the basis of the individual, dietary-genomic situation among patients (Chaudhary et al., 2025; de Roos B., 2013; Ramos-Lopez et al., 2017). This focus can increase the specificity of appropriate strategies aimed at facilitating polybiome immune interactions but, more crucially, highlight the associations among personalized nutrition methods and the ultimate efficacy of the maximum number of monoclonal antibodies in accurate medical frameworks (Wiertsema et al., 2021).

This convergence is not only additive but also paradigm shift. Research on polybiomes combined with immunotherapy and artificial intelligence is redefining our approach to investigating disease causal mechanisms, developing pharmacotherapies, and measuring clinical outcomes. It supports a cycle of feedback that is rich in feedback, with real-world patient data driving the narrowing of the therapeutic process, hence becoming more accurate and adaptive in clinical practice. The development of an interdisciplinary scope requires interdisciplinary cooperation among microbiology, immunology, computational biology, systems medicine and clinical practice. This overlap of these domains outlines a new era in therapeutic creation, whereby monoclonal antibodies and immunomodulation are utilized not only as interventions but also as an elaborate means of decoding and redefining the language of the human polybiome, hence offering a holistic explanation of the complex internal interactions of the human body (Wiertsema et al., 2021). This theoretical integration is represented in Figure 1, which encapsulates the multidimensional framework of the chapter that combines artificial intelligence, monoclonal antibodies, nutrigenomics, and the polybiome within the context of precision medicine.

Figure 1.

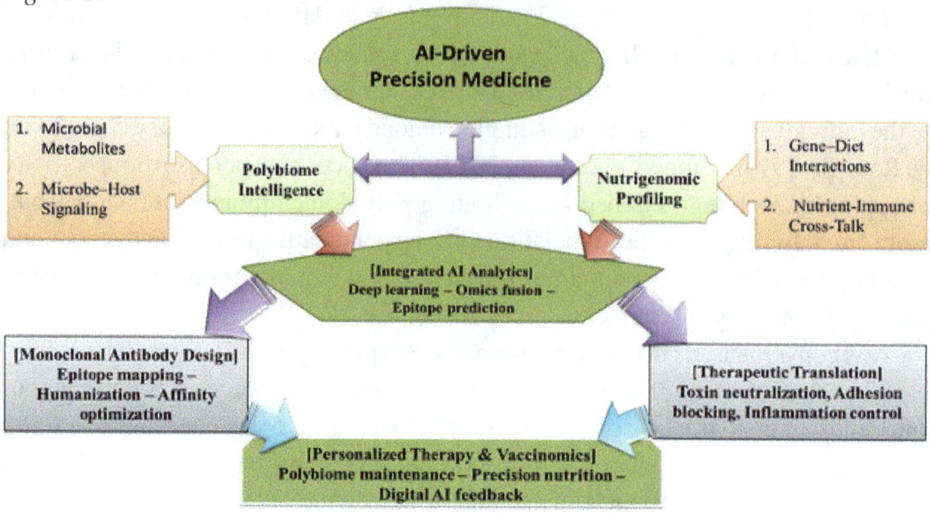

The figure illustrates how artificial intelligence, polybiome science, nutrigenomics, and immunotherapy intersect in a multidisciplinary way, revealing the translational journey through the integration of omics data into personalized therapy design.

2. MONOCLONAL ANTIBODIES IN POLYBIOME PROFILING AND DETECTION

2.1. Principles of mAb Production and Specificity

Monoclonal antibodies are highly specialized types of immunoglobulines that are formed on a single clone of B cells and are intended to bind to a specific epitope of a particular antigen (Hanack et al., 2016). They have high accuracy and can be broadly used, which makes them very useful in diagnostic, treatment, and investigative settings. Their production and specificity are controlled by determinants, the understanding of which is vital to their ability to maximize functional performance and clinical efficacy (Tabll et al., 2015).

1 Antigen selection and epitope targeting

The initial stages of developing a monoclonal antibody include the finding of a suitable target antigen, usually a protein, peptide or carbohydrate structure of a pathogen; a cancer-causing cell; or an unregulated host element (Zhou et al., 2022).

This choice can include bacterial surface proteins, virulence factors or host receptors that are adjusted to microbial metabolites in polybiome-related research. The antigen should have acceptable immunogenicity, be accessible to the antibody, and be distinct so that it can be exposed to a minimum of off-target effects (Li, 2022).

2 Hybridoma Technology

The concept of hybridoma technology, which was developed by Kohler and Milstein in 1975 (Köhler & Milstein, 1975), continues to be the key to the conventional production of monoclonal antibodies. It involves immunization of a mouse or any other suitable host via the target antigen. The fusion of B cells with immortal myeloma cells is then achieved by first isolating B cells in the spleen of the immunized animal and then fusing them (El Abd et al., 2022).

The screening of such hybridomas determines the clones that secrete the intended antibody and then multiplies to yield clinically relevant levels (Zola and Brooks, 2018). Despite the fact that it is implemented to a great extent, the validity of the method may be restricted by its species specificity and possible immunogenicity when used in human therapy (Mitra and Tomar 2021).

Current studies have surpassed the conventional idea of immunology and used artificial intelligence to predict immunodominant epitopes (Acar et al., 2024), incorporating polybiome data into the calculation of antigens that mediate host–microbe interactions (Wiertsema et al., 2021) and interrogating the nutrigenomic signatures that tune antigenicity in pharmacological contexts. These improvements make monoclonal antibodies more specific and enable innovations in the form of personalized immunotherapy depending on polypeptide structure and dietary intake (David & Lev-Ari 2024).

3 Recombinant Antibody Technology

Modern methods have replaced classical hybridoma methods. Recombinant approaches include the cloning and isolation of immunoglobulin genes via B lymphocytes, which may then be produced via a prokaryotic, yeast, or mammalian apparatus. The goals of altering the constant domains of the antibody are to reduce immunogenicity through humanization or to enhance effector functions (Basu et al., 2019). Such methodologies enable higher yield levels, higher internal accuracy, and patient-centered treatment (Sifniotis et al., 2019).

In addition to the structural and production advantages recorded by recombinant antibody methodologies, there have continued to be advances in contemporary platforms through the integration of artificial intelligence (AI) workflows. Such AI-based pipelines are expected to maximize sequence architecture, improve prediction of antigen-binding properties, and tolerate epitope differences elicited by the polybiome or specific nutrigenomic programmes. The completely integrated structure that follows significantly increases the design abilities of monoclonal antibodies. Thus, the generated antibodies are not simply structurally optimized but also personalized according to the polybiome immune and dietary genomic environment of personhood. It is expected that such high-precision therapeutic constructs will have greater efficacy and safety in targeted biotherapeutics (Grewal, 2024; Wiertsema et al., 2021; Ramos-Lopez et al., 2017).

4 Humanization and Fully Human mAbs

Frequently, the complementarity-determining regions (CDRs) involved in antigen binding are grafted onto human antibody scaffolds to humanize murine monoclonal antibodies and lower immunogenicity in recipients (Ministro et al., 2020). Alternative sources of fully human antibodies are miRNA transgenic mice, which are engineered to express human immunoglobulin loci, and phage and yeast libraries, which display the entire human immunoglobulin repertoire. Humanized and fully human monoclonal antibodies have become more common in contemporary clinical practice and are described via machine learning-enhanced epitope selection regimes, whereby personalized genomic, microbial, and nutrigenomic datasets can be incorporated to achieve improved specificity of therapy.

5 Specificity and Affinity

The greatest specificity of monoclonal antibodies is their remarkable specificity, which is caused by the unique variable domains of the heavy and light chains that make up the region that binds the antigen. Long target antigen interactions are critical to resist target dissociation and are improved by FEV technology methods, including somatic hypermutation or in vitro directed evolution (Kothari et al., 2024). Specificity can also reduce off-target interactions, which is highly important in the polybiome setting, where different microbial taxa tend to have similar conserved molecular motifs (Bondad-Reantaso et al., 2023). Second-generation treatment plans progressively combine AI-based affinity maturation and profiling, facilitating the emergence of treatments based on individual polybiome and nutrigenomic environments (Dewaker et al., 2025).

6 Functional Considerations

Synthetic monoclonal antibodies have the potential to inhibit interactions between a receptor and a ligand, such as an immune checkpoint or a cytokine receptor

(Stone et al., 2024). Immunoefficacy is activated through Fc interactions, such as those that induce complement activation or antibody-dependent cellular cytotoxicity (ADDC) (Ochoa et al., 2017). Potential active interventions targeted to slow down the polybiome include the neutralization or prevention of microbial enzymes or inflammatory cue secretions of harmful bacteria (Mishra et al., 2023). Recent developments in precision medicine have built on artificial neural network-based functional profiling to develop monoclonal antibodies predictive of polybiome-associated immune conditions and consistent with patient-specific nutrigenomic information to allow intoxicated adaptive biology to be selectively modulated and tailor responses to patients (Dewaker et al., 2025; Ke et al., 2024; Lam et al., 2025)

Table 1. AI-driven monoclonal antibody design workflow

Workflow Phase	Omics & Metadata Inputs	Computational Methods	Translational Breakthroughs	References
Epitope Discovery	Metagenomic & metatranscriptomic profiles	Deep learning–based epitope mapping	Identification of disease-specific antigenic regions	Hu et al., 2025
Affinity Optimization	Antibody repertoire sequencing & structural datasets	Machine learning–driven affinity maturation	High-affinity mAb variants with improved binding kinetics	Matsunaga et al., 2025
Functional Profiling	Proteomic, metabolomic & cytokine response datasets	Predictive modeling of antibody–antigen–pathway interactions	Selection of mAbs that modulate key inflammatory or microbial pathways	Vebr et al., 2023
Adaptive Vaccine Formulation	Multi-omic integration (HLA typing, nutrigenomics, polybiome)	In silico docking & neural network immunogenicity scoring	Personalized multi-epitope vaccine constructs tailored to host context	Guo et al., 2025

A preceding table outlines the main four steps of a typical AI-based pipeline in monoclonal antibody (mAb) development and indicates the resulting data inputs to computational methods and the development of therapeutics via these methods at each step.

2.2. Targeted identification of microbial species via monoclonal antibodies

Monoclonal antibodies (mAbs) are powerful instruments that are able to selectively identify and characterize specific microbial taxa in complex biological environments by covalently attaching themselves to specific molecular moieties (Vacca et al., 2022). In polybiome studies, mAbs act as a means of understanding and monitoring computed or illness-causing strains or as LEDs to identify hosts and microbes (Helmy et al., 2023; Zhang et al., 2021).

1 Antigen-specific recognition principle.

The antigenic determinants, or epitopes, that monoclonal antibodies discriminate are found on the microbial cell surface. These epitopes can be proteins, polysaccharides (e.g., lipopolysaccharides or capsular antigens), teichoic acids or any other constituent of the cell wall that varies among species and intraspecies (Nagy et al., 2017). This specificity allows distinguishing between closely related microorganisms that would otherwise be indistinguishable via traditional culture methods or molecular techniques (Posner et al., 2019).

2 Applications in Microbial Detection

Monoclonal antibodies can be used in various forms to recognize and identify microbial taxa. The use of fluorophore- or enzyme-conjugated mAbs allows the detection and quantification of microorganisms in a tissue or environmental matrix (Su et al., 2020). mAbs can be used to detect microbial antigens within bodily fluids or tissue extracts (Tabll et al., 2024). The direct surface antigens of microbial populations can be rapidly sorted and enumerated with fluorescently labeled monoclonal antibodies through a flow cytometry method (Sliwa-Dominiak et al., 2025; Wang et al., 2025). Monoclonal antibodies, such as those against *Clostridium difficile*, Salmonella spp., and *Helicobacter pylori*, are used in the development of point-of-care tests to detect antigens in clinical samples via rapid lateral flow assays (Lin et al., 2015). Recent developments in multiplex profiling to include immunoglobulin off-target effects have led to the introduction of new monoclonal antibody selection and engineering protocols. With artificial intelligence, which is applied to host immune and dietary–genomic molecular signatures, researchers can now tailor monoclonal antibody design, thus promoting the establishment of next-generation, personalized diagnostics and treatment interventions.

3 Advantages over Molecular Techniques

Despite the abundance of taxonomic resolution via DNA-based approaches, such as 16S rRNA sequencing and metagenomics (Einarsson and Boutin, 2019), monoclonal antibodies yield many benefits, including the possibility of identifying phenotypically viable microbial cells (Vacca et al., 2022). It is a fast and cost-efficient method without dependence on a sequencing infrastructure, which is its means of identification (Spjuth et al., 2016). The lack of binding to nucleic acids improves the ability to detect in samples in which the DNA is either damaged or nonexistent. The notion of functional specificity includes the identification of virulence factors, toxins, or certain forms of structure (Vashisht et al., 2023).

4. Challenges and Considerations

The purification of effective monoclonal antibodies that are able to recognize microbial taxa requires that the antigens of interest are defined, preferably in an exclusive manner to the organism under interest, and are decoratable on the surface (Raj et al., 2021). The antigens adopted should demonstrate minimum cross-reactivity with nonharmful or nonpathogenic microorganisms, which means that the antibodies would not bind friendly species (Augustyniak et al., 2017). Additionally, the production of these antigens should be consistent and predictable, especially in microorganisms with phenotypic plasticity or environmental adaptation, where antigenic alteration should not be one of the factors that obstruct antibody specificity (Georgieva et al., 2018). These are particularly important in the human polybiome, which contains many different species, many of which have overlapping antigenic characteristics.

The recovery of microbiological targets has been enabled by the rapid generation of monoclonal antibodies via recent developments in the recombinant engineering of antibodies and phage display libraries (Davydova, 2022; Roth et al., 2021). Moreover, AI-based antigen discovery pipelines are able to interrogate genomic and proteomic data to model species-specific epitopes, hence allowing faster production of diagnostic antibodies. An approach involving the combination of monoclonal-antibody-based detection with high-throughput systems, such as microfluidics or multiplexed bead-based systems (Yeh et al., 2024), allows monitoring of multiple microbial species in a single sample in a more informative manner of polybiome dynamics. The use of monoclonal antibody protocols has high potential in clinical diagnostics, microbial forensics, and polybiome studies (Wang et al., 2022). With the current development of precision medicine and AI-focused discovery, monoclonal antibodies will become more instrumental in identifying not only the presence but also the functional and pathological role of bacteria in local ecosystems (Jana and Chiang, 2024; Sulaiman, 2024; Tsitou et al., 2024). New methodologies involved in diagnosis have evolved into AI-driven systems, which incorporate polybiome profiles into multidimensional omics data appended to nutrigenomic signatures (Verma et al., 2016). These types of integration contribute to subsequent monoclonal antibody design, increased functional specificity (Chen et al., 2024), and the design of individualized detection and intervention strategies. This intersection represents a new wave of polybiome-based diagnostics and tailor-made therapeutic interventions (Makowski et al., 2023).

2.3. Distinguishing Pathogens from Commensals in Complex Polybiomes

One of the main issues in the study of polybiomes is the proper distinction between pathogenic and commensal microorganisms related to the presence of the same species within a host (Tian et al., 2024). Such delineation is imperative in the detection of infections, the development of specific therapeutics, and healthy communities of microorganisms. The value of monoclonal antibodies has been identified as a promising option, as they can detect pathogens at an epitope-specific level and are designed with extremely high specificity, as well as being easily tailored in design, as required by the dense and taxonomically diverse human microbiota (Relman and Lipsitch, 2018).

1 Targeting Virulence-Specific Antigens

In contrast to commensal organisms, most infectious agents have discrete virulence factors, e.g., toxins, adhesion molecules, or invasion proteins (Leitao, 2020). Monoclonal antibodies can be specific for identifying pathogenic determinants and thus prevent cross-reactivity with nonpathogenic proteins, such as the cholera toxin of *Vibrio cholerae*, the shiga toxin of *Escherichia coli*, or the clumping factor of *Staphylococcus aureus*. By enabling clinicians and scientists to determine not only the taxon but also its potential pathogenicity, this antigen-level resolution enables this arms and allows it to identify not only its taxon but also its potential pathogenic potential. The more recent technologies in biomedical engineering utilize artificial-intelligence-driven epitope mapping and nutrigenomic contextual profiling to produce monoclonal antibodies calibrated on virulence determinants of a wide range of polybiomes (Dewaker et al., 2025; Nazir et al., 2025). Specificity is increased with such integrative therapeutics, and collateral damage to commensal organisms is minimized (Shaman, 2024).

2. Phenotypic Differentiation within Shared Taxa

Pathogens and commensals are often members of the same genus or species but differ in terms of gene expression or posttranslational modifications. Moreover, both harmless intestinal flora and highly virulent strains of *Escherichia coli*, such as enterohemorrhagic *Escherichia coli* (EHEC) or urinary tract-associated *Escherichia coli* (UPEC), can occur (Gipson et al., 2020). Monoclonal antibodies can be designed to be used in the selective recognition of structural or secreted antigens specific to pathogenic variants, e.g., lipopolysaccharide O-antigen variants or particular types of fimbria, resulting in selective interference in genetic homologous populations (Vacca et al., 2022; Yin et al., 2024).

To identify weak agiotypic differences between organisms, artificial intelligence-based methodologies that regulate multiomics data, particularly microbial transcriptomes in conjunction with host nutrigenetic metadata, have become essential to the process of developing monoclonal antibodies with selective binding within a common taxonomic consortium (Mahdavi et al., 2025; Jia et al., 2024).

3. Functional Profiling of Microbial Behavior

It is possible to use monoclonal antibodies to assess microbial activity, including the secretion of proinflammatory mediators and biomarkers of biofilm-associated proteins (Kharga et al., 2023). This applicability is especially well observed in the case of polymicrobial communities, where the pathogenesis of the disease is not dictated by one microorganism but by the interaction of two or more microorganisms and the environmental conditions of the host (Hajishengallis & Lamont, 2021). Monoclonal antibodies identify functionally valuable molecules, differentiating between bacteria actively involved in disease-causing processes and those involved in only chance (Vacca et al., 2022). Investigations performed recently have harmonized functional antibody positioning and machine learning algorithms trained on proteomic and metabolomic clusters, which can, accordingly, delineate operationally polybiome-dependent pathogenic activity much more precisely (Baiges-Gaya et al., 2023).

4. Nondestructive Profiling for Preserving Commensals.

Another strength of antibody-based detection is its ability to perform nonlethal profiling. In contrast to antibiotics, monoclonal antibodies can be used either diagnostically or therapeutically in a more selective manner. This particularity is critical to polybiome-saving methods, where commensal diversity is required to maintain immunologic homeostasis and avoid secondary complications such as opportunistic infection or dysbiosis (Alagiakrishnan et al., 2024). Modern antibody engineering methods that integrate artificial intelligence and host nutrigenomics can enable pathogen-dependent neutralization and maintain the polybiome, promoting polybiome-saving therapeutic goals (Dewaker et al., 2025; Lynch et al., 2025).

5. Integration with AI for Improved Target Identification

Machine learning and artificial intelligence approaches are now used to query large omics datasets to suggest putative pathogenic signs (Liu et al., 2024). Such computational methods can identify differential patterns of expression, antigenic motifs, or protein structures, which are pathogen-specific in polybiome samples (Rawal et al., 2021). It is then possible to use the resulting targets to produce monoclonal antibodies for either diagnostic or therapeutic purposes, thus forming a cohesive pipeline between in silico and functional immunologic reagents (Rawal et al., 2021). Existing advanced AI approaches, together with the use of high-throughput omics data, are applied in designing microbe–immune interaction targets and streamlining monoclonal antibody development models in the context of personalized nutrition and genomics through pipelines. This method can successfully generate high-quality predictive results, even with relatively small sample sizes, in large part because it incorporates domain-specific knowledge and uses rich data representations and thoroughly structured optimization algorithms (Vasudevan et al., 2025).

6. Clinical and translational implications

Clinically, monoclonal antibodies are under development to detect pathogens quickly in conditions without clear separation between colonization and infection, inflammatory bowel disease, sepsis, and respiratory infection. Rapid mAb-based methods can enable the differentiation of pathogenic and commensal flora in transplant medicine and can be used to optimize the accuracy of immunosuppressive or antibacterial treatments (Raj, et al., 2021). Current computational models guide the increasingly used application of targeted monoclonal antibodies in current therapeutics by incorporating polybiome and nutrigenomic information about both individual patients, thus refining the diagnostic accuracy and medicinal treatment of infectious and immune-related illnesses (Vecchietti et al., 2025; Reyed, 2023a).

3. FUNCTIONAL ANALYSIS OF MICROBIAL INTERACTIONS VIA MABS

3.1. Antimicrobial activity of microbial metabolites and toxins.

The human polybiome includes microorganisms that produce various compounds and toxins with the ability to significantly change the physiology of the host (Ogunrinola et al., 2020). Although many metabolites help in the immunology and metabolic homeostasis of the body, a specific group of metabolites, especially those produced by pathological or dysbiological bacteria, can enhance disease pathology, such as inflammation, neoplasia, neurotoxicity, and tissue damage (Zhao et al., 2023). Monoclonal antibodies represent a select immunotherapeutic modality that is capable of neutralizing harmful microbial products and maintaining the entire microbial community (Vacca et al., 2022). One recent method is artificial intelligence-augmented integrative metabolomics, which can increase our understanding of the complex interactions between microbes and their hosts and fluidly characterize both the chemical repertoires of isolated polybiomes (Sabry et al., 2025). Such datasets with analytics can reveal pathogenic patterns that cannot be detected via traditional assays. In addition, machine learning systems involving the inflammatory profiles of hosts and nutrigenomics can be used to predict the presence of certain microbial metabolites that exacerbate disease when the host is exposed to certain genetic and nutritional conditions. These insights will help to create monoclonal antibodies that can logically be applied selectively to disproportionately harmful metabolites in specific groups of patients (Zafar et al., 2024). This potential holds with the potential, in particular, to reduce chronic pathology that occurs due to polybiome-produced metabolites, including those associated with metabolic syndrome and colorectal cancer (Reyed, 2023b). Together, these strategies further the pursuit of microbial–nutrigenomic interaction profiling on the basis of individuals.

3.2. The bacterial exotoxin method was used.

Bacterial exotoxins are highly eligible protein toxins that either destroy or disrupt the cellular responses of host tissues. These toxins can be neutralized by monoclonal antibodies engineered in a highly specific manner (Li et al., 2022). Models to explain paralysis caused by botulinum toxins, tetanus toxins, and neuronal inhibition of Shiga toxin, toxin A, and endothelial damage are mitigated by monoclonal antibodies (Rasetti-Escargueil and Popoff, 2019), and the FDA-approved monoclonal antibody bezlotoxumab neutralizes the toxin A of *Clostridioides difficile* (Navalkele & Chopra 2018(. Antibody research based on a repertoire and machine learning that is positioned at the host-polybiome-nutrigenomic interface is producing new therapeutic platform gateways that adopt antibody-mediated antigenicity of bacterial exotoxins with great specificity and decreased off-target actions, thus laying the basis for personalized clinical therapy (Sakari et al., 2022; Mazzocco et al., 2021).

3.3. Neutralizing inflammatory metabolites.

The polybiome, which can be defined as the interrelated network of all microbial habitats, produces a catalog of metabolites, including secondary bile acids, lipoteichoic acid, and derivatives of short-chain fatty acids, capable of causing chronic inflammatory or metabolic disturbances in distal tissues (Jenkins et al., 2019). These molecules may be captured and neutralized by monoclonal antibodies, preventing their production of proinflammatory effects. The existence of lipopolysaccharides (LPS) from gram-negative bacteria that can induce diarrhea requires proper identification, e.g., through skin sampling, before antibodies against exogenous LPS are exchanged, which renders gram-negative bacterial *C. perfringens* responses insignificant (Page et al., 2022). The same has been explored regarding the use of anti-flagellin-based antibodies and their ability to suppress gut inflammation in inflammatory bowel disease, as flagellin is recognized abnormally by the immune system (Morgan et al., 2022). Inflammatory disease and metabolic disease neutralization have already proven the efficiency of the systematics of monoclonal antibodies applied in response to particular nutrigenomic systems that simplify the prediction of treatment effects and drug doses (Linares-, 2022). Such a multidimensional approach can be regarded as a way to personalize the therapy of such maladies.

3.4. Microbe-Host Signaling Intrusion.

Some microbial metabolites mimic the action of immune responses and physiological actions by functioning as structural homologs of host signaling molecules. These microbe–host interactions can be disrupted with the help of monoclonal antibodies. For example, phazenamides that are produced in the course of infection with *Pseudomonas aeruginosa* change the redox level of airway and pulmonary tissue immunopathology (Lew et al., 2025). This could be improved by antibody-mediated neutralization of such compounds and help to diminish tissue bronchial tissue injury in long-term pulmonary infections. The indole and tryptophan metabolites interact with aryl hydrocarbon receptors (AhRs); antibodies induced by immunogenicity to specific indole-benzoic acid derivatives have the potential to alter AhR signaling (Miao et al., 2024). Multiscale in silico methodologies have the ability to simulate complicated biochemical networks containing feedback loops based on microbe–host signaling to discover therapeutic targets via patient-derived gastrointestinal data (Reyed, 2023b). Modern studies integrate nutrigenomic data to explain the effects of food consumption on receptor–receptor communication, thereby guiding the design of monoclonal antibodies.

3.5. The advantages of antibody-mediated neutralization.

Compared with antibiotics and small molecules, monoclonal antibodies offer a higher specificity level and reduce off-recognition but do not affect or kill beneficial microorganisms (MacNair et al., 2024). They have a very long half-life, which allows neutralization to be maintained over toxins and metabolites. Dysbiosis and the risk that microbes develop resistance are prevented by monoclonal antibodies, thus banning microbial extermination, which is also considered essential when introducing polybiome-based therapy (Dadgar et al., 2023). In a precision nutrition model, diet-based biomarkers together with genomic indicators under the control of artificial intelligence-directed functional assays can support the tailoring of antibody-mediated pathogen neutralization without altering the commensal population (Greathouse & Choudhury, 2025).

3.6. Future Directions and AI Integration

Artificial intelligence (AI) and machine learning (ML) can be used to predict immunogenic epitopes on microbial toxins and metabolites, optimize antibody affinity, and model interactions between hosts, microbes, and metabolites (Alawam & Alwethaynani, 2024). These computational techniques hasten the discovery and creation of next-generation classes of monoclonal antibodies with the ability to target microbial products previously inaccessible. Moreover, therapeutic agents and diagnostic markers that are conjugated with neutralizing monoclonal antibodies provide theragnostic opportunities in which both the neutralization and detection of dangerous microbial compounds can occur in real time (Kothari et al., 2024).

Recent developments in deep learning have generated integrative schemes that combine genomic, metabolomic, and microbiomic methods with the aim of accelerating the process of selecting the best monoclonal antibody (mAb) candidates. The resulting systems automate labor-intensive experimental processes, which used to require years; thus, the timeline has been reduced to a matter of weeks, not months or years. In addition, the models integrate nutrigenomic variables to predict the effects of the nutritional host on antigen expression and provide recommendations for the redesign of mAbs.

This is because these conjugated adaptive antibodies can be used to perform real-time monitoring and treatment in conjugation with diagnostic or therapeutic payloads, a concept that is often described as theragnostic medicine (Sergeeva et al., 2025). The further development of these integrative platforms expects a paradigm shift, in which monoclonal antibodies are not used as passive therapeutic agents but rather as active components of individualized health maintenance regimens, which will transform the interface between immunotherapy, precision nutrition, and polybiome science.

4. BLOCKING HOST MICROBEAD ADHESION AND COLONIZATION

The capacity of bacteria, viruses and fungi to bind to the tissues of the host represents a major initial step in microbial pathogenesis (Casadevall 2025). Adhesion supports colonization, immunological evasion, and infection initiation and defines the site specificity of microbial invasion in dermal, respiratory, urinary, and gastrointestinal tissues (Holmes et al., 2021).

Owing to their high level of specificity and unique binding properties, monoclonal antibodies (mAbs) are efficient tools for interfering with the interactions between early hosts and microbes, ultimately leading to the elimination of microbial implantation and pathogenesis (Sauvat et al., 2024). Novel integrative paradigms have used AI models that are blinded to large datasets of interactions between hosts and microbes, including host nutrigenomic signatures and polybiome adhesion signatures, to identify clinical situations when interference in polybiome–host adhesion can be beneficial (D'Urso & Broccolo, 2024).

These studies demonstrate individual patient vulnerabilities and genetic polymorphisms in mucosal boundaries and dietary habits, which control receptor expression. As a result, patients who are at high risk of adhesion-related complications can be accurately treated with monoclonal antibody-related interventions in the clinic (Aboul-Ella et al., 2024).

Computational biology, immunotherapy and nutrigenomics integration thus form the core of adhesion interference to provide personalized control strategies, such as microbe management.

4.1. Monoclonal Antibodies as Adhesion Blockers

Several methods for producing monoclonal antibodies that block such adhesion pathways exist (Villamor et al., 2003). Monoclonal antibodies (mAbs) react with adhesin proteins, which consequently inhibits their ability to bind with host receptors in a specific manner (Speziale and Pietrocola, 2021). To avoid microbial interactions, mAbs can also be focused on the host receptor. The formation of biofilms has also been inhibited (Vacca et al., 2022). MAbs have been demonstrated to neutralize key surface elements in the context of working with *Staphylococcus aureus* and *Pseudomonas aeruginosa* to inhibit the initial aggregation of microbes or the formation of biofilms (Speziale & Pietrocola, 2021). This particular interference, by significantly impairing the capacity of infections to maintain themselves in host tissues, allows the immune system to eliminate de-epithelialized microbial invaders (Thakur et al., 2019).

In recent years, the exploration of adhesin proteins has been dominated by eptope-prediction pipelines based on artificial intelligence to map the key binding domains that mediate bacterial adhesion (Yu et al., 2025). Polybiome population analysis via machine-learning approaches also reveals strain-specific adhesion behaviors, thus guiding the choice of mAbs that can prevent pathogenic adhesion without affecting mutualistic symbionts. The synthesis of these developments will allow a paradigm shift in patient care to occur and the adoption of more specific adhesion blockade techniques to replace more generalized antiadhesive techniques, which will decrease the possible negative consequences, including unintended disruption of the ecological balance of a patient. This personalized strategy is also effective in treatment and is safe in preventing the risk of infection.

4.2. Applications in Polybiome and Infectious Disease Contexts

Anti-fimbrial mAbs prevent colonization of the gastrointestinal tract by *Escherichia coli*, Salmonella spp., and Campylobacter spp. (Corcionivoschi et al., 2023). The use of mAbs targeting *E. coli* type 1 and P. pili to prevent recurring infections is promising in the context of urinary tract infections (UTIs) (Zhou et al., 2023).

In terms of respiratory diseases, blocking virus–host cell attachment is possible through the administration of antibodies targeting viral fusion proteins, including the respiratory syncytial virus (RSV) F protein and influenza hemagglutinin (Battles & McLellan, 2019).

Adsorptive interactions between opportunistic pathogens, such as *Candida albicans* and *Propionibacterium* acnes, can also be disrupted by monoclonal antibodies, which reduce their proliferation in dysbiotic polybiomes (Asadi et al., 2019).

All these agents work together and do not act directly with a bactericidal effect, thus minimizing the collateral effects on the commensal flora and the pressure to increase selective pressure, leading to antimicrobial resistance (Vacca et al., 2022).

Innovative uses of monoclonal antibodies in the gut, urinary, and respiratory polybiomes increasingly make use of simulations supported by artificial intelligence to anticipate the consequences of targeted therapeutic interventions to a defined bacterial community (Zhang et al., 2022).

These computational tools combine multimodal data on local microbiota composition, host immune genotypes, and diet and use simulations to discover conditions where antiadhesion mAbs do not harm the local microbiota but rather which commensal organisms are spared. The models also assist in prioritizing the individual mAb intervention of these genetically susceptible patients with recurrent infections or nutrigenomic profiles that are prone to pathogen colonization (Matute, 2021). As such, the application of mAbs in these polybiome microenvironments is shifting toward a data-supported individualized therapeutic model.

This specificity minimizes secondary microbiota loss and positively affects the efficiency of treatment, in general (Shen et al., 2025).

4.3. Advantages over conventional antimicrobials

There is also a need to avoid the use of broad-spectrum antimicrobial agents to preserve the structural and functional integrity of polybiomes (Alm & Lahiri,

2020). Strain or virulence-type selectivity allows preferential targeting of pathogens within microbial environments (O'Brien et al., 2022). The most critical factor is preventive efficacy, especially in immunocompromised patients or during surgical interventions, chemotherapy, and transplantation (Coccolini et al., 2021). When adhesion mechanisms are not targeted, this does not cause much evolutionary pressure and, therefore, reduces the development of resistance (Christodoulou et al., 2022). The recent wave of monoclonal antibody (mAb)-based therapeutics, along with polybiome effect model types based on machine learning, is aimed not at the indiscriminate elimination of microbes as opposed to traditional antibiotics but instead at maintaining the diversity of microbes (Liang et al., 2022). These calculation applications combine and examine host nutritional conditions, genetic aspects, and pathogenic routes to assess the resilience of the microbiota against intervention to develop a more balanced approach to pathogen inhibition (Wei et al., 2025). Moreover, machine learning-based risk assessment provides clinicians with the ability to identify patients with the highest probability of therapeutic gain through antibiotic prophylaxis, especially in the perioperative and immunosuppressed groups (de la Cuesta-Zuluaga et al., 2024). These types of frameworks reduce selective pressures that promote antimicrobial resistance, which can be aggravated by therapeutic practices with high therapeutic value coupled with limited ecological disturbance. In this respect, the combination of specific immunotherapy, AI-driven analytics, and polybiome science opens the way for a new paradigm of infection prevention (Zhao et al., 2023).

4.4. AI and epitope prediction for adhesin-targeting mAbs

Machine learning has also been increasingly used to simulate microbial surfaces, forecast preserved or naturally useful epitopes on adhesins and screen potential antibody constructs (Kim et al., 2024).

This will help speed up the creation of next-generation adhesion-blocking monoclonal antibodies that can be tailored to specific pathogens or even to polybiomes in one patient who is distinctive to another (Gribonika et al., 2025).

It is possible to map adhesion structural models and host receptor profiles simultaneously with modern deep learning models and, thus, prioritize antibody-based interventions at immunogenic and functionally important epitopes (Liu et al., 2025).

To provide even greater refinements to these forecasts, nutrigenomic datasets will also be factored into the equation, hence allowing inference when the expression of the adhesin is regulated and when the localization of the receptors under individual nutritional regimens is studied.

The resulting system is dynamically adaptive, as it possesses the ability to create monoclonal antibodies that are universally suppressive of adhesion but are calibrated to patient-specific polybiome and patient-specific metabolic settings (Matsunaga & Tsumoto, 2025).

This reactive design has the potential to turn adhesion blockade into a heterogeneous intervention into a platform technology for targeted polybiome-sensitive therapeutics. Therefore, the active combination of artificial intelligence, immunotherapy, and personalized health data represents an important breakthrough in combating microbial colonization (Ajibade & Madu, 2025).

5. INVESTIGATING IMMUNE MODULATION AND INFLAMMATORY PATHWAYS VIA MONOCLONAL ANTIBODIES IN THE POLYBIOME

A major definition is that the human polybiome is an important determinant of both local and systemic immunity. Microbe-derived signals regulate the process of immunological cell maturation, cytokine production, and an inflammatory tone and maintain homeostasis under physiological conditions or initiate dysbiosis (Ding et al., 2025).

In addition to their already known therapeutic and diagnostic uses, monoclonal antibodies are versatile research tools that can be used to characterize the complex immunological processes induced by interactions with microbes in the host. Host receptors, immune mediators, or certain microbial ligands are used to target the polybiome in a specific way via mAbs to dissect the role of the polybiome in immune reactions and inflammatory diseases (Malik and Ghatol, 2023).

More recent methodological developments have attempted to utilize artificial intelligence for single-cell and spatial transcriptomics in conjunction with polybiome profiling to outline the role of microbial taxa in specific immune cell subsets in vivo. By being combined with nutrigenomic targets, such datasets can be used to predict the effects of dietary components, which signal to modulate microbial signaling pathways that are biased to either pro- or anti-inflammatory phenotypes. The use of subsequent monoclonal antibodies may be used to query and disrupt the identified pathways, thus allowing functional validation of AI-derived microbial targets.

This method involves identifying patient-specific cytokine–microbe interaction nodes with respect to the level of inflammatory bowel disease and metabolic inflammation. Finally, organoids and polybiome members can be cultured together with adaptive mAb libraries, which allows real-time screening of immune modulation in the context of personalized dietary conditions. Optimality with AI is also used to optimize antibody affinity and isotype selection, matching specific inflammatory profiles and avoiding unwanted immunomodulation.

Taken together, these approaches open a new era of accurate immunoprofiling and antibody treatment via the microbial–diet–host nexus.

6. AI-DRIVEN ANTIGEN DISCOVERY FOR POLYBIOME-TARGETED VACCINES

The human polybiome is very complex and variable, which makes strategies for immunotherapeutic development difficult and promising. The ability to query large quantities of data and identify small but clinically relevant molecular footprints is the factor that can identify specific antigens in a microbial community that can be used as a vaccine or monoclonal antibody (Varade et al., 2021). The field of artificial intelligence is pushing antigen discovery limits, particularly machine-learning and deep-learning methods. Therefore, it is possible to create immunizations that target polybiome elements with precision and effectiveness, which has never been achieved before (Olawade et al., 2024).

Multiomic datasets, such as metagenomics, metatranscriptomics, and proteomics datasets, can be analyzed via AI-based platforms to identify one or more microbial genes, proteins, and metabolic products that serve as viable antigens (Sauceda et al., 2022). These algorithms use high-throughput sequencing data to identify conserved and surface-expressed microbial epitopes in a variety of patient groups, determine their possible immunogenicity, and predict their interactions with host or commensal organisms (Khanum et al., 2022). Examples of this paradigm are reverse-vaccinology 2.0 and deep-learning-based antigenicity predictors, including Vaxign-ML and DeepVacPred, which autonomously select candidates on the basis of their structural, functional and immunological properties (Alashwal et al., 2025).

Artificial intelligence tools are essential in the most recent advances in antigen discovery. A new AI-based pipeline has been constructed by researchers to trace the pathology between antigen discovery and its future treatment use (see Figure 2). Artificial intelligence (AI) is crucial in empowering researchers to differentiate between harmless and aggressive microbial strains, therefore allowing the creation of vaccines or monoclonal antibodies (mAbs), which specifically target the harmful elements of the polybiome without inducing harm to commensal populations (Alsulimani et al., 2024). It is highly beneficial, especially in chronic disorders, such as inflammatory bowel disease, metabolic syndrome, and several other neuroimmune conditions, in which dysbiosis is a primary etiological factor and not an acute infection (Vebr, et al., 2023).

Figure 2.

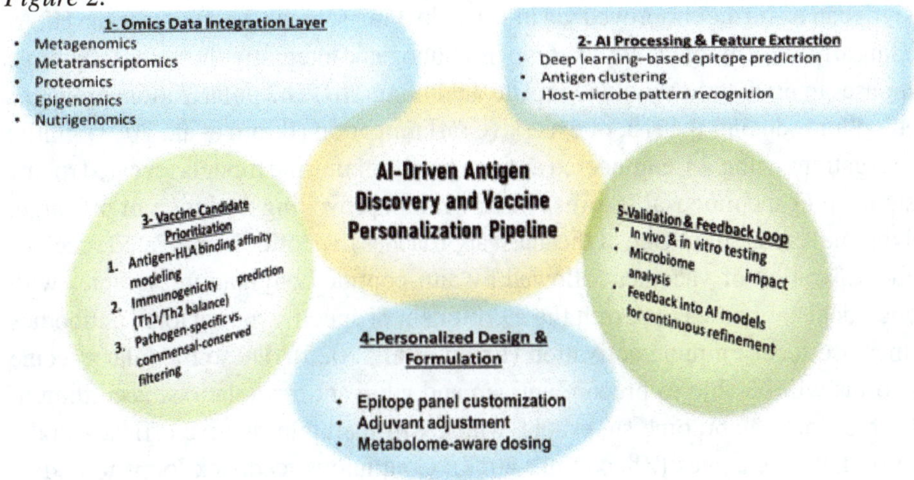

This figure shows the computational process by which benign and pathogenic microbial strains can be distinguished on the basis of the integration of multiomic data. This study shows how deep-learning models use metagenomic, transcriptomic, proteomic, and nutrigenomic signatures to recognize viable antigenic targets, allowing the design of precision-engineered monoclonal antibodies and vaccines without disturbing the commensal microbiota.

AI-based detection of microbial antigens can be used to design immunizations to prevent infection or induce a tolerance response to dysregulated microbial components (Ajibola et al., 2025; Wong et al., 2023). This is necessary in the development of prophylactic and therapeutic vaccines. Microorganisms harboring antigens can also be used to prepare microorganism-specific mAbs that neutralize toxin effects, prevent microbial adhesion, or suppress microbiota-regulated immune signaling networks alongside monoclonal antibody platforms (Raj et al., 2021). With the development of AI abilities, it is anticipated that they will find their way into polybiome studies and immunotherapeutic design to allow the creation of a new generation of tailored vaccines and antibody-based therapies and interventions based on polybiome factors (Odoom et al., 2025). Recent AI systems have been able to utilize metagenomic, metatranscriptomic, proteomic, and host epigenomic data to discover subclinical antigenic patterns of selected polybiome spectra (Sun et al., 2024). These multiomic antigenic situations are contextualized using dispositional nutrigenomic signatures to prioritize vaccine candidates because of the likelihood of producing protective immunity among different dietary-genetic environments (Shannon et al., 2020). Data collection and deep-learning models trained on established protective, tolerogenic and neutralizing antigens can be used to help construct multiepitope constructs that best enable optimal pathogen elimination and intestinal homeostasis. New vaccine designs can be further improved via in silico docking screens that increase the HLA-binding affinity of antigens. Predictors of immunogenic integrative host transcriptomic responses to predict the cytokine profile, including Th1/Th2 polarization, provide a method for reducing the risk of excessive inflammation (Ikram et al., 2023). Initial investigations using AI-engineered antigen panels of murine models revealed robust systemic IgG and mucosal IgA responses without provoking dysbiosis or off-target polybiome remodeling. One of the latest approaches to peptide-conjugated vaccines is the separation of vaccines followed by anti-peptide conjugation therapies with monoclonal antibodies to permit the recruitment of specific monoclonal antibodies to increase local immune activation (Imani et al., 2025). These modular vaccine platforms will be able to prepare antigen and adjuvant formulations according to host intestine metabolome snapshots, which will maintain relative efficacy in the face of dietary changes (Zhao et al., 2017). Continuous feedback loops to update formulations through dynamic reoptimization of booster formulations are motivated by changing polybiome responses to vaccinations reintroduced into AI models (El Arab et al., 2025). The operation of this feedback mechanism is based on the principles of precision nutrition combined with dietary recommendations, which contributes to improved vaccine-induced immune responses and polybiome resilience. This feedback loop aligns with the principles of precision nutrition and the broader concept of the polybiome (Reyed, 2023b).

It integrates dietary and immune regulation to enhance the body's vaccine response and maintain polybiome stability. Artificial intelligence applications have already shown tangible outcomes and are ready to reshape healthcare paradigms through monoclonal antibody–integrated vaccines, personalized diets and live biotherapeutics. Collectively, these innovations aim to provide comprehensive, patient-centered protection (Olawade et al., 2024).

7. CONCLUSION: A CONVERGING FUTURE FOR PERSONALIZED POLYBIOME MEDICINE

This chapter provides the contextual framework of science, tracing the conceptual advancement of science beginning with the principles of biological inquiry and continuing into the multiomics innovations that have formed the basis of current personalized health approaches. It describes a clear-cut transformation in the conceptualization and design of monoclonal antibodies (mAbs). mAbs are no longer passive therapeutic agents; they are dynamic, specifically designed molecules with the ability to interact with host biology in a far more complicated way than previous generations of immunotherapy ever permitted. With the development of structural biology, protein engineering, and computational modeling, mAbs have become modular systems with adaptability, optimality, and context-response capabilities.

One of the main topics that emerges as a result of this work is the combination of artificial intelligence and whole multiomic data. The integration of layered information provided to AI models (polybiome, nutrigenome, and metabolome) can help them conduct simultaneous and interdependent analyses that could not be performed by humans previously. This integrative potential will enable extremely sophisticated epitope identification; specific discrimination between pathogenic and commensal microbial signals; and antibody isotype tailoring, glycosylation patterns, affinities, and optimal therapeutic benefits. The therapeutic decision-making process therefore ceases to be a one-size-fits-all approach and evolves into an interactive, feedback-driven system that is capable of responding to biological changes in a responsive manner.

This trend is also reinforced by experimental evidence. Monoclonal antibodies have been shown to neutralize toxins, homoaggregate proinflammatory metabolites, prevent the attachment of pathogens to host tissues, and selectively maintain the beneficial microbial species necessary for metabolic and immune homeostasis. Combined with AI-assisted antigen recognition and in silico docking technology, these efforts are driving the design of next-generation vaccines and theragnostic systems. It is hoped that such systems will be able to adjust in real time to alterations in diet, genotype, and polybiome composition, making personalized therapy no longer an impressive dream but a realistic expectation.

This system-based model reconfigures the therapeutic strategy as an ongoing adaptation process. Intensive optimization of mAb structure and dosing, along with individualized nutritional recommendations, follows polybiome profiling subsequent to the administration of antibodies. The model focuses on ecological complexity, predicts emerging resistance patterns, and highlights the significant role of lifestyle factors in influencing therapeutic success. All of these elements create a platform for a new set of generation context-based biomedical interventions.

Moving ahead, the area of clinical translation will require successful and well-planned studies that demonstrate the multidimensional character of the suggested model. Digital health pipelines, longitudinal polybiome surveillance, and real-time data feedback loops are involved in randomized trials. Implementing the combination of precision nutrition, artificially intelligent immune engineering, and polybiome-conscious antibody therapeutics is merely the beginning of interventions as personalized as the patients themselves. The implications are far broader than those of mAbs; they signify a shift toward data-intensive, ecologically informed, patient-specific therapeutic ecosystems.

One major question arises: how does one translate proof-of-concept into working in the clinical reality? One practical roadmap includes building an end–end pilot platform that integrates patient polybiome sequencing with a secure, cloud-based AI engine. Clinicians can enter multiomic profiles and receive personalized recommendations for therapy, such as optimized mAb constructs and nutritional plans, on the basis of the ecological and genetic backgrounds of individual patients. This type of platform represents the culmination of the entire vision of precision immunotherapy, where the environment, nutrition, host inheritance, and gut microbial ecosystems are all incorporated into a single decision support system.

The chapter thus positions monoclonal antibodies at the intersection of polybiome science, nutrigenomics, and AI-driven design, serving as a stepping stone in the development of therapeutic technology. It lays the conceptual groundwork for a new therapeutic paradigm—one that understands biological complexity, exploits data-centric intelligence, and continuously adapts to the evolving environments presented by the human body.

REFERENCES

Aboul-Ella, H., Gohar, A., Ali, A. A., Ismail, L. M., Mahmoud, A. E. E., Elkhatib, W. F., & Aboul-Ella, H. (2024). Monoclonal antibodies: From magic bullet to precision weapon. *Molecular Biomedicine*, *5*(1), 47. DOI: 10.1186/s43556-024-00210-1 PMID: 39390211

Acar, D. D., Witkowski, W., Wejda, M., Wei, R., Desmet, T., Schepens, B., De Cae, S., Sedeyn, K., Eeckhaut, H., Fijalkowska, D., Roose, K., Vanmarcke, S., Poupon, A., Jochmans, D., Zhang, X., Abdelnabi, R., Foo, C. S., Weynand, B., Reiter, D., & Vandekerckhove, L. (2024). Integrating artificial intelligence-based epitope prediction in a SARS-CoV-2 antibody discovery pipeline: Caution is warranted. *EBioMedicine*, *100*, 104960. DOI: 10.1016/j.ebiom.2023.104960 PMID: 38232633

Ajibade, V. M., & Madu, C. S. (2025). The Integration of Artificial Intelligence Into Precision Medicine for Neuro-Oncology: Ethical, Clinical, and Nursing Implications in Immunotherapy Care. *Cureus*, *17*(5), e85024. DOI: 10.7759/cureus.85024 PMID: 40443837

Ajibola, F. O., Nwojiji, E. C., Lawal, M., Barakat, S. T., Temitayo, I. A., Ali, V. E., & Bakare-Abidola, T. (2025). The role of the gut microbiome in immune modulation: Implications for autoimmune diseases and cancer therapy. *World Journal of Biology Pharmacy and Health Sciences.*, *22*(1), 564–590.

Alagiakrishnan, K., Morgadinho, J., & Halverson, T. (2024). Approach to the diagnosis and management of dysbiosis. *Frontiers in Nutrition*, *11*, 1330903. DOI: 10.3389/fnut.2024.1330903 PMID: 38706561

Alam, M. Z., Maslanka, J. R., & Abt, M. C. (2023). Immunological consequences of microbiome-based therapeutics. *Frontiers in Immunology*, *13*, 1046472. DOI: 10.3389/fimmu.2022.1046472 PMID: 36713364

Alashwal, H., Kochunni, N. P., & Hayawi, K. (2025). Application of machine learning and deep learning techniques on reverse vaccinology – a systematic literature review. *Soft Computing*, *29*(1), 391–403. DOI: 10.1007/s00500-025-10480-8

Alawam, A. S., & Alwethaynani, M. S. (2024). Construction of an aerolysin-based multiepitope vaccine against *Aeromonas hydrophila*: An in silico machine learning and artificial intelligence-supported approach. *Frontiers in Immunology*, *15*, 1369890. DOI: 10.3389/fimmu.2024.1369890 PMID: 38495891

Alm, R. A., & Lahiri, S. D. (2020). Narrow-Spectrum Antibacterial Agents-Benefits and Challenges. *Antibiotics (Basel, Switzerland)*, *9*(7), 418. DOI: 10.3390/antibiotics9070418 PMID: 32708925

Alsulimani, A., Akhter, N., Jameela, F., Ashgar, R. I., Jawed, A., Hassani, M. A., & Dar, S. A. (2024). The Impact of Artificial Intelligence on Microbial Diagnosis. *Microorganisms, 12*(6), 1051. DOI: 10.3390/microorganisms12061051 PMID: 38930432

Asadi, A., Razavi, S., Talebi, M., & Gholami, M. (2019). A review on anti-adhesion therapies of bacterial diseases. *Infection, 47*(1), 13–23. DOI: 10.1007/s15010-018-1222-5 PMID: 30276540

Augustyniak, D., Majkowska-Skrobek, G., Roszkowiak, J., & Dorotkiewicz-Jach, A. (2017). Defensive and Offensive Cross-Reactive Antibodies Elicited by Pathogens: The Good, the Bad and the Ugly. *Current Medicinal Chemistry, 24*(36), 4002–4037. DOI: 10.2174/0929867324666170508110222 PMID: 28482788

Baiges-Gaya, G., Iftimie, S., Castañé, H., Rodríguez-Tomàs, E., Jiménez-Franco, A., López-Azcona, A. F., Castro, A., Camps, J., & Joven, J. (2023). Combining Semi-Targeted Metabolomics and Machine Learning to Identify Metabolic Alterations in the Serum and Urine of Hospitalized Patients with COVID-19. *Biomolecules, 13*(1), 163. DOI: 10.3390/biom13010163 PMID: 36671548

Battles, M. B., & McLellan, J. S. (2019). Respiratory syncytial virus entry and how to block it. *Nature Reviews. Microbiology, 17*(4), 233–245. DOI: 10.1038/s41579-019-0149-x PMID: 30723301

Bondad-Reantaso, M. G., MacKinnon, B., Karunasagar, I., Fridman, S., Alday-Sanz, V., Brun, E., & Caputo, A. (2023). Review of alternatives to antibiotic use in aquaculture. *Reviews in Aquaculture, 15*(4), 1421–1451.

Casadevall, A. (2025). Pathogenicity and virulence of *Cryptococcus neoformans* from an environmental perspective. *Virulence, 16*(1), 2547090. DOI: 10.1080/21505594.2025.2547090 PMID: 40810603

Chang, L., Liu, J., Zhu, J., Guo, S., Wang, Y., Zhou, Z., & Wei, X. (2025). Advancing precision medicine: The transformative role of artificial intelligence in immunogenomics, radiomics, and pathomics for biomarker discovery and immunotherapy optimization. *Cancer Biology & Medicine, 22*(1), 33–47. PMID: 39749734

Chaudhary, D., Guleria, D., Aggarwal, H., Mishra, V., Chauhan, A., Dufossé, L., & Joshi, N. C. (2025). Nutrigenomics and personalized diets: Tailoring nutrition for optimal health. *Applied Food Research, 5*(1), 100980. Advance online publication. DOI: 10.1016/j.afres.2025.100980

Chen, H. C., Pan, Y. L., Chen, Y., Yang, T. H., Hsu, E. T., Huang, Y. T., & Chiang, M. H. (2024). Monoclonal Antibodies as a Therapeutic Strategy against Multidrug-Resistant Bacterial Infections in a Post-COVID-19 Era. *Life (Basel, Switzerland), 14*(2), 246. DOI: 10.3390/life14020246 PMID: 38398755

Christodoulou, M. S., Villa, F., Pinto, A., & Cappitelli, F. (2022). Correlation between Perturbation of Redox Homeostasis and Antibiofilm Capacity of Phytochemicals at Non-Lethal Concentrations. *Antioxidants (Basel, Switzerland), 11*(12), 2451. DOI: 10.3390/antiox11122451 PMID: 36552659

Coccolini, F., Improta, M., Cicuttin, E., Catena, F., Sartelli, M., & Bova, R., De' Angelis, N., Gitto, S., Tartaglia, D., Cremonini, C., Ordonez, C., Baiocchi, G. L., & Chiarugi, M. (2021). Surgical site infection prevention and management in immunocompromised patients: A systematic review of the literature. *World Journal of Emergency Surgery; WJES, 16*(1), 33. DOI: 10.1186/s13017-021-00375-y PMID: 34112231

Corcionivoschi, N., Balta, I., Butucel, E., McCleery, D., Pet, I., Iamandei, M., Stef, L., & Morariu, S. (2023). Natural Antimicrobial Mixtures Disrupt Attachment and Survival of *E. coli* and *C. jejuni* to Non-Organic and Organic Surfaces. *Foods, 12*(20), 3863. DOI: 10.3390/foods12203863 PMID: 37893756

D'Urso, F., & Broccolo, F. (2024). Applications of Artificial Intelligence in Microbiome Analysis and Probiotic Interventions—An Overview and Perspective Based on the Current State of the Art. *Applied Sciences (Basel, Switzerland), 14*(19), 8627. DOI: 10.3390/app14198627

Dadgar, N., Edlukudige Keshava, V., Raj, M. S., & Wagner, P. L. (2023). The Influence of the Microbiome on Immunotherapy for Gastroesophageal Cancer. *Cancers (Basel), 15*(18). Advance online publication. DOI: 10.3390/cancers15184426 PMID: 37760397

Dakal, T. C., Xu, C., & Kumar, A. (2025). Advanced computational tools, artificial intelligence and machine-learning approaches in gut microbiota and biomarker identification. *Frontiers in Medical Technology, 6*, 1434799. DOI: 10.3389/fmedt.2024.1434799 PMID: 40303946

David, A., & Lev-Ari, S. (2024). Targeting the Gut Microbiome to Improve Immunotherapy Outcomes: A Review. *Integrative Cancer Therapies, 23*, 15347354241269870. DOI: 10.1177/15347354241269870 PMID: 39223798

Davydova, E. K. (2022). Protein Engineering: Advances in Phage Display for Basic Science and Medical Research. *Biochemistry. Biokhimiia (Moscow, Russia), 87*(Suppl 1), S146–S110. DOI: 10.1134/S0006297922140127 PMID: 35501993 de la Cuesta-Zuluaga, J., Müller, P., & Maier, L. (2024). Balancing act: Counteracting adverse drug effects on the microbiome. *Trends in Microbiology*, *32*(1), 10–22. DOI: 10.1016/j.tim.2023.10.002 PMID: 39395850

De Roos, B. (2013). Personalised nutrition: Ready for practice? *The Proceedings of the Nutrition Society*, *72*(1), 48–52. PMID: 23199675

Dewaker, V., Morya, V. K., Kim, Y. H., Park, S. T., Kim, H. S., & Koh, Y. H. (2025). Revolutionizing oncology: The role of Artificial Intelligence (AI) as an antibody design, and optimization tools. *Biomarker Research*, *13*(1), 52. DOI: 10.1186/s40364-025-00764-4 PMID: 40155973

Ding, G., Yang, X., Li, Y., Wang, Y., Du, Y., Wang, M., Ye, R., Wang, J., Zhang, Y., Chen, Y., & Zhang, Y. (2025). Gut microbiota regulates gut homeostasis, mucosal immunity and influences immune-related diseases. *Molecular and Cellular Biochemistry*, *480*(4), 1969–1981. DOI: 10.1007/s11010-024-05077-y PMID: 39060829

Do Rosario, N. A. D., Gokhale, D., & Gore, M. (2025). Exploring the future of nutrigenomics: Dietitians' perceptions on integration in Indian practice. *Genes & Nutrition*, *20*(1), 7. DOI: 10.1186/s12263-024-00760-7 PMID: 40097997

Einarsson, G., & Boutin, S. (2019). The Lung Microbiome - ERS Monograph: Techniques: culture, identification and 16S rRNA gene sequencing. In Cox, M. J., Ege, M. J., & von Mutius, E. (Eds.), *ERS Monograph: The Lung Microbiome* (Vol. 83, p. 18). ERS Monograph. DOI: 10.1183/2312508X.10000819

El Abd, Y., Tabll, A., Smolic, R., & Smolic, M. (2022). Mini-review: the market growth of diagnostic and therapeutic monoclonal antibodies–SARS CoV-2 as an example. *Human antibodies, 30*(1), 15-24. .DOI: 10.3233/HAB-211513

El Arab, R. A., Alkhunaizi, M., Alhashem, Y. N., Al Khatib, A., Bubsheet, M., & Hassanein, S. (2025). Artificial intelligence in vaccine research and development: An umbrella review. *Frontiers in Immunology*, *16*, 1567116. DOI: 10.3389/fimmu.2025.1567116 PMID: 40406131

Georgieva, M., Kagedan, L., Lu, Y. J., Thompson, C. M., & Lipsitch, M. (2018). Antigenic Variation in *Streptococcus pneumoniae* PspC Promotes Immune Escape in the Presence of Variant-Specific Immunity. *mBio*, *9*(2), e00264–e18. DOI: 10.1128/mBio.00264-18 PMID: 29535198

Gipson, K. S., Nickerson, K. P., Drenkard, E., Llanos-Chea, A., Dogiparthi, S. K., Lanter, B. B., Hibbler, R. M., Yonker, L. M., Hurley, B. P., & Faherty, C. S. (2020). The Great ESKAPE: Exploring the Crossroads of Bile and Antibiotic Resistance in Bacterial Pathogens. *Infection and Immunity*, *88*(10), e00865–e19. DOI: 10.1128/IAI.00865-19 PMID: 32661122

Gleeson, P. J., Camara, N. O. S., Launay, P., Lehuen, A., & Monteiro, R. C. (2024). Immunoglobulin A Antibodies: From Protection to Harmful Roles. *Immunological Reviews*, *328*(1), 171–191. DOI: 10.1111/imr.13424 PMID: 39578936

Greathouse, K. L., & Choudhury, A. (2025). Precision nutrition and the gut microbiome: Harnessing AI to revolutionize cancer prevention and therapy. *Cell Host & Microbe*, *33*(6), 766–776. DOI: 10.1016/j.chom.2025.05.011 PMID: 40505617

Grewal, S., Hegde, N., & Yanow, S. K. (2024). Integrating machine learning to advance epitope mapping. *Frontiers in Immunology*, *15*, 1463931. DOI: 10.3389/fimmu.2024.1463931 PMID: 39403389

Gribonika, I., Band, V. I., Chi, L., Perez-Chaparro, P. J., Link, V. M., Ansaldo, E., Oguz, C., Bousbaine, D., Fischbach, M. A., & Belkaid, Y. (2025). Skin autonomous antibody production regulates host-microbiota interactions. *Nature*, *638*(8052), 1043–1053. DOI: 10.1038/s41586-024-08376-y PMID: 39662506

Guo, W., Wang, X., Hu, J., Zhang, B., Zhao, L., Zhang, G., Qi, J., Wei, Z., Bao, Y., Tian, M., & Wang, S. (2025). In silico design of a multiepitope vaccine against *Mycobacterium avium* subspecies paratuberculosis. *Frontiers in Immunology*, *16*, 1505313. DOI: 10.3389/fimmu.2025.1505313 PMID: 39935480

Hajishengallis, G., & Lamont, R. J. (2021). Polymicrobial communities in periodontal disease: Their quasiorganismal nature and dialog with the host. *Periodontology 2000*, *86*(1), 210–230. DOI: 10.1111/prd.12371 PMID: 33690950

Hanack, K., Messerschmidt, K., & Listek, M. (2016). Antibodies and Selection of Monoclonal Antibodies. *Advances in Experimental Medicine and Biology*, *917*, 11–22. DOI: 10.1007/978-3-319-32805-8_2 PMID: 27236550

Helmy, Y. A., Taha-Abdelaziz, K., Hawwas, H. A. E., Ghosh, S., AlKafaas, S. S., Moawad, M. M. M., Saied, E. M., Kassem, I. I., & Mawad, A. M. M. (2023). Antimicrobial Resistance and Recent Alternatives to Antibiotics for the Control of Bacterial Pathogens with an Emphasis on Foodborne Pathogens. *Antibiotics (Basel, Switzerland)*, *12*(2), 274. DOI: 10.3390/antibiotics12020274 PMID: 36830185

Holmes, C. L., Anderson, M. T., Mobley, H. L. T., & Bachman, M. A. (2021). Pathogenesis of Gram-Negative Bacteremia. *Clinical Microbiology Reviews*, *34*(2), e00234–e20. DOI: 10.1128/CMR.00234-20 PMID: 33692149

Hu, R.-S., Gu, K., Ehsan, M., Abbas Raza, S. H., & Wang, C.-R. (2025). Transformer-based deep learning enables improved B-cell epitope prediction in parasitic pathogens: A proof-of-concept study on Fasciola hepatica. *PLoS Neglected Tropical Diseases*, *19*(4), e0012985. DOI: 10.1371/journal.pntd.0012985 PMID: 40300022

Ikram, A., Alzahrani, B., Zaheer, T., Sattar, S., Rasheed, S., Aurangzeb, M., & Ishaq, Y. (2023). An In Silico Deep Learning Approach to Multi-Epitope Vaccine Design: A Hepatitis E Virus Case Study. *Vaccines*, *11*(3), 710. DOI: 10.3390/vaccines11030710 PMID: 36992295

Imani, S., Li, X., Chen, K., Maghsoudloo, M., Jabbarzadeh Kaboli, P., Hashemi, M., Khoushab, S., & Li, X. (2025). Computational biology and artificial intelligence in mRNA vaccine design for cancer immunotherapy. *Frontiers in Cellular and Infection Microbiology*, *14*, 1501010. DOI: 10.3389/fcimb.2024.1501010 PMID: 39902185

Jana, M. K., & Chiang, M.-H. (2024). Monoclonal antibodies: A promising weapon against the silent pan-demic of multidrug-resistant bacteria. In Microbes of Medical Importance (IIP Series, Vol. 4, pp. 546–568). IIP. https://doi.org/DOI: 10.58532/nbennurmmch23

Jenkins, T. P., Fryer, T., Dehli, R. I., Jürgensen, J. A., Fuglsang-Madsen, A., Føns, S., & Laustsen, A. H. (2019). Toxin Neutralization Using Alternative Binding Proteins. *Toxins*, *11*(1), 53. DOI: 10.3390/toxins11010053 PMID: 30658491

Jia, D., Wang, Q., Qi, Y., Jiang, Y., He, J., Lin, Y., Sun, Y., Xu, J., Chen, W., Fan, L., Yan, R., Zhang, W., Ren, G., Xu, C., Ge, Q., Wang, L., Liu, W., Xu, F., Wu, P., & Wang, L. (2024). Microbial metabolite enhances immunotherapy efficacy by modulating T cell stemness in pancancer. *Cell*, *187*(7), 1651–1665.e21. DOI: 10.1016/j.cell.2024.02.022 PMID: 38490195

Jiang, Y., & Li, Y. (2024). Nutrition Intervention and Microbiome Modulation in the Management of Breast Cancer. *Nutrients*, *16*(16), 2644. DOI: 10.3390/nu16162644 PMID: 39203781

Kapellou, A., Salata, E., Vrachnos, D. M., Papailia, S., & Vittas, S. (2025). Gene–Diet Interactions in Diabetes Mellitus: Current Insights and the Potential of Personalized Nutrition. *Genes*, *16*(5), 578. DOI: 10.3390/genes16050578 PMID: 40428400

Ke, S., Kil, H., Roggy, C., Shields, T., Quinn, Z., Quinn, A. P., Small, J. M., Towne, F. D., Brooks, A. E., & Brooks, B. D. (2024). Potential Therapeutic Targets for Combination Antibody Therapy Against *Staphylococcus aureus* Infections. *Antibiotics (Basel, Switzerland), 13*(11), 1046. DOI: 10.3390/antibiotics13111046 PMID: 39596740

Khanum, S., Carbone, V., Gupta, S. K., Yeung, J., Shu, D., Wilson, T., Parlane, N. A., Altermann, E., Estein, S. M., Janssen, P. H., Wedlock, D. N., & Heiser, A. (2022). Mapping immunogenic epitopes of an adhesin-like protein from Methanobrevibacter ruminantium M1 and comparison of empirical data with in silico prediction methods. *Scientific Reports*, *12*(1), 10394. DOI: 10.1038/s41598-022-14545-8 PMID: 35729277

Kharga, K., Kumar, L., & Patel, S. K. S. (2023). Recent Advances in Monoclonal Antibody-Based Approaches in the Management of Bacterial Sepsis. *Biomedicines*, *11*(3), 765. DOI: 10.3390/biomedicines11030765 PMID: 36979744

Köhler, G., & Milstein, C. (1975). Continuous cultures of fused cells secreting antibody of predefined specificity. *Nature*, *256*(5517), 495–497. DOI: 10.1038/256495a0 PMID: 1172191

Kothari, M., Wanjari, A., Acharya, S., Karwa, V., Chavhan, R., Kumar, S., Kadu, A., & Patil, R. (2024). A Comprehensive Review of Monoclonal Antibodies in Modern Medicine: Tracing the Evolution of a Revolutionary Therapeutic Approach. *Cureus*, *16*(6), e61983. DOI: 10.7759/cureus.61983 PMID: 38983999

Lagoumintzis, G., & Patrinos, G. P. (2023). Triangulating nutrigenomics, metabolomics and microbiomics toward personalized nutrition and healthy living. *Human Genomics*, *17*(1), 109. DOI: 10.1186/s40246-023-00561-w PMID: 38062537

Lam, H. N., Lin, S.-P., Nguyen, D. H. N., Chen, C.-M., Su, C.-T., Fang, T.-C., & Li, S.-C. (2025). Integrative Roles of Functional Foods, Microbiotics, Nutrigenetics, and Nutrigenomics in Managing Type 2 Diabetes and Obesity. *Nutrients*, *17*(4), 608. DOI: 10.3390/nu17040608 PMID: 40004938

Leitão, J. H. (2020). Microbial Virulence Factors. *International Journal of Molecular Sciences*, *21*(15), 5320. DOI: 10.3390/ijms21155320 PMID: 32727013

Lew, S. Q., Chong, S. Y., & Lau, G. W. (2025). Modulation of pulmonary immune functions by the *Pseudomonas aeruginosa* secondary metabolite pyocyanin. *Frontiers in Immunology*, *16*, 1550724. DOI: 10.3389/fimmu.2025.1550724 PMID: 40196115

Li, M., Mei, S., Yang, Y., Shen, Y., & Chen, L. (2022). Strategies to mitigate the on- and off-target toxicities of recombinant immunotoxins: An antibody engineering perspective. *Antibody Therapeutics*, *5*(3), 164–176. DOI: 10.1093/abt/tbac014 PMID: 35928456

Liang, H., Jo, J. H., Zhang, Z., MacGibeny, M. A., Han, J., Proctor, D. M., Taylor, M. E., Che, Y., Juneau, P., Apolo, A. B., McCulloch, J. A., Davar, D., Zarour, H. M., Dzutsev, A. K., Brownell, I., Trinchieri, G., Gulley, J. L., & Kong, H. H. (2022). Predicting cancer immunotherapy response from gut microbiomes using machine learning models. *Oncotarget*, *13*, 876–889. DOI: 10.18632/oncotarget.28252 PMID: 35875611

Lin, Z., Cheng, S., Yan, Q., Liu, X., Zheng, W., Hu, X., & Zhang, J. (2015). Development of an immunochromatographic lateral flow device for rapid detection of *Helicobacter pylori* stool antigen. *Clinical Biochemistry*, *48*(18), 1298–1303. DOI: 10.1016/j.clinbiochem.2015.08.004 PMID: 26256542

Liñares-Blanco, J., Fernandez-Lozano, C., Seoane, J. A., & López-Campos, G. (2022). Machine Learning Based Microbiome Signature to Predict Inflammatory Bowel Disease Subtypes. *Frontiers in Microbiology*, *13*, 872671. DOI: 10.3389/fmicb.2022.872671 PMID: 35663898

Liu, J., Bao, C., Zhang, J., Han, Z., Fang, H., & Lu, H. (2024). Artificial intelligence with mass spectrometry-based multimodal molecular profiling methods for advancing therapeutic discovery of infectious diseases. *Pharmacology & Therapeutics*, *263*, 108712. DOI: 10.1016/j.pharmthera.2024.108712 PMID: 39241918

Liu, X., Fu, H., Yang, Y., & Zhang, J. (2025). Bio-Inspired Mamba for Antibody-Antigen Interaction Prediction. *Biomolecules*, *15*(6), 764. DOI: 10.3390/biom15060764 PMID: 40563406

Lynch, L. E., Lahowetz, R., Maresso, C., Terwilliger, A., Pizzini, J., Melendez Hebib, V., Britton, R. A., Maresso, A. W., & Preidis, G. A. (2025). Present and future of microbiome-targeting therapeutics. *The Journal of Clinical Investigation*, *135*(11), e184323. DOI: 10.1172/JCI184323 PMID: 40454480

MacNair, C. R., Rutherford, S. T., & Tan, M. W. (2024). Alternative therapeutic strategies to treat antibiotic-resistant pathogens. *Nature Reviews. Microbiology*, *22*(5), 262–275. DOI: 10.1038/s41579-023-00993-0 PMID: 38082064

Mahdavi, S., Anthony, N. M., Sikaneta, T., & Tam, P. Y. (2025). Perspective: Multiomics and Artificial Intelligence for Personalized Nutritional Management of Diabetes in Patients Undergoing Peritoneal Dialysis. Advances in nutrition (Bethesda, Md.), 16(3), 100378. DOI: 10.1016/j.advnut.2025.100378

Makowski, E. K., Chen, H. T., & Tessier, P. M. (2023). Simplifying complex antibody engineering using machine learning. *Cell Systems*, *14*(8), 667–675. DOI: 10.1016/j.cels.2023.04.009 PMID: 37591204

Malik, B., & Ghatol, A. (2025). Understanding how monoclonal antibodies work. In *StatPearls*. StatPearls Publishing. https://www.ncbi.nlm.nih.gov/books/NBK572118/

Matsunaga, R., & Tsumoto, K. (2025). Accelerating antibody discovery and optimization with high-throughput experimentation and machine learning. *Journal of Biomedical Science*, *32*(1), 46. DOI: 10.1186/s12929-025-01141-x PMID: 40346589

Matute, J., Finander, B., Pepin, D., Ai, X., Smith, N., Li, J., Edlow, A., Villani, A., Lerou, P., & Kalish, B. (2021). Single-cell immunophenotyping of the fetal immune response to maternal SARS-CoV-2 infection in late gestation. Research square, rs.3.rs-311000. https://doi.org/DOI: 10.21203/rs.3.rs-311000/v1

Mazzocco, G., Niemiec, I., Myronov, A., Skoczylas, P., Kaczmarczyk, J., Sanecka Duin, A., Gruba, K., Król, P., Drwal, M., Szczepanik, M., Pyrc, K., & St pniak, P. (2021). AI aided design of epitope based vaccine for the induction of cellular immune responses against SARS CoV 2. *Frontiers in Genetics*, *12*, 602196. Advance online publication. DOI: 10.3389/fgene.2021.602196 PMID: 33841493

Miao, H., Wang, Y. N., Yu, X. Y., Zou, L., Guo, Y., Su, W., Liu, F., Cao, G., & Zhao, Y. Y. (2024). Lactobacillus species ameliorate membranous nephropathy through inhibiting the aryl hydrocarbon receptor pathway via tryptophan-produced indole metabolites. *British Journal of Pharmacology*, *181*(1), 162–179. DOI: 10.1111/bph.16219 PMID: 37594378

Ministro, J., Manuel, A. M., & Goncalves, J. (2020). Therapeutic Antibody Engineering and Selection Strategies. *Advances in Biochemical Engineering/Biotechnology*, *171*, 55–86. DOI: 10.1007/10_2019_116 PMID: 31776591

Mishra, R., Mohanty, S., Mahapatra, S., & Prasad, P. (2023). Overview of microbial therapeutics in immunological disorders. In Chauhan, N. S., & Kumar, S. (Eds.), *Microbiome therapeutics* (pp. 289–353). Academic Press. DOI: 10.1016/B978-0-323-99336-4.00008-2

Mitra, S., & Tomar, P. C. (2021). Hybridoma technology; advancements, clinical significance, and future aspects. *Journal of Genetic Engineering and Biotechnology*, *19*(1), 159. DOI: 10.1186/s43141-021-00264-6 PMID: 34661773

Morgan, N. N., Duck, L. W., Wu, J., Rujani, M., Thomes, P. G., Elson, C. O., & Mannon, P. J. (2022). Crohn's Disease Patients Uniquely Contain Inflammatory Responses to Flagellin in a CD4 Effector Memory Subset. *Inflammatory Bowel Diseases, 28*(12), 1893–1903. DOI: 10.1093/ibd/izac146 PMID: 35931421

Nagy, E., Nagy, G., Power, C. A., Badarau, A., & Szijártó, V. J. R. A. f. I. D. (2017). Antibacterial monoclonal antibodies. Adv Exp Med Biol.1053:119-153 119-153.

Navalkele, B. D., & Chopra, T. (2018). Bezlotoxumab: An emerging monoclonal antibody therapy for prevention of recurrent *Clostridium difficile* infection. *Biologics, 12*, 11–21. DOI: 10.2147/BTT.S127099 PMID: 29403263

Nazir, A., Hussain, F. H. N., Nadeem Hussain, T. H., Al Dweik, R., & Raza, A. (2025). Therapeutic targeting of the host-microbiota-immune axis: Implications for precision health. *Frontiers in Immunology, 16*, 1570233. DOI: 10.3389/fimmu.2025.1570233 PMID: 40364844

O'Brien, T. J., Figueroa, W., & Welch, M. (2022). Decreased efficacy of antimicrobial agents in a polymicrobial environment. *The ISME Journal, 16*(7), 1694–1704. DOI: 10.1038/s41396-022-01218-7 PMID: 35304578

O'Riordan, K. J., Moloney, G. M., Keane, L., Clarke, G., & Cryan, J. F. (2025). The gut microbiota immune brain axis: Therapeutic implications. *Cell Reports Medicine, 6*(3), 101982. DOI: 10.1016/j.xcrm.2025.101982 PMID: 40054458

Ochoa, M. C., Minute, L., Rodriguez, I., Garasa, S., Perez-Ruiz, E., Inogés, S., Melero, I., & Berraondo, P. (2017). Antibody-dependent cell cytotoxicity: Immunotherapy strategies enhancing effector NK cells. *Immunology and Cell Biology, 95*(4), 347–355. DOI: 10.1038/icb.2017.6 PMID: 28138156

Odoom, A., Osman, A. H., & Dzuvor, C. K. O. (2025). Recent Advances in Immunotherapeutic and Vaccine-Based Approaches for the Treatment of Drug-Resistant Bacterial Infections. *ACS Infectious Diseases, 11*(6), 1366–1402. DOI: 10.1021/acsinfecdis.5c00001 PMID: 40315159

Ogunrinola, G. A., Oyewale, J. O., Oshamika, O. O., & Olasehinde, G. I. (2020). The Human Microbiome and Its Impacts on Health. *International Journal of Microbiology, 8045646*. Advance online publication. DOI: 10.1155/2020/8045646 PMID: 32612660

Olawade, D. B., Teke, J., Fapohunda, O., Weerasinghe, K., Usman, S. O., Ige, A. O., & Clement David-Olawade, A. (2024). Leveraging artificial intelligence in vaccine development: A narrative review. *Journal of Microbiological Methods, 224*, 106998. DOI: 10.1016/j.mimet.2024.106998 PMID: 39019262

Olawade, D. B., Teke, J., Fapohunda, O., Weerasinghe, K., Usman, S. O., Ige, A. O., & Clement David Olawade, A. (2024). Leveraging artificial intelligence in vaccine development: A narrative review. *Journal of Microbiological Methods, 224*, 106998. Advance online publication. DOI: 10.1016/j.mimet.2024.106998 PMID: 39019262

Page, M. J., Kell, D. B., & Pretorius, E. (2022). The Role of Lipopolysaccharide-Induced Cell Signaling in Chronic Inflammation. *Chronic Stress (Thousand Oaks, Calif.), 6*, 24705470221076390. DOI: 10.1177/24705470221076390 PMID: 35155966

Phugat, S., & Goel, P. (2025). Review on Advancement of AI in Nutrigenomics. *Methods in Molecular Biology (Clifton, N.J.), 2952*, 429–444. DOI: 10.1007/978-1-0716-4690-8_23 PMID: 40553346

Posner, J., Barrington, P., Brier, T., & Datta-Mannan, A. (2019). Monoclonal Antibodies: Past, Present and Future. *Handbook of Experimental Pharmacology, 260*, 81–141. DOI: 10.1007/164_2019_323 PMID: 31820172

Raj, G. M., Priyadarshini, R., Murugesan, S., & Adhimoolam, M. (2021). Monoclonal Antibodies Against Infectious Microbes: *So Long and Too Little! Infectious Disorders Drug Targets, 21*(1), 4–27. DOI: 10.2174/187152652066620031215 4649 PMID: 32164518

Ramos-Lopez, O., Milagro, F. I., Allayee, H., Chmurzynska, A., Choi, M. S., Curi, R., De Caterina, R., Ferguson, L. R., Goni, L., Kang, J. X., Kohlmeier, M., Marti, A., Moreno, L. A., Pérusse, L., Prasad, C., Qi, L., Reifen, R., Riezu-Boj, J. I., San-Cristobal, R., & Martínez, J. A. (2017). Guide for current nutrigenetic, nutrigenomic, and nutriepigenetic approaches for precision nutrition involving the prevention and management of chronic diseases associated with obesity. *Journal of Nutrigenetics and Nutrigenomics, 10*(1–2), 43–62. DOI: 10.1159/000477729 PMID: 28689206

Rasetti-Escargueil, C., & Popoff, M. R. (2019). Antibodies and Vaccines against Botulinum Toxins: Available Measures and Novel Approaches. *Toxins, 11*(9), 528. DOI: 10.3390/toxins11090528 PMID: 31547338

Rawal, K., Sinha, R., Abbasi, B. A., Chaudhary, A., Nath, S. K., Kumari, P., Preeti, P., Saraf, D., Singh, S., Mishra, K., Gupta, P., Mishra, A., Sharma, T., Gupta, S., Singh, P., Sood, S., Subramani, P., Dubey, A. K., Strych, U., & Bottazzi, M. E. (2021). Identification of vaccine targets in pathogens and design of a vaccine using computational approaches. *Scientific Reports, 11*(1), 17626. DOI: 10.1038/s41598-021-96863-x PMID: 34475453

Relman, D. A., & Lipsitch, M. (2018). Microbiome as a tool and a target in the effort to address antimicrobial resistance. *Proceedings of the National Academy of Sciences*, 115(51), 12902-12910. doi: DOI: 10.1073/pnas.1717163115

Reyed, R. M. (2023a). Focusing on individualized nutrition within the algorithmic diet: An in depth look at recent advances in nutritional science, microbial diversity studies, and human health. *Food & Health*, *5*(1), 5. DOI: 10.53388/FH2023005

Reyed, R.M. (2023b). Simulate the dynamic interactions across nutritional supplements, the polybiome, and the biochemical approach dependent on patient derived gastrointestinal knowledge. *Journal of Clinical Epidemiology & Toxicology*, *4*, 133. DOI: 10.47363/JCET/2023(4)133

Reyed, R.M., & Prabhakar, P. K. M. (2025). Polybiome Derived Therapies and Fermentobiometrics: Novel Strategies for Blocking Cancer Pathways and Progression. In Cancer Etiology, Diagnosis and Treatments (pp. 145-212). Nova Medicine and Health/Pathways to Healing: Advancements in Cancer Therapy. DOI: 10.52305/JKMU3682

Roth, K. D. R., Wenzel, E. V., Ruschig, M., Steinke, S., Langreder, N., Heine, P. A., Schneider, K. T., Ballmann, R., Führer, V., Kuhn, P., Schirrmann, T., Frenzel, A., Dübel, S., Schubert, M., Moreira, G. M. S. G., Bertoglio, F., Russo, G., & Hust, M. (2021). Developing Recombinant Antibodies by Phage Display Against Infectious Diseases and Toxins for Diagnostics and Therapy. *Frontiers in Cellular and Infection Microbiology*, *11*, 697876. DOI: 10.3389/fcimb.2021.697876 PMID: 34307196

Sabry, Z., Arora, H. S., Chandrasekaran, S., & Wang, Z. (2025). AI driven drug discovery and repurposing using multi omics for myocardial infarction and heart failure. *Exploration of Medicine*, *6*, 1001340. Advance online publication. DOI: 10.37349/emed.2025.1001340 PMID: 41522213

Sakari, M., Laisi, A., & Pulliainen, A. T. (2022). Exotoxin targeted drug modalities as antibiotic alternatives. *ACS Infectious Diseases*, *8*(3), 433–456. DOI: 10.1021/acsinfecdis.1c00296 PMID: 35099182

Sauceda, C., Bayne, C., Sudqi, K., Gonzalez, A., Dulai, P. S., Knight, R., Gonzalez, D. J., & Gonzalez, C. G. (2022). Stool multi-omics for the study of host-microbe interactions in inflammatory bowel disease. *Gut Microbes*, *14*(1), 2154092. DOI: 10.1080/19490976.2022.2154092 PMID: 36503356

Sauvat, L., Verhoeven, P. O., Gagnaire, J., Berthelot, P., Paul, S., Botelho-Nevers, E., & Gagneux-Brunon, A. (2024). Vaccines and monoclonal antibodies to prevent healthcare-associated bacterial infections. *Clinical Microbiology Reviews*, *37*(3), e0016022. DOI: 10.1128/cmr.00160-22 PMID: 39120140

Sergeeva, O. V., Luo, L., & Guiseppi-Elie, A. (2025). Cancer theragnostics: Closing the loop for advanced personalized cancer treatment through the platform integration of therapeutics and diagnostics. *Frontiers in Bioengineering and Biotechnology*, *12*, 1499474. DOI: 10.3389/fbioe.2024.1499474 PMID: 39898278

Shaman, J. A. (2024). The Future of Pharmacogenomics: Integrating Epigenetics, Nutrigenomics, and Beyond. *Journal of Personalized Medicine*, *14*(12), 1121. DOI: 10.3390/jpm14121121 PMID: 39728034

Shannon, C. P., Blimkie, T. M., Ben-Othman, R., Gladish, N., Amenyogbe, N., Drissler, S., Edgar, R. D., Chan, Q., Krajden, M., Foster, L. J., Kobor, M. S., Mohn, W. W., Brinkman, R. R., Le Cao, K.-A., Scheuermann, R. H., Tebbutt, S. J., Hancock, R. E. W., Koff, W. C., Kollmann, T. R., & Lee, A. H.-Y. (2020). Multi-Omic Data Integration Allows Baseline Immune Signatures to Predict Hepatitis B Vaccine Response in a Small Cohort. *Frontiers in Immunology*, *11*, 578801. DOI: 10.3389/fimmu.2020.578801 PMID: 33329547

Shen, Y., Fan, N., Ma, S. X., Cheng, X., Yang, X., & Wang, G. (2025). Gut Microbiota Dysbiosis: Pathogenesis, Diseases, Prevention, and Therapy. *MedComm*, *6*(5), e70168. DOI: 10.1002/mco2.70168 PMID: 40255918

Sifniotis, V., Cruz, E., Eroglu, B., & Kayser, V. (2019). Current Advancements in Addressing Key Challenges of Therapeutic Antibody Design, Manufacture, and Formulation. *Antibodies (Basel, Switzerland)*, *8*(2), 36. DOI: 10.3390/antib8020036 PMID: 31544842

Singar, S., Nagpal, R., Arjmandi, B. H., & Akhavan, N. S. (2024). Personalized Nutrition: Tailoring Dietary Recommendations through Genetic Insights. *Nutrients*, *16*(16), 2673. DOI: 10.3390/nu16162673 PMID: 39203810

Śliwa-Dominiak, J., Czechowska, K., Blanco, A., Sielatycka, K., Radaczyńska, M., Skonieczna-Żydecka, K., Marlicz, W., & Łoniewski, I. (2025). Flow Cytometry in Microbiology: A Review of the Current State in Microbiome Research, Probiotics, and Industrial Manufacturing. *Cytometry. Part A*, *107*(3), 145–164. DOI: 10.1002/cyto.a.24920 PMID: 40028773

Speziale, P., & Pietrocola, G. (2021). Monoclonal Antibodies Targeting Surface-Exposed and Secreted Proteins from Staphylococci. *Vaccines*, *9*(5), 459. DOI: 10.3390/vaccines9050459 PMID: 34064471

Spjuth, O., Bongcam-Rudloff, E., Dahlberg, J., Dahlö, M., Kallio, A., Pireddu, L., Vezzi, F., & Korpelainen, E. (2016). Recommendations on e-infrastructures for next-generation sequencing. *GigaScience*, *5*, 26. DOI: 10.1186/s13742-016-0132-7 PMID: 27267963

Stone, C. A.Jr, Spiller, B. W., & Smith, S. A. (2024). Engineering therapeutic monoclonal antibodies. *The Journal of Allergy and Clinical Immunology, 153*(3), 539–548. DOI: 10.1016/j.jaci.2023.11.018 PMID: 37995859

Sulaiman, I. M. (Ed.). (2024). *Diagnosis of Pathogenic Microorganisms Causing Infectious Diseases* (1st ed.). CRC Press., DOI: 10.1201/9780429001192

Sun, Y., Gan, Z., Wang, X., Liu, J., Zhong, W., Zhang, Z., Zuo, J., Zhong, H., Huang, X., Yan, Z., & Cao, Q. (2024). Integrative metagenomic, transcriptomic, and proteomic analysis reveal the microbiota-host interplay in early-stage lung adenocarcinoma among nonsmokers. *Journal of Translational Medicine, 22*(1), 652. DOI: 10.1186/s12967-024-05485-0 PMID: 38997719

Tabll, A., Abbas, A. T., El-Kafrawy, S., & Wahid, A. (2015). Monoclonal antibodies: Principles and applications of immmunodiagnosis and immunotherapy for hepatitis C virus. *World Journal of Hepatology, 7*(22), 2369–2383. DOI: 10.4254/wjh.v7.i22.2369 PMID: 26464752

Tabll, A. A., Shahein, Y. E., Omran, M. M., Hussein, N. A., El-Shershaby, A., Petrovic, A., Glasnovic, M., Smolic, R., & Smolic, M. (2024). Monoclonal IgY antibodies: Advancements and limitations for immunodiagnosis and immunotherapy applications. *Therapeutic Advances in Vaccines and Immunotherapy, 12*, 25151355241264520. DOI: 10.1177/25151355241264520 PMID: 39071998

Thakur, A., Mikkelsen, H., & Jungersen, G. (2019). Intracellular Pathogens: Host Immunity and Microbial Persistence Strategies. *Journal of Immunology Research, 1356540*. Advance online publication. DOI: 10.1155/2019/1356540 PMID: 31111075

Theis, B. F., Park, J. S., Kim, J. S. A., Zeydabadinejad, S., Vijay-Kumar, M., Yeoh, B. S., & Saha, P. (2025). Gut Feelings: How Microbes, Diet, and Host Immunity Shape Disease. *Biomedicines, 13*(6), 1357. DOI: 10.3390/biomedicines13061357 PMID: 40564077

Tian, K., Jing, D., Lan, J., Lv, M., & Wang, T. (2024). Commensal microbiome and gastrointestinal mucosal immunity: Harmony and conflict with our closest neighbor. *Immunity, Inflammation and Disease, 12*(7), e1316. DOI: 10.1002/iid3.1316 PMID: 39023417

Tsigalou, C., Konstantinidis, T., Aloizou, A. M., Bezirtzoglou, E., & Tsakris, A. (2022). Future Therapeutic Prospects in Dealing with Autoimmune Diseases: Treatment Based on the Microbiome Model. In Dwivedi, M. K., Sankaranarayanan, A., Kemp, E. H., & Shoenfeld, Y. (Eds.), *Role of Microorganisms in Pathogenesis and Management of Autoimmune Diseases*. Springer. DOI: 10.1007/978-981-19-4800-8_25

Tsitou, V. M., Rallis, D., Tsekova, M., & Yanev, N. (2024). Microbiology in the era of artificial intelligence: Transforming medical and pharmaceutical microbiology. *Biotechnology, Biotechnological Equipment, 38*(1). Advance online publication. DOI: 10.1080/13102818.2024.2349587

Vacca, F., Sala, C., & Rappuoli, R. (2022). Monoclonal Antibodies for Bacterial Pathogens: Mechanisms of Action and Engineering Approaches for Enhanced Effector Functions. *Biomedicines, 10*(9), 2126. DOI: 10.3390/biomedicines10092126 PMID: 36140226

Varadé, J., Magadán, S., & González-Fernández, Á. (2021). Human immunology and immunotherapy: Main achievements and challenges. *Cellular & Molecular Immunology, 18*(4), 805–828. DOI: 10.1038/s41423-020-00530-6 PMID: 32879472

Vashisht, V., Vashisht, A., Mondal, A. K., Farmaha, J., Alptekin, A., Singh, H., Ahluwalia, P., Srinivas, A., & Kolhe, R. (2023). Genomics for Emerging Pathogen Identification and Monitoring: Prospects and Obstacles. *BioMedInformatics, 3*(4), 1145–1177. DOI: 10.3390/biomedinformatics3040069

Vasudevan, K. T. D., Hebbar, S. R., Selvam, P. K., Rambabu, M., Anbarasu, K., & Rohini, K. (2025). Multiomics and AI-driven immune subtyping to optimize neoantigen-based vaccines for colorectal cancer. *Scientific Reports, 15*(1), 19333. DOI: 10.1038/s41598-025-01680-1 PMID: 40456769

Veal, D. A., Deere, D., Ferrari, B., Piper, J., & Attfield, P. V. (2000). Fluorescence staining and flow cytometry for monitoring microbial cells. *Journal of Immunological Methods, 243*(1-2), 191–210. DOI: 10.1016/s0022-1759(00)00234-9 PMID: 10986415

Vebr, M., Pomahačová, R., Sýkora, J., & Schwarz, J. (2023). A Narrative Review of Cytokine Networks: Pathophysiological and Therapeutic Implications for Inflammatory Bowel Disease Pathogenesis. *Biomedicines, 11*(12), 3229. DOI: 10.3390/biomedicines11123229 PMID: 38137450

Verma, M., Hontecillas, R., Abedi, V., Leber, A., Tubau-Juni, N., Philipson, C., Carbo, A., & Bassaganya-Riera, J. (2016). Modeling-Enabled Systems Nutritional Immunology. *Frontiers in Nutrition, 3*, 5. DOI: 10.3389/fnut.2016.00005 PMID: 26909350

Villamor, N., Montserrat, E., & Colomer, D. (2003). Mechanism of action and resistance to monoclonal antibody therapy. *Seminars in Oncology, 30*(4), 424–433. DOI: 10.1016/s0093-7754(03)00261-6 PMID: 12939711

Waldmann, H. (2019). Human Monoclonal Antibodies: The Benefits of Humanization. *Methods in Molecular Biology (Clifton, N.J.), 1904*, 1–10. DOI: 10.1007/978-1-4939-8958-4_1 PMID: 30539464

Wang, H., Zhang, W., & Tang, Y. W. (2022). Clinical microbiology in detection and identification of emerging microbial pathogens: Past, present and future. *Emerging Microbes & Infections, 11*(1), 2579–2589. DOI: 10.1080/22221751.2022.2125345 PMID: 36121351

Wang, X.-D., Ma, B.-Y., Lai, S.-Y., Cai, X.-J., Cong, Y.-G., Xu, J.-F., & Zhang, P.-F. (2025). High-throughput strategies for monoclonal antibody screening: Advances and challenges. *Journal of Biological Engineering, 19*(41). Advance online publication. DOI: 10.1186/s13036-025-00513-z PMID: 40340930

Wei, Y., Qiu, T., Ai, Y., Zhang, Y., Xie, J., Zhang, D., Luo, X., Sun, X., Wang, X., & Qiu, J. (2025). Advances of computational methods enhance the development of multi-epitope vaccines. *Briefings in Bioinformatics, 26*(1), bbaf055. Advance online publication. DOI: 10.1093/bib/bbaf055 PMID: 39951549

Wiertsema, S. P., van Bergenhenegouwen, J., Garssen, J., & Knippels, L. M. J. (2021). The Interplay between the Gut Microbiome and the Immune System in the Context of Infectious Diseases throughout Life and the Role of Nutrition in Optimizing Treatment Strategies. *Nutrients, 13*(3), 886. DOI: 10.3390/nu13030886 PMID: 33803407

Wong, F., de la Fuente-Nunez, C., & Collins, J. J. (2023). Leveraging artificial intelligence in the fight against infectious diseases. *Science, 381*(6654), 164–170. DOI: 10.1126/science.adh1114 PMID: 37440620

Yeh, R. M. M., Taha, B. A., Bachok, N. N., Sapiee, N. M., Othman, A. R., Abd Karim, N. H., & Arsad, N. (2024). Advancements in detecting porcine-derived proteins and DNA for enhancing food integrity: Taxonomy, challenges, and future directions. *Food Control, 161*, 110399. DOI: 10.1016/j.foodcont.2024.110399

Yin, X., Shan, J., Dou, L., Cheng, Y., Liu, S., Hassan, R. Y. A., Wang, Y., Wang, J., & Zhang, D. (2024). Multiple bacteria recognition mechanisms and their applications. *Coordination Chemistry Reviews, 517*, 216025. DOI: 10.1016/j.ccr.2024.216025

Yu, T., Chae, M., Wang, Z., Ryu, G., Kim, G. B., & Lee, S. Y. (2025). Microbial Technologies Enhanced by Artificial Intelligence for Healthcare Applications. *Microbial Biotechnology, 18*(3), e70131. DOI: 10.1111/1751-7915.70131 PMID: 40100535

Zafar, A., Khan, M. J., Abu, J., & Naeem, A. (2024). Revolutionizing cancer care strategies: Immunotherapy, gene therapy, and molecular targeted therapy. *Molecular Biology Reports, 51*(1), 219. DOI: 10.1007/s11033-023-09096-8 PMID: 38281269

Zhang, Y., Thompson, K. N., Branck, T., Yan, Y., Nguyen, L. H., Franzosa, E. A., & Huttenhower, C. (2021). Metatranscriptomics for the Human Microbiome and Microbial Community Functional Profiling. *Annual Review of Biomedical Data Science, 4*, 279–311. DOI: 10.1146/annurev-biodatasci-031121-103035 PMID: 34465175

Zhang, Y., Zhou, L., Xia, J., Dong, C., & Luo, X. (2022). Human Microbiome and Its Medical Applications. *Frontiers in Molecular Biosciences, 8*, 703585. DOI: 10.3389/fmolb.2021.703585 PMID: 35096962

Zhao, M., Chu, J., Feng, S., Guo, C., Xue, B., He, K., & Li, L. (2023). Immunological mechanisms of inflammatory diseases caused by gut microbiota dysbiosis: A review. Biomedicine & pharmacotherapy = Biomedecine & pharmacotherapie, 164, 114985. DOI: 10.1016/j.biopha.2023.114985

Zhao, Q., Chen, Y., Huang, W., Zhou, H., & Zhang, W. (2023). Drug-microbiota interactions: An emerging priority for precision medicine. *Signal Transduction and Targeted Therapy, 8*(1), 386. DOI: 10.1038/s41392-023-01619-w PMID: 37806986

Zhou, J., Chen, J., Peng, Y., Xie, Y., & Xiao, Y. (2022). A Promising Tool in Serological Diagnosis: Current Research Progress of Antigenic Epitopes in Infectious Diseases. *Pathogens (Basel, Switzerland), 11*(10), 1095. DOI: 10.3390/pathogens11101095 PMID: 36297152

Zhou, Y., Zhou, Z., Zheng, L., Gong, Z., Li, Y., Jin, Y., Huang, Y., & Chi, M. (2023). Urinary Tract Infections Caused by Uropathogenic *Escherichia coli*: Mechanisms of Infection and Treatment Options. *International Journal of Molecular Sciences, 24*(13), 10537. DOI: 10.3390/ijms241310537 PMID: 37445714

Zola, H., & Brooks, D. (2018). Techniques for the production and characterization of monoclonal hybridoma antibodies. In Hurrell, J. G. R. (Ed.), *Monoclonal hybridoma antibodies: Techniques and applications* (pp. 1–57). CRC Press. DOI: 10.1201/9781351074827

8 ABBREVIATIONS

Antibody-Dependent Cellular Cytotoxicity: ADCC
Artificial Intelligence: AI

Aryl Hydrocarbon Receptors: AhR
Complementarity-Determining Regions: CDRs
Inflammatory Bowel Disease: IBD
Monoclonal Antibodies: mAbs

Chapter 5
Algorithmic Nutrient Engineering via Dynamic Microbial Bioprocesses:
Omics Translation for Precision Nutrient Design

Irem Kılınç
http://orcid.org/0000-0002-3398-8532
Tokyo University of Marine Science and Technology, Japan

Sherein Ismail Abd El-Moez
http://orcid.org/0000-0003-3620-8930
National Research Centre, Egypt

ABSTRACT

This chapter describes the mechanistic approach used to translate nutrigenomic data into clinically meaningful nutritional formulations through artificial intelligence (AI) and precision fermentation (PF). This highlights the central challenge of personalized nutrition: converting complex genomic and metabolic signals into consistent, actionable nutrient specifications. This chapter introduces an AI-driven inverse-design system that generates digital nutrient specifications (DNSs) that define the molecular composition and production parameters. PF is a biomanufacturing platform capable of executing these specifications through engineered microbial systems and real-time process control to produce genotype-aligned peptides, vitamins, and metabolic cofactors. The chapter also outlines the analytical pipeline linking multiomic interpretation to fermentation optimization and provides examples relevant to metabolic and neurodevelopmental applications, concluding with key regulatory and ethical considerations for implementing AI-based personalized nutrition.

DOI: 10.4018/979-8-3373-3121-8.ch005

1. INTRODUCTION

The modern food system is experiencing an immediate epistemic break. The constraints of conventional agricultural and bioprocessing paradigms are becoming particularly obvious as climate instability, ecological depletion, and demographic expansion continue to increase. However, it is not only the pressure of the environment that dominates modern food science; it is also the biological pressure. The new understanding that metabolic wellness, disease predisposition, and nutritional effectiveness are tightly connected to genomic variation, microbiome structure, and multiomic biochemical profiles has upset the long-held belief that food can be just a population-level good. Instead, food is becoming a programmable biological interface that should react correctly to the metabolic architecture of individuals (Rouskas et al., 2025).

In this context, precision fermentation can no longer be viewed as an incremental innovation in food production but as a molecular manufacturing regime that has the potential to generate nutritional agents of interest with unprecedented fidelity. Introduced microorganisms, including yeasts, fungi, and bacteria, have been repurposed into controlled microfactories that can produce caseins, whey proteins, enzymatic complexes, structured lipids, and designer peptides with the same reproducibility that is not possible through conventional fermentation (Mirsalami & Mirsalami, 2025). Nonetheless, this molecular accuracy requires equally accurate computational intelligence. The microbial genetics and thermodynamics of fermentation and biochemical flux networks combine to produce a design space that is too complex to be explored by human intuition alone, necessitating the implementation of artificial intelligence systems capable of modeling, predicting, and optimizing microbial behavior (Leon-Buitimea et al., 2022; Lateef, 2025).

Artificial intelligence is thus not an auxiliary addition but rather the interpretive and translational drive of the whole biochemical pipeline. Inverse-design architectures based on AI learn high-dimensional biological pathologies, such as changes in epigenomic regulation, proteomic perturbations, and gut metabolite pathologies, into machine-readable production specifications that can be printed by a microbial chassis. Such an algorithmic basis of biological complexity into manufacturable molecular products is the primary mechanistic contribution, as presented in Section 4.1, in which the AI is formalized as the computational interface between genotype and nutrient phenotype. Section 5.1 builds upon this basis by examining how these computational frameworks unlock clinically actionable nutrition via metabolic modeling and real-time feedback loops.

Combining AI-enhanced fermentation and microbiome-informed dietary design will advance the field to a new area of biomedical exploration. Algorithms are algorithmically designed to adjust gut microbial ecology to promote the health of

Lactobacillus and Bifidobacterium via prebiotics, postbiotics, and tailor-made bioactive peptides (Zhang et al., 2024; Elhalis et al., 2023). Machine-learned models are able to detect molecular interactions that reinforce immune modulation or promote metabolic resilience (Lopez-Moreno et al., 2023). Protein structure predictions and reengineering epitopes have also allowed the computational reduction of allergenicity and inflammatory potential and the development of next-generation nutritional analogs for people with lactose intolerance, celiac disease, or immune sensitivity (Achinna &Deshpande, 2024; Yang et al., 2025). Such abilities are indications of a shift in population nutrition to genotype-calibrated dietary therapeutics.

Nutrigenomics goes deeper into this paradigm. The incorporation of genotype-inspired predictive models into fermentation processes allows AI to enable the development of nutrient formulations that match individual genomic disposition and metabolic needs (Bogueva & Danova, 2024). These formulations have implications for precision prevention programmes against metabolic syndrome, obesity, and micronutrient deficiencies (Onyeaka et al., 2023). Notably, the reasoning of the system is not based on general dietary intuitions but rather on the computational explanation of individual biological information, translated into manufacturable specifications, a principle that underlies the mechanistic explanation given in Section 4.1.

In addition to the concept of personalization, the environmental and industrial design of AI-directed fermentation presents fresh dimensions of sustainable biomanufacturing. AI-powered life-cycle assessment models can project energy requirements, emissions, and water consumption and thus design environmentally efficient operating regimes (Mirzabaev et al., 2019). agro-industrial byproducts (molasses, whey permeate, and spent grains) can be reused in fermentation feeds through AI-optimized metabolic flux analyses (Sigalla et al., 2021), making precision fermentation an enabler of circular bioeconomies.

These systems are further enhanced by parallel developments in the area of digitalization (Dakhia et al., 2025). IoT-connected, smart bioreactors with real-time sensors produce continuous data streams that drive machine-learning models that can autonomously control pH, dissolved oxygen, biomass growth, and metabolite formation to achieve GMP- and HACCP-compliant production quality (Terefe, 2022). With this integration, fermentation becomes a nonexperimental craftsmanhip, merely a manufacturing infrastructure governed by self-regulating algorithms (Augustin et al., 2024).

All of these convergent directions—AI-based molecular design, omics-based personalization, microbiome-based modulation, and sustainable biomanufacturing—constitute the conceptual basis of a new generation of adaptive, genotype-inspired approaches to nutrition (Kong et al., 2022). The architecture of this integrated framework is depicted in Figure 1, which not only redefines the technical boundaries of food production but also develops the mechanistic logic of the translational patterns that have been established over the course of this chapter.

Figure 1. Systems-level integration of machine learning–driven bioprocesses for next-generation functional food design.

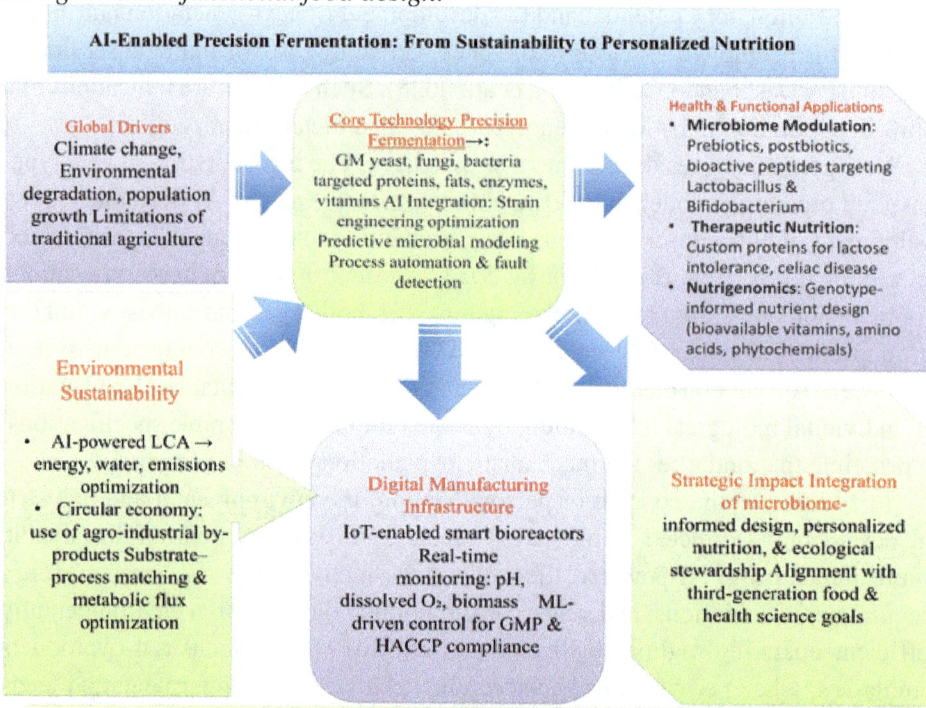

2. TECHNOLOGICAL FOUNDATIONS OF PRECISION FERMENTATION RELATED TO FOOD SCIENCE

Precision fermentation is a critical innovation at the intersection of synthetic biology, systems biotechnology, and artificial intelligence (AI), presenting an innovative paradigm of molecular food system reengineering. In precision fermentation, however, rationally designed microorganisms, often genetically engineered via CRISPR/Cas9 gene editing and pathway engineering, are exploited to create specific biocompounds of nutritional or functional value (Reyed et al., 2026, Reyed, & Prabhakar, 2025). Unlike conventional fermentation, which utilizes natural microbial activity with a relatively small repertoire of specificity, this molecular precision targets the increasing popularity of clean-label, allergen-free, and functional enrichment of food ingredients in personalized nutrition as well as therapeutic applications (Boukid et al., 2023).

The essence of this technological breakthrough is genetic circuit engineering, in which biosynthesis-related gene clusters are converted into organismal hosts—typically *Saccharomyces cerevisiae* (Liu et al., 2021), *Escherichia coli* (Bilyk et al., 2016), or *Yarrowia lipolytica*, also known as *Candida lipolytica* (Cao et al., 2023)—to generate proteins, enzymes, and vitamins in a well-regulated fashion. Artificial intelligence platforms are invaluable in streamlining these pathways: deep learning and constraint-based modeling make prior predictions of enzyme performance, codon optimization, and metabolic bottlenecks, contributing to the considerably faster cycle of design-build-test-learn (DBTL) in microbial engineering (Noll and Henkel, 2020).

Precision fermentation stands out in food science because of its ability to combine AI-based metabolic flux analysis and bioinformatics (Vinestock et al., 2024). These tools facilitate time-dependent analysis of genomic and transcriptomic data, modeling microbial growth, response to stress, and product yields under changing fermentation conditions. Here, AI is not just an optimization tool but rather a cognitive coworker in strain selection, pathway reconfiguration, and real-time quality assurance. An example is AI-driven multiomics integration, which allows correlations between gene expression profiles and product production to inform the selection of knock-in or knock-out genetic events to optimize metabolic performance.

A strategic assessment of the modern scene in precision fermentation reveals that the major challenge lies in the scalability and standardization of AI-based technologies in the industrial environment. Biofoundries, as well as computational frameworks, are often created in an academic setting, which makes them insufficiently robust industrially and transferable to real applications (Asin-Garcia et al., 2025). In addition, the predictive power of such models is highly dependent on relatively limited training datasets that are still limited by limited diversity or operating within implicit biases toward model organisms. This subsequently leads to an urgent need to create universal, open-access repositories of microbes along with broader adoption of standardized processes that can enable the realization of results and increase the extensiveness of their application in food-grade systems (Reyed, 2023). One notable research gap is the compatibility of the microbiome with precision-fermented ingredients. Even though the produced products might translate nutritional similarity to animal nature-based analogs, the extended effects on the host microbiome ecosystem, accessibility to bioavailability, and host metabolism are poorly researched. For example, recombinant casein or omega-3 s synthesized by yeast fermentation could have different interactions with existing intestinal flora than natural casein or omega-3 s due to structural differences as well as byproducts. This knowledge gap should be addressed by integrated in vitro gut modeling, AI-assisted metabolomics, and bioengineering of microbiomes to predict downstream health impacts, i.e., the transition between bioproduction and nutrigenomic safety (Elhalis et al., 2023).

Personally, an analytical viewpoint on the future direction of this field should be human-focused systems biology, where AI and synthetic biology will be combined to develop not only foodstuff but also nutritional therapeutics matching the host genomics and microbiome ecosystems. To facilitate such integration, a multilayered approach would be needed: to include predictive models of AI, the simulation of the healthy microbiome-gut, and the evaluation of the functional outcomes of that combination, shifting the paradigm of a technology-driven approach to one based on universal health—solicitation—and ethical biodesign (Bogueva & Danova, 2024).

Precision fermentation is an extremely ambitious area of research in the broader study of AI-driven food innovation: the scientific and technological basis is interesting, but as it relates to individualized nutrition and microbiome science, its applicability is challenging. To demonstrate the full potential of precision fermentation, promoting transdisciplinary cooperation, robust clinical testing, and other bioethically sound measures will be necessary to ensure alignment among new technology, population health, and disciplined science (Vinestock, et al., 2024).

2.1 Bioinformatics as a Catalyst for Microbial Performance and Sustainability

Bioinformatics presents one of the most critical foundations of the ripening of AI-infused precision fermentation, transforming the very way in which microbial systems are designed, refined, and implemented in the context of sustainable food production (Saleem et al., 2024). By utilizing the ability to process and integrate large volumes of genomic, transcriptomic, proteomic, and metabolomic data through often automated methods, bioinformatics can offer a systems-level understanding of microbial physiology and the functionality of metabolism. Compared with conventional fermentation paradigms, which are still rooted in empirical screening and heuristic optimization, AI-enhanced bioinformatics frameworks provide predictive systems that can guide strain selection, the reconstruction of metabolic pathways, and the determination of fermentation parameters with impeccable precision. For example, genome annotation algorithms allow the identification of stress tolerance genes, nutrient absorption genes, and secondary metabolite biosynthesis genes—features that are vital for strong industrial performance. Furthermore, the integration of multiomics data enhances the detail of metabolic flux models, enabling scientists to predetermine how engineered organisms respond to diverse substrate conditions or distressed operating environments (Kanwal et al. 2025). The versatility of such capabilities is particularly important in precision fermentation products that require microbial strains to reliably produce highly specific products, such as recombinant milk proteins, plant-based heme, and functional peptides, in many production runs.

One of the most significant contributions of bioinformatics lies in its ability to maximize microbial performance while simultaneously contributing to ecological sustainability. In food systems characterized by resource shortages, climatic uncertainty, and rising world demand, bioinformatics allows for the creation of strains that utilize low-cost renewable feedstocks such as lignocellulosic biomass, food-processing wastes, and farm wastes with high efficiency (Ye et al., 2025). Modeling supported by AI can redirect metabolic pathways to target end products while encountering maximum byproduct suppression, accompanied by reduced energy and carbon dioxide emissions. Metabolic network analysis packages, especially COBRApy and OptFlux, predict carbon redistribution patterns caused by the deletion of genes or changes in promoters, which provides practical ideas about how to coordinate the behavior of microorganisms with the goals of a circular bioeconomy.

Driven by bioinformatics, precision fermentation would act not only as a paradigm of production but also as a tool that has the potential to catalyze sustainable restructuring of the agri-food sector. Despite the great progress that has been made, there are several obstacles. Of most urgent concern are the constraints in the entire characterization of industrially relevant microbial genomes, especially in nonmodel organisms.

The metabolism of less conventional hosts, such as *Corynebacterium glutamicum* and *Pichia pastoris*, as well as probiotic strains such as *Lactobacillus plantarum*, has been poorly explored despite the availability of significant genomic resources in widely studied organisms such as *Escherichia coli* and *Saccharomyces cerevisiae*. Underexplored genomes result in a limited design space to produce food-grade compounds, particularly in cases where strain selection must comply with food safety, regulatory, and allergen-free requirements (Sturme et al., 2025).

Another limitation is the lack of access to high-quality, annotated datasets to train machine learning models used in predictive bioinformatics. Numerous existing sources include fragmented or inconsistent metadata standards or are biased toward particular applications, thereby limiting the generalizability of trained AI systems. In addition, the dynamic nature of microbial metabolism, which is influenced by environmental variations, host–microbe interactions, and posttranslational control, is highly complex, with current computational systems equipped to handle it only currently.

To counter such disparities, future research should focus on creating open-access, standardized, and interoperable bioinformatics platforms that can manage multisource biological information with high-throughput model validation. A combination of efforts among academic institutions, biotechnology businesses, and regulatory authorities will be critical to building a shared infrastructure of AI-powered microbial engineering (Bakkeren et al., 2025)

These platforms are expected to incorporate microbiome-guided design rules to allow predictions of the interactions of fermentation-derived chemicals with the human gut microbiome. The simulation of intestinal microbial metabolism via AI should be integrated into strain design processes to ensure that functional ingredients, such as post-biotics and synthetic fibers, do not have disruptive effects on human health.

The intersection between computational biology and nutrigenomics will not only improve products to optimize efficacy but also prevent undesirable effects such as dysbiosis, inflammation, or uncharacterized metabolic perturbations (Asar et al., 2025).

One of the most promising directions lies in the utilization of digital twin technology in modeling bioprocessing. Virtual representations of physical fermentation systems (digital twins) collect omics data, real-time sensor outputs, and past process history to improve estimates of strain performance.

With the addition of AI, such digital twins can anticipate the effects of genetic or environmental changes on the production of metabolites, allowing for active control of processes and adaptive optimization planning. These real-time feedback loops significantly lower the costs, time, and uncertainty associated with scaling-up moves between laboratory fermenters and industrial fermenters (Vallée, 2025).

Furthermore, in conjunction with IoT-powered bioreactors, such systems form a smart fermentation ecosystem where the role of bioinformatics is an intelligence layer that coordinates the variables involved in the biological, chemical, and operational aspects of fermentation to achieve an optimal result. At a broader health scale, issues of bioinformatics integration in microbial engineering should also cover food safety, allergenicity, and personalized nutrition.

Predictive models can be created to test whether given microbial pathways produce byproducts that are of toxicological interest or that are immunogenic (Yee, et al., 2025). Additionally, consumer genomic data enable tuning of fermentation-derived food products to specific metabolomic consumer profiles, dietary allergies, or microbiome compositions—potentializing the dream of truly personalized functional foods. To achieve this goal, however, there is a need to design bioinformatics pipelines that are scalable, efficient, and compliant with regulations while ensuring transparency and reproducibility. Multidisciplinary governance models are required to address ethical issues related to instances of algorithmic bias and data privacy and custody, which must provide ethical oversight of equitable access to AI-enabled nutrition technologies. Overall, bioinformatics is not a secondary computational technology that accompanies precision fermentation; instead, it is a building block of microbial performance, nutritional specificity, and ecological sustainability. Transforming biological complexity into actionable predictive intelligence, bioinformatics forms the foundation for reshaping nutritious and accessible food systems through the lens of improved productivity and personalization, as well as an environmentally friendly planet (Namli, et al., 2024). Nevertheless, pharmaceutical innovations need to be explored further beyond technical optimization into the sociobiological challenges inherent in health systems, microbiome ecology, and global food justice. The future of functional food innovations will build on adaptable, interoperable, and ethically based AI-bioinformatics infrastructures, thereby bringing such advances closer to adhering to human health and stewardship of the environment (Simon et al., 2025).

2.2 Key differences between precision and traditional fermentation

As an extension of maximizing the potential of microbial strains with the help of bioinformatics, precision fermentation contrasts with conventional manufacturing models in terms of its integrated nature, scalability, and sustainability. Typical methods, which are based on empirical, organism-based manipulation, frequently utilize uncontrolled microbial consortia and environmental factors to transform materials into food products such as yogurt, bread, beer, and tempeh.

In contrast, precision fermentation transcends this empirical model, employing synthetic biology, systems-level biotechnology, and artificial intelligence to design microbial hosts to manufacture nutritional or functional biooptimized biomolecules in a targeted manner. Instead of relying on native metabolic pathways, precision fermentation designs reprogrammed strains, e.g., *Saccharomyces cerevisiae*, *Pichia pastoris*, or *Escherichia coli*, whose genomes are optimized to produce casein, ovalbumin, lipase, phytase, vanillin, and omega-3 fatty acids in a highly precise and reproducible manner (Teng et al., 2021). Such selectivity is achievable owing to the application of CRISPR-based genome editing, modular pathway engineering, and AI-aided metabolic optimization, which collectively optimize intracellular flux distributions without driving undesired side effects by competing reactions. The result is a series of bioproduction platforms, each capable of producing bioidentical food compounds without allergens, antimicrobials, or zoonotic contamination, potentially offering improved nutritional specificity (Binczarski et al., 2025). Figure 2 elucidates a theoretical evaluation between traditional fermentation models and AI-enhanced precision fermentation, emphasizing dissimilarities in control, specificity, sustainability, and production scalability.

Another positive aspect of precision fermentation is the flexibility of its feedstock and its ecological sustainability. Unlike conventional systems that typically rely on sugar-rich feedstocks produced through energy-intensive monocultures, precision fermentation can utilize a wide range of low-cost and renewable substrates, including agricultural residues, lignocellulosic biomass, food processing waste streams, and methane-derived carbon feedstocks. This flexibility enables the decoupling of production from crop-based inputs and reduces dependence on sugar-rich resources that are often unsustainable and energy intensive (Knychala et al., 2024). Moreover, AI-powered metabolic modeling platforms—such as flux balance analysis (FBA) and machine learning-based yield prediction algorithms—allow researchers to simulate substrate uptake kinetics, carbon utilization pathways, and byproduct formation under different scenarios. This adaptability not only lessens environmental burdens and reduces agricultural input requirements but also supports the development of circular economy frameworks, wherein waste streams from one process serve as nutrient sources in another fermentation process (Khanna et al., 2024). In doing so, precision fermentation contributes directly to resource conservation, waste valorization, and greenhouse gas mitigation—areas where traditional fermentation remains inherently limited owing to its reliance on batch-scale, monoculture-driven systems that lack adaptive control (Qiu et al., 2023).

A second defining distinction between traditional fermentation and precision fermentation lies in the scalability and modularity of the production infrastructure. Conventional fermentation typically requires large-scale, centralized facilities involving complex downstream processing, high water consumption, and significant energy inputs. These systems are characterized by long processing times, susceptibility to contamination, and a limited ability to adapt to changes in demand or environmental conditions. In contrast, precision fermentation employs compact, automated bioreactors that can be deployed in decentralized, modular units (Yee et al., 2025). Equipped with Internet of Things (IoT)-enabled sensors and AI-driven control systems, these bioreactors monitor real-time variables—such as temperature, pH, dissolved oxygen, and biomass accumulation—and adjust operational parameters dynamically to optimize yields and ensure batch-to-batch consistency. This degree of process intelligence not only reduces operational costs and resource use but also enhances responsiveness and resilience, which are critical to ensuring food security in the face of climate variability, geopolitical instability, and supply chain disruptions (Rather et al., 2024).

From a technological perspective, the integration of AI and synthetic biology within precision fermentation enables a level of metabolic control and product customization that is unattainable in conventional systems (Tsalas et al., 2024). AI algorithms are increasingly applied to genomic and phenotypic datasets to predict enzyme efficiencies, forecast strain performance under varying conditions, and shorten strain development timelines while reducing experimental failure. Coupled with FBA and similar AI-powered modeling tools, these methods allow researchers to predict substrate uptake rates, carbon flux distributions, and byproduct profiles across diverse input conditions (Zhang et al., 2025). In parallel, synthetic biology enables the de novo design of biosynthetic pathways for compounds that are otherwise difficult or prohibitively expensive to produce naturally. Examples include human milk oligosaccharides, collagen peptides, and heme analogs for plant-based meat. These biomolecules can be tailored to meet specific nutritional, sensory, or functional needs, allowing manufacturers to produce sustainable food products that align with the principles of personalized nutrition. In contrast, conventional fermentation is constrained by the inherent limits of natural metabolic pathways and microbial physiology, restricting both metabolic flexibility and conversion efficiency in modern food systems.

Despite these advantages, precision fermentation has several drawbacks. A major barrier includes the excessive capital and operational costs associated with constructing bioreactors, developing strains, and obtaining regulatory authorization. Precision-fermented products are thus more expensive than traditional fermented foods are (Ramírez Rojas et al., 2022), hindering their widespread use, especially among low- and middle-income groups. Moreover, the acceptability of fermentation-generated food ingredients among consumers, particularly those based on genetically modified organisms, varies significantly across cultures and regulatory authorities. Labeling transparency, thorough safety verification, and educational campaigns are critical steps toward establishing trust in and acceptance of these new developments (Cet al., 2025). Additionally, the understanding of the long-term health effects of recombinant compounds, especially their interactions with the human microbiome, is limited. Even though molecular mimicry is achievable via precision fermentation, concerns exist regarding whether targeted compounds are products of metabolism under the same conditions or whether their uptake could disrupt microbial homeostasis or trigger immune reactions in unpredictable ways. The existing uncertainties underscore the need to develop a more multisided approach where microbiome-informed design, clinical validation, and postmarket surveillance are at the center of precision fermentation products. From a scientific perspective, future development must focus on making precision fermentation processes more affordable, accessible, and biologically friendly. The highest priority must be given to the development of AI-trained open-source strain libraries on the basis of comprehensive and synergistic biology datasets and interoperable digital fermentation platforms that enable efficient transfer of technology between regions. Effective cross-disciplinary coordination between synthetic biologists, microbiome experts, AI developers, and food scientists will be invaluable for scaling up laboratory achievements to solutions that can be used worldwide. At the same time, ethical concerns related to patenting microorganisms, data ownership, and food sovereignty must be considered to ensure access to new precision fermentation in equal terms. At an individual analytical level, precision fermentation serves as a technological invention and a paradigm shift in the way food is perceived, produced, and consumed as part of health policies. Its ability to create tailor-made, functional, and sustainable food ingredients makes it a keystone of the food system of the future. However, achieving this goal will require a balance between highly detailed and accurate biological design and open, diverse, and morally aware innovation processes.

Figure 2. Precision fermentation vs. traditional fermentation

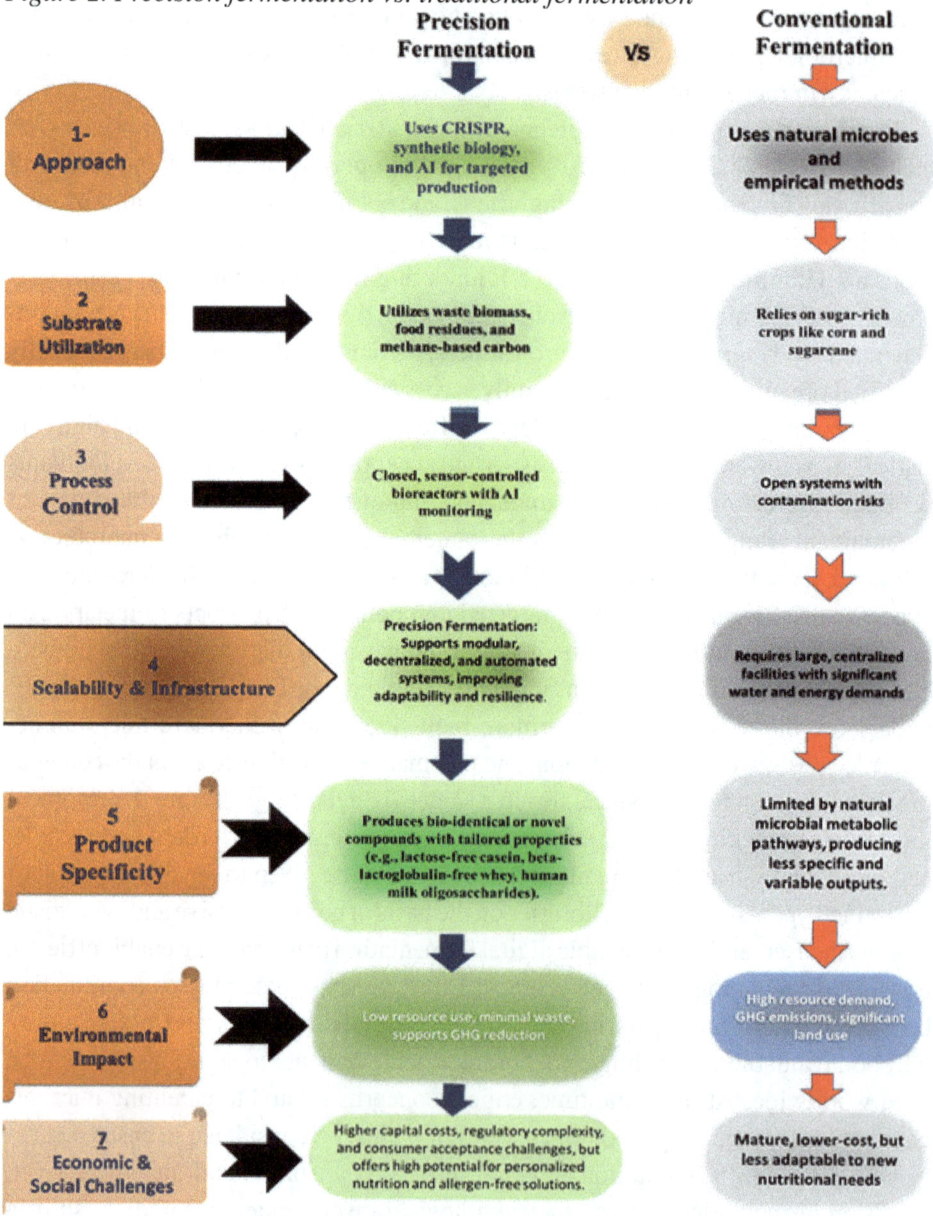

Analytical Summary:

Research shows that precision fermentation can be used to turn food systems into reproducible, customizable, and sustainable food systems. It streamlines the ability of flexibility in feedstock selection, allows the development of compounds without allergies, and matches production with customized nutrition aspirations.

Residual Gaps:

- High costs
- Fragmented regulations
- Consumer acceptance issues
- lack of safety information on microbiomes.

Proposed Solutions:

- Build open-source strain libraries connected to compatible electronic platforms.
- Identify technoeconomic applications through artificial intelligence to reduce costs and enhance manufacturing.
- Perform strict clinical trials and postmarketing microbiome tracking.
- Establish end-to-end ethical regulations facilitating data openness and equity.

3-INTERSECTIONS OF PRECISION FERMENTATION, FOOD SAFETY, AND PUBLIC HEALTH

Precision fermentation is not merely a technological advancement; it is the systematic change in both food safety and allergen management and food nutrition as an element of the social and cultural fabric of well-being (Fig. 3). The most significant power is the greatest extent of control over the processes: closed and hermetically sealed bioreactors are sterile, the temperature and pH are precisely maintained, and opportunistic pathogens such as Salmonella and Listeria are removed. These abilities transform food production into a computationally controlled process instead of a process of probability. The crux of the issue discussed here is that there are risks associated with the imperviousness of traditional systems of production to contamination, fluctuations in harvested nutritional products, and insufficient real-time confirmation of safety status (Tsalas et al., 2024). AI-based monitoring is one of the solutions because quality control can be turned into an active process. Rather than allowing contamination or deviation, real-time sensor networks are used to predict risks in place to intervene prior to their occurrence. This not only minimizes the losses in practice during batches but also greatly fosters consumer confidence by proving that safety information is not a one-point-only checkpoint. This is written, but the dilemma is how to incorporate such advanced monitoring in systems that may be resource-limited owing to constraints in infrastructure, power constant, and maintenance of sensors. Such a solution is proactive, that is, the creation of low-cost, modular, AI-driven biosurveillance kits capable of scenario deployment in decentralized plants so that high safety standards would not be those of only well-funded manufacturing sites. The other important dimension is that of manufacturing hypoallergenic and inclusive foodstuffs. Recombinant caseins, whey proteins, and gluten analogs can be regarded as direct reactions to increasing global health mismatch: increasing the number of food allergies and intolerances that affect quality of life and nutrition. Precision fermentation is highly specific because it eliminates immunogenic epitopes without compromising functional capabilities (e.g., gelation and emulsification).

This remains, however, not the complete answer to the problem unless long-term clinical validation is performed regarding the interaction of these modified molecules with the human immune system and gut microbiota in the long term. The biological issue concerns not only the challenge but also regulatory aspects; other jurisdictions can place IT items under different categories, which may cause delayed entry into the market. The suggested direction is an integrated framework involving the integration of in silico allergenicity awareness, gut-on-chip meal digestion, and global population-based microbiome data to provide evidence of security and mixability, which are globally accepted. Another urgent need involves micronutrient fortification by engineered microbes; micronutrient fortification is a hidden hunger and nutrient deficiency that disproportionately afflicts children, pregnant women, and elderly individuals. Pilot-scale production is already technically possible, but yield continuity in variable environments continues to be a bottleneck. The result is AI-based feedstock monitoring intertwined with the need to implement induction logistics and predictive demand forecasting, since the introduction of AI-controlled monitoring must be corrected by adjusting the type of production to the density of the population in real time. The vision of the future is a responsive system, a dynamic and adaptive one where epidemiologic trends can automatically result in fermentation runs that are generated and shipped at the precise point and time when they are to be consumed as fortified foods. Another overlay effect is shelf-life extension via the use of bacteriocin-producing probiotics, which reduce waste levels through the transformation of food and facilitate sustainable distribution. Inline spectroscopy is coupled with machine-learning algorithm tools, suggesting the occurrence of spoilage in advance, which allows process modifications to preserve the quality of the product. The effect it may have is considerable: the carbon footprint of food supply chains decreases when the shelf life is longer, and resilience to disruptions is also enhanced. The only gap that exists is the lack of understanding of the interactions of these bacteriocins with beneficial gut microbes and how such interactions affect the structure of the gut ecosystem with increased long-term ingestion. Future approaches should involve combining multiomics profiling technology with microbiome stability tests to ensure that preservation methods do not inadvertently disrupt host–microbe symbiosis.

All these factors lead to a scenario whereby precision fermentation is a component of preventive population health. Nevertheless, to actualize the vision, a number of systemic issues must be addressed.

First, infrastructure equity: the scaled fermentation units using solar power and decentralized facilities should address settings with low resource usage.

Second, regulatory harmonization: AI-assisted safety dossiers need to be accepted internationally so that approval can be achieved across jurisdictions.

Third, social acceptance requires open communication, codesign with locals, and culturally sensitive product development. The prognosis of long-term work is that the food produced will cease to be a commodity-oriented business and become a precision health service. This vision of the future is characterized by bioreactors hosting health platforms, AI as a nutritional planner, and each food batch optimized, not only to taste but also for microbiome compatibility, metabolic health, and individualized nutritional TFs. This marks a paradigm shift: food will cease to be just a source of calories and will be a focused intervention aimed at functioning as a preventive measure against disease, as an immune-boosting solution, and at improving the quality of life where nutrition and medicine meet.

Figure 3. AI-enabled precision fermentation workflow and its public health and nutritional applications. The figure shows an example of the integrated decision-support workflow, which involves the linkage of AI-based optimization of the microbial strain, predictive safety modeling, and real-time bioprocess analysis and improves food safety validation, micronutrient accuracy, and nutritional interventions for the population.

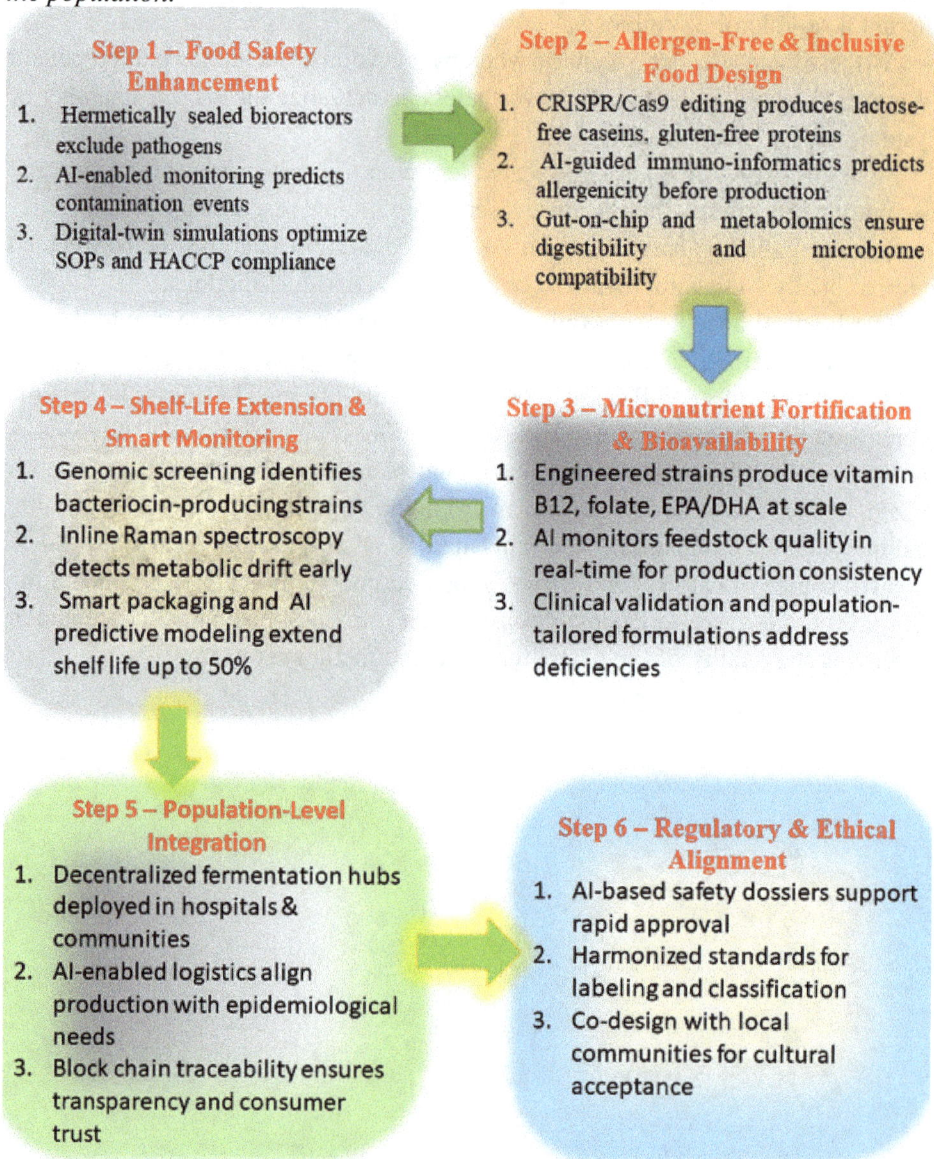

4. INTEGRATIVE TRANSLATIONAL FRAMEWORK: INTERRELATIONSHIP BETWEEN NUTRIGENOMICS AND AI-CONTROLLED PRECISION FERMENTATION

4.1. Overview and Rationale

The combination of nutrigenomics and AI-based precision fermentation represents a significant paradigm shift in the development of personalized nutrition. In this new model, food becomes redefined to be no longer a traditional source of caloric food but rather a genotype-calibrated biomedical intervention with profound multiomics intelligence. The main challenge in translational efforts is the development of an ultraconstructive mechanistic system that will convert patient-customized genomic and metabolic patterns into extremely exact tailor-made nutrient compounds. This is achieved by microbial biofactories coordinated by sophisticated AI-based regulatory systems that ensure faithfulness, security, and biochemical selectivity at clinically usable concentrations (Iqbal et al., 2025).

The implementation of nutrigenomics would require the capacity to translate multifaceted multiomic signals into high-fidelity biomanufacturing specifications. This conversion has been accomplished by means of AI-based inverse-design systems that can take complicated biological manufacturing deficits and render them machine-read production blueprints (Karbalaei-Heidari & Budisa, 2024). Here, precision fermentation is the only manufacturing system that can bring these AI-produced blueprints to the physical reality of molecular products that are physically realizable and cover patients at the level of their specificity. The conventional chemical or scale biomanufacturing strategies are insufficient since they are based on stagnant, one-size-fits-all architectures incapable of fulfilling quick reparameterization rates according to the comparable conditions of an individual genome or metabolism. In contrast, precision fermentation involves a modular, reprogrammable microbial chassis with metabolic pathways that can be rapidly reconfigured to generate custom nutritional or therapeutic molecules. Artificial intelligence is used as the translational interface in this pipeline, as a complex biological signature is transformed into specific operative parameters needed by the fermenter: genomic variation, proteomic imbalances, and metabolic deficits. By providing feedback and closed-loop optimization, AI converts patient-derived biological data into real-time biomanufacturing instructions, thus operationalizing personalized nutrition at production-scale resolution.

4.2. The Specification Challenge: Signaling Genomic Knowledge into manufacturing Logic.

4.2.1 Multiomic Fidelity and Functional Interpretation.

An intensive nutrigenomic investigation has gone far beyond individual single-nucleotide polymorphisms (SNPs) of the types MTHFR C677T, FADS2 C174575, or BCMO1 C12934922. These genetic phenotypes interact with metabolic processes in cooperation with functional biomarkers obtained through the use of transcriptomics, proteomics, metabolomics, and microbiome profiling (Thilakar, 2025; Loric & Conti, 2022). Federated multiomic learning structures are needed to capture these complex system interactions by ensuring cross-data layer security. To obtain a single-step expression, such a comprehensive system yields a functional deficiency score (FDS), a powerful quantitative evaluation of the metabolic effects of genotype–phenotype interactions (Calvino, et al.,2024).

4.2.2 AI Specification Engine

The AI specification engine then processes the FDS, which uses a Bayesian inverse-design model to transform the score into a digital production specification that is detailed. This specification defines the biologically optimal ranges of concentrations of the biologically required nutrient with a special focus on the required molecular structure (e.g., the biologically active form of folate versus its precursor, folic acid) (Queiroz, et al.,2024). In addition, the specification places strict requirements on purity, allergenicity and physicochemical stability, which consists of stability to thermal, oxidative and pH changes. It also specifies the characteristics of bioenhancing matrices of material necessary to achieve the maximum intake of nutrients. The generated digital specification file regulates and controls the totality of microbial engineering and subsequent fermentation procedures (qbal, et al., 2025).

4.3. Breaking the Barrier: Data to Mechanical Control.

4.3.1 Strain engineering and metabolic flux optimization

To achieve the high-fidelity outputs specified by the AI specification, microbial chassis (e.g., *Saccharomyces cerevisiae*, *Pichia pastoris* or microalgae) manipulations are highly complex. These biofactories are designed by CRISPR-mediated pathway rewiring, promoter optimization and a specific redirection of flux toward the desired metabolites. One of the most significant safety interventions involves the eradication of any actively movable genetic component in the host strain. Constant metabolic flux analysis (MFA) and constraint-based modeling (CBM) ensure yield and stability and optimize the performance of metabolic organisms under specified bioprocessing conditions (lin et al., 2019).

4.3.2 Safety-by-Design Architecture.

Individualized nutrition requires a degree of clinical-grade biosafety, to which the manufacturing process is inherently integrated. The components of this architecture include strict removal of all known host proteins that are allergens, removal of residual host DNA to nondetectable levels, and extensive endotoxin clearing procedures. The system is also provided with advanced safety features, including fixed-point orthogonal kill switches (Pohling et al. 2024), which are used to guarantee absolute biological containment and compliance with regulations.

4.3.3 AI-Managed Precision Fermentation: Online Cororectal Fidelity.

A Cyber-Physical Integrated Architecture controls the implementation phase (Zeng, 2019). This topology implements a one-way, real-time connection between the tangible bioreactor and its predictive digital twin, thus permitting real-time perturbation analysis and prediction of metabolic change states. The deep learning-based control systems act with a prescriptive control paradigm to control key processes of the process: gaseous mass-transfer coefficients, redox potential, and microscale substrate dosing to be guided by the metabolic flux pathway toward the genomic blueprint but controlled by a system that aims to maintain the genomic blueprint as a strict constraint on its behavior. Production fidelity will be compared with high, clinical-grade performance goals (Shariatifar et al., 2025), such as testing concentration variability within the tolerance envelope, zero tolerance to metabolic flux drift and real-time bioprocess fidelity of the final bioactive nutrient. The precision fermentation workflow shown in Figure 4 illustrates how genomic and multiomic inputs can be computationally converted via specification modeling, microbial engineering, and controlled bioprocessing to generate personalized nutrient formulations.

Figure 4. Integrative Workflow Linking Genomic Profiling, AI Specification, Microbial Engineering, and Precision Fermentation in Personalized Nutrition.

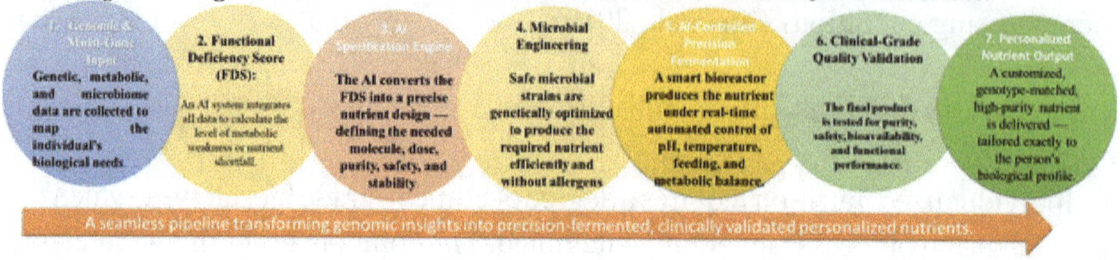

The diagram presents the full translational pathway from genomic input to final precision-fermented nutrient output, highlighting the computational, biological, and quality-control stages involved.

4.4. Application and Translational Use Cases.

4.4.1 The following example is used: 4.1 MTHFR-C677T and precision-fermented 5-MTHF.

Carriers of the MTHFRC677T polymorphism demonstrate reduced folic acid to its active form 5-methyltetrahydrofolate (5-MTHF) enzymatic transformation, making them susceptible to cardiometabolic and neuropsychiatric conditions (Andrade et al., 2024). AI-precision fermentation responds to this by producing the active metabolite 5-MTHF, which is already ready to be used without involving the process of human-equivalent catalysis, thus overcoming the damaged host pathway. The manufacturing specifications of this personalized product are highly detailed: the active metabolite should be synthesized at a therapeutically relevant concentration, a high purity level should be achieved, and an allergen-free formulation with increased physicochemical stability should be included. The efficacy of the final products is confirmed by thorough incorporation activities on cells in tests of biological activity. The entire operation loop is present in this broad-scale genotype–phenotype-specific engineered biosynthesis-validated outcome pathway (Ciaglia et al., 2025).

4.4.2 Translational Use Cases

This framework is widely applicable to a continuum of multifaceted chronic illnesses caused by genetic–metabolic dysregulation. An example is the ability to modify variants of lipid metabolism related to cardiometabolic disorders through establishing APOE-specific omega-3 lipid ratios and controlling the production of phytosterols (Clark et al., 2015; AlQahtani et al., 2025). Similarly, interventions for autism spectrum disorder address abnormalities in methylation and oxidative stress, which entail highly accurate fermentation of active 5-MTHF and methyl-vitamin B12. Fermented chromium picolinate and the use of probiotics promoting the production of specific fatty acids taken as short chains could be used to manage type 2 diabetes and insulin resistance (Belardo, et al., 2019).

5. CONCLUSIONS AND FUTURE PROSPECTS

5.1 Future Direction and Research Omissions

Although this open-system translational framework is robust, the discipline faces a number of strategic challenges that need to be overcome to enable its prevalently broad clinical implementation: realization of full multiomic integration, especially deeper transcriptomic and metabolomic profiling; conducting large-scale, genotype-stratified clinical trials; and alignment of regulatory requirements across the traditionally separate disciplines of food, biotechnology and medicine (Mohr, et al., 2024). In addition, special work is needed to make bioreactor engineering more cost-effective and introduce strong social-ethical models that can guarantee the fair distribution of personalized nutrient products worldwide (Murdock, et al., 2024).

5.2 Concluding Rationale

This integrative framework, which is discussed in the following section, supports the overall goal of the manuscript, i.e., to combine artificial intelligence, the human microbiome, and nutrigenomic personalization (Kunal et al., 2025). AI-mediated precision fermentation is defined as a fabrication process but a key functional principle that is at the core of this new look at personalized food. Even though methodological and empirical gaps are present in the framework, interdisciplinary overlap is expected to greatly enable the restructuring of nutrition worldwide as a means of therapeutically customized and genotype-specific alimentary systems (Priyadharshini, et al., 2025).

6. APPLICATIONS AND CASE STUDIES OF AI-ENHANCED PRECISION FERMENTATION IN THE FOOD INDUSTRY

Innovative preservation and production technologies are driving the pace at which AI-powered precision fermentation is being used in the development of functional foods. These technologies combined with synthetic biology enable programmable microbial systems to de novo create bioidentical or enhanced ingredient-free dairy proteins, structured flavors, tailor-made fats, and optimized nutrients with postbiotic potential with substantially diminished environmental impact by integrating real-time AI monitoring, adaptive shelf-life extension, and bacteriocin-based protection (Knychala et al., 2024). Companies such as Perfect Day, Every Company, and Formo are already taking proven engineered microbial strains to AI-optimized bioreactors, with demonstrated reductions of up to 91% in GHG emissions and 98% in water usage compared with livestock agriculture (Kanwal et al., 2025).

The use of AI in fermentation also involves the identification of fine chemicals and aroma compounds, thus replacing volatility in agricultural supply chains and lessening pressure on deforestation. Examples of functional drinks include kombucha; therefore, AI modeling of microbial consortia has the potential to maximize flavor, nutritional, and probiotic stability (Priyadharshini et al., 2025).

The key industrial issues include laboratory scale-up for manufacturing, oxygen transfer, pH gradient, foaming, and data and equipment standardization. The concepts of digital twins via AI and the interoperability of fermentation are thus key, as are the need to combine production with circular bioeconomic pathways that incorporate waste-derived substrates and AI-based supply chain maximizers (Kanwal et al., 2025).

In the future, precision-fermented foods are expected to be made possible by individualizing to omics-based consumer characteristics such that microbatch manufacturing of nutrient-specific foods can be feasible in clinical practice (Priyadharshini et al., 2025). Data, bias in algorithms, and equal access to them in terms of algorithms must be handled by ethical and regulatory frameworks to prevent the spread of nutritional inequality.

Through integrated strategies of multimodal scaling, digital standardization, regulatory alignment, and structured access, precision fermentation in the form of AI-driven precision fermentation can become a crux of resurgent, personalized, and sustainable international food systems (Knychala et al., 2024).

6.1 Production of Alternative Proteins and Functional Ingredients

Precision fermentation powered by artificial intelligence is transforming the landscape of alternative proteins, functional food ingredients, fermented beverages, and the unlocking of more uses of biomass through sustainable and scalable microbial biomanufacturing (Paz Cedeno et al., 2025). These companies, including Perfect Day and Every Company (Perfect Day LCA. 2024), utilize machine-learning pipelines to rationally design microbial strains that produce casein, ovalbumin, and collagen analogs of the same functional parity as their animal-derived products but also result in a reduction in land, water, and greenhouse gas burdens (Wood & Tavan, 2022). The magnification of the screening of enzyme variants and pathway structure, which can be seen through recent developments increasing the production of vanillin, highlights the rigor of precision and yield enhancement, which can be achieved through AI tools (Tazon et al., 2024). However, scaling is still an issue; bottlenecks of oxygenation and metabolic stress are common problems that run into the industrial-scale bottleneck. In turn, process simulations guided by AI, control via reinforcement learning, and adaptive networks of biosensors per se are not optional components of safe functioning (Teknoscienze, 2024). In the context of a fermented drink, omics-driven community modeling enables the design of high-performance probiotic consortia. However, to ensure reproducibility throughout the microbial complexity and variability of raw materials, the integration of real-time microbiome monitoring and AI-based contamination prevention is vital. The usefulness of AI is extrapolated to sustainable biomaterials where biomaterial options include PHAs and chitosan packaging, and a predictive model is useful in the design of polymer properties. However, operations at the downstream stage will follow a hustling process, demanding creativity in solid-state fermentation and extraction according to an optimized AI. Among the major obstacles are unregulated or regulatory uncertainty, a mismatch of social-economic understanding of consumer acceptance (especially in emerging markets) and the absence of harmonized ontologies that facilitate cross-platform AI application. The vision of personalized, health-integrated, sustainably produced, and globally scalable food and material production can be made around future models of modular, convergent production hubs, which combine and integrate digital twins, demand-responsive microbial routing, and health-integrated feedback systems (Kuhl, 2025).

6.2 Flavor and Fragrance Development

The use of AI-assisted precision fermentation is changing the dynamics of flavor and fragrance production with a radical trend. Instead of harvesting volatile molecules from plants or synthesizing them chemically, programmed microorganisms are engineered to produce them in fermentation vats. Machine learning algorithms can decipher the metabolic pathways that lead to naturally occurring flavor compounds, including vanillin, citrus terpenes, and cocoa aroma molecules, and define the genetic alterations required to produce them (Manikandan et al., 2023). By mapping sensory properties to the actual composition of molecules, AI applications can also create flavors on the basis of specific cultural or regional taste variations, further enhancing adaptability within the market (Yang et al., 2025). However, there are still important research challenges. The creation of high-resolution standardized sensory datasets should be prioritized because only with these datasets can it be possible to train AI models that can predict subtle flavor perception.

Second, microbially produced flavor compounds must undergo thorough safety and compositional reviews to pass regulatory controls and achieve consumer acceptance.

Last, the regulatory systems in different jurisdictions that govern the AI-driven development of flavors and fragrances via bioproduction are not harmonized, which constitutes a market entry and scale-up challenge.

Closing these knowledge gaps will require synergistic work between sensory scientists, bioprocess engineers, data scientists, and regulatory experts to fulfill the promise of precision fermentation powered by AI in the flavor and fragrance sectors.

6.3 Fermented Beverages and Functional Foods: Personalized Health through Microbial Engineering

There has been a paradigm shift in traditional fermented beverages, including kombucha, kefir, and artisanal beers; these beverages are no longer just cultural artifacts, and scientific advancements place them as finely tuned functional foods. Current advances in microbiome science and computational modeling now allow researchers and food technologists to optimize these products for their desired sensory qualities and measurable health benefits. In concert with the wider ambition of personalized nutrition, this transition augments typical themes across this book, especially in the linking of nutrigenomics, microbial ecology, and personalized wellness regimens.

The microbial communities that naturally form as part of fermented drinks are by definition quite complex, but when carefully controlled, they can create significant benefits, such as improved intestinal health, metabolic health, and immune functions. Recent research underlines the important influence of the variability in fermentation factors, such as the composition and pH of the substrate, temperature, and inoculum ratios, on final products in terms of their bioactive metabolome (Du et al., 2022; Imamoglu, 2024). In turn, computational tools have become invaluable in the prediction of microbial behavior as well as the optimization of coculture interactions and ensuring reproducible performance in production. These attempts are based on human-directed design, which is dependent on experimental confirmation and continuous data-driven improvement, as opposed to caprice.

The movement toward functional customization additionally affords the integration of nutrigenomic profiles into fermentation strategies. An example is the selection of beverages with specific probiotic strains that may target host-specific genomic markers related to lactose intolerance, inflammatory responses, or microbiota composition. Such interventions are illustrative of how targeted microbial intervention can be used as an adjunct to dietary personalization, which is one of the most fundamental tenets of precision medicine (Abbaspour, 2024).

Modern breakthroughs in fermented beverage science have broadened the field of research, but various conceptual knowledge gaps remain. The first constraint is in phenotypic and sensory characterization: measurements available today do not meet the spatial and temporal resolution needed to detect minor changes in flavor or bioactivity during fermentation under batch variables. The second weakness is our limited understanding of postingestion metabolic processes. Despite the current acceptance that functional metabolites may contribute to the beneficial effects of a probiotic, the specific mechanisms involved remain to be determined, making claims regarding its putative antioxidant or therapeutic properties very speculative. Finally, differences in regulatory bodies result in regulatory fragmentation that adds complexity to commercialization and consumer adoption, highlighting the need to consider the standardization of functional labeling and safety validation. The opportunity to bridge these gaps requires an integrative research plan. This goal can be fulfilled only when the fields of nutritional genomics and microbial ecology are integrated with biotechnological innovation to develop models that can explain genetic, dietary, and environmental diversity at the population level (Keskin & Yıldırım, 2025). This broad approach can be used to ensure that the interventions are technically viable and still clinically and culturally applicable. This point of view does not conflict with the idea presented throughout this book that AI-powered microbiome-based systems will turn nutrition into a personalized, clinically anchored, and sustainable paradigm. This new vision represents a paradigm change in the manufacturing of fermented beverages, a shift in approach that is no longer based on the traditional noncontrolled manner of producing fermented beverages but a method in which a person regulates and can personalize. The bioengineering of microbial consortia has now reached the point where nutritional and functional goals can be precisely controlled, fully in line with the fundamentals of personalized medicine. To demonstrate the scope of this material transformation, Table 1 summarizes the scientific and methodological distinctions between the conventional fermentation methods and the individualized fermentation methods.

Comparative Table (1). Traditional vs. human-guided personalized fermented beverages

Aspect	Traditional Fermented Beverages	Human-Guided Personalized Fermented Beverages	Scientific Analysis and Methodological Perspective
Fermentation Microbiota	Naturally occurring communities with mixed wild strains and inherent variability (Park et al., 2025)	Curated consortia selected for targeted metabolic outputs and specific health functions (Chai, 2022)	Personalized systems enable reproducibility and targeted benefits but require careful balance to preserve beneficial microbial diversity.
Health Functionality	Provides general probiotic effects without tailoring to individual metabolic or immunological profiles (Shi et al., 2020)	Designed for host-specific outcomes such as anti-inflammatory effects and improved digestion, based on individual nutritional genomics (Petrariu et al., 2024)	Enhanced functionality increases relevance and efficacy for diverse populations; however, comprehensive clinical validation remains critical for translational applications.
Nutritional Output	Metabolite profiles vary with fermentation conditions and raw materials, leading to inconsistent nutritional outputs (Postigo et al., 2025)	Optimized and standardized metabolite production achieved through integrated genomic and metabolic modeling, ensuring alignment with dietary needs (Baimakhanova, et al.., 2025)	Improved nutritional predictability allows for better alignment with personalized dietary interventions, although further studies are necessary to validate long-term health benefits.
Flavor and Sensory Profile	Rich sensory complexity influenced by traditional techniques and terroir, but high variability limits standardization (Shi et al., 2025)	Tailored flavor profiles based on sensory mapping and cultural taste preferences, while preserving authenticity (Chaudhari, 2025)	Integration of sensory data with genetic and biochemical profiling enhances market adaptability and consumer satisfaction without compromising traditional characteristics.
Process Consistency	High batch-to-batch variability is common, often resulting in unpredictable outcomes (Lee et al., 2024)	Achieves robust reproducibility through controlled fermentation parameters and standardized protocols, facilitating scalable production (Reffai & Fechtali, 2025)	Consistency is crucial for commercial viability and clinical reliability, making quality control and systematic validation indispensable for widespread adoption.
Cultural and Ethical Relevance	Deeply rooted in local traditions and established production practices, but may lack adaptability to modern nutritional demands (Ghosh et al., 2023)	Adaptation of traditional methods is achieved with respect to cultural heritage, complemented by targeted nutritional enhancements for specific populations (Alqurashi et al., 2025)	A blended approach that marries tradition with precision nutrition is essential for ensuring both cultural acceptance and improved health outcomes, emphasizing the need for ethical and culturally sensitive innovation.

continued on following page

Comparative Table (1). Continued

Aspect	Traditional Fermented Beverages	Human-Guided Personalized Fermented Beverages	Scientific Analysis and Methodological Perspective
Clinical Relevance	Limited empirical evidence regarding specific health outcomes due to inherent variability and lack of standardized methodologies (Jilcha, 2025)	Supported by increasing clinical studies that assess bioavailability, metabolic benefits, and consumer safety of tailored fermented products (Park et al., 2025)	Rigorous clinical validation is imperative to confirm functional claims, thereby bridging the gap between laboratory research and real-world application in personalized nutrition.
Regulatory Challenges	Typically, classified as traditional foods, subject to conventional safety standards but less scrutinized regarding functional claims (Bolesławska et al., 2024)	Faces complex regulatory hurdles requiring harmonized standards for functional labeling, safety assessment, and microbial strain authentication (Gravante, 2025)	Clarity in regulatory definitions and standards is necessary to navigate the translational pathway for innovative functional foods, emphasizing the need for interdisciplinary collaboration among scientists, clinicians, and policymakers.

6.4 Scientific Analysis and Personal Methodological Commentary

The current table indexes a fundamental change in the landscape of fermented drinks by targeting the emergence of human-controlled personalized fermentation as an undeniable step forward past traditionally organized procedures that remain, in their specific and crystallized cultural forms, highly changeable and unamenable to large-scale adaptation and modification by user control, context, or terminal nutritional outcome. In contrast, human-controlled personalized fermentation is a stringently regulated, data-driven approach that allows the synergistic design of microbial consortia and fermentation reactions to align directly with unique health needs and cultural preferences. This level of algorithmic precision enhances product consistency and therapeutic potency and adds to the general principles of personalized nutrition integrated into the concept of precision medicine. However, translational research requires a delicate balance: although tailored microbial cocktails may generate positive effects such as gut health, metabolic balance, and immunomodulation, one must be careful not to lose the fundamental complexity of the wild microbial environment that underpins the sensory and functional diversity of fermented products. Therefore, in the process of bringing personalized fermented foods to the market, stringent clinical trials and the adoption of coordinated regulation criteria become necessary conditions to ensure consumer safety and market confidence. Overall, the shift to personalized fermentation driven by human biology has transformed nutritional genomics, microbial ecology, and precision biotechnology into a powerful example of the transformative capacity of interdisciplinary research, categorizing traditional food activities into scalable, innovative tools for evolving functional foods and refining nutritional resilience.

6.5 Bioplastic and Edible Packaging Solutions: Sustainable Innovation through AI-Guided Microbial Fermentation

The ongoing convergence of microbial engineering and artificial intelligence, which has already been reported to have been taking place with respect to the context of personalized fermented foods, is now being applied to the context of sustainable packaging. Instead of being just another straightforward lateral application, this extension supports the main message demonstrated by the personalized fermentation model, which is the integration of biology-driven solutions with computationally informed systems to achieve precision, consistency, and sustainability) González-Rojo et al., 2024).

In the edible packaging field, microbial fermentation of supplies, including agricultural waste or lignocellulosic biomass, produces bioplastics, including polyhydroxyalkanoates (PHAs). This path is similar to historic developments in personalized nutritional fermentation, in which undefined and inconsistent results are replaced with specific and objective biosynthesis. AI has a similar effect in these two applications by optimizing strain selection, modeling metabolic pathways, and controlling dynamic processes (Ibrahim et al., 2025).

Despite these developments, a number of fundamental barriers remain between innovative discoveries and transfer to scalable, validated solutions.

The lack of standardized safety/biodegradability platforms echoes the subpar harmonization of labeling and microbial strain validation existing in functional foods. ISO-grade standards and FDA-aligned pathways are unavoidable requirements in the two arenas for ensuring reproducibility, safety, and customer trust (Shi et al., 2025).

Similar to the case of personalized nutrition, where some clinical validation gap still remains, economic scalability remains a roadblock in the case of edible packaging. Full-process economic case studies of the pilot studies are therefore mandatory to fill the laboratory potential with industrial viability (Upadhyay et al., 2024).

In both scenarios, trade-offs to performance achieved in functionality, such as biodegradability versus barrier performance in the case of alternatives or microbial diversity versus reproducibility, and therefore, functional hybrids using multiple components are required to accommodate the balanced performance requirements.

The lack of data in industrial-scale fermentation, which is also a common constraint, requires the application of industrial data sharing to reinforce AI models and predictive confidence.

In conclusion, the verification of functional claims (be they in terms of micronutrient delivery of food films or probiotic functionality of fermented beverages) requires the integration of in vitro/in vivo pipelines connecting mechanistic believability to clinical outcomes.

The identified research gap, which is spread down the line of various sectors in our case, requires a comprehensive integrated framework to fill it thoroughly and to present it in the light that is least limited in any given sphere. Essentially, this gap can be attributed to the absence of a coherent translational framework or the absence of a unified combination of processes between microbial biotechnology, artificial intelligence optimization, regulatory standardization, and clinical validation. To drive such an integrated framework within food and packaging, there is the need for the creation of a multidomain innovation ecosystem where both food and packaging technologies simultaneously occur within interlocking pipelines, taking advantage of common standards, common data infrastructures, and common clinical validation approaches (Liyanapathiranage et al., 2023).

In the case that such an integrated framework came to fruition, it would increase the reduction of gaping research niches present in a given sector beyond merely narrowing their availability and opening the door to a new breed of nutritionally tuned, environmentally corrective, industrially replicable bioengineered products through a convergence of ethos that positions the exact nature of personalized fermentation with that of the sustainability dictum that defines edible packaging (Liyanapathiranage et al., 2023).

6. 6 Overarching Research Gaps and Implementation Roadmaps

Although precision fermentation is progressing faster and is expected to lead to considerable transformations into edible packaging, alternative proteins, and functional bioactives, the existing ecosystem has structural inefficiencies that are likely to limit innovation and adoption unless addressed in the future. However, these three main weaknesses, namely, the absence of homogenous data standards, regulatory byzantine, and lack of socioeconomic awareness, are not simply procedural or logistical challenges but rather systemic blind spots that hinder the movement of lab-scale innovations to feasible, culturally embedded differences operating globally.

Implementation issues at the data infrastructure level, lacking harmonized metadata ontologies and interoperable sensor control protocols, already rule out cross-domain learning, which is essential for scaling AI-aided strain engineering and fermentation optimization (Helleckes et al., 2023). In the absence of this interoperability, models can only be trained on single-sector datasets, meaning that their results would be overfitting to extremely narrow specifics in the experimental setting and lose their predictive capacity in real-world manufacturing. The solution is the development of modular FAIR-compliant (Findable, Accessible, Interoperable, Reusable) data frameworks codesigned with research consortia, industry, and regulators so that nutritional metabolomics, volatile flavor compound profiling, and PHA biosynthesis trial data can be combined in real time.

One more constraint is regulatory divergence. Incompatibility in the interpretation of definitions of novel food, toxicological limits, and claims on functional labels among jurisdictions destroys economies and delays market presence (Liyanapathiranage et al., 2023). The potential answer in this case would be to use a sandbox regulatory scheme, a multiagency environment where a comprehensive approach to dossier features representing safety as well as environmental impact and clinical validation could be tested. The frameworks would go a long way in eliminating redundancy, minimizing review time, and accelerating international uptake.

The socioeconomic focuses of the innovation story in precision fermentation are also predominantly urban, with a nominal ethnographic and cultural understanding of rural, low-income, and mobile populations, who may benefit disproportionately both in terms of the low carbon footprint of precision fermentation (compared with traditional animal products) and the high nutritional content of precision fermentation outputs, which may have significant value to these demographics. By integrating field-based participatory research during the development process, during which iterative prototypes are optimized on the basis of sensory acceptance, cultural alignment, and nutritional effects, any final product will maintain both viability (technologically) and relevancy (socially) (Lee et al., 2024).

The tri-pronged approach of 1) standardized and AI-ready data systems, 2) globally harmonized and adaptive regulations, and 3) culturally integrated deployment models is not just a policy envisioning but a technical requirement in shifting isolated innovation islands into a continuous learning environment in the context of precision fermentation. Under this paradigm, a self-optimizing network ensures that every fermentation batch, packaging prototype, and consumer feedback loop are in the mix, further forming an accelerating discovery and deployment process with the central positioning you did, i.e., societal relevancy.

7. CHALLENGES AND FUTURE PERSPECTIVES

7.1 Challenge of Regulatory and Consumer Acceptance

Precision-fermented goods presently occupy a gray area between GMOs and novel foods, making them the source of definition and pathway orthogonality that extends approval processes and atomizes markets. An intermediate solution is a common lexicon and perhaps interoperable, sandboxed approval algorithms (mutual-dossier components of safety and exposure, allergenicity, and environmental fate) to abbreviate review times and provide limited jurisdiction-specific acknowledgment. It is therefore based on mutually negotiated transparency instruments such as QR-enabled dynamic labels, which represent and display real-time fermentation conditions, model outputs, and LCA overviews that are tested via behavioral tools to fine-tune framing and cultural acceptability (Zollman et al., 2022). As a technical matter, explaining ability and liability blueprints (model cards, audit trails, fault trees) should accompany ML-optimized bioprocesses to distribute responsibility among platform vendors and producers (Sturme et al., 2025).

7.2 Costs of Scale-Up and Technical Barriers

Scale-up reveals bottlenecks in oxygen transfer, nutrient gradients, foaming, risk of contamination, and batch-to-batch variation, particularly in the case of eukaryotic hosts at high cell density. Digital twins aided by AI and self-optimizing control (closed-loop set points: kLa, feed, pH, antifoam) have the potential to preempt drift and decrease failure rates, although standardized sensors, soft-sensor calibrations, and longitudinal data are needed. To facilitate comparisons among organisms and sites, studies are encouraged to use modular, open protocols; to report benchmark TEA/LCA values; and to publish scale-down models that can be reproduced to predict scale-up behavior (Singh et al., 2024).

7.3 Infrastructure and information requirements

Cross-domain learning is hampered by heterogeneous schemas and siloed historians. Deploy FAIR-compliant data fabrics, often with shared ontologies (process, strain, metabolome, and sensory), time-synchronized edge-to-cloud pipelines, and access-controlled data trusts to open federated model training. A common metadata scheme for design-build-test learning cycles and model governance (versioning, performance envelopes, drift monitors) is needed to transition to site-transferable industrial AI (Helleckes et al., 2023).

7.4 Ethical, Social, and Environmental Considerations

The main point of the discourse that exists today regarding precision food is ethical, social, and environmental concerns. The three areas where ethical risk is focused are data rights (process versus personal health), algorithmic bias, and IP concentration/food sovereignty, as well as AI energy costs. The proposed guiding paradigm is ethics-by-design, where things such as clear data ownership contracts, the distinction between explicit consent between population analytics and personalized nutrition, bias audits performed on diverse cohorts, place-based fermentation ecologies that value local strains and operationalize benefit sharing, and transparent AI/compute LCAs with consistent product claims should be part of the equation. Some governance functions should entail the use of model cards, external auditing, community ethics boards, and combined reviews of liability, therefore making the consequent benefits regenerative and equitable and not animatedly efficient (Wood & Tavan, 2022).

Addressing the above-outlaid gaps in precision fermentation requires a coordinated effort that incorporates innovation in regulators, technical scalability, ethical liability, and societal conformity. By implementing adaptive governance models, implementing artificial intelligence in process optimization and transparency enhancement, and developing community-based fermentation ecosystems, precision fermentation can become a mainstream technology in sustainable, equitable global nutrition and smart packaging technologies (Shi, et al., 2025).

8. CONCLUDING INSIGHTS AND FUTURE DIRECTIONS

This chapter outlines an elaborate integrative translational model that significantly enhances the operational potential of nutrigenomics by setting AI-guided precision fermentation as the central pathway for the transfer of genomic and multiomics signals to clinically relevant nutritional interventions. Unlike traditional personalized nutrition models based on the single-nucleotide probabilistic interpretation of single-nucleotide polymorphisms or generalized dietary personal recommendations, the new framework provides that Bayesian inverse-design architectures combined with federated multiomic datasets are capable of producing highly specific, reproducible and clinically relevant nutrient specifications as determined by the metabolic phenotype of an individual. This study represents a new contribution to the field, filling a longstanding bottleneck: the lack of a strictly mechanistically based methodology that could be used to translate the numerous complicated genotype–phenotype interactions into biologically validated nutrient preparations (Helleckes et al., 2023; Sturme et al., 2025). Building upon the mechanistic and computational analyses presented throughout this chapter, the following synthesis integrates these contributions into the broader scientific narrative of the book.

8.1 The article generally has a broader scientific impact and aligns with the mission of the book.

The paradigm proposed below supports the overall purpose of the text, which is to combine the field of artificial intelligence with the field of microbiome science as well as the field of genotype-informed nutrition. Through the implementation of safety-by-design, concept, metabolic flux, real-time cyber-physical bioprocess monitoring, and digital-twin modeling, this chapter provides the computational and biomanufacturing infrastructure needed to transform theoretical knowledge into scalable, customizable nutrient solutions (Chaudhari et al., 2025). Furthermore, it supports the conceptual paradigm shift described in earlier sections: how food may transition to become more actively a biotherapeutic system of precision engineered mechanistic biology and computational intelligence instead of generalizations about a population (Lee et al., 2024). This placement places the chapter as one of the methodological leaders supporting the scientific consistency of the whole volume.

8.2 Strategic Research Priorities and Future Directions

The ability to translate this paradigm into broad clinical and industrial applications requires a series of studies in several areas of strategic focus. First, the field needs to reach complete multiomic convergence to include additional layers of epigenomics, metabolomics, and spatial transcriptomics in predictive models to obtain a more credible picture of metabolic activity. These developments should be tested with large-scale, genotype-stratified clinical trials capable of rigorously testing the therapeutic efficacy of such developments in various genetic structures. Moreover, global standards of digital nutrient specification (DNS) should be established to interoperate between decentralized fermentation platforms and AI ecosystems. This attempt has to be supported by powerful data-governance mechanisms that protect privacy in federated multiomic settings. Moreover, the harmonization of their regulations across the food, medical and biotechnology domains and the creation of ethical-equity models to ensure worldwide access to tailored nutrient therapeutics are long-term successes. Together, these priorities lead to a course of sustainable, scalable, and socially responsible implementation of precision-engineered nutrition (Jilcha, 2025; Sturme et al., 2025; Helleckes et al., 2023).

8.3 Closing Perspective

The interplay of nutrigenomics and AI-managed precision fermentation represents an indelible wave of evolution in the face of human nutrition. The form of dietary admonition, which is used to serve as a means of descriptive instruction, is contracted back into its original form as a genotype specific, clinically validated, biotherapeutic discipline with the ability to treat metabolic inefficiency at its interactive core in a mechanistic manner. With the scientific community driving forward into more substantial multiomic integration, more adaptive AI architecture, and more equitable regulatory approaches, personalized nutrition will become a dream, a reality, and a transformative health technology worldwide. This chapter provides the structural and conceptual framework on which this new era of preventive, exact and customized nutritional biomedicine can be achieved in its full potential. Taken together, this chapter operationalizes the scientific and translational foundation for a new era of protective, clear-cut, and completely personalized nutritional biomedicine.

REFERENCES

Abbaspour, N. (2024). Fermentation's pivotal role in shaping the future of plant-based foods: An integrative review of fermentation processes and their impact on sensory and health benefits. *Applied Food Research*, *4*(2), 100468. DOI: 10.1016/j.afres.2024.100468

Achinna, P., & Deshpande, H. (2024). Gluten-free synbiotic pseudocereals pasta: Evaluation of technological, nutritional, sensory, and probiotic characteristics. *Journal of Culinary Science & Technology*, •••, 1–27. DOI: 10.1080/15428052.2024.2354217

AlQahtani, S. N., AlGubaisi, S., AlHaffaf, F. A., Makki, R. J., Alohali, E. A., AlMadani, R. O., AlSagiheer, H. M., Al-Otaibi, M. M., & Mohammed, H. T. (2025). Nutrition Support Therapy for Hospitalized Children with Malnutrition: A Narrative Review. *Healthcare (Basel, Switzerland)*, *13*(5), 497. DOI: 10.3390/healthcare13050497 PMID: 40077059

Alqurashi, R. M., Abdalla, S. M., Bin Ammar, A., Shatwan, I. M., Alsayegh, A. A., Alnasser, A. N., Alfadhliah, J. T., Alnoubi, A. A., Fallata, G. A., Alhumaidan, O. A., & Bawazeer, N. M. (2025). The most popular local and traditional food dishes in different regions of the Kingdom of Saudi Arabia and their cultural significance. *Frontiers in Nutrition*, *12*, 1590522. DOI: 10.3389/fnut.2025.1590522 PMID: 40463474

Andrade, P., Santamarina, A. B., de Freitas, J. A., Marum, A. B. R. F., & Pessoa, A. F. M. (2024). Personalized nutrition and precision medicine in perimenopausal women: A minireview of genetic polymorphisms COMT, FUT2, and MTHFR. *Clinics (Sao Paulo, Brazil)*, *80*, 100549. DOI: 10.1016/j.clinsp.2024.100549 PMID: 39642577

Asar, R., Erenler, S., Devecioglu, D., Ispirli, H., Karbancioglu-Guler, F., Ozturk, H. I., & Dertli, E. (2025). Understanding the Functionality of Probiotics on the Edge of Artificial Intelligence (AI) Era. *Fermentation (Basel, Switzerland)*, *11*(5), 259. DOI: 10.3390/fermentation11050259

Asin-Garcia, E., Fawcett, J. D., Batianis, C., & Martins Dos Santos, V. A. P. (2025). A snapshot of biomanufacturing and the need for enabling research infrastructure. *Trends in Biotechnology*, *43*(5), 1000–1014. DOI: 10.1016/j.tibtech.2024.10.014 PMID: 39592270

Augustin, M. A., Hartley, C. J., Maloney, G., & Tyndall, S. (2024). Innovation in precision fermentation for food ingredients. *Critical Reviews in Food Science and Nutrition*, *64*(18), 6218–6238. DOI: 10.1080/10408398.2023.2166014 PMID: 36640107

Baimakhanova, B. B., Sadanov, A. K., Ratnikova, I. A., Baimakhanova, G. B., Orasymbet, S. E., Amitova, A. A., Aitkaliyeva, G. S., & Kakimova, A. B. (2025). In Silico Modeling of Metabolic Pathways in Probiotic Microorganisms for Functional Food Biotechnology. *Fermentation (Basel, Switzerland), 11*(8), 458. DOI: 10.3390/fermentation11080458

Bakkeren, E., Piskovsky, V., & Foster, K. R. (2025). Metabolic ecology of microbiomes: Nutrient competition, host benefits, and community engineering. *Cell Host & Microbe, 33*(6), 790–807. DOI: 10.1016/j.chom.2025.05.013 PMID: 40505619

Belardo, A., Gevi, F., & Zolla, L. (2019). The concomitant lower concentrations of vitamins B6, B9 and B12 may cause methylation deficiency in autistic children. *The Journal of Nutritional Biochemistry, 70*, 38–46. DOI: 10.1016/j.jnutbio.2019.04.004 PMID: 31151052

Bilyk, O., Sekurova, O. N., Zotchev, S. B., & Luzhetskyy, A. (2016). Cloning and heterologous expression of the grecocycline biosynthesis-related gene cluster. *PLoS One, 11*(7), e0158682. DOI: 10.1371/journal.pone.0158682 PMID: 27410036

Binczarski, M. J., Zuberek, J., Fraczyk, J., Kolesinska, B., Radojčin, M., Pavkov Wiktorowska-Sowa, E., Piotrowski, J., Kaminski, Z. J., & Witonska, I. A. (2025). Application of Microorganisms for the Valorization of Side-Products of Rapeseed De-Oiling. *Biomolecules, 15*(7), 917. DOI: 10.3390/biom15070917 PMID: 40723789

Bogueva, D., & Danova, S. (2024). Comparing Precision Fermentation and Traditional Fermentations: Consumer Views, Springer Books. In Bogueva, D. (Ed.), *Consumer Perceptions and Food* (pp. 563–588). Springer., DOI: 10.1007/978-981-97-7870-6_27

Bolesławska, I., Górna, I., Sobota, M., Bolesławska-Król, N., Przysławski, J., & Szymański, M. (2024). Wild Mushrooms as a Source of Bioactive Compounds and Their Antioxidant Properties-Preliminary Studies. *Foods (Basel, Switzerland), 13*(16), 2612. DOI: 10.3390/foods13162612 PMID: 39200539

Boukid, F., Ganeshan, S., Wang, Y., Tülbek, M. Ç., & Nickerson, M. T. (2023). Bioengineered Enzymes and Precision Fermentation in the Food Industry. *International Journal of Molecular Sciences, 24*(12), 10156. DOI: 10.3390/ijms241210156 PMID: 37373305

Boukid, F., Hassoun, A., Zouari, A., Tülbek, M. Ç., Mefleh, M., Aït-Kaddour, A., & Castellari, M. (2023). Fermentation for designing innovative plant-based meat and dairy alternatives. *Foods, 12*(5), 1005. DOI: 10.3390/foods12051005 PMID: 36900522

Calvino, G., Peconi, C., Strafella, C., Trastulli, G., Megalizzi, D., Andreucci, S., Cascella, R., Caltagirone, C., Zampatti, S., & Giardina, E. (2024). Federated Learning: Breaking Down Barriers in Global Genomic Research. *Genes*, *15*(12), 1650. DOI: 10.3390/genes15121650 PMID: 39766917

Cao, L., Li, J., Yang, Z., Hu, X., & Wang, P. (2023). A review of synthetic biology tools in *Yarrowia lipolytica*. *World Journal of Microbiology & Biotechnology*, *39*, 129. Advance online publication. DOI: 10.1007/s11274-023-03557-9 PMID: 36944859

Chai, Y., Ye, N., Zhang, W., & Li, Z. (2022). From spontaneous to precision fermentation: Advances in control and consistency in fermented products. *Trends in Food Science & Technology*, *128*, 1–12. DOI: 10.1016/j.tifs.2022.03.001

Chaudhari, S., Rane, A., & Kumar, A. R. (2025). Personalizing nutrition and recipe recommendation using attention mechanism with an ensemble model. Clinical eHealth, 8, 66–77. https://doi.org/DOI: 10.1016/j.ceh.2025.03.002

Ciaglia, T., Napolitano, V., Miranda, M. R., La Gioia, D., Musella, S., Schiano Moriello, A., Merciai, F., Di Sarno, V., De Vita, S., Colarusso, E., Smaldone, G., Di Matteo, F., Sommella, E. M., Kumar, P., Allarà, M., Ligresti, A., Gomez-Monterrey, I. M., Bifulco, G., Lauro, G., & Bertamino, A. (2025). Untargeted Diversity-Oriented Synthesis for the Discovery of New Antitumor Agents: An Integrated Approach of Inverse Virtual Screening, Bioinformatics, and Omics for Target Deconvolution. *Journal of Medicinal Chemistry*, *68*(15), 16483–16517. DOI: 10.1021/acs.jmedchem.5c01344 PMID: 40705771

Clark, V. R., Perkins, P., Carson, B. L., Boyd, K., & Jefferson, T. M. (2015). Fasting serum glucose and cholesterol as predictors of cardiovascular reactivity to acute stress in a sample of African American college students. *Ethnicity & Disease*, *25*(2), 175–179. PMID: 26118145

Dakhia, Z., Russo, M., & Merenda, M. (2025). AI-enabled IoT for food computing: Challenges, opportunities, and future directions. *Sensors (Basel)*, *25*(7), 2147. DOI: 10.3390/s25072147 PMID: 40218659

Du, Y. H., Wang, M. Y., Yang, L. H., Tong, L. L., Guo, D. S., & Ji, X. J. (2022). Optimization and scale-up of fermentation processes driven by models. *Bioengineering (Basel, Switzerland)*, *9*(9), 473. DOI: 10.3390/bioengineering9090473 PMID: 36135019

Elhalis, H., See, X. Y., Osen, R., Chin, X. H., & Chow, Y. (2023). Significance of fermentation in plant-based meat analogs: A critical review of nutrition and safety-related aspects. *Foods*, *12*(17), 3222. DOI: 10.3390/foods12173222 PMID: 37685155

Ghosh, S., Meyer-Rochow, V. B., & Jung, C. (2023). Embracing Tradition: The Vital Role of Traditional Foods in Achieving Nutrition Security. *Foods (Basel, Switzerland)*, *12*(23), 4220. DOI: 10.3390/foods12234220 PMID: 38231593

González-Rojo, S., Paniagua-García, A. I., & Díez-Antolínez, R. (2024). Advances in Microbial Biotechnology for Sustainable Alternatives to Petroleum-Based Plastics: A Comprehensive Review of Polyhydroxyalkanoate Production. *Microorganisms*, *12*(8), 1668. DOI: 10.3390/microorganisms12081668 PMID: 39203509

Gravante, F., Sacchini, F., Mancin, S., Lopane, D., Parozzi, M., Ferrara, G., Sguanci, M., Morales Palomares, S., Biondini, F., Marfella, F., Cangelosi, G., Caggianelli, G., & Petrelli, F. (2025). Preventing Microorganism Contamination in Starting Active Materials for Synthesis from Global Regulatory Agencies: Overview for Public Health Implications. *Microorganisms*, *13*(7), 1595. DOI: 10.3390/microorganisms13071595 PMID: 40732104

Helleckes, L. M., Hemmerich, J., Wiechert, W., von Lieres, E., & Grünberger, A. (2023). Machine learning in bioprocess development: From promise to practice. *Trends in Biotechnology*, *41*(6), 817–835. DOI: 10.1016/j.tibtech.2022.10.010 PMID: 36456404

Ibrahim, R., Aranjani, J., & Prasanna, N.. (2025). Production, isolation, optimization, and characterization of microbial PHA from *Bacillus australimaris*. *Scientific Reports*, *15*, 8395. DOI: 10.1038/s41598-025-92146-x PMID: 40069246

Imamoglu, E. (2024). Artificial intelligence and/or machine learning algorithms in microalgal bioprocesses. *Bioengineering (Basel, Switzerland)*, *11*(11), 1143. DOI: 10.3390/bioengineering11111143 PMID: 39593803

Iqbal, Z., Al-nuaim, L., Algarni, A., Alanzazi, N., AlMukhaylid, S., Al Qahtani, S., Shammas, M., Naeem, R., Taleb, Y., Karrar, T., Aleem, A., Mahmood, A., & Saglio, G. (2025). *Multi-Omics and AI-/ML-driven integration of nutrition and metabolism in cancer: A systematic review, meta-analysis, and translational algorithm*. medRxiv. https://doi.org/DOI: 10.1101/2025.07.29.25332402

Jilcha, K. (2025). Identifying existing research challenges and enhancing outcomes through the development of standardized methodologies. *Humanities & Social Sciences Communications*, *12*, 187. DOI: 10.1057/s41599-024-04269-7

Kanwal, N., Zhang, M., & Khan, S.. (2025). AI-Driven Precision Fermentation: From Restaurant Food Waste to Sustainable Protein Production. *Food and Bioprocess Technology*. Advance online publication. DOI: 10.1007/s11947-025-03981-9

Karbalaei-Heidari, H. R., & Budisa, N. (2024). Advanced and safe synthetic microbial chassis with orthogonal translation system integration. *ACS Synthetic Biology*, *13*(9), 2992–3002. DOI: 10.1021/acssynbio.4c00437 PMID: 39151168

Keskin, E., & Yildirim, İ. O. (2025). Cultivating precision: Comparative analysis of sensor-based yogurt fermentation monitoring techniques. arXiv preprint arXiv:2501.08781. https://doi.org//arXiv.2501.08781DOI: 10.48550

Khanna, M., Zilberman, D., & Hochman, G. (2024). An economic perspective of the circular bioeconomy in the food and agricultural sector. *Communications Earth & Environment*, *5*, 507. DOI: 10.1038/s43247-024-01663-6

Knychala, M. M., Boing, L. A., Ienczak, J. L., Trichez, D., & Stambuk, B. U. (2024). Precision fermentation as an alternative to animal protein, a review. *Fermentation (Basel, Switzerland)*, *10*(6), 315. DOI: 10.3390/fermentation10060315

Knychala, M. M., Boing, L. A., Ienczak, J. L., Trichez, D., & Stambuk, B. U. (2024). Precision Fermentation as an Alternative to Animal Protein, a Review. *Fermentation (Basel, Switzerland)*, *10*(6), 315. DOI: 10.3390/fermentation10060315

Kong, F. C., Ng, K. R., Samarasiri, M., & Chen, W. N. (2022). Precision fermentation to advance fungal food fermentations. *Current Opinion in Food Science*, *47*, 100881. DOI: 10.1016/j.cofs.2022.100881

Kuhl, E. (2025). *AI for food: accelerating and democratizing discovery and innovation. npj Science of Food, 9,* Article 82. https://doi.org/DOI: 10.1038/s41538-025-00441-8

Kunal, D. M., Kumari, S., & Singh, M. (2025). The Human Microbiome: An Invisible Key to Unlocking Wellness. *Journal of Pure & Applied Microbiology*.

Lateef Junaid, M. A. (2025). Artificial intelligence driven innovations in biochemistry: A review of emerging research frontiers. *Biomolecules and Biomedicine*, *25*(4), 739–750. DOI: 10.17305/bb.2024.11537 PMID: 39819459

Lee, C. H., Ahn, J., & Son, H. S. (2024). Ethnic fermented foods of the world: An overview. *Journal of Ethnic Foods*, *11*, 39. DOI: 10.1186/s42779-024-00254-2

León-Buitimea, A., Balderas-Cisneros, F. J., Garza-Cárdenas, C. R., Garza-Cervantes, J. A., & Morones-Ramírez, J. R. (2022). Synthetic biology tools for engineering microbial cells to fight superbugs. *Frontiers in Bioengineering and Biotechnology*, *10*, 869206. DOI: 10.3389/fbioe.2022.869206 PMID: 35600895

Lin, W. R., Tan, S. I., Hsiang, C. C., Sung, P. K., & Ng, I. S. (2019). Challenges and opportunity of recent genome editing and multiomics in cyanobacteria and microalgae for biorefinery. *Bioresource Technology*, *291*, 121932. DOI: 10.1016/j.biortech.2019.121932 PMID: 31387837

Liu, Z., Nielsen, J., & Lin, Z. (2021). Expression of fungal biosynthesis-related gene clusters in *Saccharomyces cerevisiae* for natural product discovery. *Synthetic and Systems Biotechnology*, *6*(1), 1–10. DOI: 10.1016/j.synbio.2021.01.003 PMID: 33474503

Liyanapathiranage, A., Dassanayake, R. S., Gamage, A., Karri, R. R., Manamperi, A., Evon, P., Jayakodi, Y., Madhujith, T., & Merah, O. (2023). Recent Developments in Edible Films and Coatings for Fruits and Vegetables. *Coatings*, *13*(7), 1177. DOI: 10.3390/coatings13071177

López-Moreno, A., Langella, P., Martín, R., & Aguilera, M. (2023). Microbiota analysis for risk assessment of xenobiotic exposure and the impact on dysbiosis: Identifying potential next-generation probiotics. *EFSA Journal*, *21*(S1), e211010. DOI: 10.2903/j.efsa.2023.e211010 PMID: 38047127

Loric, S., & Conti, M. (2022). Versatile Functional Energy Metabolism Platform Working From Research to Patient: An Integrated View of Cell Bioenergetics. *Frontiers in Toxicology*, *3*, 750431. DOI: 10.3389/ftox.2021.750431 PMID: 35295105

Manikandan, S., Vickram, S., Sirohi, R., Subbaiya, R., Krishnan, R. Y., Karmegam, N., Sumathijones, C., Rajagopal, R., Chang, S. W., Ravindran, B., & Awasthi, M. K. (2023). Critical review of biochemical pathways to transformation of waste and biomass into bioenergy. *Bioresource Technology*, *372*, 128679. DOI: 10.1016/j.biortech.2023.128679 PMID: 36706818

Mirsalami, S. M., & Mirsalami, M. (2025). Advances in genetically engineered microorganisms: Transforming food production through precision fermentation and synthetic biology. *Future Foods : a Dedicated Journal for Sustainability in Food Science*, *11*, 100601. DOI: 10.1016/j.fufo.2025.100601

Mirzabaev, A., Wu, J., Evans, J., Garcia-Oliva, F., Hussein, I. A. G., . . . Weltz, M. (2019). Desertification. In P. R. Shukla, J. Skea, E. Calvo Buendia, V. Masson-Delmotte, … & J. Malley (Eds.), *Climate Change and Land: An IPCC special report on climate change, desertification, land degradation, sustainable land management, food security, and greenhouse gas fluxes in terrestrial ecosystems*. Intergovernmental Panel on Climate Change. https://www.ipcc.ch/srccl/chapter/chapter-3/

Mohr, A. E., Ortega-Santos, C. P., Whisner, C. M., Klein-Seetharaman, J., & Jasbi, P. (2024). Navigating Challenges and Opportunities in Multi-Omics Integration for Personalized Healthcare. *Biomedicines*, *12*(7), 1496. DOI: 10.3390/biomedicines12071496 PMID: 39062068

Murdock, L., Taylor &. Francis, & Stephenson, R., HEPI. (2024). Advancing Translational Research. *Journal of Translational Research*, *1*(1). Advance online publication. DOI: 10.1080/29947448.2024.2432753

Namli, T., Anıl Sınacı, A., Gönül, S., Herguido, C. R., Garcia-Canadilla, P., Muñoz, A. M., Esteve, A. V., & Ertürkmen, G. B. L. (2024). A scalable and transparent data pipeline for AI-enabled health data ecosystems. *Frontiers in medicine*, *11*, 1393123. DOI: 10.3389/fmed.2024.1393123 PMID: 39139784

Noll, P., & Henkel, M. (2020). History and evolution of modeling in biotechnology: Modeling and simulation, application, and hardware performance. *Computational and Structural Biotechnology Journal*, *18*, 3309–3323. DOI: 10.1016/j.csbj.2020.10.018 PMID: 33240472

Onyeaka, H., Tamasiga, P., Nwauzoma, U. M., Miri, T., Juliet, U. C., Nwaiwu, O., & Akinsemolu, A. A. (2023). Using artificial intelligence to tackle food waste and enhance the circular economy: Maximizing resource efficiency and minimizing environmental impact: A review. *Sustainability*, *15*(13), 10482. DOI: 10.3390/su151310482

Park, I., & Mannaa, M. (2025). Fermented Foods as Functional Systems: Microbial Communities and Metabolites Influencing Gut Health and Systemic Outcomes. *Foods*, *14*(13), 2292. DOI: 10.3390/foods14132292 PMID: 40647044

Paz Cedeno, F. R., Olubiyo, O. J., & Ferreira, S. (2025). From microbial proteins to cultivated meat for alternative meat-like products: A review on sustainable fermentation approaches. *Journal of Biological Engineering*, *19*, 44. DOI: 10.1186/s13036-025-00509-9 PMID: 40369620

Perfect Day, L. C. A. (2024). Life cycle assessment of animal-free dairy protein production. Perfect Day Inc. https://perfectday.com/blog/life-cycle-assessment-of-perfect-day-protein/

Petrariu, O. A., Barbu, I. C., Niculescu, A. G., Constantin, M., Grigore, G. A., Cristian, R. E., Mihaescu, G., & Vrancianu, C. O. (2024). Role of probiotics in managing various human diseases, from oral pathology to cancer and gastrointestinal diseases. *Frontiers in Microbiology*, *14*, 1296447. DOI: 10.3389/fmicb.2023.1296447 PMID: 38249451

Pohling, J., Ramakrishnan, V. V., Hossain, A., Trenholm, S., & Dave, D. (2024). Optimization of Enzymatic Deproteination of Northern Shrimp (Pandalus borealis) Shell Chitin Using Commercial Proteases. *Marine Drugs*, *22*(10), 445. DOI: 10.3390/md22100445 PMID: 39452853

Postigo, V., García, M., Crespo, J., Canonico, L., Comitini, F., & Ciani, M. (2025). Bioactive Properties of Fermented Beverages: Wine and Beer. *Fermentation (Basel, Switzerland)*, *11*(5), 234. DOI: 10.3390/fermentation11050234

Priyadharshini, D., Muthuvel, I., Saraswathy, S., & Jegadeeswari, V. (2025). Precision to plate: AI-driven innovations in fermentation and hyperpersonalized diets. *Frontiers in Nutrition*, *12*, 1659511. DOI: 10.3389/fnut.2025.1659511 PMID: 40977977

Priyadharshini, D., Muthuvel, I., Saraswathy, S., Kavitha, P. S., & Jegadeeswari, V. (2025). Precision to plate: AI-driven innovations in fermentation and hyperpersonalized diets. *Frontiers in Nutrition*, *12*, 1659511. DOI: 10.3389/fnut.2025.1659511 PMID: 40977977

Qiu, S., Yang, A., & Zeng, H. (2023). Flux balance analysis-based metabolic modeling of microbial secondary metabolism: Current status and outlook. *PLoS Computational Biology*, *19*(8), e1011391. DOI: 10.1371/journal.pcbi.1011391 PMID: 37619239

Queiroz, L. P., Nogueira, I. B. R., & Ribeiro, A. M. (2024). Flavor Engineering: A comprehensive review of biological foundations, AI integration, industrial development, and sociocultural dynamics. Food research international (Ottawa, Ont.), 196, 115100. https://doi.org/DOI: 10.1016/j.foodres.2024.115100

Ramírez Rojas, A. A., Swidah, R., & Schindler, D. (2022). Microbes of traditional fermentation processes as synthetic biology chassis to tackle future food challenges. *Frontiers in Bioengineering and Biotechnology*, *10*, 982975. DOI: 10.3389/fbioe.2022.982975 PMID: 36185425

Rather, M. A., Ahmad, I., Shah, A., Hajam, Y. A., Amin, A., Khursheed, S., Ahmad, I., & Rasool, S. (2024). Exploring opportunities of artificial intelligence in aquaculture to meet increasing food demand. *Food Chemistry: X*, *22*, 101309. DOI: 10.1016/j.fochx.2024.101309 PMID: 38550881

Reffai, Y. M., & Fechtali, T. (2025). A Critical Review on the Role of Lactic Acid Bacteria in Sourdough Nutritional Quality: Mechanisms, Potential, and Challenges. *Applied Microbiology*, *5*(3), 74. DOI: 10.3390/applmicrobiol5030074

Reyed, R. M. (2023). Focusing on individualized nutrition within the algorithmic diet: An in-depth look at recent advances in nutritional science, microbial diversity studies, and human health. *Food Health*, *5*(5). Advance online publication. DOI: 10.53388/FH2023005

Reyed, R. M., Patel, R., Patel, S., & Prajapati, B. G. (2026). Application of synthetic biology in genetic disorder. In Prajapati, B. G., Basu, B., Dutta, A., & Chatterjee, A. (Eds.), *Synthetic biology and its consequences in present healthcare* (pp. 201–213). Academic Press., DOI: 10.1016/B978-0-443-33649-2.00001-0

Reyed, R. M., & Prabhakar, P. K. (2025). Personalized Antiviral Therapy TailoringTreatments to Genetic Environmental and Lifestyle Factors. In Aslam, M., Yingngam, B., Javed, T., & Riaz, S. (Eds.), *Global Perspectives on Antiviral Drug Development* (pp. 63–138). IGI Global Scientific Publishing., DOI: 10.4018/979-8-3693-9276-8.ch003

Rouskas, K., Guela, M., Pantoura, M., Pagkalos, I., Hassapidou, M., Lalama, E., Pfeiffer, A. F. H., Decorte, E., Cornelissen, V., Wilson-Barnes, S., Hart, K., Mantovani, E., Dias, S. B., Hadjileontiadis, L., Gymnopoulos, L. P., Dimitropoulos, K., & Argiriou, A. (2025). The Influence of an AI-Driven Personalized Nutrition Program on the Human Gut Microbiome and Its Health Implications. *Nutrients*, *17*(7), 1260. DOI: 10.3390/nu17071260 PMID: 40219016

Sadekar, O., Budamagunta, M., Sreejith, G. J., Jain, S., & Santhanam, M. S. (2021). An infectious diseases hazard map for India based on mobility and transportation networks. *Current Science*, *121*(9), 1208–1215. https://www.jstor.org/stable/27310784

Saleem, A., Anwar, S., Nawaz, T., & Cheema, N. (2024). Securing a sustainable future: The climate change threat to agriculture, food security, and sustainable development goals. *Journal of Umm Al-Qura University for Applied Sciences*. https://doi.org/DOI: 10.1007/s43994-024-00177-3

Shariatifar, M., Salimian Rizi, M., Sotudeh-Gharebagh, R., Zarghami, R., & Mostoufi, N. (2025). On digital twins in bioprocessing: Opportunities and limitations. *Process Biochemistry*, *156*, 274–299. DOI: 10.1016/j.procbio.2025.05.023

Shi, X., Cui, L., Xu, C., & Wu, S. (2025). Next-Generation Bioplastics for Food Packaging: Sustainable Materials and Applications. *Materials (Basel)*, *18*(12), 2919. DOI: 10.3390/ma18122919 PMID: 40573049

Sigalla, O. Z., Tumbo, M., & Joseph, J. (2021). Multistakeholder platform in water resources management: A critical analysis of stakeholders' participation for sustainable water resources. *Sustainability*, *13*(16), 9260. DOI: 10.3390/su13169260

Simon, B. O., Nnaji, N. D., Anumudu, C. K., Aleke, J. C., Ekwueme, C. T., Uhegwu, C. C., Ihenetu, F. C., Obioha, P., Ifedinezi, O. V., Ezechukwu, P. S., & Onyeaka, H. (2025). Microbiome-Based Interventions for Food Safety and Environmental Health. *Applied Sciences (Basel, Switzerland)*, *15*(9), 5219. DOI: 10.3390/app15095219

Singh, A., Ram, S., & Chandra, R.. (2024). A district-level geospatial analysis of anemia prevalence among rural men in India, 2019-21. *International Journal for Equity in Health*, *23*, 9. DOI: 10.1186/s12939-023-02089-w PMID: 38243230

Singh, V. K., Jiménez Del Val, I., Glassey, J., & Kavousi, F. (2024). Integration Approaches to Model Bioreactor Hydrodynamics and Cellular Kinetics for Advancing Bioprocess Optimization. *Bioengineering (Basel, Switzerland)*, *11*(6), 546. DOI: 10.3390/bioengineering11060546 PMID: 38927782

Sturme, M., van der Berg, J. P., & Kleter, G. (2025). *Precision fermentation – With a focus on food safety* (FAO Report No. CD/4448 EN). Food and Agriculture Organization of the United Nations (FAO). https://doi.org/DOI: 10.4060/cd4448en

Tazon, A. W., Awwad, F., Meddeb-Mouelhi, F., & Desgagné-Penix, I. (2024). Biotechnological Advances in Vanillin Production: From Natural Vanilla to Metabolic Engineering Platforms. *Biochem*, *4*(4), 323–349. DOI: 10.3390/biochem4040017

Teknoscienze. (2024). AI – Harnessing AI for alternative protein development: promising advances with realistic challenges. Agro Food Industry Hi-Tech, 35(3).

Teng, T. S., Chin, Y. L., Chai, K. F., & Chen, W. N. (2021). Fermentation for future food systems: Precision fermentation can complement the scope and applications of traditional fermentation. *EMBO Reports*, *22*(5), e52680. DOI: 10.15252/embr.202152680 PMID: 33908143

Terefe, N. S. (2022). Recent developments in fermentation technology: Toward the next revolution in food production. In *Food Engineering Innovations Across the Food Supply Chain* (pp. 89–106). Springer., DOI: 10.1016/B978-0-12-821292-9.00026-1

Thilakar, S. J. (2025). Exploring artificial intelligence and gene–diet interaction mapping to drive the formulation of evidence-based functional foods tailored to individual genetic profiles and metabolic health for addressing age-old health conditions. *International Journal of Food Science Research and Development*, *3*(1), 1–8. https://iaeme.com/MasterAdmin/Journal_uploads/IJFSRD/VOLUME_3_ISSUE_1/IJFSRD_03_01_001.pdf

Tsalas, N., Golfinopoulos, S. K., Samios, S., Katsouras, G., & Peroulis, K. (2024). Optimization of Energy Consumption in a Wastewater Treatment Plant: An Overview. *Energies*, *17*(12), 2808. DOI: 10.3390/en17122808

Upadhyay, A., Agbesi, P., Arafat, K. M. Y., Urdaneta, F., Dey, M., Basak, M., Hong, S., Umeileka, C., & Argyropoulos, D. (2024). Biobased smart packaging: Fundamentals and functions in sustainable food systems. *Trends in Food Science & Technology, 145*, 104369. DOI: 10.1016/j.tifs.2024.104369

Vallée, A. (2025). Digital twins for personalized medicine require epidemiological data and mathematical modeling: Viewpoint. *Journal of Medical Internet Research, 27*, e72411. DOI: 10.2196/72411 PMID: 40762974

Venugopal, G., Khan, Z. H., Dash, R., Tulsian, V., Agrawal, S., Rout, S., Mahajan, P., & Ramadass, B. (2023). Predictive association of gut microbiome and NLR in anemic low middle-income population of Odisha- a cross-sectional study. *Frontiers in Nutrition, 10*, 1200688. DOI: 10.3389/fnut.2023.1200688 PMID: 37528994

Vinestock, T., Short, M., Ward, K., & Guo, M. (2024). Computer-aided chemical engineering research advances in precision fermentation. *Current Opinion in Food Science, 58*, 101196. DOI: 10.1016/j.cofs.2024.101196

Wood, P., & Tavan, M. (2022). A review of the alternative protein industry. *Current Opinion in Food Science, 47*, 100869. DOI: 10.1016/j.cofs.2022.100869

Yang, H., Jiao, W., Zouyi, L., & Li, R. (2025). Artificial intelligence in the food industry: Innovations and applications. *Discover Artificial Intelligence, 5*, 60. DOI: 10.1007/s44163-025-00296-8

Ye, Y., Cai, Y., Wang, F., He, Y., Yang, Y., Guo, Z., Liu, M., Ren, H., Wang, S., Liu, D., Xu, J., & Wang, Z. (2025). Industrial Microbial Technologies for Feed Protein Production from Non-Protein Nitrogen. *Microorganisms, 13*(4), 742. DOI: 10.3390/microorganisms13040742 PMID: 40284579

Yee, C. S., Zahia-Azizan, N. A., Abd Rahim, M. H., Mohd Zaini, N. A., Raja-Razali, R. B., Ushidee-Radzi, M. A., Ilham, Z., & Wan-Mohtar, W. A. A. Q. I. (2025). Smart fermentation technologies: Microbial process control in traditional fermented foods. *Fermentation (Basel, Switzerland), 11*(6), 323. DOI: 10.3390/fermentation11060323

Zeng, A. P. (2019). New bioproduction systems for chemicals and fuels: Needs and new development. *Biotechnology Advances, 37*(4), 508–518. DOI: 10.1016/j.biotechadv.2019.01.003 PMID: 30639927

Zhang, C., Singh, R. P., Yadav, P., Kumar, I., Kaushik, A., Roychowdhury, R., Mubeen, M., Singh, S. K., Kumar, A., & Wang, J. (2025). Recent advances in biotechnology and bioengineering for efficient microalgal biofuel production. *Fuel Processing Technology, 270*, 108199. DOI: 10.1016/j.fuproc.2025.108199

Zhang, Q., Li2, G., Zhao, W., Wang, X., He, J., Zhou, L. et al. (2024). Efficacy of *Bifidobacterium animalis* subsp. lactis BL-99 in the treatment of functional dyspepsia: A randomized placebo-controlled clinical trial. *Nature Communications, 2024*(15), 227. DOI: 10.1038/s41467-023-44292-x PMID: 38172093

Zhu, Y., Wang, K., Jia, X., Fu, C., Yu, H., & Wang, Y. (2024). Antioxidant peptides, the guardian of life from oxidative stress. *Medicinal Research Reviews, 44*(1), 275–364. DOI: 10.1002/med.21986 PMID: 37621230

Zollman Thomas, O., Dillard, C., Bowman, D., & Le Roy, B. (2022). Framing the futures of animal-free dairy: Using focus groups to explore early-adopter perceptions of the precision fermentation process. *Frontiers in Nutrition, 9*, 997632. DOI: 10.3389/fnut.2022.997632 PMID: 36263302

Chapter 6
Microbiome-Driven Food Allergies and Precision Therapeutic Innovations

Shweta Thusoo
http://orcid.org/0000-0002-0316-3193
Dr. D.Y. Patil School of Allied Health Sciences, Dr. D.Y. Patil Vidyapeeth, Pune, India

Rohini Vipin Adepwar
http://orcid.org/0009-0006-3753-4200
Dr. D.Y. Patil School of Allied Health Sciences, Dr. D.Y. Patil Vidyapeeth, Pune, India

ABSTRACT

Food allergies have become a growing concern in the general population, and an increase of evidence is pointing to gut microbiome disbalances as the cause of immune malfunctions and allergic reactions. Altered filtering between the microbes may cause defects in immune tolerance and elevated allergicity. Conventional therapies tend to deal with symptoms and not the cause. The novel developments in the metagenomics, microbiome sequencing, and AI derived analytics have further elucidated that the gut microbes have the ability to regulate immunity. Microbial biomarkers can now be detected with the help of these technologies, and personalized interventions involving probiotics, prebiotics, fecal microbiota transplantation (FMT) and engineered microbial consortia can be developed. In spite of the promise, there are still challenges of the variability in the individual microbiome, regulatory issues, and long-term safety. In this chapter of the book, the author offers a detailed account of the role of the microbiome in food allergies, the development of AI-enhanced research instruments, and the development of microbiome-based therapeutics. It

DOI: 10.4018/979-8-3373-3121-8.ch006

also bridges the gap between science, technology, and clinical implementation by pointing out the possibility of precision medicine changing the way allergy is treated into a more focused and sustainable method of healthcare provision.

INTRODUCTION:

The gut microbiome plays a critical role in the formation and protection of allergies to foods (Yang et al., 2025; Chernikova et al., 2022). Disproportion of intestinal microorganisms is frequently the antecedent of food allergy and the onset of the disease (Mishra et al., 2025; Nance et al., 2020). Environmental factors, such as diet, mode of delivery, and the administration of antibiotics, are also the causes of dysbiosis (Faiyazuddin & Ashique, 2025). Regulations of gut microbiome composition and the induction of immune tolerance will be carried out with the help of new treatments, including dietary changes, probiotics, and microbiome-based biotherapies (Chernikova et al., 2022; Taico Oliva et al., 2025). Even though a clinical trial on prebiotics and probiotics has been conducted, no specifications are presented on how these treatment modalities can be used on a daily basis. The promise lies in the interventions e.g. fecal microbiota transplantation and post-biotic interventions grounded on microbial metabolites (Nance et al., 2020). The benefit of the gut microbiome is that it offers a good pathway of food allergy prevention and control in both children and adults (Niechcial et al., 2025). Clinically, the immune tolerance and allergen sensitivity is regulated by dysbiosis (Zhang et al., 2025; Gupta et al., 2024). Among the factors predisposing a person to allergies, there are caesarean delivery and the lack of breastfeeding (Tuli et al., 2025). Early prophylactic treatments can also be given by the beneficial bacteria that contributes to the induction of T cells regulatory differentiation (Taico Oliva et al., 2025; Mishra et al., 2025).

Immune response in relation to allergy is regulated with microbial products, mainly short-chain fatty acids (SCFAs), which improve gut barrier integrity and cause immune tolerance, but may also induce inflammation. It is essential to protect the gut barrier against translocation of allergens, and probiotics can help to recover this system (Faiyazuddin & Ashique, 2025; Obukohwo et al., 2024). Other new treatments, including oral, sublingual, and epicutaneous immunotherapies, are expected to alleviate allergen tolerance regardless of the side effects (Faiyazuddin & Ashique, 2025; Tuli et al., 2025). The combination of existing treatment plans with microbiome-based approaches has a bright future.

1. THE ROLE OF THE MICROBIOME IN FOOD ALLERGIES:

The microbiome is also instrumental in the formation and treatment of the food allergies as they affect the immune responses and intestinal homeostasis. The Dysbiosis (gut microbial composition imbalance) makes one vulnerable to food allergies (Yang et al., 2025; Chernikova et al., 2022). These allergies are caused when the immune system reacts to harmless food proteins i.e. peanuts, milk, eggs, or shellfish as threats resulting in IgE-mediated responses. Food allergies such as peanut and tree nut allergies have been on the increase at a high rate, and environmental and dietary influence have been identified as causal factors (Tuli et al., 2025).

The intestinal microbiome is known to promote immune tolerance, that is, to distinguish harmless antigens and harmful ones (Faiyazuddin & Ashique, 2025). Conversely, the immune-mediated diseases caused by dysbiosis include asthma, atopic dermatitis and food allergies. The presence of beneficial bacteria such as *Bifidobacterium* and *Lactobacillus* can prevent allergies whereas the presence of pathogens such as *Staphylococcus aureus* can aggravate the allergies. Diet, mode of delivery and antibiotic use are some of the factors that greatly influence microbiome composition (Berni Canani et al., 2019). Some of the interventions that are promising to restore the balance and immune tolerance are dietary modulation interventions, probiotics and transplantation of fecal microbiota. The hygiene hypothesis advocates early microbial exposure to prevent allergies, and more research on the use of microbiomes to prevent and treat should be conducted.

Case Study 1: Early Microbiome Disruption, Risk of allergy and AI-led Intervention. (Boggio Marzet et al., 2022; Liu et al., 2019; Abavisani et al., 2024)

Clinical Scenario:

A baby born through Caesarean section and fed on formula since birth obtained eczematous and subsequently, demonstrated a demonstration of allergic reaction to eggs at the age of one. At four months the infant was given broad-spectrum antibiotics due to a respiratory infection. Such a sequence is consistent with clinical findings that associate the disruption of microbiome in early life with allergic phenotypes.

Mechanistic Interpretation:

Loss of vertical microbial transmission (through bypassing the birth canal), the absence of human milk oligosaccharides (essential to the maturation of *Bifidobacteria*), and the loss of commensals due to antibiotics all interfered with intestinal colonization. This led to decreased abundance of immunomodulatory taxa such as *Bifidobacterium* and *Lactobacillus*, which inhibits the integrity of the mucosal barrier and regulatory T-cell (Treg) differentiation. Therefore, the development of immune tolerance did not occur, which encouraged a Th2-biased reaction and sensitization to dietary antigens (e.g., ovalbumin).

Intervention and Outcome:

Lactobacillus rhamnosus GG (LGG), an agent demonstrated to have evidence-based effectiveness in allergy prevention, reversed microbial-immune crosstalk, improved Treg activity and inhibited inflammatory mechanisms. This was associated with healed eczema and developed tolerance to egg antigens, which proves the potential of microbiome-targeted therapy.

Relevance and Integration of AI:

This case highlights the predisposition by early abiotic influences on allergy through microbiome disruption (C-section, formula feeding, antibiotics). It also points to the therapeutic potential of probiotics. Nonetheless, such interventions can be radically refined with the help of AI-based precision medicine:

Stratification of Risk (Predictive):

Genetic information (e.g., polymorphisms in FILAGGRIN that predict eczema risk, IL4/IL13 genotypes that predict atopy) in combination with microbiome data (e.g., low *Akkermansia,* high *Staphylococcus*) and clinical history (mode of delivery, antibiotic use) can be used to predict high-risk infants before allergy develops with AI models.

Selection of Probiotics Strains:

Instead of generic LGG supplementation, AI can compute specific microbiome functional impairments (e.g., the loss of butyrate production, disturbed bile acid metabolism), and can be paired with probiotic strains whose genetically encoded functions address those deficits (e.g., *Bifidobacterium longum subsp.* infantis to use HMOs).

Supervision and Non-isolating Adjustment:

Longitudinal metagenomic sequencing can monitor microbiome recovery following interventions in machine learning algorithms, and changes in taxonomic abundance (i.e. more *Faecalibacterium*) can be associated with cytokines or allergen-specific IgE. This enables probiotic/prebiotic regimens to be modified in real time.

Polygenic Risk Score (PRS):

AI is able to integrate polygenic risk scores of allergic illnesses and indices of microbial dysbiosis to compute cumulative risk and tailor the intervention severity (e.g., extend the probiotic course in cases of high genetic and high microbiome risk).

Conclusion:

This example of the hygiene hypothesis in action, however, demonstrates the shortcomings of universal probiotics. AI enables a transition toward proactive to reactive care, and combines genetic susceptibility, microbiome ecology, and environmental exposures to direct focused, timely microbial therapies. Such a unified treatment is the key to restoring immune homeostasis and preventing the allergy pandemic.

a. Key Microbial Species in Allergy Development

The role of gut microbiota and the immune system connection in the development of allergenic diseases has become rather important, in particular, food allergies. Recent researches have established that some of the healthy gut bacteria are essential in conditioning the immune system to tolerate harmless antigens, thus avoiding hypersensitive immune responses, e.g. allergies.

Table 1. Key Gut Microbes and Their Health Roles

Microbial Species	Primary Functions	Health Relevance
Bifidobacterium longum	Ferments oligosaccharides; secretes lactic acid, increases barrier integrity in the gut	Present in high levels in breastfed infants; facilitates immune tolerance and prevents allergies from developing.
Lactobacillus rhamnosus GG	Biosynthesizes antimicrobial peptides; adjusts dendritic cell activity	Widely investigated in the treatment and prevention of eczema and food allergies; proven to enhance epithelial cell functionalities
Faecalibacterium prausnitzii	Produce butyrate; act as anti-inflammatory agents on the gut epithelium	Decrease in IBD and food hypersensitivity patients; makes the intestines immune-balanced

continued on following page

Table 1. Continued

Microbial Species	Primary Functions	Health Relevance
Akkermansia muciniphila	Breaks down mucin; increased tight junction protein synthesis	Linked to a healthy mucosal lining and defense against metabolic and inflammatory diseases
Clostridium clusters IV & XIVa	Induce regulatory T cells (Tregs); produce SCFAs like butyrate	Immune tolerance key; depletion is associated with the greater chances of allergic diseases
Bacteroides fragilis	Programs T cell outcomes through polysaccharide A; alters cytokinome	Involved in host-gut-intestinal immuno-homeostasis; distortion associated with systemic inflammation
Escherichia coli (commensal)	Aids in the production of vitamin K2; fills gut niches to prevent pathogens	Non-pathogenic isolates lead to microbial homoeostasis; abundance could indicate dysbiosis

Table no. 1 provided above outlines some of the key microbial players with substantial implications in the regulation of the immune system, inflammatory reduction and the development of novel treatment strategies that include: *Bifidobacterium, Lactobacillus, Clostridium, Akkermansia muciniphila, Faecalibacterium prausnitzii and Bacteroides fragilis* (Zmora et al., 2019; Wan et al., 2022).

The gut microbiome is very crucial in digestion, immunity, gut integrity and vitamin synthesis. It breaks down dietary fibers into short-chain fatty acids (SCFA), enhances the epithelial barrier, removing xenobiotics, and balancing the immune cells (Wan et al., 2022; Liu et al., 2021). The composition of intestinal microbiota in early life plays a major role in shaping an individual towards allergy; low levels of *Bifidobacteria* and *Enterococci* and high levels of *Clostridia* and *Staphylococcus aureus* are associated with allergy (Gupta et al., 2024; Kumar et al., 2025). Beneficial microbes have a protective effect and allergic symptoms are usually preceded by dysbiosis. Atopic eczema can also be reduced through *Lactobacillus* supplementation that increases the production of IL-10 and TGF-b (Xie et al., 2023). Animal and human research indicates that early exposure to microbes prevents allergies, which has confirmed the hygiene hypothesis, which attributes decreased contact with microbes during childhood to dysregulation of the immune system and rising allergies. The preventive interventions involve the change in the maternal diet, antibiotic stewardship, and the reduction of elective cesarean births (Aguilera et al., 2020).

b. Mechanisms of Influence of Microbiome on Allergies

The recent literature shows much importance of gut microbiota in food allergies (Chernikova et al., 2022; Faiyazuddin & Ashique, 2025). Low *Lactobacilli* and *Bifidobacterium* accompanied by high levels of *Staphylococcus aureus* are associated with allergy to food such as eggs and milk (Meneses do Rêgo & Araaujo-Filho, 2025). The microbiome differences in the gut affect immune tolerance and IgE-associated cross-reactivity observed in the pollen-food syndrome (Wang & Hu, 2023; Taico

Oliva et al., 2025). Early exposure to microorganisms activates macrophage-produced IL-12 which encourages Th1 responses and diminishes Th2 ones whereas the short-chain fatty acids (SCFA) can regulate immunity by stimulating regulatory T cells and suppressing inflammation (Mishra et al., 2025; Gupta et al., 2024). Nevertheless, SCFAs are capable of causing pro- and anti-inflammatory responses (Chernikova et al., 2022; Faiyazuddin & Ashique, 2025).

c. Immune Response and Dysbiosis

Exposure to favorable microbes at early age is important in immune tolerance development since the immune system learns to distinguish between antigens that are harmful and those that are harmless (Faiyazuddin & Ashique, 2025). *Bifidobacteria* and *Lactobacillus* help keep the immune system balanced by promoting vaginal delivery and breastfeeding to increase the production of regulatory T cells (Tregs) and immunoglobulin A (IgA). When this early colonization is disrupted it may result in dysbiosis or augmenting intestinal permeability or leaky gut, permitting bacterial by-products to initiate allergy (Gupta et al., 2024). Cesarean section, the absence of breast milk, antibiotics, unhealthy diet is some of the causes of dysbiosis. It can be proposed that the risk of allergy can be decreased by microbiome-specific interventions in the early stages before clinical manifestation (Taico Oliva et al., 2025; Homayouni Rad et al., 2021).

2. AI-DRIVEN ADVANCEMENTS IN MICROBIOME RESEARCH:

Artificial intelligence (AI) is transforming microbiome studies by allowing the study of complicated communities of microbes and their effects on health. Recent progress in metagenomics and treatment based on AI has provided further insights into microbe-interaction and immune-regulation (Wang et al., 2022). Taxonomic profiling, functional annotation, and prediction effects of antimicrobial resistance, and examination of diagnostic data can be achieved with the help of machine learning (ML) and deep learning (DL). Multi-omics data is also incorporated into AI to discover microbial biomarkers, and it can be used to design individual nutrition (Chen et al., 2024). Naive Bayes and Support Vector Machines are algorithms, which help in the prediction of drugs targets and pathogen classification. It has been applied in microbial therapeutics, predictions of the disease and precision medicine (Abdelwahab et al., 2024). Nonetheless, the issues such as the heterogeneity of data and ethical issues still exist (Filardo et al., 2024). In addition, AI-assisted enzyme design (Huo & Wang, 2024) and multi-omics integration in his forensics illustrate that AI has the potential to transform and disrupt the understanding of the nature of microbial systems (health innovation) (Wu et al., 2024).

a. AI in biomarker identification via microbes

AI has become an effective technology of describing microbial biomarkers, improving disease management and diagnosis with great accuracy and performance. Through high complexity computational algorithms, AI can serve to analyze the difficult microbiome data in order to recognize the important microbial signatures with regard to health conditions. It aids in pathogen detection, predicting the antibiotic resistance, and enhanced diagnostics (Li et al., 2025). AI understands the genomic and Next-Generation sequencing (NGS) data to provide personal treatment and lab diagnostics (Gatla, 2024). Machine learning allows the classification and functional annotation of microbes as well as prediction of metabolic paths (Huo & Wang., 2024). The latest such methods as WSGMB apply weighted signed graph neural networks to accurately predict biomarkers in colorectal cancer and Crohn's disease (Pan et al., 2023), and laser tweezers Raman spectroscopy combined with AI can distinguish microbes down to the single-cell scale (Özyaman & Yılmaz, 2022). They are useful in the study of the human microbiome and isolation of bacterial biomarkers to predict and diagnose diseases and cancer (Xiao & Zhao, 2024).

b. Biomarker ID methods of AI

i. **Interpretable Deep Learning:** Microbial biomarkers are greatly improved through AI because it assists in better image analysis, genomic sequencing, and natural language processing to find and detect early infections and provide personalized treatments and enhance the overall health of the people. Due to the drawbacks of linear methods, MIIDL package relies on convolutional neural networks and interpretability algorithms to discover microbial biomarkers in high-dimensional data, which is not limited by linear methods (Jiang, 2021). It is a stronger line of investigation usable compared to the conventional linear approaches which are less interpretable. The goal of studies is to discover microbial biomarkers of Inflammatory Bowel Disease (IBD) and Colorectal Cancer (CRC) based on Deep Feedforward Neural Networks (DFNN), representing the model of microbiological interactions within a microbial interaction network (MIN) (Abdelwahab et al., 2024; Gupta et al., 2024).

ii. **Graph Neural Networks:** Graph Embedding Deep Feedforward Networks (GEDFN) are the deep learning-based microbial interaction network incorporating graph embeddings, which can further improve feature selection, classification, and are not constrained by the traditional information source, as well as identify biologically relevant markers (Malakar et al., 2024). WSGMB approach enhances graph embeddings, which increases disease-related microbes detection. Weighted signed graph models are also used at RGGWSGMB to study microbial co-occurrence networks and identify disease-related biomarkers that are accurate and tested against literature databases (Pan et al., 2023).

iii. **Personalized Interventions Based on AI:** Personalized interventions in healthcare and education processes through AI enhance the quality of individualized treatment, learning, and disease risk prediction, in addition to improving outcome and decreasing disparities (Ahmed et al., 2025; Zheng et al., 2023). In a healthcare context, AI allows customizing dietary and probiotic interventions with regard to the individual microbiome profiles and combine multi-omics data to learn about gut microbiome-host interactions on the level of metabolism (Chen et al., 2024). Pilot studies demonstrate that AI-guided diets decrease the IBS symptoms and expand the beneficial microbes (Karakan et al., 2021). Dietary strategies of individualized microbiota are being refined in silico, in vitro, and in vivo. The ethical aspects, data privacy, bias, and interdisciplinary collaboration are still significant in the responsible AI use in the personalized interventions.

3. METHODOLOGY: DATA-TO-INSIGHT PIPELINE

In AI-based multi-omics research, consistency of the analysis procedure is a major requirement of transparency and reliability. It involves a combination of the methods of data collection, pre-processing, computational modeling and biological explanation into one model. The systematic workflow to extract meaningful information using microbiome and o0ther omics data is as illustrated in the steps below:

a. **Multi-Omics Data acquisition and Integration:** The recent developments in the high throughput sequencing and metabolomics platforms have allowed the in-depth characterization of microbial as well as host biological responses. Dispersed datasets being 16S rRNA profiles, shotgun metagenomes, genomic variants of hosts, metabolomic profiles, and related clinical data are found in public archives, including the NCBI Sequence Read Archive (SRA), MG-RAST, and the EMBL-EBI. The collection of such heterogeneous data will be in favor of the system-level perception of the host-microbe interactions, particularly in terms of the food-allergic phenotypes (Arıkan & Muth, 2023).

b. **Preprocessing and Engineering of Data:** Processing entails raw sequence reads filtration of succinct quality filtration, contaminant removal and taxonomic and functional annotation. Samples are normalized by normalization processes that normalize species and metabolic properties of samples. The feature engineering is used to design synthessible biological features, including alpha/beta diversity measures, metabolites profiles, microbe-microbe interactions (including LUPINE), SNPs of the host, and diet-microbiome interactions (Kodikara & Lê Cao, 2025). They are advanced features that enhance the precision of the analysis and render the latter modeling of the downstream easier.

c. **Predictive Modeling by Algorithms of Machine learning:** Machine learning is applied to give future and future insight. In the microbiome data, the latent structures occur when unsupervised algorithms, e.g., clustering and dimensionality reduction, are applied. Some of the examples of machine learning methods used in the classification of individuals based on whether they are allergic or not include the following: Random Forest, XGBoost, SVMs, and neural networks that can be used to identify the discriminative microbial or metabolic signatures (Sutanto & Fetarayani, 2025). The hyperparameter optimization also strengthens the model along with ensuring the reproducibility (Topcuoglu et. al., 2020).

d. **Model Performance Evaluation and Validation:** K-fold cross-validation is also applicable to the reliability assessment where the overfitting procedure should be minimized to find out the generalizability of the samples (Wong & Yeh, 2020). Predictive control The model predictive power is measured with the help of conventional performance measurements, including, though not limited to, accuracy, precision, recall, F1 score, and area under the ROC curve (AUC). The recent multi-omics works suggest that trained combined models can provide AUCs on the order of 0.72-0.83 that suggests high diagnostic and predictive relevance (Vänni et al., 2023).

e. **Explainable Artificial Intelligence Biological Interpretation:** Interpretability can be enhanced with explainable AI techniques particularly SHAP (SHapley Additive explanations) which effectively measures the strength of features in prediction numerically. Such methods may prove helpful in the disclosure of mechanistic interrelationships between microbial communities, metabolic pathways and clinical outcome, to contribute to the unveiling of biomarkers and precision-guided treatment (Przymus et al., 2025; Probul et al., 2024).

f. **Evidence of Practical Applicability:** The evidence of the practical applicability can be seen through an analysis in which an open-access list of food allergies is recognized. Taxa such as *Clostridia*, *Ruminococcaceae* and butyrate-producing are likely to be the key predictors in such a re-analysis. These metabolic processes such as bile-acid metabolism and tryptophan metabolism are commonly employed in the classification processes which once again demonstrates the biological importance of these processes. To depict the potential of AI-driven multi-omics synthesis, this image depicts the opportunity of identifying with risk factors and identifying the decision on a particular treatment regimen (Maaskant et al., 2025).

The other problems that face the study are compositional bias, loss of data and model interpretation. Nevertheless, new techniques, including Bayesian multitask modeling, multi-omics time-series integration, can offer a possibility to increase the amount of analytical accuracy and personalized diet and microbiome-based therapy (Ruiz-Perez et al., 2021).

4. CURRENT MICROBIOME-TARGETED THERAPIES:

Microbiome-based therapeutics provide some insight in the manner in which the gut microbiome relates with the immune system, particularly with the increase in food allergies around the world. Allergic sensitization can occur because of changes in the environment to the extent of interfering with the gut microbial composition. One of the roles of the gut microbiota is an immune defense, which is generated by producing short-chain fatty acids (SCFAs), promoting thematization of regulatory T cells, and affecting cytokine-mediated inflammation (Faiyazuddin & Ashique, 2025; Gupta et al., 2024).

Probiotics and prebiotics have the potential to optimally modify intestinal microbiota and improve immunity. Research on germ-free mice demonstrates that Anaerostipes caccae is a species that inhibits allergic reactions and Th2 antibodies and cytokines, whereas Lactobacillus rhamnosus enhances gut barriers and decreases allergen sensitivity (Probul et al., 2024). Immune homeostasis is also restored through fecal microbiota transplantation (FMT). The future study must optimize microbiome-specific interventions, and AI-authoritative interventions can be used with safety and ethical concerns. Collectively, the innovations have the potential of being transformative in the prevention and treatment of food allergies.

a. Prebiotics and Probiotics

Prebiotics and probiotics are important to gut health, digestive and immune health.

Probiotics refer to live organisms associated with health promotion after taking them in proper quantities. By controlling phagocytosis, as well as controlling the production of pro-inflammatory cytokines, they can prevent or alleviate food allergies. The most common are fermented foods (yogurt, kefir, miso, and sauerkraut), dietary supplements (Zhang et al., 2025; Wei et al., 2025; Leiva Alaniz et al., 2025). The immune protection, gastrointestinal health, and absorption of nutrients are improved through probiotic strains such as *Bifidobacteria* and *Lactobacilli*, which increases the diversity of the gut microorganisms. Multicenter randomized trials and other clinical studies indicate that probiotics have the ability to enhance the effect of oral immunotherapy in children with peanut allergy. The evidence is not consistent, however, as *Lactobacillus rhamnosus GG* also may prevent IgE sensitization, whereas *Lactobacillus acidophilus* might enhance atopic susceptibility, so the further strain-specific studies are necessary (Faiyazuddin & Ashique, 2025).

Prebiotics are nondigestible dietary fibers that promote the growth of beneficial microbes and enhance gut microbial diversity. Found in foods like garlic, onions, bananas, asparagus, oats, and flax seeds, prebiotics are fermented into short-chain fatty acids (SCFAs), which activate regulatory T cells (Treg) and may help prevent allergic reactions (Faiyazuddin & Ashique, 2025; Taico Oliva et al., 2025; Wang & Hu, 2023). Together, probiotics and prebiotics are emerging as dietary strategies to modulate the gut microbiome and manage food allergies, though long-term effects require further study.

b. *Synbiotics* are prebiotics and probiotics that are used together to maximize the benefits to the gut, which may reduce the chances of allergies and generally promote the overall health of the gut (Al-Habsi et al., 2024).
c. **Engineered Microbial Consortia:** Engineered microbial communities are synergistic microbial communities that have been engineered to improve metabolic processes, environmental cleanup, and sustainable agriculture (Hu et al., 2022). These consortia enhance stability of microbes, resource use and performance through mutualistic or commensal interactions. Engineered consortia can be used in medicine to personalize the gut microbiome and produce therapeutic molecules, as well as to regulate immune response (Duncker et al., 2021; Mkilima, 2025). Experiments have shown that they have the potential to alleviate food allergy symptoms by restoring immune tolerance and decreasing the allergic sensitization of animal models (Chernikova et al., 2022; Gelli et al., 2025). There are still such obstacles as stable colonization, geographically specific treatments, and safety measures. In order to test efficacy and safety in humans, clinical trials are required (Mkilima, 2025).
d. *Adjuvant Therapies* used as a supplement to probiotics and prebiotics in the management of food allergies. Oral immunotherapy (OIT) should be used in combination with pharmacotherapy comprising of antihistamines and biologics, which enhances safety and decreases adverse reactions. An example is that omalizumab has improved the safety profile of OIT and still, aluminium salts serve as adjuvants by reducing the dispersion of allergens and raising the local immunity exposure (Dantzer & Wood, 2021).
e. **Dietary Interventions:** Eating habits to boost fiber intake are vital in increasing the production of short-chain fatty acids (SCFA) which is fundamental in immune modulation (Taico Oliva et al., 2025). Gut microbiota-derived SCFAs

suppress pro-inflammatory cytokines at the expense of anti-inflammatory mediators and promote immune tolerance to harmless antigens (Gupta et al., 2024; Wei et al., 2025; Korczak et al., 2020). The gut microbiota composition can also be achieved through diets high in fruits, vegetables, whole grains, and legumes (Taico Oliva et al., 2025; Tuli et al., 2025; Zhang et al., 2022).

Combined, prebiotics, probiotics, synbiotics, engineered microbial consortia, adjuvant therapies, and dietary interventions form a multifaceted strategy for boosting gut health, immune responses, and managing food allergies. Yet, more research is needed to fine-tune their clinical application.

In Figure 1, a conceptual model proposed by Dr. Reyed M. Reyed outlines the center of the editorial analytical framework of this work. The graph depicts a connected microbiome-AI circuit, where the colonization of the gut during early life triggers gut ecosystem maturation, influences immune signaling mechanisms, and coordinates metabolite homeostasis. The AI layer separates these host-microbe interactions to provide personalized therapeutic feedback, establishing a self-regulating feedback loop that signals accurate interventions and fosters long-term immune and metabolic homeostasis.

Figure 1. The Integrative Microbiome-AI Diagnostic and Therapeutic Loop. Note: This figure was specifically developed and curated by Reyed M. Reyed (Editor) to provide a tailored conceptual framework that aligns with the specialized content of this chapter.

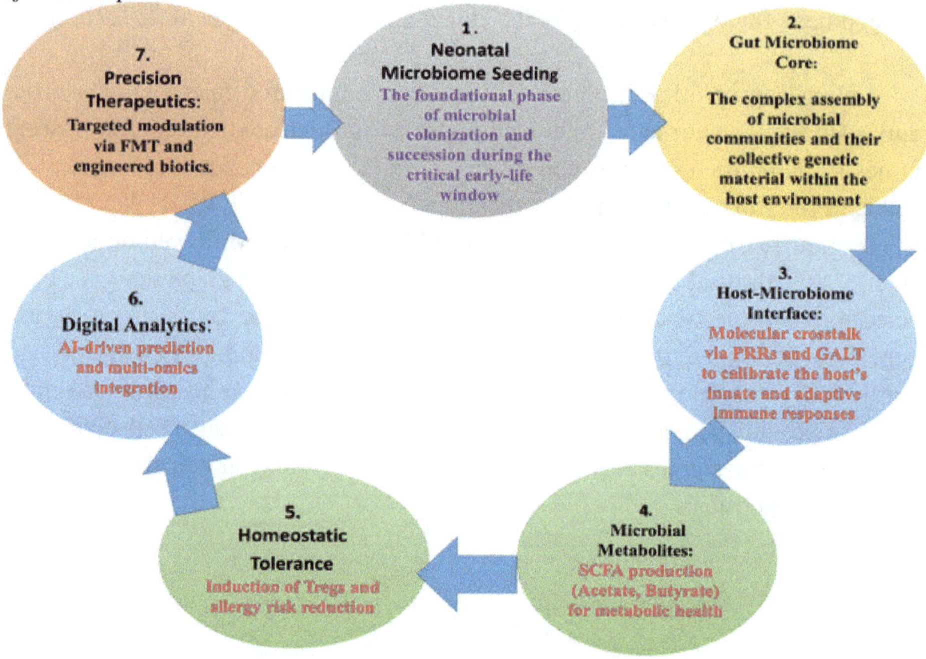

Case Study 2: AI-Directed Fecal Microbiota Transplant in a Chronic Food Sensitivity Case (Dantzer & Wood, 2021; Haring et al., 2021; Huang et al., 2022; Wang et al., 2022; Abavisani et al., 2024)

Scenario:

A young adult was introduced with the persistent bloating, abdominal pains, and new sensitivities to the food substances like gluten and dairy. IBS has a low response to standard therapy. Diagnostics based on AI-assisted analysis and a combination of gut microbiome sequencing and host genetic profiling were further advanced. The outcome of the combined data demonstrated the decreased microbial diversity, the severe loss of anti-inflammatory species, such as Faecalibacterium prausnitzii, and genetic polymorphisms related to the insufficiency of the gut barrier functions and immune responsiveness.

Interpretation:

A fecal microbiota transplant (FMT) was conducted personally with the donor being matched both on microbiome and immune-genetic markers as recommended by AI algorithms to rectify the dysbiosis and address underlying genetic predispositions. The patient had rapid recovery of symptoms and progressive food tolerance after FMT. The AI also suggested an individualized post-FMT diet as it is enriched with certain prebiotic fibers depending on the microbiome and genotype of a patient, which guarantees the resilience of the microbiome in the long term.

Relevance:

This example highlights the new position of artificial intelligence in precision gut health where multi-omics data, such as microbiome and host genetics, can be integrated to improve the response to treatment. The concept of AI in the welfare research is potentially disruptive because it allows the customization of the microbiota-based treatment of chronic gastrointestinal and immune-related diseases.

5. NUTRIGENOMICS: UNVEILING THE DIALOGUE BETWEEN GENES AND DIET

The gut microbiome is an active determinant of health but the genetic blueprint of the host offers the context on which dietary interventions have an effect. Nutrigenomics is an interdisciplinary science that studies the two-way relationship between nutrition and genetics, such as the way genetic differences affect food reactivity, and vice versa (Kumar et al., 2025). Nutrigenomics is a field combining molecular biology, genomics, and nutrition, investigating the interaction between genes and environmental factors in case of changes of health outcomes and disease susceptibility (Kumar et al., 2025; Agrawal et al., 2024). Nutrigenomics examines two processes—DNA methylation and histone modification as the means to control the expression of the genes, and nutrigenetics studies how genetic differences influence individual response to the components of the food (Agrawal et al., 2024). The end vision is to create an individualized nutrition plans upon genetic makeup, metabolic phenotype, and the intestinal microbiome composition so that chronic conditions can be prevented and well-being enhanced (Mishra et al., 2022). The extended nutrigenomics are modern approaches, considering the interaction of three genomes: the host, diet, and gut microbiota, which will enable making relevant decisions regarding the diet and therapy (Agrawal et al., 2024). This mixture of genetic, nutrition, and microbiome research is a radical shift of personal health and illness prevention.

a. Genetic Variations Influence Response to Diet:

Human genetic variations play a major role in reaction to dietary intervention. Some 50 percent of the variance in plasma cholesterol and 30 to 60 percent of the variance in blood pressure is genetically predetermined. Metabolic genes have SNPs that control reactions to lifestyle and dietary adjustment, including body weight, lipid profiles, glucose homeostasis, and blood pressure (Fenwick et al., 2019). An example of gene-diet interaction is the Apolipoprotein E (ApoE) variants ApoE4 carriers incur higher risk of cholesterol and coronary artery disease compared to ApoE3/2 women who gain minimal benefit by consuming polyunsaturated fats. Interindividual differences in nutrient response are caused by SNPs in enzymes that control lipid metabolism, glycolysis and insulin. To give an example, SNPs in folate metabolism and choline metabolism genes define the tendency towards fatty liver and muscle damage during low-choline diets (Gentile et al., 2020). Such genetic differences show the need to use specific dietary interventions that are applicable in an individual genetic profile.

b. Role of Dietary Components in Gene Expression:

Bioactive dietary molecules serve as signalling molecules that control gene expression through chromatin alterations, transcriptional factors and ligand-binding to nuclear receptors (Mierziak et al., 2021; Micheletti et al., 2023). Fatty acids, glucose, cholesterol, amino acid, and glucose metabolism transcription factors are PPARs and ChREBPs, respectively, whereas lxr and FXR are fatty acid transcription factor (Micheletti et al., 2023). These processes allow metabolism to be strictly controlled and its malfunction to lead to obesity, diabetes, and cardiovascular diseases.

Nutrigenomics offers an empirical basis of individualized nutrition beyond the statistical approach of providing population-based one-size-fits-all rules to the individual perspective based on genetic inclination. Individualized dietary prescriptions have the potential of preventing obesity, type 2 diabetes, cardiovascular disorders and some types of cancers, which are classified as chronic diseases related to nutrition. Nutrigenomics shows that nutrients have a genetic dialogue with our cells and that provides the untold possibilities of very specific health intervention. By combining artificial intelligence with nutrigenomics, it is possible to obtain advanced analysis of both genetic and dietary data, which will result in nutrition advice. Nutrigenetics and nutrigenomics are complementary concepts as they study the variability of metabolism of nutrients and diet-gene interactions. Such issues as data privacy, algorithmic biases and access inequity to nutrigenomic services remain (Marcum, 2020).

In general, nutrigenomics is one of the foundations of personalized medicine, utilizing both genetic and dietary data to maximize their health outcomes and avoid chronic diseases in a wide range of populations.

c. Clinical Relevance of Nutrigenetics

The clinical relevance of nutrigenetics is premised on the fact that the nutritional history of a person has a major impact on the present and future health conditions of the individual.

i. **LCT Gene and Lactose Intolerance:** Well-defined interactions among genes and diet are clinically relevant since they have direct effects of nutrigenetics on health and disease. One of the well-known examples is the genetic variant affecting dietary reactions in LCT gene and lactose intolerance. Lactase persistence is a relatively rare hereditary disorder that causes lactase enzyme activity to remain high in adulthood to allow lactose to be digested (Cohen et al., 2025). Important ones are the -13910 C>T and -22018 G SNP in an MCM6 intron that is common in dairy farming-related populations. Those who do not have such variants undergo lactase downregulation, which leads to bloating, gas, diarrhea, and abdominal pain and requires specific dietary approaches (JanssenDuijghuijsen et al., 2024).
ii. **MTHFR Gene and Folate Metabolism:** MTHFR C677T polymorphism, which is present in 5-15% of the population and inhibits folate metabolism by lowering the enzyme activity. Homozygotes have a higher level of homocysteine and a lower level of folate, which puts them at risk of cardiovascular diseases and neural tube defects. A1298C polymorphism also reduces the activity of the enzyme, which underlines the genotype-specific folate demand (Raghubeer & Matsha, 2021).
iii. **APOA2 Gene and Saturated Fat Response:** The effect of saturated fat on the risk of obesity is changed by an APOA2-265 T>C (rs5082) polymorphism. It has been demonstrated that CCH homozygotes become more obese and overweight in diets with high saturated fat content, mediated by the change in eating habits, increased ghrelin, and epigenomic modifications, such as altered APOA2 promoter methylation (Abaj et al., 2022; Kim et al., 2023).

6. MULTI-OMICS DATA AND AI INTEGRATION:

Integration of multi-omics data and AI has transformed precision medicine. Convolutional neural networks, deep neural networks and transformers can be used to study intricate interactions in genomics, transcriptomics, proteomics, and metabolomics. SNPs are represented as categorical or continuous additive scores of AI models, thousands of SNPs are summarized as polygenic risk scores (PRS), which are quantitative features and improve the performance of the model (Elgart et al., 2022). Similarly, non-linear interactions are represented in the microbial risk scores (MRS) which enhances prediction of diseases like diabetes type 2 and glucose increment (Wang et al., 2022). The combination of AI with electronic health records also allows the most effective use of therapeutic effects and minimizes adverse events through real-time drug-gene interaction. There are still difficulties with standardization of data, interpreting and applying it clinically.

Table 2. Multi-Omics Nutrigenomics Framework

Omics Layer	Role in Nutrigenomics	Relevance to Food Allergy
Genomics	Identifies diet-responsive SNPs (e.g., LCT, MTHFR, APOA2, IL-10, TLR4)	Predicts susceptibility to food sensitivities; explains inter-individual variation in allergen tolerance.
Epigenomics	DNA methylation, histone remodeling influenced by nutrients	Allergens and nutrients modulate FOXP3, IL-4, IL-13 expression affecting Th1/Th2 balance.
Transcriptomics	Diet-regulated gene expression via nutrient signaling	Helps decode immune activation pathways that lead to IgE overexpression.
Metabolomics	Identifies nutrient-derived and microbe-derived metabolites	SCFAs (butyrate) promote Treg development; dysbiosis reduces tolerogenic metabolites.
Microbiomics	Microbial gene expression shifts with diet	High-fiber diets support bacteria producing anti-allergic metabolites (propionate, acetate).

Table 2 summarizes how different omics layers illustrate the influence of diet on immune tolerance and allergic reactions (Micheletti et al., 2023, Elgart et al., 2022; Rubas et al., 2025; Lagoumintzis & Patrinos, 2023).

a. Data Convergence: Genetics, Microbiome and Metabolomics

Complementary health information is offered by host genetics, gut microbiome and metabolomics. Although genomic data are fixed and inability to forecast disease, the gut microbiome is an indicator of dynamic metabolic potential, and individual species such as Prevotella copri predicts postprandial glucose metabolism and the composition of the entire microbiome informs cardiometabolic markers (Asnicar et al., 2021). Metabolomics represents the functional results of diet, genetics, and microbial activity in order to be able to assess the risk of diseases comprehensively (Chen et al., 2022).

b. Artificial Intelligence Driven Synthesis in Precision Nutrition

Complex interactions between diet, microbiome, and host metabolism can be decoded using AI and machine learning, which non-AI approaches are unable to do (Abavisani et al., 2024). Microbiome, genetic and metabolic data are combined with algorithms such as the Random Forest and Gradient Boosting to provide personalized and accurate dietary interventions, which have been shown to be effective in treating obesity and diabetes. The implementation should be successful and should address the issues of data privacy, algorithm bias, ethics, and regulatory frameworks (Ferreira et al., 2025).

On the whole, nutrigenetics, multi-omics, and AI convergents will present an effective model of precision nutrition to create customized diet in the play of particular illnesses, maximize the game of health and enhance preventive and curative treatments to the greatest number of diseases (Ferreira et al., 2025).

c. Contributing to Patient-Tailored Therapeutic Strategies

A personalized medicine is a paradigm shift and involves the application of therapeutic interventions on patient features on a patient-specific basis through multi-omic approaches (Verma et al., 2025). This paradigm is based on integrating clinicopathological indices with genomics, transcriptomics, proteomics and metabolomics to come up with diagnostic, prognostic and therapeutic modalities that are unique to a given patient. Pharmacogenomics also allows the Personalization of drug therapy on genetic composition, and biomarker signatures of gene and protein expression also increase the ability to predict how the drug therapy will respond (Shaman et al., 2024; Su et al., 2024).

i. *Precision Prevention* is based on genetic information to determine the high-risk individuals. As an example, Celiac disease is closely linked to HLA-DQ2 and DQ8 alleles and 95 percent of the patients carry HLA-DQ2 allele (Aboulaghras et al., 2023). The degree of risk is associated with the dosage of the gene, in which homozygous DQ2.5, and heterozygous DQ2.2/DQ2.5 individuals are more prone to showing anti-endomysial antibodies (Del Pozzo et al., 2021). This is further enhanced by combining HLA and non-HLA alleles to identify individuals at risk to apply preventive practices (Aboulaghras et al., 2023; Brown et al., 2019).
ii. *Precision Therapeutics* and Nutrition are concerned with a personalized approach to interventions instead of a universal diet. The interaction between genetics and diet affects the risk of cardiovascular and metabolic responses necessitating a combination of genetics, metabolism, and environment to customize the

nutritional phenotype (Desjardins & Vohl, 2023; Laddu & Hauser, 2019). More sophisticated panomic tools such as metabolomics and microbiomics can be used to further phenotype patients and offer a personalized intervention.

iii. *Dynamic Optimization* is an AI-based algorithm to combine longitudinal microbiome and metabolome data, which will subsequently allow the personalization of interventions in a continuous cycle (Oh & Li, 2024; Mukhopadhyay et al., 2025). New artificial intelligence platforms exist to predict what happens through the use of computer-generated gut twins, which are used to model nutrient, multi-omic, and clinical interactions to optimize individualized dietary plans (Saxena et al., 2024; Greathouse & Choudhury, 2025).

7. CHALLENGES IN MICROBIOME-BASED ALLERGY THERAPEUTICS:

Microbiome-based therapies in allergy treatment are not yet at the advanced stage, and have a number of challenges. These treatments demonstrate possibilities of immune response regulation and reducing the symptoms of allergy but need further optimization (Gelli et al., 2025; Haring et al., 2021). The key problems are the stability of the microbial colonization of the transplantation, custom-made treatment, and safety issues. Bacterial engraftment is so heavily dependent on immunostatus of the recipients,whereas technical and regulatory barriers remain an obstacle (Alam et al., 2023). The presence of such strategies as dietary changes, the use of prebiotics, control over the metabolism of microbiota, and live biotherapeutics are now available (Chernikova et al., 2022). Nevertheless, these treatments could ultimately offer patient-specific, sustained allergy care, which is better than standard treatments (Yadav & Chauhan, 2021).

Nevertheless, dysbiosis can easily decrease the efficacy of therapy and deteriorate allergic disorders (Mishra et al., 2025). Probiotics have different effects because of the host-microbial interactions (Nance et al., 2020; Berni Canani et al., 2019). The combination of probiotics and prebiotics may potentially become more effective, although the most effective combinations are not yet known. Recognizing the biomarkers of microbiome might be a personalized treatment option, but the diversity of microbes makes diagnosis more difficult (Chernikova et al., 2022).

The microbiota is dynamic, and it is hard to correlate particular microbes with allergies because the populations are affected by the environment and the diet (Berni Canani et al., 2019). Although probiotics, prebiotics, and synbiotics have a bright future, the impact of preventing allergy remains doubtful. Placebo-controlled trials are crucial in order to perfect the therapeutic strategies. Custodial therapies on the basis of genetics, initial microbiota, and food exposures are essential to enhance outcomes (Faiyazuddin & Ashique, 2025). Further research and investment would be required to overcome biological and engineering complications and define microbiome-immune relationships (Chernikova et al., 2022).

a. Engineering Challenges

The difficulties in engineering include the stability and functionality of metabolically interdependent consortia of microbes. These microbes have to work effectively and perform desired tasks especially in such important applications of life-support machines. Monitoring and optimization of interactions in the consortia involves technical solutions, which are not simple, such as functional redundancy, where the same module ensures the availability of the biochemical circuit in the event of failure (Chen et al., 2024).

b. Biological Considerations

The species of microorganisms must be adaptive to environmental changes biologically. The alterations in the population numbers can result in the destabilization of the metabolic fluxes. The selection of functional redundancy of the species (such as *Nitrosomonas* or *Nitrobaster*) helps to maintain the biochemical activity in the changing conditions (Ramond et al., 2025).

c. Operational Limitations

Operation complexities such as dynamic interaction of microbes are also present. Communities of microbes can also be relatively stable over long durations with change of dominant species/strains change over short time periods. The external interference can influence such changes, and, therefore, to make a successful consortium work, it is necessary to understand and regulate these forces. Moreover, the reduction of complexity of species within synthetic consortia as compared to natural ones expresses concern about the entire community stability itself and maintenance of its functions over time. Lastly, the mathematical and ecological models, which are applied to advance and simulate these consortia to design them, tend to need considerable data that can render them accurate and valid. Such models are also complex, and thus challenging to actualize, which explains why continuous optimization and improvement of such models is required to better represent microbial interactions (Prosser & Martiny, 2020).

d. Complexity and Variability

Human microbiome is very dynamic as it depends on the environment, diet and genetic makeup. It is the variability that makes it hard to come up with universal health markers. Although probiotics, prebiotics, and synbiotics have the potential in allergy prevention, there is no sound evidence, so it is necessary to conduct stronger clinical trials. Dietary interventions are essential because the microbiomes differ across geography, diet and lifestyle (Redondo-Useros et al., 2020). Individual modification of microbes can influence immunity, metabolism and stability of microbiomes. The gut health largely depends on the presence of microbial diversity and resilience, which are influenced by antibiotics and other perturbations. Fecal microbiota transplantation and customized dietary approaches have potential but need to be considered individually as patients react differently to them (Fassarella et al., 2021; Ilan, 2019). This complexity needs to be comprehended to put the microbiome science into practical clinical application (Jain, 2020).

e. Transformation into Clinical Practice

Microbiome healthcare roles are generally not well known among healthcare professionals and the general population. There is the need to conduct educational campaigns to support microbiome-based interventions in preventing and treating allergies (Daniel et al., 2021; Butler et al., 2019).

f. Microbiome Therapeutics Issues

The inconsistency of gut systems, and the interaction between different microbes, makes the development of universal treatment difficult. Colonization of administered bacteria is not an easy thing because most of them do not survive after oral administration. Studies are still underway to determine the microbial species that can safely regulate the immune system and decrease the occurrence of allergies (Xiao et al., 2025 ; Yadav & Chauhan, 2021).

i. AI struggles in Microbiome Research.

Use of artificial intelligence (AI) in microbiome studies has a number of challenges and constraints that need to be overcome to be successfully applied in microbiome studies. These shortcomings are mainly based on interpretability, quality of data, model assessment, and management of imbalanced types of data (Barredo et al., 2019).

ii. Interpretability

AI models and especially deep learning are frequently black boxes, and the rationale of their clinical decisions is hard to understand. Distrust and accountability can be minimized by the lack of transparency. Prior knowledge is also beneficial because it enhances interpretability and confidence in AI-supported decisions (Malakar et al., 2024; Norgeot et al., 2020).

iii. Data Quality

The quality and representative data are essential. Unreliable predictions are obtained using poor or biased data (Kumar et al., 2025; Li et al., 2025). The key to a durable analytical pipeline is data preprocessing and cleaning, as well as the feature selection (Faiyazuddin & Ashique, 2025; Jiao et al., 2022).

iv. Model Evaluation and Selection

Good performance metrics should be able to address the task constraints, the imbalance in the classes, and the impact of various errors. The outcomes of hyper-parameter tuning have an important impact on the results of the model, and thus, the evaluation process is not simple (Wojciuk et al., 2024).

v. Imbalanced Datasets

Biased data sets are capable of influencing AI projections. If these issues are tackled with techniques like data augmentation or class reduction, and it is difficult to perfectly balance it (Delgado et al., 2022).

vi. Regulatory and Ethical Concerns

Customized microbiome interventions present some ethical issues in terms of data privacy, consent, and possibly discrimination. Genomic information is distinct and unalterable; it is advisable to use secure methods of calculating data such as federated learning. Dynamic consent systems provide the possibility to control further use of data by the participant. The issue of algorithmic biases should be tackled in order to avoid health inequity, with diverse training data. Accessibility and equity in precision nutrition is also important to prevent the aggravation of inequality on a global scale. The fecal microbiota transplantation (FMT) and donor screening is continuing its standardization (Jain, 2020). Factors such as patient confidentiality, risks of safety, and accountability are ethical issues at hand. Regulatory loopholes exist, yet new omics technologies and AI can eliminate risks (Grigoryan et al., 2020; Ejtahed et al., 2023).

vii. Long-Term Efficacy and Safety

Recent studies demonstrate the importance of the gut microbiome in the development of the immune system and long-term health effects (Gupta et al., 2024; Obukohwo et al., 2024). The initial three years are especially pivotal when it comes to the development of a varied and operative microbiome (Nunez et al., 2025). The main variables that impact this process are the maternal transmission of microbes, breastfeeding, and early child exposures (Liu et al., 2024). Even such dietary interventions as fiber-rich foods and fermented foods consumption have been associated with altering the gut microbiota and influencing immunity in adulthood (Wastyk et al., 2021). Nevertheless, challenges in this field exist, such as technical and methodological barriers in the study of microbiome (Nunez et al., 2025). To go further in the study of microbiome-immune interactions, scientists recommend taking the priority of longitudinal intervention research in human beings and studying non-industrialized people to determine the features of human microbial identity that have been lost (Yuan et al., 2021). Such methods can result in better microbiome-based interventions for enhancing immune health and preventing chronic diseases.

8. FUTURE RESEARCH DIRECTIONS AND CLINICAL APPLICATIONS:

Gut microbiome is a collection of trillions of microorganisms that are very important in immune control and physiological functioning. The imbalance in the microbial composition of the gut, known as dysbiosis, can lead to the inability of the immune system to differentiate harmful pathogens and harmless food proteins, which causes allergies to food (Berni Canani et al., 2019). The complex interaction of the microbiome and immune system can only be understood through prospective, randomized, placebo-controlled trials. Childhood diet and antibiotic exposures have a strong effect on the formation of the microbiome and could be the cause of the increasing rate of allergy diseases. The objective of food allergy immunotherapy (FA-AIT) is to regulate the gut microbiome to enhance oral tolerance and outcome of treatment. The efficacy of microbiome-based interventions (prebiotics and probiotics) is yet to be well determined though. Further studies must determine individual microbial strains and metabolites that control allergy-mediated immune responses and determine their therapeutic value. The balance of the immune system, a healthy gut microbiome, a lower number of allergies, and overall well-being and quality of life are crucial to maintaining a healthy gut microbiome (Nance et al., 2020).

a. Research Priorities

Studies on microbiomes have established a new emphasis on host-microbiota interactions, in particular, food allergies. The effects of various microbial mixtures on immunotropic reactions and allergy intensity should be studied in the future. Specific focus is to be placed on testing the preventive and therapeutic potential of the interventions involving probiotics, prebiotics, and synbiotics (Nance et al., 2020).

b. Improving Data Integration Methods

There will be an increase in AI-supported studies of microbiomes using highly developed data integration techniques. Multi-omics can be used to have a multi-faceted microbial ecosystem and their effects on host health. It is possible to convert the data of omics data into intermediate forms: graph structures or latent representations to enhance predictive accuracy and better understand microbe-associated biomarkers (Wang et al., 2024; Xin et al., 2024). Variational autoencoders and graph neural networks are some of the techniques that can be used to refine such integrations to improve model training and understanding of complex interactions among microbes.

c. Optimizing a machine learning pipeline

Machine learning (ML) pipeline optimization is one of the areas that are still developing. Efficient model training and evaluation can be done through methods such as hyperparameter optimization (HPO) and automated machine learning (AutoML) framework, such as JADBio. More complex models that are more consistent and interpretable can be generated using advanced search strategies like the Bayesian optimization and evolutionary algorithms when exploration of high-dimensional model spaces is required (Wang et al., 2024; Xin et al., 2024).

d. Ethical Considerations

The use of AI in the field of microbiome suggests serious ethical issues, especially when it comes to fair access and utilization of data. In low- and middle-income countries (LMICs), such problems as data scarcity, cost-efficiency, and technological colonization may exist. Other ethical concerns in LMICs include not following ethical standards in genomic databases, lack of consent procedures, and insufficient benefits of research (Yu & Zhai, 2024; Mubarak & Ashraf, 2024). Microbiome research ethics consider the four principles of Do Good, Don't Harm, Respect, and Act Justly, and advise to treat gut microbiome data as a global heritage (Lange et al., 2022). Privacy, algorithmic bias, transparency, and professional responsibility should be comprehensively integrated to ensure the access to responsible AI use, responsible treatment of patients, and fair treatment of patients. Genomic data should be shared ethically, as advocated by organizations like WHO, which will make biases a thing of the past and enhance inclusivity.

e. Promoting the Interdisciplinary Collaboration

The interdisciplinary cooperation of microbiologists, data scientists, and health-care workers will be useful in future AI-driven microbiome studies. This type of collaboration will be able to maximize the application of AI technology to enhance diagnostic precision and develop individualized therapy and patient outcomes in the case of microbiome-associated conditions (Wang et al., 2024).

f. Personalized Medicine Applications

Personalized medicine overlaps with the study of the microbiome with AI. Individual microbiome profiles can be analyzed by big data to identify unique microbial health signatures, which can be used to prescribe individual treatment to an individual. The future of AI is a potential of incorporating microbiome data into patient-centered treatment, with the microbiomes being used to improve care (Xin et al., 2024).

g. Personalized Nutrition

Personalized nutrition is also possible with the help of AI which can analyze microbiome, metabolome, and genomic data to create dietary interventions to improve gut health and prevent chronic inflammatory diseases, such as allergies. It is necessary to conduct research on the translation of such personalized nutrition strategies into clinical practice to enhance patient outcomes in the future (Wang et al., 2024).

h. Community Engineering and Synthetic Ecology

Synthetic biology provides the possibilities to design the microbial community with optimized interactions that could be applied in healthcare, agriculture, and waste management. Graphing cellular interactions has the potential to generate more stable and competent consortia that can be useful in reducing problems with natural microbial populations (Yang et al., 2021). The role of genetic engineering tools can increase the capabilities of microbial consortia and extend the limits of microbial applications (Li et al., 2023).

i. New Technology in Production Methods

The second step in research to follow should be on the process of manufacturing microbial inoculants in high quantities to enable their use in practice. There is the need to shift to mass production systems so as to tackle the needs of agriculture and clinical needs. Small-scale production units that can be supported by the government initiatives are present in rural areas that can enhance the shelf-life and availability of microbial products. Automation would also help in ensuring efficiency in production since it is possible to prepare sterile, stable and viable microbial inoculants (Domeignoz-Horta et al., 2024).

j. Extended Impact and Sustainable Outcomes

Microbial consortia can be engineered to be environmentally sustainable and add to economic growth and development through facilitating the use of a circular economy by reducing waste and reusing resources. Synthetic microbial systems should be implemented in an ethical and sustainable way that is focused on environmental goals and human health (Wijerathna et al., 2024).

9. NEW THERAPEUTIC APPROACHES:

New treatments like fecal microbiota transplantation (FMT) and live biotherapeutics can replace the good intestinal microbiota and improve the tolerance in the immunity. The research must assess safety, effectiveness, and long-term effects, particularly in patients with food allergies, and examine the potential to be used in combination with dietary interventions to achieve maximum positive results (Kim et al., 2024).

a. Probiotic Interventions

The concept of probiotics, pre/synbiotics has potential in terms of managing balances in the microbiome and can be deployed prophylactically, complementary to immunotherapy. Although the initial results are promising, additional research based on larger, better-characterized cohorts must be conducted to help define the contribution of the microbiome in the allergic diseases (Tan-Lim & Esteban-Ipac, 2025).

b. Therapeutic Implications

The gut microbiome has a positive effect of preventing allergies to food. Atopic sensitization to frequent allergens has proven to be reduced by oral supplementation with specific types of probiotics, including *Ligilactobacillus salivarius* and *Bifidobacterium bifidum* (Vijayan et al., 2024). Profiling personal microbial antigens that induce or exacerbate IgE reactions is a possible application of desensitization therapies that will be tailored personally (Taico Oliva et al., 2025).

c. Clinical Trials

To determine the safety and effectiveness of microbiome based interventions, it is necessary to consider clinical trials to test the intervention. Clinical trials assessing new strategies and interventions, including epicutaneous immunotherapy patches, are also expected to desensitize all patients with minimal side effects (Hervé et al., 2023; Anagnostou & Greenhawt, 2025). Identification of specific microbial strain functional roles will enable the tailoring of separate therapeutic approaches (NCT02960074, 2024), potentially leading to the increase of the effectiveness of allergy treatment overall (Berni Canani et al., 2019).

d. Meta-Analyses and Systematic Reviews

There have been systematic reviews on the effectiveness of probiotics in the management of food allergies. Thirteen studies were included according to the inclusion criteria in accordance with PRISMA as it is vital to consider the long-term post-therapeutic outcomes. These reviews offer important feedback in future studies of long-term microbiome-based interventions (Tan-Lim & Esteban-Ipac, 2025; Hussein et al., 2025).

10. CLINICAL APPLICATIONS:`

a. Immunotherapy Approaches

Incentives in newer technologies in allergy treatment have resulted in several immunotherapy interventions, especially in food allergies. Oral immunotherapy (OIT) and sublingual immunotherapy (SLIT) have been used to perform clinical trials involving allergens like peanuts, eggs, and milk in the last 20 years (Wang et al., 2022; Siddiqui & Cresci, 2021). OIT is the process of desensitizing by adding more and more of the allergen to the body, which causes the development of tolerance, and SLIT is less aggressive and can be done at home Integration of biologic, including Omalizumab and Dupilumab, could lead to better desensitization and a decrease in adverse effects (Nance et al., 2020).

The multi-targeted effects of phytopharmaceuticals used in cancer treatment through the gut microbiota and metabolites have indicated that they may have application in immune-course conditions (Kwon et al., 2023; Li et al., 2023). Another edge is engineered microbiomes, which can be used to provide customized interventions to right a particular microbiota imbalance in immune-related allergies (Murali & Mansell, 2024).

b. Oral Immunotherapy (OIT)

OIT is used to cause tolerance by exposure to allergen in a gradual manner. Though successful in desensitization and the long-term creation of unresponsiveness, OIT is linked to gastrointestinal side effects, rigid schedules, and anxiety associated with the possibility of an allergic response, which can decrease the quality of life (Meneses do Rêgo & Araaujo-Filho, 2025).

c. Sublingual Immunotherapy (SLIT)

SLIT involves the low dose of allergens into the tongue. It is not as intrusive as OIT, and in most cases, has less adverse reactions, yet it is less effective at attaining complete desensitization (Jiao et al., 2022).

d. Epicutoan Pharyngeal Insulin Therapy (EPIT)

EPIT involves the delivery of allergenic proteins with the help of skin patches. There is some evidence that it may be a safe and effective intervention, but additional research is needed to determine long-term effectiveness and safety (Jiao et al., 2022; Hervé et al., 2023; Anagnostou & Greenhawt, 2025).

11. POTENTIAL RISKS OF GENERATIVE AI IN CLINICAL PRACTICE

Generative AI is fast to generate information and is susceptible to hallucinations, where models generate confident and false or fabricated information. The clinical context of such outputs can result in false interpretation, reference, or unfounded clinical recommendations. It is particularly problematic in the context where AI-generated answers are regarded as the official medical information in need of the proper verification.

Another issue is the absence of transparency and openness to audit. The vast majority of large language models are black-box, which means that they do not tell much about the process of decision-making and the data that affected the output. This form of obscurity presents regulatory validation, medico-legal responsibility, and clinician trust of issue, especially within a high-stakes diagnostic or therapy environment.

Also, generative models can reproduce clinical misinformation and bias in cases where they are trained on data sets that do not represent sufficiently diverse populations. Unequal performance within demographic groups is dangerous as it may strengthen disparities in healthcare provision. Therefore, to ensure patient safety and ethical practice, clinical adoption has to be vigorously validated, explained, and monitored in real-time and with human supervision (Chustecki, 2024; Tung et al., 2025).

CONCLUSION

The increasing awareness of the gut microbiome in food allergies will become another significant change of allergy science and management. Recent development of metagenomics, microbiome sequencing and artificial intelligence (AI) have elucidated the multidimensional connection between microbial imbalance and immune auto-regulation, leading to the possibility of custom therapeutic interventions. The new therapies of probiotics, prebiotics, fecal microbiota transplantation (FMT) and engineered microbial consortia are also interested in restoring the microbial balance and preventing allergic reactions at the origin instead of acting only on the symptoms.

Although it promises, it is difficult to translate these findings into clinical practice. Microbiome variations in the personalities of the individuals, compliance problems in patients and regulatory obstacles are barriers to large scale implementation. Moreover, an ethical issue, such as the privacy of data, the transparency of the algorithm, and fair access supports the necessity of strict supervision and recognition of predictive models based on AI. These obstacles will only be overcome through interdisciplinary efforts of clinicians, microbiologists, bioinformaticians, and policy makers.

The key to microbiome-based precision medicine is in the future when various biological systems are integrated. Genomic information can be used to indicate predisposition differences at the individual level, whereas profiles of gut microbiome can indicate the current state of metabolism and flexibility. The role of AI is the computational interface, which transforms massive multi-omic data into practical knowledge. Such integration allows specific actions to be taken, which deal with the underlying causes of metabolic and immune dysfunction. As an example, AI models can be used to discover relationships between genetic polymorphisms and microbial metabolic pathways to predict the responses of individuals to particular diets and assist in very personalized treatment schemes.

Nevertheless, this great integration comes with a big burden. To ensure fairness and trust, it is essential that ethical and legal aspects like informed consent, minimization of algorithmic bias, and safeguarding of data be ensured. Precision healthcare led by microbiomes lies in the ability to develop a wide range of datasets, cost-effective technologies, and open global policies that can benefit not only the privileged groups.

In the future, interdisciplinary collaboration will be mandatory. The upcoming studies are needed to further explain the dynamics between host-microbe-diet and run high-quality clinical trials to prove the recommendations provided by AI-based tools really have health outcomes. The training of medical personnel in the analytics of data, genomics, and microbiome science will play a crucial role in introducing this knowledge into daily clinical practice.

Finally, the combination of microbiome, AI and clinical medicine is a paradigm shift to patient-centered, predictive and preventive medicine. Through the convergence of science with moral leadership, food allergy management and the management of the immune and metabolic health in general can shift its focus towards the reactive management of symptoms to the proactive, customized and globally available management.

REFERENCES

Abaj, F., Esmaeily, Z., Naeini, Z., Rafiee, M., & Koohdani, F. (2022). Dietary acid load modifies the effects of ApoA2-265 T > C polymorphism on lipid profile and serum leptin and ghrelin levels among type 2 diabetic patients. *BMC Endocrine Disorders*, *22*(1), 190. DOI: 10.1186/s12902-022-01083-7 PMID: 35883173

Abavisani, M., Khoshrou, A., Foroushan, S. K., Ebadpour, N., & Sahebkar, A. (2024). Deciphering the gut microbiome: The revolution of artificial intelligence in microbiota analysis and intervention. *Current Research in Biotechnology*, *7*, 100211. DOI: 10.1016/j.crbiot.2024.100211

Abdelwahab, S., Taha, M., Jerah, A., Farasani, A., Abdullah, S., Aljahdali, I., Ibrahim, R., Oraibi, O., Oraibi, B., Alfaifi, H., Alzahrani, A., & Babiker, Y. (2024). Artificial intelligence and microbiome research: Evolution of hotspots, research trends, and thematic-based narrative review. *Cellular and Molecular Biology*, *70*(10), 182–192. DOI: 10.14715/cmb/2024.70.10.24 PMID: 39605107

Aboulaghras, S., Piancatelli, D., Taghzouti, K., Balahbib, A., Alshahrani, M. M., Al Awadh, A. A., Goh, K. W., Ming, L. C., Bouyahya, A., & Oumhani, K. (2023). Meta-Analysis and Systematic Review of HLA DQ2/DQ8 in Adults with Celiac Disease. *International Journal of Molecular Sciences*, *24*(2), 1188. DOI: 10.3390/ijms24021188 PMID: 36674702

Agrawal, P., Kaur, J., Singh, J., Rasane, P., Sharma, K., Bhadariya, V., Kaur, S., & Kumar, V. (2024). Genetics, Nutrition, and Health: A New Frontier in Disease Prevention. *Journal of the American Nutrition Association*, *43*(4), 326–338. DOI: 10.1080/27697061.2023.2284997 PMID: 38015713

Aguilera, A. C., Dagher, I. A., & Kloepfer, K. M. (2020). Role of the microbiome in allergic disease development. *Current Allergy and Asthma Reports*, *20*(9), 44. DOI: 10.1007/s11882-020-00944-2 PMID: 32548788

Ahmed, M. M., Okesanya, O. J., Olaleke, N. O., Adigun, O. A., Adebayo, U. O., Oso, T. A., & Lucero-Prisno, D. E.III. (2025). Integrating digital health innovations to achieve universal health coverage: Promoting health outcomes and quality through global public health equity. *Health Care*, *13*(9), 1060. DOI: 10.3390/healthcare13091060 PMID: 40361838

Al-Habsi, N., Al-Khalili, M., Haque, S. A., Elias, M., Olqi, N. A., & Al Uraimi, T. (2024). Health Benefits of Prebiotics, Probiotics, Synbiotics, and Postbiotics. *Nutrients*, *16*(22), 3955. DOI: 10.3390/nu16223955 PMID: 39599742

Alam, M. Z., Maslanka, J. R., & Abt, M. C. (2023). Immunological consequences of microbiome-based therapeutics. *Frontiers in Immunology*, *13*, 1046472. DOI: 10.3389/fimmu.2022.1046472 PMID: 36713364

Amália Cinthia Meneses do Rêgo, & Araújo-Filho, I. (2025). The role of gut microbiota in tumor recurrence and therapy resistance in colorectal cancer: Molecular mechanisms and clinical implications. *Journal of Surgery and Postoperative Care*, *4*(1). Advance online publication. DOI: 10.58489/2836-8657/014

Anagnostou, A., & Greenhawt, M. (2025). Epicutaneous immunotherapy: A review of safety and efficacy. *Pediatric Allergy and Immunology*, *36*(4), e70096. DOI: 10.1111/pai.70096 PMID: 40285330

Arıkan, M., & Muth, T. (2023). Integrated multi-omics analyses of microbial communities: A review of the current state and future directions. *Molecular Omics*, *19*(8), 607–623. DOI: 10.1039/d3mo00089c PMID: 37417894

Asnicar, F., Berry, S. E., Valdes, A. M., Nguyen, L. H., Piccinno, G., Drew, D. A., Leeming, E., Gibson, R., Le Roy, C., Khatib, H. A., Francis, L., Mazidi, M., Mompeo, O., Valles-Colomer, M., Tett, A., Beghini, F., Dubois, L., Bazzani, D., Thomas, A. M., & Segata, N. (2021). Microbiome connections with host metabolism and habitual diet from 1,098 deeply phenotyped individuals. *Nature Medicine*, *27*(2), 321–332. DOI: 10.1038/s41591-020-01183-8 PMID: 33432175

Barredo Arrieta, A., Herrera, F., Chatilaf, R., Benjaminsh, R., Molina, D., Gil-Lopez, S., . . . Díaz-Rodríguez, N. (2019). Explainable artificial intelligence (XAI): Concepts, taxonomies, opportunities and challenges toward responsible AI. Artificial Intelligence Review. https://doi.org//arXiv.1910.10045DOI: 10.48550

Berni Canani, R., Paparo, L., Nocerino, R., Di Scala, C., Della Gatta, G., Maddalena, Y., Buono, A., Bruno, C., Voto, L., & Ercolini, D. (2019). Gut microbiome as target for innovative strategies against food allergy. *Frontiers in Immunology*, *10*, 191. DOI: 10.3389/fimmu.2019.00191 PMID: 30828329

Boggio Marzet, C., Burgos, F., Del Compare, M., Gerold, I., Tabacco, O., & Vinderola, G. (2022). Approach to probiotics in pediatrics: the role of Lactobacillus rhamnosus GG. Abordaje de los probióticos en pediatría: el rol de Lactobacillus rhamnosus GG. *Archivos Argentinos de Pediatria*, *120*(1), e1–e7. DOI: 10.5546/aap.2022.eng.e1 PMID: 35068121

Brown, N. K., Guandalini, S., Semrad, C., & Kupfer, S. S. (2019). A Clinician's Guide to Celiac Disease HLA Genetics. *The American Journal of Gastroenterology*, *114*(10), 1587–1592. DOI: 10.14309/ajg.0000000000000310 PMID: 31274511

Butler, M. I., Mörkl, S., Sandhu, K. V., Cryan, J. F., & Dinan, T. G. (2019). The gut microbiome and mental health: What should we tell our patients? *Canadian Journal of Psychiatry, 64*(11), 747–760. DOI: 10.1177/0706743719853169 PMID: 31530002

Chen, J., Zhao, T., Li, H., Xu, W., Maas, K., Singh, V., Chen, M. H., Dorsey, S. G., Starkweather, A. R., & Cong, X. S. (2024). Multi-omics analysis of gut microbiota and host transcriptomics reveal dysregulated immune response and metabolism in young adults with irritable bowel syndrome. *International Journal of Molecular Sciences, 25*(6), 3514. DOI: 10.3390/ijms25063514 PMID: 38542485

Chen, L., Zhernakova, D. V., & Kurilshikov, A.. (2022). Influence of the microbiome, diet and genetics on inter-individual variation in the human plasma metabolome. *Nature Medicine, 28*, 2333–2343. DOI: 10.1038/s41591-022-02014-8 PMID: 36216932

Chen, Y. C., Destouches, L., Cook, A., & Fedorec, A. J. H. (2024). Synthetic microbial ecology: Engineering habitats for modular consortia. *Journal of Applied Microbiology, 135*(7), lxae158. Advance online publication. DOI: 10.1093/jambio/lxae158 PMID: 38936824

Chernikova, D. A., Zhao, M. Y., & Jacobs, J. P. (2022). Microbiome therapeutics for food allergy. *Nutrients, 14*(23), 5155. DOI: 10.3390/nu14235155 PMID: 36501184

Chustecki, M. (2024). Benefits and Risks of AI in Health Care: Narrative Review. *Interactive Journal of Medical Research, 13*, e53616. DOI: 10.2196/53616 PMID: 39556817

Cohen, C. E., Swallow, D. M., & Walker, C. (2025). The molecular basis of lactase persistence: Linking genetics and epigenetics. *Annals of Human Genetics, 89*(5), 321–332. DOI: 10.1111/ahg.12575 PMID: 39171584

Daniel, N., Lécuyer, E., & Chassaing, B. (2021). Host/microbiota interactions in health and diseases—Time for mucosal microbiology! *Mucosal Immunology, 14*(5), 1006–1016. DOI: 10.1038/s41385-021-00409-y PMID: 33772148

Dantzer, J. A., & Wood, R. A. (2021). Omalizumab as an adjuvant in food allergen immunotherapy. *Current Opinion in Allergy and Clinical Immunology, 21*(3), 278–285. DOI: 10.1097/ACI.0000000000000736 PMID: 33605614

Del Pozzo, G., Farina, F., Picascia, S., Laezza, M., Vitale, S., & Gianfrani, C. (2021). HLA class II genes in precision-based care of childhood diseases: What we can learn from celiac disease. *Pediatric Research, 89*(2), 307–312. DOI: 10.1038/s41390-020-01217-4 PMID: 33122841

Delgado, R., & Núñez-González, J. D. (2022). Bayesian network-based over-sampling method (BOSME) with application to indirect cost-sensitive learning. *Scientific Reports, 12*(1), 8724. DOI: 10.1038/s41598-022-12682-8 PMID: 35610323

Desjardins, L., & Vohl, M. (2023). Precision Nutrition for Cardiovascular Disease Prevention. *Lifestyle Genomics, 16*, 73–82. DOI: 10.1159/000529054

Domeignoz-Horta, L. A., Cappelli, S. L., Shrestha, R., Gerin, S., Lohila, A. K., Heinonsalo, J., Nelson, D. B., Kahmen, A., Duan, P., Sebag, D., Verrecchia, E., & Laine, A. L. (2024). Plant diversity drives positive microbial associations in the rhizosphere enhancing carbon use efficiency in agricultural soils. *Nature Communications, 15*(1), 8065. DOI: 10.1038/s41467-024-52449-5 PMID: 39277633

Duncker, K. E., Holmes, Z. A., & You, L. (2021). Engineered microbial consortia: Strategies and applications. *Microbial Cell Factories, 20*(1), 211. DOI: 10.1186/s12934-021-01699-9 PMID: 34784924

Ejtahed, H. S., Parsa, M., & Larijani, B. (2023). Ethical challenges in conducting and the clinical application of human microbiome research. *Journal of Medical Ethics and History of Medicine, 16*, 5. DOI: 10.18502/jmehm.v16i5.13313 PMID: 37753524

Elgart, M., Lyons, G., Romero-Brufau, S., Kurniansyah, N., Brody, J. A., Guo, X., Lin, H. J., Raffield, L., Gao, Y., Chen, H., de Vries, P., Lloyd-Jones, D. M., Lange, L. A., Peloso, G. M., Fornage, M., Rotter, J. I., Rich, S. S., Morrison, A. C., Psaty, B. M., & Sofer, T. (2022). Non-linear machine learning models incorporating SNPs and PRS improve polygenic prediction in diverse human populations. *Communications Biology, 5*(1), 856. DOI: 10.1038/s42003-022-03812-z PMID: 35995843

Faiyazuddin, M., & Ashique, S. (Eds.). (2025). *The Role of Probiotics in Cancer Management* (1st ed.). Apple Academic Press., DOI: 10.1201/9781003610748

Fassarella, M., Blaak, E. E., Penders, J., Nauta, A., Smidt, H., & Zoetendal, E. G. (2021). Gut microbiome stability and resilience: Elucidating the response to perturbations in order to modulate gut health. *Gut, 70*(3), 595–605. DOI: 10.1136/gutjnl-2020-321747 PMID: 33051190

Fenwick, P. H., Jeejeebhoy, K., Dhaliwal, R., Royall, D., Brauer, P., Tremblay, A., Klein, D., & Mutch, D. M. (2019). Lifestyle genomics and the metabolic syndrome: A review of genetic variants that influence response to diet and exercise interventions. *Critical Reviews in Food Science and Nutrition, 59*(13), 2028–2039. DOI: 10.1080/10408398.2018.1437022 PMID: 29400991

Ferreira, D. D., Ferreira, L. G., Amorim, K. A., Delfino, D. C. T., Ferreira, A. C. B. H., & Souza, L. P. C. E. (2025). Assessing the Links Between Artificial Intelligence and Precision Nutrition. *Current Nutrition Reports*, *14*(1), 47. DOI: 10.1007/s13668-025-00635-2 PMID: 40087237

Filardo, S., Di Pietro, M., & Sessa, R. (2024). Current progresses and challenges for microbiome research in human health: A perspective. *Frontiers in Cellular and Infection Microbiology*, *14*, 1377012. DOI: 10.3389/fcimb.2024.1377012 PMID: 38638832

Gatla, T. R. (2024). An innovative study exploring revolutionizing healthcare with AI: Personalized medicine, predictive diagnostic techniques and individualized treatment. *International Journal of Advanced Research and Interdisciplinary Scientific Endeavours*, *1*(2), 61–70. DOI: 10.61359/11.2206-2406

Gelli, H. P., Vazquez-Uribe, R., Buckley, S. T., Andersen, J. T., & Alexander Sommer, M. O. (2025). Advanced microbiome therapeutics for oral delivery of peptides and proteins: Advances, challenges, and opportunities. *Advanced Drug Delivery Reviews*, *222*, 115603. DOI: 10.1016/j.addr.2025.115603 PMID: 40349728

Gentile, F., Doneddu, P. E., Riva, N., Nobile-Orazio, E., & Quattrini, A. (2020). Diet, Microbiota and Brain Health: Unraveling the Network Intersecting Metabolism and Neurodegeneration. *International Journal of Molecular Sciences*, *21*(20), 7471. DOI: 10.3390/ijms21207471 PMID: 33050475

Greathouse, K. L., & Choudhury, A. (2025). Precision nutrition and the gut microbiome: Harnessing AI to revolutionize cancer prevention and therapy. *Cell Host & Microbe*, *33*(6), 766–776. DOI: 10.1016/j.chom.2025.05.011 PMID: 40505617

Grigoryan, Z., Shen, M. J., Twardus, S. W., Beuttler, M. M., Chen, L. A., & Bateman-House, A. (2020). Fecal microbiota transplantation: Uses, questions, and ethics. *Medicine in Microecology*, *6*, 100027. DOI: 10.1016/j.medmic.2020.100027 PMID: 33834162

Gupta, S., Kumar, D., Negi, R., Singh, R., Ansari, M. J., Kashyap, S., & Mehrotra, S. (Eds.). (2024). *Genomic intelligence: Metagenomics and artificial intelligence*. CRC Press., DOI: 10.1201/9781003570233

Haring, E., Zeiser, R., & Apostolova, P. (2021). Interfering With Inflammation: Heterogeneous Effects of Interferons in Graft-Versus-Host Disease of the Gastrointestinal Tract and Inflammatory Bowel Disease. *Frontiers in Immunology*, *12*, 705342. DOI: 10.3389/fimmu.2021.705342 PMID: 34249014

Hervé, P. L., Dioszeghy, V., Matthews, K., Bee, K. J., Campbell, D. E., & Sampson, H. A. (2023). Recent advances in epicutaneous immunotherapy and potential applications in food allergy. *Frontiers in Allergy, 4*, 1290003. DOI: 10.3389/falgy.2023.1290003 PMID: 37965375

Homayouni Rad, A., Aghebati Maleki, L., Samadi Kafil, H., & Abbasi, A. (2021). Postbiotics: A novel strategy in food allergy treatment. *Critical Reviews in Food Science and Nutrition, 61*(3), 492–499. DOI: 10.1080/10408398.2020.1738333 PMID: 32160762

Hu, H., Wang, M., Huang, Y., Xu, Z., Xu, P., Nie, Y., & Tang, H. (2022). Guided by the principles of microbiome engineering: Accomplishments and perspectives for environmental use. *mLife, 1*(4), 382–398. DOI: 10.1002/mlf2.12043 PMID: 38818482

Huang, H. L., Zhu, J. Q., Yang, L. S., Wu, Q., Shou, D. W., Chen, H. T., Ma, J., Li, Y. Q., Xu, H. M., & Zhou, Y. J. (2022). Fecal microbiota transplantation combined with a low FODMAP diet for the treatment of irritable bowel syndrome with predominant diarrhea. *Oxidative Medicine and Cellular Longevity, 5121496*. Advance online publication. DOI: 10.1155/2022/5121496 PMID: 36187337

Huo, D., & Wang, X. (2024). A new era in healthcare: The integration of artificial intelligence and microbial. *Medicine in Novel Technology and Devices, 23*, 100319. DOI: 10.1016/j.medntd.2024.100319

Hussein, M., Essam, S., & Mekky, M. E.. (2025). Safety profile of probiotics as an adjuvant to oral immunotherapy for food allergies: A meta analysis of randomized controlled trials. *Probiotics and Antimicrobial Proteins*. Advance online publication. DOI: 10.1007/s12602-025-10544-z PMID: 40320506

Ilan, Y. (2019). Why targeting the microbiome is not so successful: Can randomness overcome the adaptation that occurs following gut manipulation? *Clinical and Experimental Gastroenterology, 12*, 209–217. DOI: 10.2147/CEG.S185317 PMID: 31190948

Jain, N. (2020). The Need for Personalized Approaches to Microbiome Modulation. *Frontiers in Public Health, 8*, 144. DOI: 10.3389/fpubh.2020.00144 PMID: 32411648

JanssenDuijghuijsen, L., Looijesteijn, E., van den Belt, M., Gerhard, B., Ziegler, M., Ariens, R., Tjoelker, R., & Geurts, J. (2024). Changes in gut microbiota and lactose intolerance symptoms before and after daily lactose supplementation in individuals with the lactase nonpersistent genotype. *The American Journal of Clinical Nutrition, 119*(3), 702–710. DOI: 10.1016/j.ajcnut.2023.12.016 PMID: 38159728

Jiang, J. (2021). MIIDL: A Python package for microbial biomarkers identification powered by interpretable deep learning. arXiv preprint. https://arxiv.org/abs/2109.12204

Jiao, L., Su, C. W., Cao, T., Zheng, S., Walker, W. A., & Shi, H. N. (2022). Maternal influences and intervention strategies on the development of food allergy in offspring. *Frontiers in Immunology*, *13*, 817062. DOI: 10.3389/fimmu.2022.817062 PMID: 35281070

Karakan, T., Gundogdu, A., Alagözlü, H., Ekmen, N., Ozgul, S., Hora, M., Beyazgul, D., & Nalbantoglu, O. U. (2021). Artificial intelligence-based personalized diet: A pilot clinical study for IBS. *medRxiv*. https://doi.org/DOI: 10.1101/2021.02.23.21251434

Kim, D. Y., Lee, S. Y., Lee, J. Y., Whon, T. W., Lee, J. Y., Jeon, C. O., & Bae, J. W. (2024). Gut microbiome therapy: Fecal microbiota transplantation vs live biotherapeutic products. *Gut Microbes*, *16*(1), 2412376. DOI: 10.1080/19490976.2024.2412376 PMID: 39377231

Kim, M., Lee, S., Cho, E., Hong, K. W., You, S. J., & Choi, H. J. (2023). CETP and APOA2 polymorphisms are associated with weight loss and healthy eating behavior changes in response to digital lifestyle modifications. *Scientific Reports*, *13*(1), 21615. DOI: 10.1038/s41598-023-48823-w PMID: 38062157

Kodikara, S., & Lê Cao, K. A. (2025). Microbial network inference for longitudinal microbiome studies with LUPINE. *Microbiome*, *13*(1), 64. DOI: 10.1186/s40168-025-02041-w PMID: 40033386

Korczak, R., Kocher, M., & Swanson, K. S. (2020). Effects of oats on gastrointestinal health as assessed by in vitro, animal, and human studies. *Nutrition Reviews*, *78*(5), 343–363. DOI: 10.1093/nutrit/nuz064 PMID: 31638148

Kumar, S., Mukherjee, R., Gaur, P., Leal, É., Lyu, X., Ahmad, S., Puri, P., Chang, C. M., Raj, V. S., & Pandey, R. P. (2025). Unveiling roles of beneficial gut bacteria and optimal diets for health. *Frontiers in Microbiology*, *16*, 1527755. DOI: 10.3389/fmicb.2025.1527755 PMID: 40041870

Kwon, C., Ediriweera, M. K., & Kim Cho, S. (2023). Interplay between phytochemicals and the colonic microbiota. *Nutrients*, *15*(8), 1989. DOI: 10.3390/nu15081989 PMID: 37111207

Laddu, D., & Hauser, M. (2019). Addressing the Nutritional Phenotype Through Personalized Nutrition for Chronic Disease Prevention and Management. *Progress in Cardiovascular Diseases*, *62*(1), 9–14. DOI: 10.1016/j.pcad.2018.12.004 PMID: 30553801

Lagoumintzis, G., & Patrinos, G. P. (2023). Triangulating nutrigenomics, metabolomics and microbiomics toward personalized nutrition and healthy living. *Human Genomics*, *17*(1), 109. DOI: 10.1186/s40246-023-00561-w PMID: 38062537

Lange, L., Berg, G., Cernava, T., Champomier-Vergès, M. C., Charles, T., Cocolin, L., Cotter, P., D'Hondt, K., Kostic, T., Maguin, E., Makhalanyane, T., Meisner, A., Ryan, M., Kiran, G. S., de Souza, R. S., Sanz, Y., Schloter, M., Smidt, H., Wakelin, S., & Sessitsch, A. (2022). Microbiome ethics, guiding principles for microbiome research, use and knowledge management. *Environmental Microbiome*, *17*(1), 50. DOI: 10.1186/s40793-022-00444-y PMID: 36180931

Leiva Alaniz, M. J., Vergara, S. C., Mestre Furlani, M. V., Vazquez, F., Mancha-Agresti, P., & Maturano, Y. P. (2025). Assessment of yeasts from winemaking environments: Biosafety and functional perspectives on probiotic capabilities. *Frontiers in Industrial Microbiology*, *3*, 1494334. DOI: 10.3389/finmi.2025.1494334

Li, R., Hao, X., Diao, Y., Yang, L., & Liu, J. (2025). Explainable Machine Learning Models for Colorectal Cancer Prediction Using Clinical Laboratory Data. *Cancer Control : Journal of the Moffitt Cancer Center*, *32*, 10732748251336417. DOI: 10.1177/10732748251336417 PMID: 40334702

Li, S., Feng, W., Wu, J., Cui, H., Wang, Y., Liang, T., An, J., Chen, W., Guo, Z., & Lei, H. (2023). A narrative review: Immunometabolic interactions of host–gut microbiota and botanical active ingredients in gastrointestinal cancers. *International Journal of Molecular Sciences*, *25*(16), 9096. DOI: 10.3390/ijms2516909 PMID: 39201782

Liu, P., Wang, Y., Yang, G., Zhang, Q., Meng, L., Xin, Y., & Jiang, X. (2021). The role of short-chain fatty acids in intestinal barrier function, inflammation, oxidative stress, and colonic carcinogenesis. *Pharmacological Research*, *165*, 105420. DOI: 10.1016/j.phrs.2021.105420 PMID: 33434620

Liu, X., Chen, M., & Zhou, W. (2024). The gut microbiome: A key player in immune system regulation and disease prevention. *Medical Insight*, *1*(4), 1–7. DOI: 10.70088/9j7wvw75

Liu, Y., Qin, S., Song, Y., Feng, Y., Lv, N., Xue, Y., Liu, F., Wang, S., Zhu, B., Ma, J., & Yang, H. (2019). The perturbation of infant gut microbiota caused by Cesarean delivery is partially restored by exclusive breastfeeding. *Frontiers in Microbiology*, *10*, 598. DOI: 10.3389/fmicb.2019.00598 PMID: 30972048

Maaskant, A., Lee, D., Ngo, H., Montijn, R. C., Bakker, J., Langermans, J. A. M., & Levin, E. (2025). AI for rapid identification of major butyrate-producing bacteria in rhesus macaques (Macaca mulatta). *Animal Microbiome, 7*(1), 39. DOI: 10.1186/s42523-025-00410-2 PMID: 40275402

Malakar, S., Sutaoney, P., Madhyastha, H., Shah, K., Chauhan, N. S., & Banerjee, P. (2024). Understanding gut microbiome-based machine learning platforms: A review on therapeutic approaches using deep learning. *Chemical Biology & Drug Design, 103*(3), e14505. DOI: 10.1111/cbdd.14505 PMID: 38491814

Marcum, J. A. (2020). Nutrigenetics/Nutrigenomics, Personalized Nutrition, and Precision Healthcare. *Current Nutrition Reports, 9*(4), 338–345. DOI: 10.1007/s13668-020-00327-z PMID: 32578026

Micheletti, C., Madeo, G., Macchia, A., Donato, K., Cristoni, S., Ceccarini, M. R., Beccari, T., Iaconelli, A., Aquilanti, B., Matera, G., Herbst, K. L., & Bertelli, M. (2023). Nutrigenomics: SNPs correlated to vitamins' deficiencies. La Clinica terapeutica, 174(Suppl 2(6)), 173–182. https://doi.org/DOI: 10.7417/CT.2023.2485

Mierziak, J., Kostyn, K., Boba, A., Czemplik, M., Kulma, A., & Wojtasik, W. (2021). Influence of the Bioactive Diet Components on the Gene Expression Regulation. *Nutrients, 13*(11), 3673. DOI: 10.3390/nu13113673 PMID: 34835928

Mishra, N., Ashique, S., & Garg, A. (Eds.). (2025). *Inflammatory bowel disease and gut microbiota: Current strategies and future prospects in nanomedicine*. CRC Press., DOI: 10.1201/9781003598336

Mishra, N., Ashique, S., & Kumar, A. (Eds.). (2025). *Gastrointestinal inflammations and gut microbiota: Pathophysiology, management, and emerging therapies*. CRC Press., DOI: 10.1201/9781003493143

Mishra, U. N., Jena, D., Sahu, C., Devi, R., Kumar, R., Jena, R., & Kumar, A. (2022). Nutrigenomics: An inimitable interaction amid genomics, nutrition and health. *Innovative Food Science & Emerging Technologies, 82*, 103196. DOI: 10.1016/j.ifset.2022.103196

Mkilima, T. (2025). Engineering artificial microbial consortia for personalized gut microbiome modulation and disease treatment. *Annals of the New York Academy of Sciences, 1548*(1), 29–55. DOI: 10.1111/nyas.15352 PMID: 40320966

Mubarak, S., & Ashraf, M. (2024). Ethics considerations for precision medicine research and genetic testing in low- and middle-income countries. Eastern Mediterranean health journal = La revue de sante de la Mediterranee orientale = al-Majallah al-sihhiyah li-sharq al-mutawassit, 30(6), 455–460. https://doi.org/DOI: 10.26719/2024.30.6.455

Mukhopadhyay, S., Ulaganathan, N., Dumpuri, P., & Aich, P. (2025). Integrative AI-Based Approaches to Connect the Multiome to Use Microbiome-Metabolome Interactive Outcome as Precision Medicine. *Methods in Molecular Biology (Clifton, N.J.), 2952*, 15–37. DOI: 10.1007/978-1-0716-4690-8_2 PMID: 40553325

Murali, S. K., & Mansell, T. J. (2024). Next generation probiotics: Engineering live biotherapeutics. *Biotechnology Advances, 72*, 108336. DOI: 10.1016/j.biotechadv.2024.108336 PMID: 38432422

Nance, C. L., Deniskin, R., Diaz, V. C., Paul, M., Anvari, S., & Anagnostou, A. (2020). The role of the microbiome in food allergy: A review. *Children (Basel, Switzerland), 7*(6), 50. DOI: 10.3390/children7060050 PMID: 32466620

Niechcial, A., Schwarzfischer, M., Wawrzyniak, P., Determann, M., Pöhlmann, D., Wawrzyniak, M., Gueguen, E., Walker, M. R., Morsy, Y., Atrott, K., Wilmink, M., Linzmeier, L., Spalinger, M. R., Holowacz, S., Leblanc, A., & Scharl, M. (2025). Probiotic administration modulates gut microbiota and suppresses tumor growth in murine models of colorectal cancer. *International Journal of Molecular Sciences, 26*(9), 4404. DOI: 10.3390/ijms26094404 PMID: 40362641

Norgeot, B., Muenzen, K., Peterson, T. A., Fan, X., Glicksberg, B. S., Schenk, G., Rutenberg, E., Oskotsky, B., Sirota, M., Yazdany, J., Schmajuk, G., Ludwig, D., Goldstein, T., & Butte, A. J. (2020). Protected health information filter (Philter): Accurately and securely de-identifying free-text clinical notes. *NPJ Digital Medicine, 3*, 57. DOI: 10.1038/s41746-020-0258-y PMID: 32337372

Nunez, H., Nieto, P. A., Mars, R. A., Ghavami, M., Sew Hoy, C., & Sukhum, K. (2025). Early life gut microbiome and its impact on childhood health and chronic conditions. *Gut Microbes, 17*(1), 2463567. DOI: 10.1080/19490976.2025.2463567 PMID: 39916516

Obukohwo, O., Joseph, G., Oyekanmi, B., Odokuma, E., & Thankgod, O. (2024). Gut microbiota and neuroinflammation: An interconnected nexus of health and neurodegenerative disease. *OBM Neurobiology, 8*, 1–28. DOI: 10.21926/obm.neurobiol.2404254

Oh, V. S., & Li, R. W. (2024). Wise Roles and Future Visionary Endeavors of Current Emperor: Advancing Dynamic Methods for Longitudinal Microbiome Meta-Omics Data in Personalized and Precision Medicine. Advanced science (Weinheim, Baden-Wurttemberg, Germany), 11(47), e2400458. https://doi.org/DOI: 10.1002/advs.202400458

Özyaman, F., & Yılmaz, Ö. (2022). Artificial intelligence, microbiology, and Raman technologies. *Artificial Intelligence Theory and Applications*, *2*(2), 51–58.

Pan, S., Jiang, X., & Zhang, K. (2023). WSGMB: Weight signed graph neural network for microbial biomarker identification. *Briefings in Bioinformatics*, *25*(1), bbad448. Advance online publication. DOI: 10.1093/bib/bbad448 PMID: 38084923

Probul, N., Huang, Z., Saak, C. C., Baumbach, J., & List, M. (2024). AI in microbiome-related healthcare. *Microbial Biotechnology*, *17*(11), e70027. DOI: 10.1111/1751-7915.70027 PMID: 39487766

Prosser James, I., & Martiny Jennifer, B. H. (2020). Conceptual challenges in microbial community ecologyPhil. *Trans. R. Soc.*, *B37520190241*. Advance online publication. DOI: 10.1098/rstb.2019.0241 PMID: 32200750

Przymus, P., Rykaczewski, K., Martín-Segura, A., Truu, J., Carrillo De Santa Pau, E., Kolev, M., Naskinova, I., Gruca, A., Sampri, A., Frohme, M., & Nechyporenko, A. (2025). Deep learning in microbiome analysis: A comprehensive review of neural network models. *Frontiers in Microbiology*, *15*, 1516667. DOI: 10.3389/fmicb.2024.1516667 PMID: 39911715

Raghubeer, S., & Matsha, T. E. (2021). Methylenetetrahydrofolate (MTHFR), the One-Carbon Cycle, and Cardiovascular Risks. *Nutrients*, *13*(12), 4562. DOI: 10.3390/nu13124562 PMID: 34960114

Ramond, P., Galand, P. E., & Logares, R. (2025). Microbial functional diversity and redundancy: Moving forward. *FEMS Microbiology Reviews*, *49*, fuae031. Advance online publication. DOI: 10.1093/femsre/fuae031 PMID: 39689915

Redondo-Useros, N., Nova, E., González-Zancada, N., Díaz, L. E., Gómez-Martínez, S., & Marcos, A. (2020). Microbiota and Lifestyle: A Special Focus on Diet. *Nutrients*, *12*(6), 1776. DOI: 10.3390/nu12061776 PMID: 32549225

Rubas, N. C., Torres, A., & Maunakea, A. K. (2025). The Gut Microbiome and Epigenomic Reprogramming: Mechanisms, Interactions, and Implications for Human Health and Disease. *International Journal of Molecular Sciences*, *26*(17), 8658. DOI: 10.3390/ijms26178658 PMID: 40943575

Ruiz-Perez, D., Lugo-Martinez, J., Bourguignon, N., Mathee, K., Lerner, B., Bar-Joseph, Z., & Narasimhan, G. (2021). Dynamic Bayesian Networks for Integrating Multi-omics Time Series Microbiome Data. *mSystems*, *6*(2), e01105–e01120. DOI: 10.1128/mSystems.01105-20 PMID: 33785573

Saxena, R., Sharma, V., Saxena, A. R., & Patel, A. (2024). Harnessing AI and gut microbiome research for precision health. Journal of Artificial Intelligence General science (JAIGS) ISSN: 3006-4023, 3(1), 74-88. https://doi.org/DOI: 10.60087/jaigs.v3i1.68

Shaman, J. A. (2024). The Future of Pharmacogenomics: Integrating Epigenetics, Nutrigenomics, and Beyond. *Journal of Personalized Medicine*, *14*(12), 1121. DOI: 10.3390/jpm14121121 PMID: 39728034

Siddiqui, M. T., & Cresci, G. A. M. (2021). The immunomodulatory functions of butyrate. *Journal of Inflammation Research*, *14*, 6025–6041. DOI: 10.2147/JIR.S300989 PMID: 34819742

Su, J., Yang, L., Sun, Z., & Zhan, X. (2024). Personalized Drug Therapy: Innovative Concept Guided With Proteoformics. *Molecular & Cellular Proteomics*, *23*(3), 100737. DOI: 10.1016/j.mcpro.2024.100737 PMID: 38354979

Sutanto, H., & Fetarayani, D. (2025). Harnessing Neural Networks for Food Allergy Diagnosis, Management, and Research: A Review of Current Applications and Future Prospects. *Current Treatment Options in Allergy*, *12*, 16. DOI: 10.1007/s40521-025-00395-y

Taico Oliva, C., Musa, I., Kopulos, D., Ardalani, F., Maskey, A., Wilson, A., Yang, N., & Li, X. M. (2025). The gut microbiome and cross-reactivity of food allergens: Current understanding, insights, and future directions. *Frontiers in Allergy*, *5*, 1503380. DOI: 10.3389/falgy.2024.1503380 PMID: 39872379

Tan-Lim, C. S. C., & Esteban-Ipac, N. A. R. (2025). Systematic review and meta-analysis on probiotics as treatment for food allergies among pediatric patients: A 2024 update. *Pediatric Allergy and Immunology*, *36*(1), e70028. DOI: 10.1111/pai.70028 PMID: 39803979

Topçuoğlu, B. D., Lesniak, N. A., Ruffin, M. T.IV, Wiens, J., & Schloss, P. D. (2020). A Framework for Effective Application of Machine Learning to Microbiome-Based Classification Problems. *mBio*, *11*(3), e00434–e20. DOI: 10.1128/mBio.00434-20 PMID: 32518182

Tuli, H. S., Kaur, G., & Kaur, S. (Eds.). (2025). *Gut microbiota: Characterization and food conversions to therapeutic implications*. CRC Press.

Tung, T., Hasnaeen, S. M. N., & Zhao, X. (2025). Ethical and practical challenges of generative AI in healthcare and proposed solutions: A survey. *Frontiers in Digital Health, 7*, 1692517. DOI: 10.3389/fdgth.2025.1692517 PMID: 41333106

Vänni, P., Tejesvi, M. V., Paalanne, N., Aagaard, K., Ackermann, G., Camargo, C. A.Jr, Eggesbø, M., Hasegawa, K., Hoen, A. G., Karagas, M. R., Kolho, K.-L., Laursen, M. F., Ludvigsson, J., Madan, J., Ownby, D., Stanton, C., Stokholm, J., & Tapiainen, T. (2023). Machine-learning analysis of cross-study samples according to the gut microbiome in 12 infant cohorts. *mSystems*, *8*(6), e0036423. DOI: 10.1128/msystems.00364-23 PMID: 37874156

Verma, A. K., Singh, K., Gupta, J. K., Kumar, S., & Jain, D. (2025). Pharmacological Approaches and Innovative Strategies for Individualized Patient Care. *Recent Patents on Biotechnology*. Advance online publication. DOI: 10.2174/0118722083359334250116063638 PMID: 39838664

Vijayan, S., Kandi, V., Palacholla, P. S., Rajendran, R., Jarugu, C., Ca, J., Pravallika, M., Reddy, S. C., & Sucharitha, A. S. (2024). Probiotics in Allergy and Immunological Diseases: A Comprehensive Review. *Cureus*, *16*(3), e55817. DOI: 10.7759/cureus.55817 PMID: 38590477

Wan, Z., Zheng, J., Zhu, Z., Sang, L., Zhu, J., Luo, S., Zhao, Y., Wang, R., Zhang, Y., Hao, K., Chen, L., Du, J., Kan, J., & He, H. (2022). Intermediate role of gut microbiota in vitamin B nutrition and its influences on human health. *Frontiers in Nutrition*, *9*, 1031502. DOI: 10.3389/fnut.2022.1031502 PMID: 36583209

Wang, C., Segal, L. N., Hu, J., Zhou, B., Hayes, R. B., Ahn, J., & Li, H. (2022). Microbial risk score for capturing microbial characteristics, integrating multi-omics data, and predicting disease risk. *Microbiome*, *10*(1), 121. DOI: 10.1186/s40168-022-01310-2 PMID: 35932029

Wang, K., & Hu, S. (2023). The synergistic effects of polyphenols and intestinal microbiota on osteoporosis. *Frontiers in Immunology*, *14*, 1285621. DOI: 10.3389/fimmu.2023.1285621 PMID: 37936705

Wang, L. J., Mu, S. C., Lin, M. I., Sung, T. C., Chiang, B. L., & Lin, C. H. (2022). Clinical manifestations of pediatric food allergy: A contemporary review. *Clinical Reviews in Allergy & Immunology*, *62*(1), 180–199. DOI: 10.1007/s12016-021-08895-w PMID: 34519995

Wang, T., Wang, X., & Liu, Y. (2024). Artificial Intelligence for Microbiology and Microbiome Research. arXiv Preprint. https://arxiv.org/abs/2411.01098

Wastyk, H. C., Fragiadakis, G. K., Perelman, D., Dahan, D., Merrill, B. D., Yu, F. B., Topf, M., Gonzalez, C. G., Van Treuren, W., Han, S., Robinson, J. L., Elias, J. E., Sonnenburg, E. D., Gardner, C. D., & Sonnenburg, J. L. (2021). Gut-microbiota-targeted diets modulate human immune status. *Cell, 184*(16), 4137–4153.e14. DOI: 10.1016/j.cell.2021.06.019 PMID: 34256014

Wei, N., Dai, S., Li, W., Zhou, J., & Chen, Y. (2025). Gut microbiota and sepsis-associated encephalopathy: Pathogenesis and precision therapies. *Frontiers in Neuroscience, 19*, 1596467. DOI: 10.3389/fnins.2025.1596467 PMID: 40697274

Wijerathna, P. A. K. C., Udayagee, K. P. P., Idroos, F. S., & Manage, P. M. (2024). Formulation of novel microbial consortia for rapid composting of biodegradable municipal solid waste: An approach in the circular economy. *Environment and Natural Resources Journal, 22*(3), 283–300. https://ph02.tci-thaijo.org/index.php/ennrj/article/view/251078

Wojciuk, M., Swiderska-Chadaj, Z., Siwek, K., & Gertych, A. (2024). Improving classification accuracy of fine-tuned CNN models: Impact of hyperparameter optimization. *Heliyon, 10*(5), e26586. DOI: 10.1016/j.heliyon.2024.e26586 PMID: 38463880

Wong, T.-T., & Yeh, P.-Y. (2020, August 1). Reliable Accuracy Estimates from k-Fold Cross Validation. *IEEE Transactions on Knowledge and Data Engineering, 32*(8), 1586–1594. DOI: 10.1109/TKDE.2019.2912815

Wu, Z., Guo, Y., Hayakawa, M., Yang, W., Lu, Y., Ma, J., Li, L., Li, C., Liu, Y., & Niu, J. (2024). Artificial intelligence-driven microbiome data analysis for estimation of postmortem interval and crime location. *Frontiers in Microbiology, 15*, 1334703. DOI: 10.3389/fmicb.2024.1334703 PMID: 38314433

Xiao, L., & Zhao, F. (2024). Exploring the frontier of microbiome biomarker discovery with artificial intelligence. *National Science Review, 11*(11), nwae325. Advance online publication. DOI: 10.1093/nsr/nwae325 PMID: 39606148

Xiao, Y., Huang, L., Zhao, J., Chen, W., & Lu, W. (2025). The gut core microbial species Bifidobacterium longum: Colonization, mechanisms, and health benefits. *Microbiological Research, 290*, 127966. DOI: 10.1016/j.micres.2024.127966 PMID: 39547052

Xie, A., Chen, A., Chen, Y., Luo, Z., Jiang, S., Chen, D., & Yu, R. (2023). Lactobacillus for the treatment and prevention of atopic dermatitis: Clinical and experimental evidence. *Frontiers in Cellular and Infection Microbiology, 13*, 1137275. DOI: 10.3389/fcimb.2023.1137275 PMID: 36875529

Xin, L., Huang, C., Li, H., Huang, S., Feng, Y., Kong, Z., Liu, Z., Li, S., Yu, C., Shen, F., & Tang, H. (2024). Artificial intelligence for Central Dogma-centric multi-omics: Challenges and breakthroughs. arXiv Preprint. https://arxiv.org/abs/2412.12668

Yadav, M., & Chauhan, N. S. (2021). Microbiome therapeutics: Exploring the present scenario and challenges. *Gastroenterology Report*, *10*, goab046. Advance online publication. DOI: 10.1093/gastro/goab046 PMID: 35382166

Yang, N., Nesme, J., Røder, H. L., Li, X., Zuo, Z., Petersen, M., Burmølle, M., & Sørensen, S. J. (2021). Emergent bacterial community properties induce enhanced drought tolerance in Arabidopsis. *NPJ Biofilms and Microbiomes*, *7*(1), 82. DOI: 10.1038/s41522-021-00253-0 PMID: 34795326

Yang, Y., Hong, Q., Zhang, X., & Liu, Z. (2025). Bifidobacterium animalis BD400 protects from collagen-induced arthritis through histidine metabolism. *Frontiers in Immunology*, *16*, 1518181. DOI: 10.3389/fimmu.2025.1518181 PMID: 39911381

Yu, L., & Zhai, X. (2024). Use of artificial intelligence to address health disparities in low- and middle-income countries: A thematic analysis of ethical issues. *Public Health*, *234*, 77–83. DOI: 10.1016/j.puhe.2024.05.029 PMID: 38964129

Yuan, X., Chang, C., Chen, X., & Li, K. (2021). Emerging trends and focus of human gastrointestinal microbiome research from 2010-2021: A visualized study. *Journal of Translational Medicine*, *19*(1), 327. DOI: 10.1186/s12967-021-03009-8 PMID: 34332587

Zhang, J., Ren, X., Li, B., Zhao, Z., Li, S., & Zhai, W. (2025). Fecal microbiota transplantation is a promising therapy for kidney diseases. *Frontiers in Medicine*, *12*, 1628722. DOI: 10.3389/fmed.2025.1628722 PMID: 40703302

Zhang, T., Gao, G., Sakandar, H. A., Kwok, L. Y., & Sun, Z. (2022). Gut Dysbiosis in Pancreatic Diseases: A Causative Factor and a Novel Therapeutic Target. *Frontiers in Nutrition*, *9*, 814269. DOI: 10.3389/fnut.2022.814269 PMID: 35242797

Zheng, S., Edney, S. M., Goh, C. H., Tai, B. C., Mair, J. L., Castro, O., Salamanca-Sanabria, A., Kowatsch, T., van Dam, R. M., & Müller-Riemenschneider, F. (2023). Effectiveness of holistic mobile health interventions on diet, and physical, and mental health outcomes: A systematic review and meta-analysis. *EClinicalMedicine*, *66*, 102309. DOI: 10.1016/j.eclinm.2023.102309 PMID: 38053536

Zmora, N., Suez, J., & Elinav, E. (2019). You are what you eat: Diet, health and the gut microbiota. *Nature Reviews. Gastroenterology & Hepatology*, *16*(1), 35–56. DOI: 10.1038/s41575-018-0061-2 PMID: 30262901

KEY TERMS AND DEFINITIONS

Adjuvant Therapy: Additional therapy provided at the end of initial treatment (surgery or chemotherapy) to reduce the chances of recurrence of the disease.

Designed Microbial Consortium: A combination of less complex organisms that have been developed or modified in a laboratory setting to act in unison toward a particular goal, such as facilitating better gut health.

Dysbiosis: The number of harmful bacteria increases in the gut, and this might cause illness.

Engineered Bacteria: The bacteria that have been modified in the laboratory to perform certain tasks, such as killing a disease or manufacturing vitamins in the body.

Fecal microbiota transplantation: A medical procedure in which health-giving bacteria of a donor feces are introduced into the intestinal tract of a person to restore the intestinal microflora to normal.

Gut Microbiota: The population of the microscopic living things (such as bacteria) that inhabit our guts. They support digestion and immunity, as well as general health.

Immune Tolerance: The capacity of the immune system to recognize the cells of its own body and non-harmful objects like foodstuffs or benign bacteria and not manipulate them.

Prebiotics: The special fiber that nurtures the good bacteria in the intestines to make them multiply and be effective.

Probiotics: Live, friendly bacteria that are present in foods or supplements and maintain the health of the digestive system.

Synbiotics: Probiotics (indices of good bacteria) and prebiotics (food sources of good bacteria) as a combination to help with improving gut health.

Chapter 7
Using AI to Profile Egyptian Lactobacillus for Next Generation Functional Foods:
Integrating AI for Probiotic Development

Sherein Ismail Abd El-Moez
http://orcid.org/0000-0003-3620-8930
Veterinary Research Institute, Egypt

Mohamed Gomaa Seadawy
Biodefense Center for Infectious and Emerging Diseases, Egypt

Mostafa Fetoh El-Hosseny
http://orcid.org/0000-0002-0860-2052
Biodefense Center for Infectious and Emerging Diseases, Egypt

Mohamed Abd Elhakim Kelany
Central Lab of Residue Analysis of Pesticides and Heavy Metals in Food, Egypt

Amr Mohamed Ageez
Independent Researcher, Egypt

Miral Gamal AbdElwahab
Faculty of Technological Industry and Energy, Thebes Technological University, Luxor, Egypt

Ahmed Zaki Ghareeb
http://orcid.org/0000-0002-6683-6631
Pharmaceutical and Fermentation Industries Development Canter, Egypt

Marwa Ashraf Zayed
Faculty of Veterinary Medicine, Damanhour University, Egypt

Doaa A. Ghareeb
Independent Researcher, Egypt

Abdelmohsen M. Soliman
http://orcid.org/0000-0001-8816-5309
Pharmaceutical Research Institute, Egypt

DOI: 10.4018/979-8-3373-3121-8.ch007

ABSTRACT

In response to the growing demand for sustainable, health-promoting foods, this chapter details an AI-powered pipeline for whole-genome sequencing and functional annotation of six Egyptian Lactobacillus isolates selected from 50 dairy-derived candidates. High-coverage Illumina MiSeq reads were quality-filtered (FastQC), assembled (PATRIC), and annotated via RASTtk, antiSMASH, BAGEL4, ResFinder, and CRISPRCasFinder. Analyses revealed four bacteriocin biosynthesis clusters, two Type I-E CRISPR arrays, prophage elements, and no antibiotic resistance or virulence genes. Phylogenomic analysis confirmed membership in the L. paracasei and L. rhamnosus groups. This in silico approach accelerates safety assessment and guides the selection of bacteriocin-rich cultures for next-generation functional foods, with alignment with Egyptian regulatory standards and recommendations for digital-twin modeling and pilot-scale validation.

1. BACKGROUND

1.1 Functional Foods and Their Health Significance

Functional foods are foods that have health-promoting qualities as well as basic nutritional value. Modern consumers are becoming more attracted to products that are aimed toward or promote well-being, disease prevention, or the activation of a particular physiological pathway (Mudondo, 2025). The world functional food industry reached USD 280.7 billion in 2021, and it is estimated that it will increase to USD 586.1 billion in 2030, promising an 8.5% compound annual growth rate (CAGR) (Grand View Research, 2024). This growth reflects increased awareness of dietary measures for health preservation and disease prevention.

A survey of 2024 revealed that 22 percent of global consumers chose foods specifically to support mood and psychological health, an increase of 3 points over the past year (Nutraceuticals World, 2024). Dairy products, beverages, and gut health, immunity, and cognitive benefits promoted by fortified snacks will continue to dominate the functional segment. These kinds of trends warrant the presence of powerful scientific frameworks that can identify and confirm the bioactive compounds in these foods.

Concurrently, the growth of chronic diseases and the emergence of antimicrobial resistance have prompted health authorities to seek solutions that minimize disease occurrence in ways that do not jeopardize the safety of the population (Osivand et al., 2025). Interventions on a diet basis are safe and scalable options, in those circumstances, as long as they are supported by credible evidence.

Probiotics live microorganisms, which are known to provide health benefits when administered in adequate quantities, hold a central place in numerous functional-food formulations. These include Lactobacillus species, which are the most prevalent owing to their long history of safety as well as their known health-supportive effects (Reyed & Prabhakar 2026).

Experimental evidence shows that *Lactobacillus spp.* promote interactions between intestinal epithelial and immune cells, cytokine production, barrier functions, and immune homeostasis (Reyed, & Prabhakar 2025). These bacteria promote balanced T-cell responses, an increase in the production of immunoglobulin A (IgA) in the gut and the activation of dendritic cells and macrophages. This regulatory process reduces inflammation and enhances immunity against bacterial and viral infections (Neurath et al., 2025).

1.2 Role of Probiotics (Lactobacillus) in Immunity and Host Health

Probiotics, which are living microorganisms that yield health benefits when present in adequate amounts, constitute the foundation of many functional food formulations. In this context, *Lactobacillus* species are dominant because of their long-term record of safety and strong evidence supporting their health-promoting effects. On the basis of empirical evidence, *Lactobacillus spp.* have direct effects on intestinal epithelial and immune cells, regulating the production of cytokines, strengthening the barrier, and maintaining immune homeostasis (Ji et al., 2025). These bacteria have the potential to activate dendritic cells and macrophages, which in turn secrete immunoglobulin A (IgA) in the gut and promote balanced T-cell responses. This rule reduces the processes of inflammation and the defense against bacteria or viruses. Probiotics can be enhanced with prebiotics, known as synbiotics, which further increase these effects by providing substrates that hasten the survival and activity of the probiotics, resulting in better clinical outcomes in metabolic, gastrointestinal, and even neurological disorders (Reyed & Prabhakar 2026). As a result, products containing *Lactobacillus* strains with immune-enhancing, antimicrobial, and gut–brain axis-regulating properties are regularly included in product formulations. Regardless of these benefits, there are distinct differences between various *Lactobacillus* strains. The specific genetic characteristics, such as bacteriocin biosynthesis clustering, adhesion elements, immunomodulatory peptides, and genomic safety keepers (lack of virulence or antibiotic resistance genes), control the functional effectiveness of these compounds. Therefore, profile screening requires high-resolution genetic characterization to select the most appropriate candidates (Abdelaziz et al., 2024).

1.3 Shifting Toward Data-Driven and AI-Enabled Approaches

The emergence of next-generation sequencing (NGS) has led to the generation of large amounts of microbial genome data. However, the value of transforming raw sequence data into actionable insights is limited. Standard bioinformatic pipelines are useful for the annotation of genes and prediction of their functions but often lack the specificity necessary to detect complex or low-abundance biosynthetic pathways (Ghoreyshi et al., 2025).

Machine learning (ML) and artificial intelligence (AI) are also viable options. Compared with rule-based approaches, AI-driven tools identify weak patterns, predict new groupings of genes, and assess safety profiles with greater sensitivity and specificity, in part owing to training on large volumes of annotated genomes (Wang et al., 2025). The RASTtk (Patel et al., 2025), BAGEL4 (Fernandez-Cantos et al., 2025), antiSMASH (Blin et al., 2021), and ResFinder (Martiny, 2022) resources incorporate ML to automate bacteriocin operons, CRISPR/Cas arrays (Reyed & Prabhakar 2025), prophage elements (Wishart et al., 2023), and resistance gene recognition.

Additionally, AI is proficient in integrative analytics, which helps in linking genomic, metabolomic, and phenotypic data to predict the functional capabilities of a strain. Digital-twin models can be used to predict strain behavior under a variety of conditions to guide pilot-scale validation and formulation design. AI-powered experiences: Another trend in the food and beverage industry is the emergence of the use of AI to produce personalized nutrition platforms where products are customized to reflect the specifics of individual microbiome profiles (Agrawal et al., 2025).

In the case of probiotic development, this change includes the following:

- High-throughput safety evaluation on the basis of robotized screening of resistance and virulence gene profiles;
- Acute functional characterization of microbial peptides and modulators;
- Regulatory conciliatory, since AI pipelines produce uniform documentation that could be coordinated with the Codex, EFSA, and WHO criteria.

With AI in hand, researchers have transferred to reproducible high-throughput genomic workflows that were previously labor-intensive, culture-based screens (Liu et al., 2025). The strategy will hasten the identification and implementation of next-generation functional foods designed with particular populations in mind, including Egyptian shoppers, whose food preferences and microbiome signatures can vary significantly from those in other areas.

1.4 Scope of the Chapter

The current chapter describes the theoretical and scientific development of an AI-empowered pipeline focused on profiling six Lactobacillus isolates acquired in Egypt. The goals include the following:

1. To describe the health and market drivers motivating functional food innovation.
2. To explore the relevant mechanisms by which probiotics affect immune and host physiology.
3. We can explore the move between traditional and AI-powered genome annotation.

The next sections will outline the process and results of dataset generation and annotation and then provide functional validation work. These approaches collectively form a pathway for developing isolates into culturally relevant, regulatory-compliant functional foods. In general, the study represents an example of an AI-guided precision nutrigenomic intervention that fits within the larger goals of Advancing Personalized Medicine with AI-Driven Microbiome and Nutrigenomics.

2. PROBLEM DEFINITION

2.1 Global Health Challenges: Chronic Diseases and Antimicrobial Resistance

The progression of chronic, noncommunicable diseases, most prominently cardiovascular diseases, type 2 diabetes, and inflammatory bowel disease, is unprecedented, causing tremendous burdens on national healthcare systems and dramatically lowering quality of life worldwide. The current type 2 diabetes incidence among Egyptian adults is 15.2%, which is higher than the global level of 9.3% and highlights a national health problem that is a matter of urgency (Farag et al., 2023). In parallel, antibiotics have been similarly overused and misused both in human medicine and animal farming, leading to antimicrobial resistance, which has resulted in decreased effectiveness of standard medicines and extended patient stays, increasing medical expenses and causing mortality. Chronic patients often need several rounds of antibiotics; thus, increasing resistance and high levels of AMR are important for ensuring the viability of the routine management of chronic complications. As a result, it has led to a feedback loop of increasing clinical demands and greater social destabilization.

2.2 Emerging Viral Pandemics and their Impacts on Food Systems

Moreover, recent outbreaks of novel viral pandemics have had an impact on the exposure of weak points in international food systems. The production–processing–distribution chains were disrupted due to COVID-19-related lockdowns and labor and transportation constraints, leading to food waste (accounting for an estimated 300,000 tons per week globally) and food insufficiency in vulnerable groups (Żurek & Rudy, 2024). Emerging respiratory viruses remain a food security risk, especially in low- and middle-income countries, where logistical and economic limitations negatively impact the acceleration of responses. Viral epidemics increase cases of malnutrition by disrupting the supply of fresh foods and distracting consumer demand toward shelf-stable yet micronutrient-poor foods. Malnourished individuals will then have more adverse disease outcomes and compromised immune defenses, further propagating a chain of health and food-security crises.

2.3 Demand for Quick, Safe Functions Food

With chronic diseases and infections rising to epidemic proportions across the globe, the need has never been stronger to develop, validate, and commercialize functional foods, as defined here as foods supplemented by bioactive components, i.e., foods supplemented by probiotic strains of bacteria, with truly spectacular speed. Achieving both aims, namely, preventing chronic diseases and being resilient to infections, would require the three criteria outlined in a report by the European Food Safety Authority (EFSA) to be met) Ma et al., 2025):

1. **High-Throughput Development**: High-throughput genomic and *in silico* screening to reduce the current five-- to 10-year bench-to-shelf timeline to less than two years.
2. **Genomic Safety Assurance**: Ensuring the safety of the proposed probiotic strains by ensuring that they do not contain transferable antimicrobial resistance or virulence genes in accordance with Codex Alimentarius and EFSA requirements (Boyte et al, 2025).
3. **Efficacy Validation:** Justification of health claims using methodologically sound biological and early clinical data showing measurable results in immune markers or pathogen suppression. The achievement of these objectives is forcing a move beyond labor-intensive and culture-based assays to integrate AI-driven pipelines that can annotate whole-genome sequences quickly, identify biosynthesis-related gene clusters (e.g., bacteriocins and antiviral peptides), and highlight any safety considerations. These strategies parallel the changing regulatory pressures and

meet the need for increasing consumer demands to see food as part of a healthy life in a postpandemic world.

The given considerations highlight the importance of the extensive and expeditious genomic profiling that becomes possible through the use of AI-based tools and methods; the following chapter presents the composition, sources, and preliminary descriptive statistics of six Egyptian Lactobacillus isolates that constitute the subject of analyses.

3. RESEARCH GAP

3.1 Methodological Limitations of Classical Probiotic Selection

Classical probiotic identification procedures have remained reliant on phenotypic experiments—such as acid and bile tolerance, adhesion to cell lines, and antimicrobial inhibition zones—along with culture-based identification techniques. Although such methods have led to the development of many safe and effective strains, there are several limitations to this process:

1. *Low throughput and time intensity*: Each candidate requires days or weeks of laboratory work, which limits the number of strains that can be screened.
2. *Inadequate Functional Insights*: Phenotypic screenings can measure only some characteristics and do not consistently translate to efficacy or toxicity in vivo, which translates to low conversion rates in downstream validation.
3. *Safety screening gaps:* Low-abundance genotypes from antibiotic resistance tests or nonexpressed antibiotic resistance genes are possible safety risks not targeted by culture-based antibiotic resistance tests.

These limitations act as barriers to innovation and exploration of microbial diversity, especially in less-studied countries such as Egypt, where dozens of locally adapted Lactobacillus isolates have never been characterized. A comprehensive selection framework that is scalable, predictive, and holistic in approach is thus urgently needed.

3.2 Lack of Integration between AI and Local Genomic Annotation

Despite their maturity, AI-based software-based platforms, such as RASTtk, antiSMASH, and CRISPRCas Finder, have not yet been fully adopted in the screening of probiotics in the region. There are three main barriers that have limited broader use:

1. *Generic training data*: Most machine learning models are trained on coarse (restricted to a single country), wide-scope training collections that underrepresent North African taxa, thus lowering sensitivity to clades of interest.
2. *Isolated annotation pipelines*: Within the majority of studies, *in silico* AI-based annotation does not occur alongside laboratory validation or regulatory reporting, slowing the process of rapid iteration between computational prediction and bench-based actions.
3. *Regulatory mismatch:* Regular annotation results do not consistently match the EFSA or Codex thresholds of safety and should be reinterpreted by specialists.

The possibilities of AI to reduce risk and speed up the established process of probiotic development, therefore, remain untapped in situations where probiotic ecosystems exist, but their microbial reservoir has yet to be explored, as in the case of the Egyptian raw milk ecosystem. The result is that a locally tuned, unified AI and wet-lab pipeline is needed to overcome this gap and achieve the potential of AI-driven technologies in probiotic screening.

3.3 Impact of the Gap on Harnessing Egyptian Microbial Diversity

The dairy culture in Egypt has a diverse microbial environment with Lactobacillus spp. This environment appears to be well adapted to different climates and feedstocks. This biodiversity is, however, untapped unless it comes with a carefully laid AI-genomic structure. This gap produces a series of complex effects:

1. *Unidentified bioactive compounds*: Novel bacteriocin clusters that are specific to Egyptian isolates or genes that produce antiviral peptides may have been left undescribed, thus limiting the variety of functional components that can be utilized to produce next-generation foods.
2. *Safety doubts:* A lack of detailed genomic vetting will reduce confidence in the ability of locally identified promising strains to progress to pilot-scale development, slowing product development.

3. *Economic opportunity costs:* There is value loss when the agri-food sector uses imported probiotic cultures that might not fit domestic diets and government regulations.

To overcome these constraints, we developed a hybrid AI-genomic pipeline specific to Egyptian Lactobacillus biodiversity:

1. *Regional model tuning*: Fine-tune machine-learning models on a curated North African Lactobacillus database to improve endemic biosynthetic cluster detection.
2. *Automated safety and function dashboards*: Design software that consumes raw assemblies, performs ResFinder and VFDB searches, and produces regulatory-compliant safety reports, as well as anticipated bacteriocin and CRISPR/Cas operon maps.
3. *In silico (AI) and* in vitro *(laboratory analyses):* Feedback from laboratory tests (e.g., directory mass spectrometry of modeled metabolites) can help improve AI predictions as a dynamic *in silico*, in vitro loop with active-learning techniques.
4. *Digital twin simulations:* Utilize agent-based modeling to predict strain behavior in a gastrointestinal environment and food processing applications as inputs to model selection before expensive pilot runs.

The proposed system involves regionalized AI training, automated reporting, and iteration-based experimental verification to scale, leading to more rapid and safe identification of viable potent probiotic candidates in Egypt because of their vast microbial flora. This approach directly addresses methodological bottlenecks, combines computational and experimental areas, and opens up economic and health opportunities for region-specific functional foods.

4. IMPORTANCE OF THE GENOMIC DATASET

The genomic data that are currently shown represent a subset of six Lactobacillus isolates selected from a larger set of sixty dairy-related samples. All the isolates represent a unique set of genetic characteristics that can provide some functional benefits for food applications. Considering six isolates allows them to perform a deep analysis and, at the same time, be feasible for experimentally validating the results. Selection was based on preliminary phenotypic selection in which acid-tolerant, bile-tolerant, and antimicrobial strains were investigated to ensure that the chosen strains were probiotic instead of random contamination in the environment.

Correct and proper documentation of spatial and temporal metadata largely adds credibility to the scientific outcome of the research work. As a part of a local Egyptian research study, a sample of five governorates—Cairo, Giza, Sohag, Monofeya, and Qalyubia—was chosen systematically over the period from September 2020 to September 2023. Currently, environmental conditions, such as temperature, feed programs, and seasonality, are captured by ensuring that each dairy isolate is clearly defined in terms of temporal and geographical provenance. Notably, heat shock proteins were also found to be highly correlated with isolates collected in relatively warm locations, providing empirical data on adaptive responses to thermal stresses. The importance of contextual detail is therefore emphasized by Rajamanickam et al. (2025). Genomic stability and possible horizontal gene transmission can be detected via longitudinal analysis, which has a temporal resolution period between 6 months and 3 years. If a bacteriocin gene cluster is unique to late-season isolates, it can indicate the conditions of storage or handling (Reuben & Torres, 2024). Full genomic retention and necessity throughout successive months would imply consistency of essential functional capacities, thus adding certainty to strain suitability. Overall, through careful selection of the six representative isolates and the incorporation of high-resolution spatial and temporal field data, the resulting genomic dataset is informative and can be analyzed rigorously in subsequent analyses.

Geographic metadata represent a crucial source for explaining microbial biogeography (Ezzat et al., 2025). The dairy history of Egypt comprises, in particular, traditional buffalo milk fermentation in Upper Egypt, large-scale cow milk production concentrated on the Nile Delta. This geographic differentiation in the population has different impacts on gene flow and microbial community assembly (Ezzat et al., 2025). Fixing isolated genomes to exact geographical locations allows researchers to chart the presence of advantageous combinations of genes, such as those that encode antimicrobial peptides (bacteriocins) and CRISPR/Cas defense systems. This geographical model informs follow-up sampling plans and can help locate regional locations of microbial variety where probiotics can be found (Fentie, et al., 2025). The annotated data will be further used as the basis of training and validation of AI-based genomic annotation pipelines (Bhattacharya, et al., 2025). The assemblies that are of high quality are used as inputs in machine-learning models such as antiSMASH and BAGEL4. These tools require the use of well-annotated reference genomes to identify biosynthesis-related gene clusters and inform their functionality. Poorly labeled datasets may result in false-negative or false-positive predictions; therefore, optimal model performance and maximized discovery opportunities by dataset pairing, typically extensive breadth along with high depth, minimize the possibility of misclassification. Having an optimal balance of positive and negative examples also benefits machine-learning models. The majority of the sixty isolates do not possess important probiotic markers. A comparison of these isolates with the six selected probiotic-active strains will enable the model to more finely define its decision boundaries, thereby increasing its prediction accuracy for untested isolates.

For example, supervised classifiers can be trained to identify motifs in sequences related to the production of nisin-like bacteriocins. The existence of both active and nonactive genomes in probiotics reduces the decision space and therefore the model precision. In addition to providing annotations, the dataset forms the foundation of predictive analytics on the basis of integrative modeling. Machine learning frameworks are able to predict strain behavior *in silico* by integrating genomic data with initial phenotypic measurements to predict the levels of metabolite production, stress tolerance, and host–protein interaction networks. Such *in silico* predictions simplify laboratory experiments by providing a priority list of the most promising candidates to be experimentally validated, thus saving time and resources.

The six-isolate genomic data published here exhibit several characteristics that enhance reproducibility and transparency and are thus consistent with the direction of current regulatory trends and business practices. The database closely follows the genomic requirements as stated by the Codex Alimentarius and the European Food Safety Authority. Therefore, the runtime in the systems would enhance the evidentiary framework in probiotic safety claims, as these organizations are demanding more genomic data to support any assertions on such scientific safety claims (Jost et al., 2025). The dataset achieves this goal by permitting open access to assemblies and annotations, as well as providing comprehensive metadata, which support independent validation and strengthen the confidence of stakeholders, namely, manufacturers, regulators, and consumers. Overall, the six-isolate genomic data, supplemented with strong spatial and temporal metadata, constitute an all-purpose resource. It connects empirically obtained phenotypic data with the highest-quality AI-based annotation, speeding the patient selection process and setting the stage for safe, effective, and locally personalized functional foods in the Egyptian precision nutrition context.

5. RESEARCH OBJECTIVES: DEEP CRITICAL ANALYSIS AND COMPARATIVE INSIGHTS

The aim of this study was to provide a transformative model of strain selection involving Egyptian Lactobacillus as a next-generation functional food ingredient. Outlined by conventional and AI-based approaches, the project highlights three interdependent goals, each of which is exposed to analytical consideration, contrasting analyses, and the expected results.

5.1. Setting and benchmarking of genomic safety requirements

Objective: To develop a stringent, multilayer standard of safety that can balance regulatory requirements with practical screening throughput.

A- Comparative Analysis:

Phenotypic resistance testing (agar diffusion, MIC assays) tends to lack the detection of low-copy or silent resistance genes and necessitates a high level of laboratory time (Yamin et al., 2023).

1. *In silico screening* (ResFinder, PlasmidFinder) is able to find cryptic or otherwise unexpressed loci but can produce false positives in the presence of homologous nonfunctional sequences (Ruzickova et al., 2025).
2. *Hybrid Approach*: This research proposes the idea of using both rapid genotypic scans and limiting phenotypic confirmation to borderline cases. This mitigates unnecessary alarms but also ensures that high-risk strains do not progress.

B. The critical conditions:

1. Antibiotic resistance genes (ARGs): Lack of mobile ARGs. The genomic search limit was 95% or greater sequence identity, reducing false positive hits.
2. Virulence factors: Zero tolerance of known toxin or adhesion genes through VFDB screening (Dong et al., 2024).
3. Plasmid-linked Traits: Plasmids are hotspots in regard to horizontal gene transfer. The project stipulates that no ARGs should be present on any plasmid contigs or virulence genes.

C. Delivery plan:

Within 3 years, a reproducible safety score (0--100) that correlates strongly with phenotypic assays (target R^2 2 = 0.85 or greater) is sufficient to allow an 80% reduction in wet-lab confirmations.

One of the main goals of the current study is to chart and prioritize functional gene clusters through a complete survey of bacterial strains with the goal of finding isolates that exhibit better antimicrobial activity with a well-developed adaptive immune mechanism.

To identify bacteriocin gene clusters, this experiment combines two complementary computational tools: antiSMASH and BAGEL4.

- *AntiSMASH*, which is perfect at identifying large peptide synthesis modules (NRPS and PKS), tends to have an underreporting bias on small class II bacteriocins (Blin et al., 2025). However,

- *BAGEL4*, which is specifically designed for class II peptides and lanthipeptides, can be highly effective in this category but can fail to detect hybrid motifs (Sugrue et al., 2024).

The two systems are also executed in parallel, and the results are filtered against a machine-learning approach developed in-house via experimentally verified clusters (F1 score 0.9 and above).

Both methods of detection and classification of the CRISPR/Cas system adhere to the same general methodology in terms of the information used, with access to two such publicly available data sources: CRISPRdex and CasFinder. After assigning typologies, the diversity of spacers within each CRISPR array is quantified, and spacers are aligned with extensive phage databases to gauge in situ resistance to lytic bacteriophages. Strains containing at least 30 different spacers are given higher priorities to enable further analysis to achieve stability) Xu & Li 2020).

After the sequences of interest are established, their functional relevance is assessed. Individual bacteriocin clusters are initially ranked in terms of spectrum (narrow vs. broad), predicted stability of the putative peptide, and novelty, with < 70% sequence identity to defined entries, i.e., using external databases such as the MIBiG.

The CRISPR/Cas systems are scored in the same way. Comparative studies have shown that strains that are bacteriocin rich and CRISPR spacer poor are very efficient killers but are potentially susceptible to disruption by phages (Karpenko et al., 2024). As a result, a balanced profile, or rather moderate bacteriocin abundance with significant CRISPR defense capacity, is sought after. Overall, it is reasonable to expect a final list of two to three strains with both broad-spectrum antimicrobial activity and stable CRISPR defenses, thus satisfying the requirements of optimized industrial usefulness.

5.2 Building and validating an AI-driven candidate selection framework

The development and testing of an AI-based algorithm to select candidates when choosing bacteriophages is a key goal of the current investigation. This framework should be transparent and flexible, where the safety and functional performance indicators are incorporated to render an executable ranking. To achieve this goal, we propose several major components.

5.2.1. MCDA: weighted multicriteria decision analysis:

The traditional MCDA methods preset the weight on each individual criterion, a practice that may hide the interdependencies between the screening criteria. Conversely, we implement a feedback process to dynamically adjust the weights of criteria on the basis of preliminary results from in vitro assays, i.e., early results from banana and lemon in vitro assays, such as the diameter of the antimycin growth inhibition zone and modulation cytokine results, are used to estimate the predictive power of genomic characteristics (Băcescu Ene et al., 2025).

5.2.3. Creation of machine-learning classifiers:

Model selection. Random forests are natural out-of-the-box interpretable models and allow feature-importance analysis; gradient-boosted trees (XGBoost) often produce slightly better accuracy. These models are compared via nested cross-validation, and the optimality of these models is measured via precision–recall curves and F1-score optimization to find the optimal classification threshold (target >0.92), which can be defined as a high-potential threshold (Cao et al., 2025).

5.2.4. User interface and decision dashboard:

Available pipelines produce annotated tables of candidate genomes. We suggest the use of a web-based dashboard for visualization:

- Scatter plots between the safety score and functionality score;
- Cluster maps superimposed on genome schematics;
- Spacer–phage match heatmaps.
- Such data representations result in better prepared researchers and regulatory reviewers having the ability to conduct exploratory what-if analyses (e.g., how will tightened ARG thresholds work).

5.2.5. Iterative active-learning loop:

As laboratory validations are carried out through coculture inhibition assays and simulated gastric fluid survival tests, the resulting data are fed back into the model, thus fine-tuning the predictive parameters. It is anticipated that this feedback loop can minimize the total volume of needed wet-lab tests by approximately 50% in correlation with standstill model pipelines.

5.2.6. Expected outcome:

A user-friendly and powerful platform that:

- Find new genomes of bacteriophages;
- Combine AI and MCDA;
- provides a ranked list of candidates, with 90% concordance or higher between *in silico* predictions and bench validations.

This research model goes beyond incremental improvement by performing systematic comparative analysis to provide the following contributions:

1. **Major Productivity Increases**: The possibility of time and cost savings up to 70% in comparison with traditional workflows.
2. **High Predictive Accuracy**: A composite score based on a safety function is highly correlated ($R^2 > 0.85$) with the empirical variables.
3. **Regulatory Compatibility**: Automated reporting templates are fully compatible with the Codex, EFSA, and WHO provisions and thus accelerate the process of approvals.
4. **Local adaptation**: Fine-tuning the model using Egyptian isolates ensures biological relevance to the target population and food matrix.

The results suggest that the framework can be readily used to characterize other regional microbiomes, thus providing a blueprint for how to unequivocally conduct rapid, reliable probiotic discovery and functional food development at any location around the globe.

6. PROPOSED METHODOLOGICAL FRAMEWORK

6.1 Sample collection procedures and selection criteria

The study described here adopted a rigorous methodological framework in terms of procurement, storage, and analysis of samples of microbial genomes, with the explicit aim of facilitating annotation and discovery efforts on the basis of AI. All the phases of the protocol followed were designed such that they guaranteed the preservation of the fundamental scientific validity of the material gathered and the representativeness of the final genomic dataset generated.

6.1.2 Source identification

To increase genetic diversity as much as possible, microbial isolates were retrieved from a wide variety of raw milk samples collected from different Egyptian species and governorates. These sources included terrestrial and aquatic environments, natural water sources, and gastrointestinal tracts.

6.1.3 Initial Selection Requirements

Strains were selected preferentially on the basis of observed phenotypic activity or described biochemical abilities, specifically in relation to the production of natural antibiotics and secondary metabolite production.

6.1.4 Conservation and Movement

All the samples were stored at 4°C in sterile vessels and transported under controlled conditions to reduce the extent of contamination and degradation.

6.1.5 Reactivation and Preprocessing

After transport, reactivation of the isolates with favorable growth media and incubation in standard laboratory conditions was performed to reach high biomass levels, and high-yield and high-purity DNA was extracted.

6.1.6 Selection Guidelines

The criteria used to select the contents were as follows:

1. Novel genomic characteristics and poorly represented groups in the available databases.
2. Known or hypothetical metabolic capacity.
3. Statement of interest in resistance or biosynthetic innovation.

Survey of artificial intelligence software (RASTtk, BAGEL4, etc.)

A variety of AI-based sophisticated tools that are optimized either to perform annotations or make predictions have been used to inform biologically relevant conclusions on the basis of the information provided in the genomic data:

1. RASTtk (Rapid Annotations using Subsystems Technology Toolkit):

AI-boosted subsystem-based models are used to provide high-resolution functional annotations of genomic sequences.

Annotated databases and machine learning models are combined to discover gene functions and metabolic pathways.

2. BAGEL4:

1. Expert in the detection of bacteriocin biosynthesis-related gene clusters through AI-based classification models.
2. This method identifies uniquely represented motifs and structural features with great specificity in predicting functional assignments.
3. AntiSMASH (antibiotics & secondary metabolite analysis shell) is an AI-enabled application developed to mine biosynthesis-related gene cluster sequences that systematically define the production of secondary metabolites.
4. DeepARG utilizes deep learning models to identify antimicrobial resistance genes far more microsaccades and is extremely accurate, even with fragmented or incomplete genomic data.

The combined results of these systems provide a multidimensional set of annotated genes, allowing this set to be cross-validated and the experimental candidates to then be prioritized (Pei et al., 2024).

6.2 Data quality standards and biosafety measures

To maintain the strength, reproducibility, and scientific soundness of the study results, a stringent data-quality scheme and biosafety procedures were designed. In particular, the research followed the following standards:

1. Assessment of DNA Quality: Assessment of DNA quality in terms of DNA concentration, purity, and integrity was performed via spectrophotometric (NanoDrop) and fluorometric (Qubit) assays (Versmessen et al., 2024).
2. Contamination screening: FastQC (Zhang & Kang, 2021). and ContEst16 (Astashyn, et al., 2024) were utilized to screen and reduce external or internal contamination in the raw reads of sequences.
3. Data Preprocessing: Only high-confidence reads were trimmed for data quality, and adapters were removed with Trimmomatic (He et al., 2020).
4. Replication and Reliability: The analyses were repeated three times per sample to obtain data stability and reduce interference caused by batch effects.

6.3. Biosafety and ethical compliance:

The samples were handled in facilities compliant with biosafety standards and under sterile conditions.

Biohazardous materials were also stored and discarded according to national and institutional environmental standards.

All digital information was secured with encryption policies and controlled repositories.

7. METHODOLOGICAL FRAMEWORK

In this work, a holistic AI-supported workflow for characterizing and possibly valorizing probiotics in raw dairy samples is discussed. The strategy combines onsite sampling procedures with sophisticated computational annotation routines and pairs them with effective data-quality control procedures, hence constituting an appropriate model of precision microbiome analyses.

Sixty samples of raw milk were collected from five governorates of Egypt, including Cairo, Giza, Sohag, Monofeya, and Qalyubia, and specific samples from the cottage industry and industrial dairy farms were selected. This large-scale sampling program, which was conducted from September 2020 to September 2023, aimed to collect a diverse range of microbial populations.

Taken together, the findings described here indeed illustrate the possibility of a robust translational pipeline that could transform raw milk into reputable probiotic candidates, further contributing to the area of precision microbiome engineering. The deliberate selection of a range of climatic and cultural regions in Egypt guaranteed that the full range of indigenous Lactobacillus types was isolated, including heat-tolerant desert isolates, as well as strains adapted to feed stocks in the Nile-Delta. The initial phenotypic screening, aimed at acid and bile tolerance and activity against Listeria and *E. coli* antagonists, provided fast selection of nonviable or poorly active isolates. Looking backward, genome-wide profiling of a small cohort of phenotypic negatives might have defined silent but powerful genetic traits; thus, there is an infrequent need for follow-up screenings to strengthen selection boundaries (see Table 1).

After the isolates passed these initial filters, raw MiSeq sequencing was performed with high coverage on the Illumina MiSeq platform, and the read results were evaluated via FastQC to verify the quality of the reads with >Q30 quality for 90% of the bases (Ward et al., 2020). PATRIC screened assemblies: In the PATRIC environment (Wattam et al., 2017), SPAdes assemblies were carefully selected by excluding contigs less than 500 bp in length or with read depths less than 10x and only assemblies with N50 values greater than 50 kb passed (Bankevich et al., 2012).

Our adherence to data integrity, with emphasis on discovering new candidate genes, is informed by the understanding of the need to establish the middle ground between data rigor and outreach to inclusivity. To achieve this balance, we used stringent interreference RASTtk and Prokka annotations via a manual concordance procedure that highlighted possible mismatches to experienced reviewers (Brettin et al., 2015; Seemann et al., 2014). Since with automated pipelines, the local reference genomes did not appear in the sequence-based reference databases, manual curation of computational results was essential in cases of erroneous assignment or overlooking of lineage-specific functions (Table 1).

The current study, formulated in the environment of a smart-research project, emphasized implementing AI-sustained analytical instruments capable of listing both base and specialized genomic properties of the Egyptian Lactobacillus community. RASTtk provided a powerful vehicle used to perform systematic subsystem annotation; however, the genericity of its database came at the cost of offering a poor performance of strain-specific genes in performing strain specificity and thus required a custom reference library that could be made in the future.

The ability of BAGEL4, with a ruleset enabling bacteriocin operon identification, presented significant value for antibacterial cluster recognition. Nevertheless, the small repertoire of rules can miss emergent families of RiPPs; thus, incorporating a machine-learning classifier trained on known Egyptian bacteriocins would be advisable. AntiSMASH 6.0 had good performance in defining large NRPS and PKS regions, but some small RiPPs were below the detection threshold.

These ignored aspects are proposed to be followed by the use of the small cluster mode of antiSMASH alongside searched peptide-motif sequences. A combination of *in silico* screening tools such as ResFinder and PlasmidFinder as well as CRISPRCasFinder revealed mobile antibiotic resistance genes and extrachromosomal elements in the study population and phage defense arrays in the study population, respectively.

The safety factors were strictly observed. To enforce such a zero-tolerance policy toward mobile resistance and virulence determinants, automated scripts cross-validated ResFinder and VFDB results and baked in conditional phenotypic testing to resolve ambiguous findings—a pragmatic compromise that ensures the safe, prioritized investigation of novel candidates without creating opportunities to inadvertently introduce unacceptable risks. This genotypic/phenotypic hybrid industry defines methodological innovation, combining the ludicrous computational swiftness of the genotypic approach with the disciplinary rigor of the phenotypic approach.

Reproducibility and transparency formed part of the foundation of this pipeline. Analytic operations will produce standardized reports according to the Codex Alimentarius and EFSA to speed up regulatory control. To enable geospatial and temporal analyses, metadata were used to comprehensively record the provenance of the samples, the date they were collected, and the feed practices. These carefully annotated data can be used in subsequent iterations of active learning to further optimize AI models once new datasets of the field are obtained.

Over the course of the development of this framework, innovation can best be achieved not by technological breakthroughs but through the coordinated and complementary actions of field-specific expertise, algorithmic expertise, and thorough quality assurance. We developed a precision nutrigenomics paradigm by introducing region-specific datasets into machine-learning training cohorts, by recursively comparing *in silico* models with targeted wet-lab validation, and by establishing adaptive scoring infrastructures capable of responding to evolutions in regulatory factors. This approach is likely to accelerate the process of discovering antiviral and antimicrobial probiotic strains, increase the specificity of functional foods to local dietary habits, and strengthen the infrastructure needed to make microbe-driven health interventions more sustainable. The second part of this project will involve the application of this methodology to a curated genomic dataset, which will be scoped in terms of its structure, diversity, and state of preparation to be judged in terms of functionality. Through this, the discourse will move beyond mere pipeline design to imminent actionable discovery.

An overview of the workflow of the process of sample collection and data integration followed by machine learning analysis is outlined in the table below. The modules are defined by their objectives, the related tools, and the methodological procedures used to safely and successfully isolate lactobacilli from Egyptian dairy.

Table 1. Summary of the Innovative Genomic Annotation Workflow and Quality Criteria

Module	Purpose	Tools/Criteria
Sample Collection	To obtain biological material, 60 dairy samples were collected across five Egyptian governorates, with consideration for species. This broad coverage is essential for capturing the full range of *Lactobacillus* species, as they are locally adapted to specific climatic conditions and feedstocks. The samples underwent a primary enrichment, cultivation and biochemical identification process using Bergey's manual of determinative bacteriology for classic phenotypic criteria (David and John, 1994; De Man et al., 1960). Hemolytic activity and Antimicrobial activity, and tolerance evaluations for temperature, salt, acid, and bile (Angmo et al., 2016; Sitdhipol, et al., 2025) to assess the ability of probiotic to withstand these conditions, typically by measuring changes in viability, which resulted in six viable isolates. **Innovative Insight**: The strategy acknowledges one of its significant shortfalls: extremities associated with phenotypic screening may choose not to detect strains that possess so-called silent genes that remain dormant in a laboratory setting. **The next strategy** is consequently a genomic profiling of phenotypic negatives in order to bring out the hidden genetic potential.	Isolation, identification, antimicrobial testing, temperature, salt, acid and bile tolerance tests
Genome Assembly QC	The raw sequencing data is subjected to strict quality control to ensure reliability. FastQC is used to question reads, with at least 90 percent of bases yielding a Q-value of 30 or more (Andrews, 2010). Genomes are subsequently built in SPAdes, and only those builds with N50 >= 50 kb and a coverage of replicates >= 10x are kept (Bankevich et al., 2012). This filtering method marks the determination of high-quality and continuous sequences that are critical to accurate annotation. To enhance the resolution of the plasmids, which are known as vectors of gene transmission, alternative assemblers including Unicycler, are also tested (Wick et al., 2017).	FastQC; SPAdes; N50 >= 50 kb; coverage >= 10x; Unicycler
Core Annotation	After a genome of high quality is obtained, an integrative method is used to carry out the task of annotation. The RASTtk is used to annotate housekeeping and metabolic genes comprehensively in a subsystem-by-subsystem manner (Brettin et al., 2015). The caveat is that a follow-up cross-validation against Prokka (Seemann, 2014) is key to determining database constraints and possible new gene functions unique to Egyptian isolates.	RASTtk; Prokka

continued on following page

Table 1. Continued

Module	Purpose	Tools/Criteria
Function Cluster Detection	A collection of specimen tools is used to identify functional characteristics: BAGEL4 is applied to identify bacteriocin operons because they are a cluster of genes that impart a broad spectrum of antimicrobial activity (van Heel et al., 2018). antiSMASH 6.0 is also used to identify big secondary metabolism gene clusters including nonribosomal peptide synthase (NRP) and polyketide synthase (PKS) (Blin et al., 2021). CRISPRCasFinder is a database that detects CRISPR–Cas systems that can confer protective immunity to phages (Couvin et al., 2018). This multitool approach can ensure sufficient coverage of a high number of functional components, as one tool will simply exclude specific types of gene families.	BAGEL4; antiSMASH 6.0; CRISPRCasFinder
Safety Screening	The last and most important step is a strict zero-tolerance policy concerning safety risks. Each strain is vetted using *in silico* screening tools: ResFinder (Bortolaia et al., 2020), PlasmidFinder (Carattoli et al., 2014), and VFDB (Chen et al., 2005) for antibiotic resistance genes (ABRs), plasmid-linked traits (PLT; plasmids), and virulence factors (VF), respectively. In uncertain cases the genotypic observations are backed up by conditional phenotypic lab assays. This combination method combines the speed of computational analysis and the strictness of the old days of laboratory experimentation, leaving only the safest of applicants to remain as potential development targets.	ResFinder; PlasmidFinder; VFDB

7.1 Concluding points of view:

The suggested methodological framework is less of a mere checklist-style mechanism and more of an evolving ecosystem where classical microbiology is deliberately recontextualized via the combination of AI, stringent quality controls, and adaptive feedback loops. The development of a new type of summary table aims to make the hidden processes visible and decision junctions explicit. With ongoing data analysis and functional validation, the above foundation makes it possible to nonmutagenically select and characterize Lactobacillus strains that are not only genetically secure but also biofunctionally indomitable. In this way, this platform shapes route planning for the development of culturally sculpted, next-generation functional food.

8. CHAPTER CONTRIBUTION TO THE BOOK'S VISION

The given chapter is a conceptual keystone of the monograph Advancing Personalized Medicine with the AI-Driven Microbiome and Nutrigenomics. It functions concurrently as both a theory and a case study example, where artificial intelligence and microbial genomics might work in unison to resolve eminent public health challenges. It addresses the concepts of AI-guided genomic profiling, the development of functional foods, and the rational combination of local microbial resources. All these aspects together are representative of an interdisciplinary mandate of the volume: to bring computational creativity together with biological knowledge and to bring the fruits of the laboratory to a world where individuals can enjoy personalized nutrition and where the general community can experience sustained health. By doing so, this chapter puts in place the conditions required when translating findings on emergent scientific discoveries to concrete benefits to individuals and communities.

9. APPLICABILITY TO BOOK GOALS

The volume aims to demonstrate how microbiome research and nutrition genomics can be changed through AI to become personalized tools of healthcare and how one can create a methodological and conceptual framework for profiling probiotics in the microbiome, particularly local reality in Egypt, and how such probiotics can be used as targeted nutritional agents; this chapter will serve this purpose to approach a personalized methodological and conceptual model of probiotic profiling in the microbiome guided by the local reality of Egypt and how it can be used as a targeted nutritional agent. By doing so, there are three mainly overlapping, although distinct, ways through which the study operationalizes the themes of the book:

9.1.Microbial Diversity to Personalized Nutrition:

This study connects microbial ecosystems that are yet to be used (e.g., dairy-derived Lactobacillus in Egypt) and individuals with customized functional foods to fit within the themes to emphasize the utilization of local microbiomes and aid in specific health issues, including type 2 diabetes, inflammation, and resistance to infections, through customized interventions.

9.2. Data collection to actionable insights:

The interpretation shows that genomic data converted into actionable insights upon integration with AI-based predictive methods such as RASTtk, BAGEL4, antiSMASH, and ResFinder could result in structured and safety-compliant insights, which are vital in regulatory harmonization and product development, another main practical goal of personalized nutrition, as understood in this book.

9.3. Framework for Field Application:

This chapter provides a practical pipeline concept of research starting with sample collection and finishing with a scoring system of limited candidates, which enables readers to find a way of taking research to a scalable application. It reflects the message of the book to shift out of the theory and to start practicing what has been called, paying particular attention to cultural, economic, and nutritional differences.

10. POSITIONING WITHIN THE RESEARCH NARRATIVE

The current chapter is a methodological pillar of a two-phased work that aims to use artificial intelligence to exploit Egyptian Lactobacillus species. The authors summarize the entire methodological framework with the help of a succinct narrative table after describing the processes of sample acquisition, high-throughput sequencing, annotation, and quality control. This architecture can also be used as an extension roadmap, and it is also replicable. The authors do not confine themselves to elaborating on the tools they use and the datasets they create in each stage of the procedure but also present the reason behind the design decisions and the theories that are projected as continuations to be made in the future stages further. This level of presentation enhances transparency and promotes flexibility, which may be helpful later for other researchers in similar microbiome or food innovation situations. The methodological pathway between the raw samples and the ranked, safety-cleared probiotic candidates is described systematically in how Table 1 is outlined. The modular design of the table allows researchers working in different geographic locations and concentrating on different genera to customize discrete steps without needing to substantially reengineer the entire pipeline.

This table will be mentioned repeatedly in Chapter 8, where the functional outputs of these methods are applied. Furthermore, it is provided as a practical guide to regulatory reviewers, food technologists, and other innovation-based teams who wish to mimic or clone the probiotic discovery pathway.

11. REFLECTION ON GROUP CONTRIBUTION AND RESEARCH DESIGN

As a part of a larger multidisciplinary study, a specific project was developed to prove the applicability of local microbial biodiversity in health and nutrition spheres after being connected to the current state-of-the-art tools of artificial intelligence. The team included microbiologists, data scientists and food technologists, and researchers of public health, allowing extensive exploration, which is incorporated into the chapter structure: The dairy samples were curated and maintained by microbiologists. Genome assembly and annotation were performed by bioinformaticians. The weighted scoring models were developed by specialists in AI.

The criteria for safety and efficacy are influenced by nutrition and public health specialists.

This cross-disciplinary approach enhances rigorous analysis since genomic discovery is connected to food safety, culture, and food preferences, as well as clinical outcomes. Genomic novelty is not considered a sufficient criterion by the team, and the safety, efficacy, and scalability thresholds of the strains must be quantifiable. In that regard, it is still important to consider tools such as CRISPRCasFinder, antiSMASH, and BAGEL4 as useful within the specified boundaries.

The use of annotation platforms with AI, such as RASTtk, BAGEL4, and antiSMASH, has significantly increased the speed and precision of microbial genomics research. However, we find that these tools are constrained by structural and epistemological as well as regulatory disincentives that postpone their more robust materialization, especially in unexplored microbial environments, including those present in high numbers in Egypt and the North African environment as a whole. A careful analysis of these limitations, which is based on empirical findings, illustrates four areas of concern that must be improved: (1) systemic bias of AI approaches toward Western taxonomies, (2) the lack of consistency between computer and experimental confirmation, (3) insufficient regulatory adaptation of permit frameworks, and (4) extreme infrastructural differences among North African laboratories.

The most notable challenge is the proven inclination toward the Western-centrism inherent to modern AI-trained annotation systems. Most of the genomes used to support such models are North American, European, and East Asian. Once these models are used on Egyptian Lactobacillus isolates, the lack of local microbial representation often triggers misannotated or undetected functional gene clusters. These consequences undermine strain characterization, as well as the potential for Egypt-specific probiotic phenotypes inhabiting dairy microbiomes to be discovered. A successful remedial plan will require the building of a North African Lactobacillus genomics reference library that reflects diverse divergent ecologies (environment and fermentation) and the subsequent training of existing annotation models on wider regional diversity. To improve these structural shortcomings, it is imperative to develop cooperative data-sharing systems between North African academic and government health research organizations.

In addition to taxonomic bias, a second important and equally severe barrier is the current and ongoing discrepancy between computational gene-prediction results and the resulting biological expression. Although computational inferences cannot be sacrificed in situations requiring quick evaluations of data, the translational bridge between predictive and experimental evidence is still significant. Rounding this gap will require vigorous support of computational predictions with experimental validation of physiological assessments.

Third, regulatory frameworks remain consistently uncompromising in Egypt and North Africa at the expense of the full utilization of AI-based genomic tools. Procedural barriers that grant access on the basis of senior-investigator status and a massive institutional infrastructure are disproportionately more burdensome to younger researchers and taxed laboratories. Solutions such as recalibration of permit schemes to focus on substantial skill sets and innovative facility design to enable wider access to leading edge technologies are some of the remedies.

Finally, structural disparities hinder genomics studies covering the region. These changes, inadequate funding to laboratory infrastructure and equipment, power grids that are too unstable and intermittent internet connectivity limit genomic research, which makes use of enormous computational capacity. These disparities would be significantly reduced by fair allocation of funding in research grants and specific investment in regional networking facilities.

Improving these four dimensions of regional bias, computational-experimental validation mismatch, regulatory inflexibility, and infrastructural severity is crucial for the more comprehensive utilization of annotations to support microbial research in North Africa.

12. CRITICAL ANALYSIS AND FUTURE DIRECTIONS

One of the primary issues in genomics involves the fact that the number of gene groupings that are typically discovered by computers is not expressed in a laboratory. For example, a tool such as BAGEL4 can identify the presence of a bacteriocin gene, but the protein encoded by the gene is produced only under special conditions, such as a specific pH or in the presence of another molecule (Teber & Asakawa, 2024). The effect of merely possessing a gene is not sufficient to indicate the usefulness of a strain.

To correct this, we propose a multivalidation mechanism. We will profile the genome at the transcriptomic level (RNA-seq) after computer analysis to determine whether these genes are actually activated. This helps us save time on the silent genes. Afterward, we will conduct dual confirmation of the presence of the strain via metabolomic profiling (LC–MS) and proteomic analysis, which will demonstrate that the strain is indeed functional.

Maintaining a track of the changing regulations adopted by organizations such as the Codex Alimentarius and the EFSA is another critical concern. Old software where scoring systems are involved can be easily obsolete (Bottini et al., 2025). The way out of this mess is to create a new generation of AI tools that are modular and can be upgraded discreetly. This will enable the system to alter its scoring regulations on the fly to comply with new guidelines (Yin et al., 2025). Transparency and delays are also facilitated by it (Adejumo & Ogburie, 2025).

Finally, the most suitable solution is to create AI models that are regional and, at the same time, fit international standards. This implies that we will invest in governance and alliances, rather than algorithms, to ensure that local knowledge and data are employed in an equitable way. This makes AI-enabled discovery of probiotics a public health tool rather than a scientific novelty.

The proposed study also supports modern thought processes in nutritional science, as it employs local variation in microbial profiles with artificial intelligence in communing to design interventions that enhance health at the individual consumer and communal levels. This examination was initiated by using 60 dairy samples from various geographical areas in Egypt. At the end of the screening, *6 promising strains* were selected, and high-resolution sequencing was carried out with FastQC and SPAdes. The genomic data were subsequently analyzed via a set of AI-based pipelines, such as RASTtk, a functional composition pipeline; BAGEL4 and antiSMASH, secondary metabolite pathways; and ResFinder and CRISPRCasFinder, safety and defense systems. Despite this speed of analysis, some weaknesses are worth highlighting, which include the fact that these tools can have some subjectivity to Western reference data and the occurrence of *in silico*-predicted results being different from the observed empirical results. As an ongoing measure against these limitations, the research team has implemented a continuous process of retraining their models against a North African-specific reference genome dataset.

The best criterion and uncompromising prerequisite of all candidate strains is safety. The study utilizes strict guidelines that involve a zero-tolerance representation against mobile antibiotic resistance and virulence determinants, as determined in the associations VFDB and PlasmidFinder, respectively. This *in silico* evaluation is validated with the aid of matching laboratory evaluations, resulting in a composite approach that combines the computerizing economy with classical empiricism.

The overall goal of the study is to introduce a combined framework that integrates safety indices, profiling functions, and consideration of context. This type of framework would guide strain selection and product design where the end goal is to encourage regional probiotic innovation hubs embedded in existing dairy cooperatives. By transferring sophisticated genomic expertise into day-to-day dairy operations, these hubs would make precision solutions in nutrition accessible to the masses.

The proposed genomic hub networks aim to integrate highly skilled genomic expertise into mainstream nutrition practices with the broader population. The idea is to democratize nutrigenomic technology by contextualizing genomic knowledge in daily nutrition. To establish future directions in microbiome science and precision nutrition, the following six strategic initiatives have been formulated, as shown in Fig. 1:

1. Graph-theory-based predictive modeling of functionalomic and gene cluster interactions and emergent behavior induced by gene clusters.
2. Explainable AI frameworks: frameworks of open-source explainable AI that can define the connection between computational predictions and explanations associated with specific sequence characteristics.
3. Incessant learning channels of study and localized genomic redispositories are needed to visualize fashions and localization.
4. Standardized validation pipelines (transcriptomics & metabolomics & targeted assays) such that the identification of computational hits at the smaller scale is validated prior to scale-up.
5. They are fair data partnerships, empowering local centers of power, allocating intellectual property ownership democratically, and becoming regional sequencing capacity.
6. Smart deployment environments that are able to incorporate genomic data, market intelligence, regulatory challenges, and consumer preferences to promote rapid iterative development.

13. TARGETED RECOMMENDATIONS AND IMPLEMENTATION PRIORITIES:

The proposed integrative framework, despite its ample scope, has some weaknesses. The most obvious of those is the risk of dataset bias, especially with genomic profiles being selected disproportionately on the basis of ecological or demographic niches (Jiang, et al., 2025). Since genomic data are dynamic by nature, the use of prediction drift can lead to a failure to adjust the computational model on a continuing basis, making AI-based outcomes steadily less consistent with new microbial ecosystems and changing regulatory environments (Yang et al., 2024). Moreover, the lack of a strong and locally controlled data-sharing policy exposes it to unequal access and weakened stakeholder commitment. Therefore, addressing these weaknesses is crucial to ensure the scientific validity and sustainability of the framework over the long-term period (Kelliher et al., 2025; Dorst et al., 2024).

The following corrective measures are recommended:

1. The development of enhanced collaboration among regional partners, with multi-institutional bivocational collaboration across a wide ecological zone to maximize the coverage of the diversity of genomes and optimize the generalizability of the lens of models (Ojewunmi & Fatumo 2025). ;
2. Continuously update the model to be implemented via an automated pipeline tailored to the regular intake of new microbial and environmental data to guarantee continued responsiveness to emergent microbial and regulatory settings (Raajaraam & Raman, 2024);
3. Build interoperable data repositories that combine open-science values with data protection, intellectual-property protection, and local ownership given precedence (Williamson & Prybutok (2024));
4. Real-time regulatory compliance can be built by integrating automated warnings into an analytical workflow to detect and act on alterations in both international and regional food safety regulations (Raajaraam & Raman, 2024).
5. A capacity-building strategy that includes training local scientists and technicians in microbiome analytics and AI applications in the region to ensure sustainability after the completion of initial project funding (Huo & Wang, 2024).

ACKNOWLEDGMENTS:

This work was supported by the National Research Centre (NRC), Dokki, Giza, Egypt, under Grant No. MP121202 (2021–2023). The authors gratefully acknowledge the NRC staff, field collaborators, and technical teams for their invaluable contributions, expertise, and dedicated assistance throughout the study.

REFERENCES

Abdelaziz, M. N. S., Zayda, M. G., Maung, A. T., El-Telbany, M., Mohammadi, T. N., Lwin, S. Z. C., Linn, K. Z., Wang, C., Yuan, L., Masuda, Y., Honjoh, K. I., & Miyamoto, T. (2024). Genetic Characterization, Antibiotic Resistance, and Virulence Genes Profiling of *Bacillus cereus* Strains from Various Foods in Japan. *Antibiotics (Basel, Switzerland)*, *13*(8), 774. DOI: 10.3390/antibiotics13080774 PMID: 39200074

Adejumo, A. A., & Ogburie, C. (2025). The role of AI in preventing financial fraud and enhancing compliance. *GSC Advanced Research and Reviews*, *22*(3), 269–282. DOI: 10.30574/gscarr.2025.22.3.0086

Agrawal, K., Goktas, P., Kumar, N., & Leung, M.-F. (2025). Artificial intelligence in personalized nutrition and food manufacturing: A comprehensive review of methods, applications, and future directions. *Frontiers in Nutrition*, *12*, 1636980. Advance online publication. DOI: 10.3389/fnut.2025.1636980 PMID: 40771216

Andrews, S. (2010). *FastQC: A quality control tool for high throughput sequence data*. http://www.bioinformatics.babraham.ac.uk/projects/fastqc/

Angmo, K., & Kumari, A. Savitri, and Bhalla, T.C. (2016). Probiotic characterization of lactic acid bacteria isolated from fermented foods and beverage of Ladakh. Lebensm. Wiss. Technol., 66, pp. 428-435. https://doi.org/DOI: 10.1016/j.lwt.2015.10.057

Astashyn, A., Tvedte, E. S., Sweeney, D., Sapojnikov, V., Bouk, N., Joukov, V., Mozes, E., Strope, P. K., Sylla, P. M., Wagner, L., Bidwell, S. L., Brown, L. C., Clark, K., Davis, E. W., Smith-White, B., Hlavina, W., Pruitt, K. D., Schneider, V. A., & Murphy, T. D. (2024). Rapid and sensitive detection of genome contamination at scale with FCS-GX. *Genome Biology*, *25*(1), 60. DOI: 10.1186/s13059-024-03198-7 PMID: 38409096

Băcescu Ene, G.-V., Stoia, M.-A., Cojocaru, C., & Todea, D. A. (2025). SMART Multi-Criteria Decision Analysis (MCDA)—One of the Keys to Future Pandemic Strategies. *Journal of Clinical Medicine*, *14*(6), 1943. DOI: 10.3390/jcm14061943 PMID: 40142750

Bankevich, A., Nurk, S., Antipov, D., Gurevich, A. A., Dvorkin, M., Kulikov, A. S., Lesin, V. M., Nikolenko, S. I., Pham, S., Prjibelski, A. D., Pyshkin, A. V., Sirotkin, A. V., Vyahhi, N., Tesler, G., Alekseyev, M. A., & Pevzner, P. A. (2012). SPAdes: a new genome assembly algorithm and its applications to single-cell sequencing. *Journal of computational biology: a journal of computational molecular cell biology, 19*(5), 455–477. https://doi.org/DOI: 10.1089/cmb.2012.0021

Bhattacharya, A., Joishy, T. K., & Khan, M. R. (2025). Exploration of the genomic potential of the probiotic bacterium *Lactiplantibacillus plantarum* JBC5 (LPJBC5). *Discovery Bacteria*, *2*, 13. DOI: 10.1007/s44351-025-00021-x

Blin, K., Shaw, S., Kloosterman, A. M., Charlop-Powers, Z., van Wezel, G. P., Medema, M. H., & Weber, T. (2021). antiSMASH 6.0: Improving cluster detection and comparison capabilities. *Nucleic Acids Research*, *49*(W1), W29–W35. DOI: 10.1093/nar/gkab335 PMID: 33978755

Blin, K., Shaw, S., Vader, L., Szenei, J., Reitz, Z. L., Augustijn, H. E., Cediel-Becerra, J. D. D., de Crécy-Lagard, V., Koetsier, R. A., Williams, S. E., Cruz-Morales, P., Wongwas, S., Segurado Luchsinger, A. E., Biermann, F., Korenskaia, A., Zdouc, M. M., Meijer, D., Terlouw, B. R., van der Hooft, J. J. J., & Weber, T. (2025). antiSMASH 8.0: Extended gene cluster detection capabilities and analyses of chemistry, enzymology, and regulation. *Nucleic Acids Research*, *53*(W1), W32–W38. DOI: 10.1093/nar/gkaf334 PMID: 40276974

Bortolaia, V., Kaas, R. S., Ruppe, E., Roberts, M. C., Schwarz, S., Cattoir, V., Philippon, A., Allesoe, R. L., Rebelo, A. R., Florensa, A. F., Fagelhauer, L., Chakraborty, T., Neumann, B., Werner, G., Bender, J. K., Stingl, K., Nguyen, M., Coppens, J., Xavier, B. B., & Aarestrup, F. M. (2020). ResFinder 4.0 for predictions of phenotypes from genotypes. *The Journal of Antimicrobial Chemotherapy*, *75*(12), 3491–3500. DOI: 10.1093/jac/dkaa345 PMID: 32780112

Bottini, M., Ryu, S.-J., Elmi Terander, A., Voglis, S., Maldaner, N., Bellut, D., Regli, L., Serra, C., & Staartjes, V. E. (2025). The ever-evolving regulatory landscape concerning development and clinical application of machine intelligence: Practical consequences for spine artificial intelligence research. *Neurospine*, *22*(1), 134–143. PMID: 40211523

Boyte, M. E., Akhtar, N., Koshy, B., & Roe, A. L. (2025). A Review of Probiotic Ingredient Safety Supporting Monograph Development Conducted by the United States Pharmacopeia (USP). *Journal of Dietary Supplements*, *22*(1), 123–161. DOI: 10.1080/19390211.2024.2314488 PMID: 38356247

Brettin, T., Davis, J., Disz, T., Edwards, R. A., Gerdes, S., Olsen, G. J., Olson, R., Overbeek, R., Parrello, B., Pusch, G. D., Shukla, M., Thomason, J. A., Stevens, R., Vonstein, V., Wattam, A. R., & Xia, F. (2015). RASTtk: A modular and extensible implementation of the RAST algorithm for building custom annotation pipelines and annotating batches of genomes. *Scientific Reports*, *5*, 8365. DOI: 10.1038/srep08365 PMID: 25666585

Cao, Y., Schartel, T. E., Houghton, D. C., Ross, J., & Infante, D. M. (2025). Evaluating machine learning algorithms for accuracy, stability, and among-predictors discriminability in modeling species-richness across ten datasets. *Ecological Informatics*, *90*, 103323. DOI: 10.1016/j.ecoinf.2025.103323

Carattoli, A., Zankari, E., García-Fernández, A., Voldby Larsen, M., Lund, O., Villa, L., Møller Aarestrup, F., & Hasman, H. (2014). *In silico* detection and typing of plasmids using PlasmidFinder and plasmid multilocus sequence typing. *Antimicrobial Agents and Chemotherapy*, *58*(7), 3895–3903. DOI: 10.1128/AAC.02412-14 PMID: 24777092

Chen, L., Yang, J., Yu, J., Yao, Z., Sun, L., Shen, Y., & Jin, Q. (2005). VFDB: A reference database for bacterial virulence factors. *Nucleic Acids Research*, *33*(suppl_1), D325–D328. DOI: 10.1093/nar/gki008 PMID: 15608208

Couvin, D., Bernheim, A., Toffano-Nioche, C., Touchon, M., Michalik, J., Néron, B., Rocha, E. P. C., Vergnaud, G., Gautheret, D., & Pourcel, C. (2018). CRISPRCasFinder, an update of CRISRFinder, includes a portable version, enhanced performance and integrates search for Cas proteins. *Nucleic Acids Research*, *46*(W1), W246–W251. DOI: 10.1093/nar/gky425 PMID: 29790974

David, H. B., & John, G. H. (1994). Bergey's manual of determinative bacteriology. 9th Ed., Baltimore, Williams and Wilkins Co., PP: 1-350.

De Man, J. C., Rogosa, M., & Elisabeth Sharpe, M. (1960). A medium for the cultivation of lactobacilli. *The Journal of Applied Bacteriology*, *23*(1), 130–135. DOI: 10.1111/j.1365-2672.1960.tb00188.x

Dong, W., Fan, X., Guo, Y., Wang, S., Jia, S., Lv, N., Yuan, T., Pan, Y., Xue, Y., Chen, X., Xiong, Q., Yang, R., Zhao, W., & Zhu, B. (2024). An expanded database and analytical toolkit for identifying bacterial virulence factors and their associations with chronic diseases. *Nature Communications*, *15*(1), 8084. DOI: 10.1038/s41467-024-51864-y PMID: 39278950

Dorst, M., Zeevenhooven, N., Wilding, R., Mende, D., Brandt, B. W., Zaura, E., Hoekstra, A., & Sheraton, V. M. (2024). FAIR compliant database development for human microbiome data samples. *Frontiers in Cellular and Infection Microbiology*, *14*, 1384809. DOI: 10.3389/fcimb.2024.1384809 PMID: 38774631

Ezzat, L., Peter, H., Bourquin, M., Busi, S. B., Michoud, G., Fodelianakis, S., Kohler, T. J., Lamy, T., Geers, A., Pramateftaki, P., Baier, F., Marasco, R., Daffonchio, D., Deluigi, N., Wilmes, P., Styllas, M., Schön, M., Tolosano, M., De Staercke, V., & Battin, T. J. (2025). Diversity and biogeography of the bacterial microbiome in glacier-fed streams. *Nature*, *637*(8046), 622–630. DOI: 10.1038/s41586-024-08313-z PMID: 39743584

Farag, H. F. M., Elrewany, E., & Abdel-Aziz, B. F.. (2023). Prevalence and predictors of undiagnosed type 2 diabetes and prediabetes among adult Egyptians: A community-based survey. *BMC Public Health*, *23*, 949. DOI: 10.1186/s12889-023-15819-0 PMID: 37231362

Fentie, E. G., Lim, K., Andargie, Y. E., Azizoglu, U., & Shin, J.-H. (2025). Globalizing the health-promoting potential of fermented foods: A culturomics pathway to probiotics. *Trends in Food Science & Technology*, *163*, 105119. DOI: 10.1016/j.tifs.2025.105119

Fernandez-Cantos, M. V., Garcia-Morena, D., Yi, Y., Liang, L., Gómez-Vázquez, E., & Kuipers, O. P. (2023). Bioinformatic mining for RiPP biosynthesis-related gene clusters in Bacteroidales reveals possible new subfamily architectures and novel natural products. *Frontiers in Microbiology*, *14*, 1219272. DOI: 10.3389/fmicb.2023.1219272 PMID: 37469430

Ghoreyshi, N., Heidari, R., & Farhadi, A.. (2025). Next-generation sequencing in cancer diagnosis and treatment: Clinical applications and future directions. *Discover. Oncology*, *16*, 578. Advance online publication. DOI: 10.1007/s12672-025-01816-9 PMID: 40253661

Grand View Research. (2024). Functional foods market size, share & trends analysis report byproduct, by application (cardiovascular health, digestive health, bone health, immunity), by region, and segment forecasts, 2024–2030.

He, B., Zhu, R., Yang, H., Lu, Q., Wang, W., Song, L., Sun, X., Zhang, G., Li, S., Yang, J., Tian, G., Bing, P., & Lang, J. (2020). Assessing the Impact of Data Preprocessing on Analyzing Next Generation Sequencing Data. *Frontiers in Bioengineering and Biotechnology*, *8*, 817. DOI: 10.3389/fbioe.2020.00817 PMID: 32850708

Huo, D., & Wang, X. (2024). A new era in healthcare: The integration of artificial intelligence and microbial. *Medicine in Novel Technology and Devices*, *23*, 100319. DOI: 10.1016/j.medntd.2024.100319

Ji, H. F., Li, M., Han, X., Fan, Y. T., Yang, J. J., Long, Y., Yu, J., & Ji, H. Y. (2025). *Lactobacilli*-Mediated Regulation of the Microbial-Immune Axis: A Review of Key Mechanisms, Influencing Factors, and Application Prospects. *Foods (Basel, Switzerland)*, *14*(10), 1763. DOI: 10.3390/foods14101763 PMID: 40428542

Jiang, Y., Aton, M., Zhu, Q., Zhang, H., Zhang, S., Wu, J., & Li, Y. (2025). Modeling microbiome-trait associations with taxonomy-adaptive neural networks. *Microbiome*, *13*, 87. DOI: 10.1186/s40168-025-02080-3 PMID: 40158141

Jost, S., Herzig, C., & Birringer, M. (2025). A Balancing Act—20 Years of Nutrition and Health Claims Regulation in Europe: A Historical Perspective and Reflection. *Foods*, *14*(9), 1651. DOI: 10.3390/foods14091651 PMID: 40361733

Karpenko, A., Mikhaylova, Y., Shelenkov, A., Tutelyan, A., & Akimkin, V. (2024). Data on the analysis of draft genome sequence of *Raoultella ornithinolytica* isolate carrying antimicrobial resistance genes, plasmid and CRISPR–Cas system. *Data in Brief*, *57*, 110973. DOI: 10.1016/j.dib.2024.110973 PMID: 39435365

Kelliher, J. M., Aljumaah, M., Bordenstein, S. R., Brister, J. R., Chain, P. S. G., Dundore-Arias, J. P., Emerson, J. B., Fernandes, V. M. C., Flores, R., Gonzalez, A., Hansen, Z. A., Hatcher, E. L., Jackson, S. A., Kellogg, C. A., Madupu, R., Miller, C. M. L., Mirzayi, C., Moustafa, A. M., Mungall, C., & Eloe-Fadrosh, E. A. (2025). Microbiome data management in action workshop: Atlanta, GA, USA, June 12–13, 2024. *Environmental Microbiome*, *20*, 40. DOI: 10.1186/s40793-025-00702-9 PMID: 40253432

Liu, Y., Zheng, H., & Wang, J.. (2025). Whole genome analysis and in vivo safety assessment of probiotic candidate *Lactobacillus acidophilus* L177. *BMC Microbiology*, *25*, 398. Advance online publication. DOI: 10.1186/s12866-025-04099-w PMID: 40604408

Ma, Z. F., Fu, C., & Lee, Y. Y. (2025). The Modulatory Role of Bioactive Compounds in Functional Foods on Inflammation and Metabolic Pathways in Chronic Diseases. *Foods*, *14*(5), 821. DOI: 10.3390/foods14050821 PMID: 40077524

Martiny, H.-M. (2022). *Analysis of global emergence and spread of antimicrobial resistance in 214k host and environmental samples*. Technical University of Denmark.

Mudondo, J., Happy, K., & Gang, R.. (2025). From nature to nutrition: Exploring the synergistic benefits of functional foods and herbal medicines for holistic health. *Applied Biological Chemistry*, *68*, 17. DOI: 10.1186/s13765-025-00985-apa

Neurath, M. F., Artis, D., & Becker, C. (2025). The intestinal barrier: A pivotal role in health, inflammation, and cancer. *The Lancet. Gastroenterology & Hepatology*, *10*(6), 573–592. DOI: 10.1016/S2468-1253(24)00390-X PMID: 40086468

Nutraceuticals World. (2024). Global consumer survey on functional food motivations.

Ojewunmi, O. O., & Fatumo, S. (2025). Driving global health equity and precision medicine through African genomic data. *Human Molecular Genetics*. Advance online publication. DOI: 10.1093/hmg/ddaf025 PMID: 40304701

Osivand, Z., Rahimi, E., Shakerian, A., & Khamesipour, F. (2025). Prevalence, antibiotic resistance, virulence and antimicrobial resistance gene profiles of Salmonella species recovered from retail beef and poultry processing environments. *BMC Microbiology*, *25*(1), 174. DOI: 10.1186/s12866-025-03871-2 PMID: 40155795

Patel, R., White, D., Kumar, A., & Uyaguari, M. (2025). Identification of *Acinetobacter radioresistens* in wastewater effluent from a First Nation reserve in Manitoba, Canada. *Microbiology Resource Announcements*, *00221-25*. Advance online publication. DOI: 10.1128/mra.00221-25 PMID: 40741752

Pei, Y., Shum, M. H., Liao, Y., Leung, V. W., Gong, Y. N., Smith, D. K., Yin, X., Guan, Y., Luo, R., Zhang, T., & Lam, T. T. (2024). ARGNet: Using deep neural networks for robust identification and classification of antibiotic resistance genes from sequences. *Microbiome*, *12*(1), 84. DOI: 10.1186/s40168-024-01805-0 PMID: 38725076

Raajaraam, L., & Raman, K. (2024). Modeling microbial communities: Perspective and challenges. *ACS Synthetic Biology*, *13*(8), 2260–2270. DOI: 10.1021/acssynbio.4c00116 PMID: 39148432

Rajamanickam, K., Visha, P., Elango, A., & Leela, V. (2025). Salivary heat shock protein 70 as a potential noninvasive biomarker of environmental thermal stress in dairy cattle. *International Journal of Biometeorology*, *69*(2), 449–467. DOI: 10.1007/s00484-024-02826-y PMID: 39556253

Reuben, R. C., & Torres, C. (2024). Bacteriocins: Potentials and prospects in health and agrifood systems. *Archives of Microbiology*, *206*(5), 233. DOI: 10.1007/s00203-024-03948-y PMID: 38662051

Reyed, R. M., & Prabhakar, P. K. (2025). Personalized Antiviral Therapy Tailoring Treatments to Genetic Environmental and Lifestyle Factors. In Aslam, M., Yingngam, B., Javed, T., & Riaz, S. (Eds.), *Global Perspectives on Antiviral Drug Development* (pp. 63–138). IGI Global Scientific Publishing., DOI: 10.4018/979-8-3693-9276-8.ch003

Reyed, R. M., & Prabhakar, P. K. (2026). Modulating Neuroinflammation and Autophagy in Psychiatric Disorders Through Mechanisms and Therapies: Therapeutic Strategies and Pathways. In Mokhtari, T., & Uludag, K. (Eds.), *Autophagy and Inflammation in Neuropsychological Disorders* (pp. 61–124). IGI Global Scientific Publishing., DOI: 10.4018/979-8-3693-5908-2.ch003

Ruzickova, M., Palkovicova, J., Papousek, I., Cummins, M. L., Djordjevic, S. P., & Dolejska, M. (2025). The presence of multiple variants of IncF plasmid alleles in a single genome sequence can hinder accurate replicon sequence typing using *in silico* pMLST tools. *mSystems, 10*(5), e0101024. DOI: 10.1128/msystems.01010-24 PMID: 40197103

Seemann, T. (2014). Prokka: Rapid prokaryotic genome annotation. *Bioinformatics (Oxford, England), 30*(14), 2068–2069. DOI: 10.1093/bioinformatics/btu153 PMID: 24642063

Sitdhipol, J., Niwasabutra, K., Chaiyawan, N., Nuankham, K., Thanagornyothin, T., Tanasupawat, S., Chanput, W. P., Phapugrangkul, P., Chaipanya, C., Phuengjayaem, S., Poothong, S., & Kingkaew, E. (2025). Evaluating the safety and efficacy of *Lacticaseibacillus paracasei* TISTR 2593 as a therapeutic probiotic for obesity prevention. *Frontiers in Microbiology, 16*, 1501395. DOI: 10.3389/fmicb.2025.1501395 PMID: 40241731

Sugrue, I., Ross, R. P., & Hill, C. (2024). Bacteriocin diversity, function, discovery and application as antimicrobials. *Nature Reviews. Microbiology, 22*(9), 556–571. DOI: 10.1038/s41579-024-01045-x PMID: 38730101

Teber, R., & Asakawa, S. (2024). *In Silico* Screening of Bacteriocin Gene Clusters within a Set of Marine *Bacillota* Genomes. *International Journal of Molecular Sciences, 25*(5), 2566. DOI: 10.3390/ijms25052566 PMID: 38473813

van Heel, A. J., de Jong, A., Song, C., Viel, J. H., Kok, J., & Kuipers, O. P. (2018). BAGEL4: A user-friendly web server to thoroughly mine RiPPs and bacteriocins. *Nucleic Acids Research, 46*(W1), W278–W281. DOI: 10.1093/nar/gky383 PMID: 29788290

Versmessen, N., Van Simaey, L., Negash, A. A., Vandekerckhove, M., Hulpiau, P., Vaneechoutte, M., & Cools, P. (2024). Comparison of DeNovix, NanoDrop and Qubit for DNA quantification and impurity detection of bacterial DNA extracts. *PLoS One*, *19*(6), e0305650. DOI: 10.1371/journal.pone.0305650 PMID: 38885212

Wang, Z., Zhu, G., & Li, S. (2025). Mapping knowledge landscapes and emerging trends in artificial intelligence for antimicrobial resistance: Bibliometric and visualization analysis. *Frontiers in medicine*, *12*, 1492709. DOI: 10.3389/fmed.2025.1492709 PMID: 39935800

Ward, C. M., To, T. H., & Pederson, S. M. (2020). ngsReports: A Bioconductor package for managing FastQC reports and other NGS related log files. *Bioinformatics (Oxford, England)*, *36*(8), 2587–2588. DOI: 10.1093/bioinformatics/btz937 PMID: 31841127

Wattam, A. R., Davis, J. J., Assaf, R., Boisvert, S., Brettin, T., Bun, C., Conrad, N., Dietrich, E. M., Disz, T., Gabbard, J. L., Gerdes, S., Henry, C. S., Kenyon, R. W., Machi, D., Mao, C., Nordberg, E. K., Olsen, G. J., Murphy-Olson, D. E., Olson, R., & Stevens, R. L. (2017). Improvements to PATRIC, the all-bacterial Bioinformatics Database and Analysis Resource Center. *Nucleic Acids Research*, *45*(D1), D535–D542. DOI: 10.1093/nar/gkw1017 PMID: 27899627

Wick, R. R., Judd, L. M., Gorrie, C. L., & Holt, K. E. (2017). Unicycler: Resolving bacterial genome assemblies from short and long sequencing reads. *PLoS Computational Biology*, *13*(6), e1005595. DOI: 10.1371/journal.pcbi.1005595 PMID: 28594827

Williamson, S. M., & Prybutok, V. (2024). Balancing Privacy and Progress: A Review of Privacy Challenges, Systemic Oversight, and Patient Perceptions in AI-Driven Healthcare. *Applied Sciences (Basel, Switzerland)*, *14*(2), 675. DOI: 10.3390/app14020675

Wishart, D. S., Han, S., Saha, S., Oler, E., Peters, H., Grant, J. R., Stothard, P., & Gautam, V. (2023). PHASTEST: Faster than PHASTER, better than PHAST. *Nucleic Acids Research*, *51*(W1), W443–W450. DOI: 10.1093/nar/gkad382 PMID: 37194694

Xu, Y., & Li, Z. (2020). CRISPR–Cas systems: Overview, innovations and applications in human disease research and gene therapy. *Computational and Structural Biotechnology Journal*, *18*, 2401–2415. DOI: 10.1016/j.csbj.2020.08.031 PMID: 33005303

Yamin, D., Usković, V., Wakil, A. M., Goni, M. D., Shamsuddin, S. H., Mustafa, F. H., Alfouzan, W. A., Alissa, M., Alshengeti, A., Almaghrabi, R. H., Fares, M. A. A., Garout, M., Al Kaabi, N. A., Alshehri, A. A., Ali, H. M., Rabaan, A. A., Aldubisi, F. A., Yean, C. Y., & Yusof, N. Y. (2023). Current and Future Technologies for the Detection of Antibiotic-Resistant Bacteria. *Diagnostics (Basel, Switzerland), 13*(20), 3246. PMID: 37892067

Yang, Y., Lin, M., Zhao, H., Peng, Y., Huang, F., & Lu, Z. (2024). A survey of recent methods for addressing AI fairness and bias in biomedicine. *Journal of Biomedical Informatics, 154*, 104646. DOI: 10.1016/j.jbi.2024.104646 PMID: 38677633

Yin, B., Tan, G., Muhammad, R., Liu, J., & Bi, J. (2025). AI-Powered Innovations in Food Safety from Farm to Fork. *Foods (Basel, Switzerland), 14*(11), 1973. DOI: 10.3390/foods14111973 PMID: 40509500

Zhang, F., & Kang, H. M. (2021). FASTQuick: Rapid and comprehensive quality assessment of raw sequence reads. *GigaScience, 10*(2), giab004. Advance online publication. DOI: 10.1093/gigascience/giab004 PMID: 33511994

Żurek, J., & Rudy, M. (2024). Impact of the COVID-19 Pandemic on Changes in Consumer Purchasing Behavior in the Food Market with a Focus on Meat and Meat Products—A Comprehensive Literature Review. *Foods, 13*(6), 933. DOI: 10.3390/foods13060933 PMID: 38540923

Chapter 8
Functional Profiling and Therapeutic Potential of Egyptian Probiotics:
Food Safety and Viral Defense Applications

Sherein Ismail Abd El-Moez
http://orcid.org/0000-0003-3620-8930
Veterinary Research Institute, Egypt

Mohamed Gomaa Seadawy
Main Chemical Warfare Laboratories, Egypt

Mostafa Fetoh El-Hosseny
http://orcid.org/0000-0002-0860-2052
Biodefense Center for Infectious and Emerging Diseases, Egypt

Doaa A. Ghareeb
Independent Researcher, Egypt

Osama Ibrahim El Batawy
http://orcid.org/0009-0007-6545-5435
Independent Researcher, Egypt

ABSTRACT

This chapter presents a practical follow-up to earlier AI-guided genomic investigations, examining the functional characterization and therapeutic benefits of Egyptian Lactobacillus strains for food safety and viral defense. Metabolomic and phenotypic assays were used to confirm bacteriocin production and viral inhibition

DOI: 10.4018/979-8-3373-3121-8.ch008

in vitro. The link between probiotic features—adhesion, immunomodulation—and host genetics has been explored to inform personalized nutrition. Key regulatory and ethical frameworks, including the Codex, EFSA, and WHO-Food Safety, are integrated. The chapter concludes with a roadmap—from laboratory validation to commercial products, preclinical trials, and cultural adaptation—promoting scalable, safe, and precision-oriented probiotic applications consistent with the thematic logic of the book.

THE APPEAL TO LOCAL MICROBES AND GLOBAL HEALTH BY HOOKING THE READER: A CREATIVE SOLUTION

What would happen if those invisible microorganisms living in traditional Egyptian dairy products or fermented grains answered some of the most difficult public health challenges facing us today—viral pandemics, antibiotic resistance, and chronic metabolic diseases? What would happen if the same organisms that had informed centuries of dietary practices were foundational to personalized medicine in the era of artificial intelligence (AI)? The answers to these questions are no longer a matter of hypothetical interest; rather, they represent an expanding body of evidence that microbial biodiversity, when translated and harnessed by means of sophisticated computational pipelines, can be converted into practical therapeutic and nutritional inventions (Bakkeren et al., 2025; Wang et al., 2024). One of the complicated challenges is that redox homeostasis is disrupted; redox reactions include reactive oxygen species (ROS) and antioxidants. Oxidative stress not only stimulates faster occurrence of cardiometabolic, neurological, and oncological conditions but also undermines immune fortitude and intestinal microbial equilibrium (Reyed & Prabhakar, 2025 a). This overlap reiterates the importance of combined interventions: antioxidant dietary supplements correct molecular homeostasis, and probiotic microorganisms reinforce the microbial ecosystems that control redox-sensitive processes. In combination, these complementary methods may offer a systems-based solution to poor health conditions that are inadequately treated by conventional interventions. The rise of AI has transformed the field of microbiome research in the last 10 years by increasing both its scope and detail. The potential to sequence complete genomes, characterize clusters of genes, and propose functional properties has put microbiology on the path toward predictive medicine. However, predictions without verification are not complete. The process by which genomic signals become measurable and repeatable is the defining challenge of the new era of translational microbiome research (Srivastava et al., 2025). That is, the power of AI-guided annotation should be balanced by the robustness of functional performance, clinical studies, and practical implementations to have any true impacts on human well-being. This is the turning point of this chapter. It is the applied form of the genomic discovery structure developed in Chapter 1, in which the capability of AI in predictive annotation was emphasized. Here, the emphasis is on practical laboratory validation, applied characterization, and translational routes. It answers the pivotal question that this book seeks to answer, which is how can AI-enabled microbiome science shift beyond the digital world into actionable results in the form of nutrigenomics, functional food-related applications, and precision medicine?

INTRODUCING THE CONTEXT: THE NECESSITY OF FUNCTIONAL VALIDATION

Transformative potential does not rely on the aggregation of data in AI-powered microbiome research but rather on the translation of that understanding into practice. Genomic annotation can be used to predict whether a lactic acid bacterium is capable of bile salt hydrolase activity or the presence of gene clusters involved in the synthesis of bacteriocins. However, all these predictions are only speculative when they have not been observed under physiologically relevant conditions. For example, it is effective to use genomic operons to predict acid resistance, but it does not determine whether a strain can work as a viable probiotic since acid resistance must be shown by survival across the gastric pH barrier during in vitro testing (Ikombayev et al., 2025).

Functional validation serves as a much-needed bridge. It translates *in silico* signatures into in vitro and in vivo measures. These results can be measured quantitatively as a reduction in *Salmonella enterica* replication in coculture systems, the suppression of pro-inflammatory cytokines in the gut epithelial system, or the production of short-chain fatty acids that are associated with host metabolic benefits. These data not only confirm or disprove computational predictions but can also improve AI models, creating a feedback loop that increases the pace of discovery and precision (Asar et al., 2025).

Examples of Egyptian probiotics demonstrate why local validation is relevant. The diets and health burdens of Egyptians differ from those of Europeans, North Americans, or East Asians. The high prevalence of hepatitis viruses and metabolic syndrome and the increased prevalence of obesity interact with a long tradition of heavy reliance on fermented dairy and grains in cuisine. This cross-section presents challenges and opportunities: on the one hand, there is the challenge of being able to shape interventions to suit local disease burden, and on the other hand, there is the opportunity to embrace existing strains that are already incorporated into cultural diets. Once the genomic and functional data support empirical evidence indicating that indigenous strains can be used to inhibit virus replication, increase the integrity of the gut barrier, or digest otherwise harmful dietary elements, not only is the scientific claim cross-examined, but there is also scientific resonance and practicality of public adoption.

Safety Level One: Preservation Of Community Trust

There is no translational route that can achieve success without ensuring safety. Although their popular image is that they are natural and safe, probiotics should be carefully tested for possible risks. Predictions can include the following safety markers: lack of transferable antibiotic resistance genes, no virulence factors, and stability under stress. Nevertheless, laboratory confirmation is essential. To build a basis of trust, hemoassays, drug resistance mapping, and genome-tropic screening against curated safety screens are needed (Amin, et al., 2025). There is also added significance in terms of establishing the safety of regionally isolated strains in the context of the Egyptian environment, where skepticism toward supplements and imported functional foods is a regular occurrence. For example, local probiotics, namely, lactic acid bacteria, which are isolated *from laboratory rayeb*, were identified as having no mobile resistance mechanism and instead produce natural antimicrobials that minimize the dependency on artificial preservatives (Barba-Ostria et al., 2025). The results generated through such studies are not only conducive to the Codex Alimentarius and EFSA safety standards but also reassuring local regulatory authorities and consumers. Safety validation also involves ethical considerations. As AI can predict new candidate strains on a large scale, issues of bioprospecting and ownership arise. Ensuring that Egyptian microbial resources are studied and commercialized on a fair-access basis prevents exploitation and could establish a sustainable platform of innovations.

Efficacy At The Second Level: Functional Potency

In addition to safety, the real acid test of a probiotic or microbe-derived product is effective. Functional validation uses such strains to evaluate whether the cancer strain does what its genome predicts it is likely to do and whether anything is biologically meaningful in healthy contexts. Egyptian isolates provide a number of enlightening examples.

Anti-pathogen production: Karish cheese (both exotic and native cheeses) is effective against *Listeria monocytogenes* and *Escherichia coli* O157:H_7, supporting genome prediction of bacteriocin operons (Manan & Musaalbakri 2025)

Antiviral activity: In early tests, lactobacilli found in traditional dairy products diminished rotavirus replication in cell cultures, which is consistent with the antiviral metabolite clusters predicted by AI (Yaqub, et al., 2025).

Metabolic control: Egyptian strains are associated with the synthesis of conjugated linoleic acid (CLA) and short-chain fatty acids (SCFAs), which reduce inflammation and alter the liver lipid profile (Fentie, et al., 2025).

Strong attachment to epithelial cells: In vitro epithelial binding is strongly correlated with genomic predictions of mucus-binding proteins, which are considered a key initial step in stable colonization of the gut and immune interactions.

Such functional outputs are not just abstract laboratory curiosities with immediate consequences for public health. Egypt, in turn, has a strong epidemiological burden of viral hepatitis and gastrointestinal infection. Probiotics with proven antiviral and pathogen-inhibiting properties may be used as supplementary measures in addition to vaccine and hygiene programs. Likewise, the increasing rates of obesity and type 2 diabetes highlight the importance of strains that can influence the levels of lipids and glucose via metabolite production.

The Third Level of Scalability: Lab Bench to Society

The most secure and efficient microbes will make minimal changes without scalability despite their safety or high efficiency. Scaling must be integrated into a form of delivery that is acceptable, accessible, and sustainable. Egyptian fermented foods already offer natural delivery vessels: laban rayeb, kishk, and fermented bread provide culturally acceptable substrates into which proven strains can become embedded. Compared with foreign capsules or powders, these matrices guarantee compliance and affordability.

Moreover, industrial and regulatory scalability has come into play. The creation of biobanks of validated Egyptian strains, pilot-scale fermentation facilities, and adherence to EFSA and WHO recommendations will be critical for translating benches to the market. Additionally, AI can increase the scalability of predictive formulations, shelf-life stability, and individual consumer connections on the basis of nutrigenomic databases (Agrawal et al., 2025; Reyed, 2023).

Scalability is also tied to the concept of precision medicine. Personalized nutrigenomics can pair probiotic strains with customized genetic make-ups, food preferences, and microbial compositions. In Egypt, such nucleation might involve the customization of such functional foods to populations that may be at risk of nonalcoholic fatty liver disease or metabolic syndrome, making microbial solutions contribute to population-wide public health efforts to prevent these disorders.

An Egyptian Roadmap Toward Translational Microbiome Science

The task of this chapter is to develop a complete roadmap for deploying AI-derived genomic discovery for practical implementation, with Egyptian probiotics as a case example. This is because of the way in which it categorizes the discussion across three levels that are connected with one another, namely, safety, efficacy, and scalability: the mode in which all of these come together because of the rational and actionable structure through which they have been organized (Yee et al., 2025; Yin, et al., 2025).

The initial business operation within such a roadmap is shifting the conceptual organization to experimental implementation, with the current steps including the phenotypic characterization of Egyptian probiotic strains.

FUNCTIONAL CHARACTERIZATION OF EGYPTIAN PROBIOTIC STRAINS

Phenotypic traits: stress tolerance and adhesion

Functional characterization shows promise in starting with evaluation of the key probiotic characteristics that define survival and activity in the gastrointestinal tract and offers a framework of reference against which to build transnational potential in functional foods and personalized nutrition applications. Among these characteristics, acid and bile tolerance are especially important, as they recreate the undesirable conditions found in the human small intestine and stomach and therefore anticipate the capacity of candidate strains to survive oral introduction and reach the intestine under functional conditions (Dong et al., 2025).

Lactobacillus plantarum EP1 and *Lactobacillus rhamnosus* EG3 are highly resistant to acidic environments but are not very consistent in laboratory studies. The strains presented good survival rates following exposure to low pH for several hours. This is in keeping with what has been observed in other studies of acid resistance, such as those found across different *L. plantarum* isolates, and this is a matter of great diversity. As an example, a significant percentage of these isolates are capable of resisting considerable durations of exposure to a highly acidic medium, reflecting an extended span of resilience in these bacteria. (Alizadeh Behbahani et al., 2025). Compared with a highly studied reference strain, such as *L. rhamnosus* GG, which possesses impressive resilience in hyperacidic environments, the introduction of sugar increased its survival rate. This finding indicates that other *L. rhamnosus* strains, including Egyptian isolates, could also share other acid resistance capabilities. On this basis, it is safe to expect that both *L. plantarum* EP1 and *L. rhamnosus* EG3 will survive long enough against the low-pH environmental conditions in the stomach so that they may be used in the orally consumed functional food. The critical evaluation of these findings points to a number of relevant considerations. The gold standard tests of acid and bile tolerance are worthwhile but are reductionist and do not properly mimic the environment of the gastrointestinal tract, where host mucus and interactions with the immune system lead to critical influences on the determination of colonization success and probiotic efficacy. Second, strain-specific variability further supports why empirical evaluation of *in silico*-predicted candidates is critical to close the gap between predicted genomic potential and outcomes of measurable functional performance (Zhang, et al., 2025).

In addition, a combination of these experimental data with AI-predicted genetic markers will be able to form a basis on how to connect genomic traits to real-world phenotyping. For example, stress–response gene combinations might be correlated with recently observed acid and bile tolerance, and this information can be used to improve the predictive power of AI algorithms and select potentially new strains with greater chances of success in the context of applied research (Sabino, et al., 2025).

This strategy bridges the important translational gap between genomic identification and laboratory characterization to confirm that only the most robust and functional isolates are further pursued for dietary or clinical applications.

In addition to survival, the potential for adhesion was measured via in vitro investigations using Caco-2 and HT-29 human intestinal epithelial cell lines. Adhesion is used as a practical surrogate of colonization potential and host interaction and is suggestive of the probability of permanent gut residence and immunomodulatory activities. Interestingly, strains with CRISPR/Cas loci also presented much greater adhesion indices on unstimulated cells and cytokine-stimulated cells than did CRISPR-negative strains, indicating possible links between adaptive immune loci and the expression of surface-mediating proteins in host interactions (Čuljak et al., 2025). This observation prompts a strong hypothesis of molecular follow-up: CRISPR-associated genes may regulate surface protein structure or the production of exopolysaccharides, thus altering adherence and mucosal colonization.

Critically, these adhesion data indicate an ongoing gap in research: data on genomic annotation sources can predict whether genes associated with adhesion are present, but phenomena related to direct contact and functional interaction with host epithelial cells remain context limited and need experimental determination (Liu et al., 2024). This gap requires an interim solution in the form of an integrated approach that combines multiomics, genomics, transcriptomics, and proteomics with cellular assays to uncover causal controlling mechanisms of adhesion phenotypes (Wu et al., 2024).

Future approaches to developing high-throughput microfluidic adhesion platforms and AI-guided predictive modeling have also been suggested by creating more realistic representations of dynamic host–microbe interactions than those observed in static assays. This cycle-based process guarantees that the accuracy of the genomic prediction is updated according to the experimental data, which enhances the selection of the most likely probiotic strains with the best colonization ability (Du Qiao et al., 2025).

In conclusion, survival and adhesion information yield a successful endpoint at which the functionality of *Lactobacillus plantarum* EP1 and *Lactobacillus rhamnosus* EG3 in the gastrointestinal tract is defined. These data are highly valuable for supporting their use in functional foods and individualizing nutrition approaches to fill the translational gap between artificial-intelligence-based genomic predictions and targeted, pharmacologically relevant probiotic effects. By combining phenotypic performance with genomic insights and suggesting the methodological advancements of the research, this framework will not only confirm the existing candidate strains but also open the way to the rational design of next-generation probiotics specific to the region and the dietary and health context (Tiwari et al., 2024).

Critical analysis: Filling the gap between in vitro and in vivo probiotic efficacy

In vitro experiments, such as adhesion tests with Caco-2 and HT-29 cell lines, are useful methods that precede results to assess the probiotic potential of Lactobacillus strains. These models recreate features of the human intestinal epithelium, providing a way to understand how microbes adhere and how they interact with host cells. Nevertheless, these models lack the ability to represent the complex physiology of the human gastrointestinal (GI) tract (Čuljak et al., 2025).

Limitations of the in vitro models

Although in vitro experiments provide a controlled setting through which to examine microbial activity, they cannot completely recapitulate the complex interactions observed in the gastrointestinal tract. The GI tract is a dynamic ecosystem, and among the most important factors contributing to the survival of the microbiota and colonization are mucus composition, host immune reactions, and the presence of diverse microbiota. For example, mucus acts as both a protective medium and a selective medium that contributes to microbial adhesion and persistence. The host immune system has the capacity to regulate the interactions between the gut microbiota and probiotics in various ways, such as by producing antimicrobial peptides and inflammatory immune cells (Kiran, et al., 2025).

Bridging the Research Gap

To overcome the challenge of the difference in results between in vitro and in vivo efficacy, it is necessary to develop more advanced models that do not ignore the complexity of the human gut environment. Improved in vitro systems, including those based on mucus-secreting HT29-MTX cells or coculture with the gut microbiota, can more faithfully reproduce microbial–host interactions (McCoy et al., 2024). On the basis of genomic and transcriptomic analysis, a combined approach provides an opportunity to discover vital attributes affecting the performance of probiotics and select strains with an increased ability to colonize.

Probiotic development implications

It is important to recognize the restraints of in vitro models in regard to developing probiotics rationally. Appreciation of the complex nature of the gut environment and consideration of this complexity in the development of experimental conditions can increase the predictability of the efficacy of probiotics (Asar et al., 2025). This practice is in line with the aims of promoting personalized nutrition and functional foods, where probiotic strains are being identified with respect to how well they can grow in the human intestine and the positive effects they have there.

In vitro assays play an irreplaceable role in probiotic studies but should be combined with more sophisticated in vivo models that can recapitulate the dynamic nature of the human gut. The field can continue to expand with the development of more effective and individualized probiotic therapies by overcoming current model limitations and integrating a holistic understanding of microbial–host interactions.

On this methodological basis, a further essential step is to focus on the practical functional activities of Egyptian probiotics, specifically their antimicrobial and antiviral properties against foodborne pathogens. Such translational emphasis has moved the discussion away from limitations built within the models and into the actual evidence of how these strains behave in real biological and food safety settings.

Antimicrobial activity against foodborne pathogens

Comparative Functional Mechanisms of Egyptian Probiotics Against Foodborne Pathogens

Competitive scope and positioning

The microbiological rationale of Egyptian probiotic strains should be recognized not only microbiologically but also in regulatory, cultural, and translational contexts.

Egyptian isolates—*Lactobacillus paracasei, Lactobacillus rhamnosus, Lactobacillus casei, Lactobacillus acidophilus, and Bacillus subtilis*—reflect microbial adaptations to artisanal fermentation conditions, which promote acid tolerance, bacteriocin diversity, and adhesion resilience. These ecological properties support both antimicrobial and antiviral feasibility in addition to posing a governance predicament of how Egypt can situate its strains in the context of the global innovation of probiotics (Zommara et al., 2023; Swanson et al., 2020).

Mechanistic Basis

The antibacterial action is based on acidification and redox changes and bacteriocin production, interference with adhesion, and ecological resistance against pathogens such as *E. coli* and *Listeria monocytogenes*. The antiviral effects are also remarkable: Egyptian LAB can reinforce epithelia, influence host immunity via PRRs and interferons, and dampen intestinal microbial responses to viral attacks, which is consistent with the body of research on LAB-provided mucosal protection (Muhammad et al., 2025; Marco et al., 2021). Indeed, recent studies have noted that nutritional interventions and the modulation of the gut microbiome are closely associated with systemic health. In particular, dysbiosis is associated with an increased risk of autoimmune diseases, metabolic syndrome, poor glucose tolerance, cerebral vascular disease, and many cancers. Microbial ecology can alter dietary intake in small but clinically significant ways, deregulating immune signaling and metabolic homeostasis (Shen et al., 2025; Reyed, 2022).

In addition, the altered microbial state of diseased imbalance, normally called microbial community pathophysiology, contributes to the morphological mishealing of DNA, unending inflammatory reactions, and the development of tumor-favoring microenvironments. These dysbiotic patterns not only decrease beneficial commensals but also enable an increase in harmful bacterial taxa that further contribute to epithelial damage and impair immune monitoring. Reyed (2022) also described how this cascade can promote malignancy progression and systemic comorbidities, supporting how diet, probiotics, and microbial ecology intersect in the nexus of redox regulation, host immunity, and disease prevention.

TRANSLATIONAL VALUE AND ONE-HEALTH ALIGNMENT: FROM BENCH DATA TO FOOD SAFETY POLICIES

Lessons from India, Brazil, and Japan are important in the case of Egypt. India shows the significance of strain-level dossiers and clinical validation, with the main issue being that it has particularly high regulatory costs that make it challenging for local innovators (Ganguly et al., 2025). Brazil incorporates probiotics within the concepts of biodiversity and benefits sharing in the General Law on Sanitary Surveillance (ANVISA) (Spacova, et al., 2024), providing ethical management of the elements, although the procedures are limited owing to bureaucratic delays. Japan is an example of how a local microbiome strain (*L. casei Shirota*) can be converted into a global health brand through the FOSHU system, which is known to have some stringent entry requirements (Naito et al., 2018).

Practical analysis of probiotic regulatory frameworks in India, Brazil, Japan, and Egypt

Examples from other countries, such as India, Brazil, and Japan, have important implications for the development of probiotics in Egypt.

India's Food Safety and Standards Authority (FSSAI): India requires strict strain-related identification and extensive safety and efficacy studies. Although this strategy provides quality assurance, it may be very expensive and a barrier to entry into the market for small firms, as was the case with the indigenous strain *Lactobacillus plantarum Lp91* (Kumar et al., 2024).

Brazil's regulatory agency (Agência Nacional de Vigilância Sanitária or ANVISA): Brazil has regulations on probiotics as new foods, which need human clinical evidence. It has also employed the use of specific biodiversity laws to facilitate equitable sharing of the benefits of local resources. This system, however, can cause complexities and delays in approvals (RRMA, 2025).

Japan (the Ministry of Health, Labor and Welfare (MHLW) and the Consumer Affairs Agency (CAA)): Japan is an extraordinary success story in the world arena of FOSHU, which focuses on early clinical trials. The indigenous strain *Lactobacillus casei Shirota* was successfully developed into a world-recognizable brand, and the research and marketing perspective of long-term investments has potential (Sutula et al., 2013; Takada et al., 2016).

Egypt (Proposed): Egypt can base out a hybrid approach that coordinates these lessons. There is a need to develop guidelines that are in tandem with international standards (such as the EFSA) with an appreciation of the potential of traditional foods with local strains. The vision is to capitalize on Egypt's microbial heritage to produce functional foods and supplements, with an efficient approval process and biodiversity conservation.

Policy Implications and Conclusions

A critical review of these experiences suggests that Egypt ought to formalize an explicit strain-approval procedure, encourage the translational public, and institutionalize benefit-sharing arrangements in accordance with the Nagoya Protocol. By integrating scientific rigor with ethical stewardship, Egypt can shift out of the academic exclusivity of individual probiotic strains and into branded consortia, developing relevance, both regionally and on a global scale. This roadmap positions Egyptian probiotics in the greater context of precision nutrition, sustainable food development, and microbiome-based healthcare (Marco et al., 2021; Swanson et al., 2020).

Mechanistic Foundations for translation

The functional characterization of Egyptian lactic acid bacteria (LAB) and bifidobacterial strains indicates that they have a considerable mechanistic basis for translational application because of the presence of antimicrobial metabolites (organic acids, bacteriocins, polyphenol-modulating enzymes, and SCFAs).

1. **Biopreservation**: LAB-loaded foods have been reliably found to reduce food-borne pathogens such as *Listeria monocytogenes* and Salmonella (Muthuvelu et al., 2023).
2. **Immunological priming**: Compared with commercial probiotics available globally, local strains with fermented dairy and traditional foods have a greater ability to regulate innate antiviral pathways (Fitsum et al., 2025).
3. **Synergistic dietary interactions**: As depicted by Shankar et al. (2017) and Reyed (2023), the Egyptian gastrointestinal structure processes polyphenols and SCFAs differently than Western populations do and thus requires contextual validation before implementation at the policy level.

These processes not only reinforce the importance of reframing probiotics as consumer wellness solutions but also illustrate why exploring new uses, as well as repurposing probiotics as proactive food safety measures, will be important issues in regulatory and policymaking systems.

ETHICAL AND SOCIOCULTURAL REQUIREMENTS

The application of indigenous probiotic strains is an issue where ethics is a concern:

1. Community governance of microbial resources: Locally isolated organisms in traditional foods should be avoided in biopiracies and benefit-shared with local producers.

2. Cultural acceptance: Defining probiotics in culturally acceptable formats of food (e.g., fermented dairy and pickled vegetables) will be culturally acceptable and allow community trust compared with capsule supplements.
3. Health equity impact: National strategies should make probiotics available not only in markets in towns but also in rural areas and to nutritionally vulnerable people.

By attending to such imperatives, more success is garnered in ensuring that the translational pathway is socially responsible and culturally sustainable.

POLICY RECOMMENDATIONS

To translate bench data into policy practice:

It is suggested that a four-pillar approach be used:

1. Regulatory Calibration:

To create a Probiotic Safety and Efficacy Dossier System as part of the Egyptian Food Safety Authority (EFSA Egypt), this system was consequently modeled after FSSAI and ANVISA.

No commercial use should be given until population-specific validation studies are conducted to validate the product.

2. Inclusion in National Nutrition Programs

School milk programs and nutritional programs should be strengthened to strengthen maternal nutrition programs using legitimized Egyptian probiotic species.

Probiotics can be utilized as a form of biopreservation to lower the nitrite and sodium contents in processed foods.

3. Ethical Governance:

Introduce a Microbial Commons Charter that ensures fair intellectual property and benefits the sharing of indigenous strains.

4. One-Health Alignment:

Probiotics have been incorporated as indirect measures to mitigate AMR in the National Action Plan on AMR in Egypt.

Industries should be encouraged to use microbiome-based approaches to redevelop products that align with SDG-3 (health and well-being) and SDG-12 (responsible consumption and production).

COMPARATIVE POLICY ANALYSIS FOR THE REGULATION OF PROBIOTICS IN EGYPT

A critical interpretation of the global and regional regulatory frameworks in probiotic governance indicates that there is a subtle difference that can guide a way forward in Egyptian regulation. Strain-specific dossiers have been developed over many years by the Codex Alimentarius (*Codex Alimentarius Commission, 2003)* and the European Food Safety Authority (EFSA) as universal yardsticks of safety and efficacy (O'Connell Motherway et al., 2024). However, their models are mostly models based on Western diets and microbiome compositions and fail to consider other populations with dissimilar nutritional and microbial environments (Sanders et al., 2019). This finding indicates that Egypt needs to regard the Codex standards not as rigid recipes but as flexible models and use them in tandem with Egyptian consumer-specific evidence (Fig. 1).

A NEW COMPREHENSIVE REGULATORY FRAMEWORK FOR EGYPTIAN PROBIOTICS

A Holistic Hybrid Regulatory Approach for Egyptian Probiotics

A major gap in the current food safety regulations in Egypt is the specific framework underlining the provision of probiotics, which can facilitate research translation in this field and industrial innovation (Merenstein et al., 2023; Ejtahed et al., 2023). To resolve this, Egypt will be able to resort to successful practices of other countries to establish a distinct hybrid framework that integrates the requirements of strict international standards with a profound understanding of microbial biodiversity that is native to the country.

This hybrid model is constructed on two fundamental pillars:

Strict Safety Dossiers: Egypt has the opportunity to incorporate the heavy bar safety and efficacy established by international organizations, such as the European Food Safety Authority (EFSA) and the Food Safety and Standards Authority (FSSAI) in India. This would mean that strain-specific dossiers would be detailed, meaning that any probiotic that enters the market has undergone intensive testing, including population-specific validation so that relevancy to a local population is established. This strategy promotes safety for the population and helps develop trust among consumers (Ejtahed, et al., 2023).

Fair Compensation: As opposed to other models, which may erect market barriers to innovators who are very local or small, the framework proposed by Egypt would borrow certain aspects of the Brazilian system. It would appreciate the importance of indigenous strains and lay down some biodiversity-specific laws that could implement the sharing of benefits in an equitable manner with local producers. This would help to deter "biopiracy" and promote the production of probiotics using traditional foods such as *laban rayeb* and *karish cheese*, whereby the economic gain of the Egyptian microbial heritage would be within and not outside the country.

A combination of the two pillars would result in Egypt establishing a regulatory pathway that is scientifically valid and socially inclusive. Such a new model would make use of AI-enabled analytics to streamline the approval process, serving as a dynamic point of connection between scientific research and national policy. This strategy would make Egypt a regional leader, converting its microbial resources into an essential element in their national approaches to nutrition and healthcare strategies.

METABOLIC PROFILING AND FUNCTIONAL OUTPUTS

Metabolic characterization and comparative insights

The characterization of indigenous Egyptian probiotic strains is a staple of valuing their capabilities, application value, and comparative advantage over other strains on the market. The targeted and untargeted metabolomics approaches allowed the characterization of individual strains *Lactobacillus plantarum EP1* and *Bifidobacterium adolescentis* EG2 and the profiling of their metabolomes on different fermentation substrates, which is indicative of the dietary diversity of traditional Egyptian food systems. Fingerprinting with mass spectrometry (MS) by gas chromatography (GC) and liquid chromatography (LC) generated a reproducible fugaceous pattern of SCFA metabolism, such as butyrate, acetate, and propionate, as well as a wide spectrum of amino acid products and products of polyphenol conversion (Ramos et al., 2022; Han et al., 2025).

The metabolic characteristic of butyrate production was also unique, at least in *L. plantarum EP1,* because the concentration of butyrate produced was greater than that of other international strains. Both the effects of butyrate on improving the integrity of epithelial barriers and promoting the differentiation of regulatory T cells and regulating downstream cascades of inflammation by blocking the action of nuclear factor-kappa B (NF-kB) provide a link between metabolomic profiles and host immunological homeostasis. *B. adolescentis* EG2 acetate has the ability to regulate the pH of the intestinal environment, which indirectly limits the growth of opportunistic pathogens (Sarita et al., 2025). These results prove that native germplasms cannot be viewed as clones of universal products but have unique biological characteristics that are directly affected by the indigenous diet and environmental circumstances. Untargeted metabolomics further demonstrated the occurrence of metabolites generated through polyphenol activity, which are characterized as antioxidants and anticarcinogens (Alexander et al., 2024). This fact is extremely consistent with Egyptian dishes that contain legume- and grain-based starches, which are rich in flavonoids and tannins. Importantly, functional annotation analysis through machine learning has integrated such metabolite profiles with artificially intelligent (AI)-inferred pathways and has revealed bacteriocin-like peptides, conjugated derivatives of linoleic acids, and signal compounds, which are credited to antimicrobial and immunomodulatory responses (Damiani et al., 2020). These systems-level creative networks allow mechanistic inferences that can steer toward translational and clinical testing and application to food safety.

CRITICAL EVALUATION OF THE METHODOLOGY

Metabolite profiling is associated with certain challenges. Reproducibility may be lost in the presence of batch-to-batch variability, caused partly by strain physiology and partly by substrate composition. Furthermore, laboratory-derived metabolite concentrations are often elevated compared with those observed in vivo, where host physiology and enzyme turnover, as well as microbial cross-feeding interactivity, heavily influence metabolic flows. This is why in vivo confirmation in area-based samples cannot be ignored in order to ascertain ecological validity. Owing to their unique dietary fiber intake and high fermented dairy consumption, Egyptian populations are indispensable testing grounds for determining whether in vitro-observed outputs of alterations in metabolism translate into clinically meaningful health outcomes.

COMPARATIVE STUDY OF COMMERCIAL STRAINS

The comparison of Egyptian isolates with well-established commercial probiotics can serve as an interesting point of view in assessing their comparative efficacy and niche capabilities. The Sandoz Linex Adult Probiotic, one of the most commonly distributed probiotic products in the known world, contains *Lactobacillus acidophilus* (Liu et al., 2025) and *Bifidobacterium animalis* subsp. lactis. These strains are adequately described in terms of their colonization stability, extensive SCFA production, and overall safety, and these features justify their acceptability on the market worldwide (Jiang et al., 2022; Rasika et al., 2021).

Egyptian isolates become instrumental in highlighting key differences when placed in direct comparisons (Tab. 1):

SCFA Profiles: Although Linex strains (*Lactobacillus acidophilus* and *Bifidobacterium animalis* subsp. lactis) produce acetate and lactate in similar proportions, *L. plantarum* EP1 increases the production of butyrate, a metabolite that has both unique cancer prevention and mucosal repair properties (Widodo et al., 2022; Ushkalova & Gushchina, 2015).

Polyphenol metabolism: Egyptian isolates also exhibit more extensive metabolism of dietary polyphenols to bioactive phenolic acids, which coincides with Egyptian foods that include lentils, fava beans, and grains (Abd Elhaliem & Zwain, 2022). This transformation is not present in the cultivated strains of Linex.

Antimicrobial Activity: The AI-predicted metabolic pathways indicate that *L. plantarum* EP1 and *B. adolescentis* EG2 possess a wider scope of bacteriocin-like peptides, specifically aimed at pathogens common to both North Africa and the Middle East, including *Salmonella enterica* (Lone et al., 2021), *Shigella flexneri,* (Jiang et al., 2022) and *Clostridium perfringens*. In comparison, Linex strains exhibit more universal antimicrobial patterns, which are applicable to wider but more pathogen-agnostic environments (Ushkalova, & Gushchina, 2015).

Host Adaptation: Egyptian isolates would experience an increase in metabolic plasticity within fiber-rich, vegetable-based fermentation environments; however, until recently, Linex strains have adapted best to the use of standardized dairy-based substrates because of their present adaptation to global, Western-based diets. (Rasika et al., 2021; Elhalik et al., 2024).

Comparative Table (1): Metabolite Profiles and Functional Outcomes

Feature	Egyptian Indigenous Strains	Commercial Strains (Sandoz *Linex Adult Probiotic:*)	Reference
SCFA Production	Elevated butyrate and acetate levels. Inhibits gut permeability and has strong anti-inflammatory properties	Balanced acetate and lactate; gut health support in general.	Widodo, et al., 2022 ; Ushkalova, & Gushchina, 2015
Polyphenol Metabolism	Efficient conversion to bioactive phenolic acids; adaptability to diet;	Limited capacity for polyphenol metabolism; poorly suited to plant-based diets	(Abd Elhaliem, & Zwain, 2022).
Antimicrobial Compounds	Theoretical predicted diversity of bacteriocins that target regionally relevant pathogens	Broad-spectrum; generic antimicrobial activity.	Ushkalova & Gushchina, 2015).
Host–Diet Adaptation	Specifically formulated for diets high in fiber & legumes, like Egyptian diets.	It is optimized for dairy-based global formulations.	Rasika et al., 2021; Elhalik, et al ., 2024
Clinical Potential	High translational impact for colorectal health, AMR prevention and food security	Extensively replicated in relation to bowel homeostasis and overall digestive health.	Merrick et al., 2023; Ushkalova & Gushchina, 2015

Integrative functional assessment and translational relevance

Metabolic profiling is a translational step of architecture in the book since it links early-stage indicators, such as survival, adhesion, and stress tolerance, and downstream functional applications, such as pathogen inhibition and viral defense. Specifically:

As indicated in the preceding sections, viability and adhesion are absolute conditions but insufficient to promote the right regular configuration needed to achieve probiotic activity.

This section presents metabolomics as the functional validation phase, which demonstrates that these strains have dynamically provable effects on host biology.

In the upcoming sections, the implications of these findings for determining real-world efficacy (e.g., protection against foodborne pathogens and improvement of antiviral immunity) will be discussed.

In a more general sense, this story arc reflects the central theme of the book of AI-enabled nutrigenomics, in which mechanistic outputs such as butyrate/indole ratios act as quantitative vehicles of precision nutrition strategies. In countries such as Egypt, where the dietary profile includes high levels of fermented dairy products, matrix-conscious metabolic metabolomics makes the conducted interventional probiotic studies not only culturally specific but also scientifically sound (Abdel Tawab et al., 2023; Hammam et al., 2020).

CRITICAL ANALYSIS

1. Ecological Validity Gap

In vitro studies tend to present SCFA concentrations that are at hyperbolic levels compared with values inside the body, failing to consider cross-feeding networks, mucus utilization, or host uptake. This misrepresentation threatens to inflate claims of efficacy. It has been demonstrated that receptor activation (e.g., GPR41, GPR43, and HCAR2) occurs at physiologic ranges rather than in absolute abundance (Duan et al., 2023).

Implication: Include dynamical models that reflect production, consumption kinetics, and occupancy of receptors in realistic intervals.

2.The World of Matrix and the Efforts of Portability:

In vivo bioanalytical results generally do not consider the food-matrix effect, through which grocery and cocultivated microbes divert the carbon current. For example, butyrate is promoted by fermented dairy, whereas acetate has been hypothesized to be promoted by plant matrices (Fadare et al., 2025).

Suggestion: All the metabolic claims must be verified in at least one matrix match condition.

3. Directionality of Signs and Sign Processing

The lack of precise peripheral measurements often results in untargeted metabolomics generating vague results, blurring the significance of isomers or confounding courses. Metabolite-level assignment of a phenotype does not consider bundled mechanisms, including the net effects of SCFAs, indoles, and peptides (Wang et al., 2024).

Implication: Focus on mechanistic bundles as opposed to single-molecule causation.

4. Methodological Variance

Differences in reproducibility among different laboratories also pose problems, and the main sources of variation lie in derivatization, extraction chemistry, and ion-source conditions. The implementation of quality assurance and control mechanisms, including pooled quality controls, lock masses, and open SOPs, is low (Roach et al., 2024).

Implication: Establish uniformity in protocols and make complete methodological metadata available.

5. Relevance to Outcomes of Antivirals

Translational claims concerning antiviral properties cannot remain at the level of metabolite titration in vitro. Animal and gut-on-a-chip models with mucus and dynamic flow are needed to confirm the fortification of barriers by CFA and the activation of AHR-mediated immune responses by indoles (Ballerini et al., 2025).

Suggestion: Prespecify antiviral mechanisms of action in metabolite validation prior to extrapolation in the clinical setting.

Strategic Implications: This comparative information highlights the strategic importance of narrowing down the development of Egyptian probiotic strains as branded next-generation products. Indigenous isolates boast localized metabolic efficacy and geographically specific pathogen resistance, although global strains such as those found in Linex cannot be neglected owing to their wide-range benefits. In the case of Egypt, this two-pronged approach of using local strains to contribute to national health and benchmarking against international brands to achieve evidence-based, microbiome-centered food safety policy development can be made more rapid. In addition, establishing indigenous strains in the worldwide probiotic market is likely to boost the Egyptian bioeconomy, as the country shifts away from consumption to innovation-led export.

Antiviral potential of Egyptian probiotics

The possibility of probiotics as antivirals has become a major area of microbiome research, nutrigenomics, and artificial intelligence. Owing to their critical role in gastrointestinal health and the control of bacterial pathogens, probiotics are increasingly thought of as a virological aid and immune modulator (Wang et al., 2022). Egyptian probiotics acquired via evidence from ancient foods provide a source of functional diversity in this emerging paradigm (Asar, 2024). The fact that they can change according to the eating habits, environmental conditions, and culture of a given area not only gives them resilience but also puts them on the front line when addressing food security challenges and virus containment uses.

Owing to the emerging paradigm of personalized antiviral therapy, recent advances have shown that probiotics cannot be considered only supplementary agents but also precision medicines. This means that the customization of antiviral therapies should consider individual personalities and factors such as genes, living conditions, and lifestyle choices to achieve suitable optimization, as noted by Reyed and Prabhakar (2025 b). In this context, probiotics, particularly those that are adapted to a certain ecological and cultural niche, such as those strains that were isolated from the Egyptian diet, can be strategically implanted into customized treatment plans. Their ability to modulate the gut–immune axis, in addition to their potential to affect the systemic inflammatory status, makes them candidates for inclusion in pharmacogenomic and AI-driven drug selection models, as well as computational frameworks with the capacity to predict resistance patterns.

In the subsequent section, the antiviral activity of Egyptian probiotics will be discussed, with a focus on shedding light on mechanisms, comparisons, and novel applications. In so doing, it contextualizes these strains as part of a broader plan to develop AI-directed probiotics, which is in accordance with the overall intent of the chapter, which is to harmonize functional characterization with therapeutic innovation in the development of probiotic strains.

COMPARATIVE PERSPECTIVES: EGYPTIAN VS. COMMERCIAL STRAINS (IMMUNITY-FIRST LENS)

A comparative analysis of *L. plantarum* EP1, *L. rhamnosus* (Egyptian isolates), and Linex Adult Probiotech AB Blend 64 probiotic strains revealed that the immunovirological orientations are complementary but not identical:

Lactobacillus plantarum EP1 (Egyptian isolate)

In *karish cheese* and *laban rayeb*, EP1 can generate a layer of antiviral immunity via pattern-recognition receptor (PRR)-linked interferon signaling, which is associated with potent stimulation of secretory IgA, preservation of epithelial tight junctions, and activation of natural killer (NK) cells (Abraham et al., 2022). The immune-epithelial strengthening of EP1 under this scenario points to its potential as a mucosal barrier, specifically in Egyptian-based diets, where high-fiber and fermented dairy products provide natural probiotic delivery vehicles. The lack of a human clinical trial that probes strains of concern with virology-relevant end points represents its central weakness, as it does not allow the translation of predictive laboratory results into clinically relevant outcomes.

Lactobacillus rhamnosus (Egyptian isolate)

By descending from conventional fermented milk, the strain can uniquely modulate the cytokine set point, balancing immune responses toward hyperinflammation and alleviating the possibility of a cytokine storm. This immuno-calibrating mechanism endows epithelial resiliency under duress, which has a bearing on both seasonal influenza-like diseases and emergent viral epidemics within Egyptian communities (Abd Elhamid et al., 2025). Its translational potential includes being an adjunct immunobiotic in the setting of viral resurgence; nonetheless, it also has low benchmarking of global standards and comparisons with global standards, lacking harmonized immune biomarker panels that would permit robust multicentric cross-comparisons with internationally validated strains.

Linex Adult Probiotec AB Blend 64 (Commercial)

Linex is a clinically validated combination *of L. acidophilus and B. animalis* subsp. lactis that is stable, reproducible, and safe. Its applicability is its general microbiome stabilization and symptom-centered central points, including enhanced gastrointestinal comfort and decreased antibiotic-related dysbiosis. These qualities make it a significant baseline comparator within probiotic studies and a possible coadministration candidate in a hybrid regimen against Egyptian strains. Nevertheless, its inability to fit well ecologically in Egyptian diet matrices and its lower diversity of metabolites restrict its potential application in specific immune modulations. Its immunological advantages are also indirect and are not very competitive in situations that require direct antiviral readiness (Ushkalova, & Gushchina, 2015).

Strategic complementarity

The comparative landscape indicates that substitution is not the appropriate business strategy; rather, it should be the incorporation of integration.

Commercial products offer clinical robustness, standardized protocols, and scalability, which are essential for regulatory approval and penetration of the market.

Egyptian isolates provide adaptive resiliency, mechanistic immunospecificity, and compatibility with local diets, which make them excellent candidates for implementing the precision immunonutrition concept (Singh et al., 2024).

The progressive way forward is through substantive immune assessment validation of Egyptian strains, benchmarking these strains against internationally recognized products. The antiviral cytokine signature, mucosal IgA spikes, and epithelial barrier biomarkers are endpoints that should be targeted in such trials. In addition, commercial strains can be used as reference comparators and, in addition, as hybrid coformulants, providing credibility and faster translation to the clinic. This dual-track policy ensures major global scientific contributions as well as local-scale public health impacts (Maldonado-Reina et al., 2025).

Analysis and personal opinion

Although these insights are encouraging, there are significant gaps. The existing studies on Egyptian isolates are still biased toward in vitro data, which lack the entire gut ecosystem and do not include mucus–layer heterogeneity, microbial cross-talk, or host genetic differences in immune reactions. Nonstandardized protocols used in isolation, characterization, and functional tests make reproducibility and meaningful cross-study comparisons difficult. In addition, a lack of immune biomarker harmonization limits the comparability of Egyptian data against international standards, thus delaying the pace of regulatory acceptance (Gunjan et al., 2023).

Critically and personally, this is not only a challenge but also a lost strategic opportunity. Egypt has rich probiotic biodiversity with little scientific identification and commercial representation to support its significant potential as a local source of population-specific nutraceuticals, viral resistance mechanisms, and food safety resilience. The utilization of this biodiversity has the potential to make Egypt a regional leader in probiotic development as long as investment in the form of a system is channeled into scientific authentication and translational pipelines (Abouelnaga et al., 2024).

CRITICAL ANALYSIS AND INNOVATIVE PATHWAYS

These barriers necessitate a multilevel roadmap to be beaten:

- **Standardized Isolation, Genomic, and Functional Assays**: Developing common standards to conduct isolation, genomic description, and evaluation of functional results will define cross-laboratory reproducibility and global comparability (Zimmer et al., 2024).
- **AI-based Multi-Omics Integration:** Artificial intelligence will be used to integrate genomics, transcriptomics, metabolomics, and proteomics to identify gene clusters and metabolic pathways associated with antiviral and immune modulatory features to inform rational prioritization of strains (Shahzad et al., 2012).
- **Targeted Clinical Trials**: It will be necessary to conduct controlled trials among Egyptian and regional populations to prove strain-specific efficacy,

with an emphasis on measuring the rates of reducing viral loads, epithelial barrier regeneration, and immune recalibration (Martinez et al., 2023).

- **Translational Embedding into Functional Foods**: Validated Egyptian strains may then be embedded into culturally relevant food matrices (e.g., laban rayeb, karish cheese, nutraceutical drinks) as well as into specific nutraceutical formulations, facilitating the interface between traditional food cultures and precision medicine (Li et al., 2019).

Concluding Perspective

By providing the functional immunovirology of Egyptian strains, this study relocates the tale of the comparison beyond description to solution-oriented translational science. Validated Egyptian isolates could be used to supplement existing branded-world commercial probiotics in offering standardized reliability along with immunological and local specificity. This paradox is consistent with the broader vision of harnessing the diversity of indigenous microbiomes, enabled by AI and integrative science, to address pressing global issues, including food security, viral resistance, and therapeutic robustness.

On the basis of this comparative immuno-virological line of reasoning, the translation of the predictive and mechanistic findings into experimental validation under controlled conditions is needed. Whereas the functional orientations defining Egyptian isolates and commercial strains underscore relative competencies, their actual translational aptitude can be untangled only by undertaking systemic screening that would examine their antimicrobial and antiviral efficiency in a controlled environment. This switch is imperative since it fills the spectrum between theoretical potential and practical utility, as we come to know not only how these strains influence host responses but also whether these actions hold true in realistic food environments and clinical scenarios. In addition to this essential experimental verification, how Egyptian probiotics chosen by AI might be harnessed in translational contexts, especially when dealing with emerging urgent topics in nutrition, such as foodborne pathogens and related precision nutrition, is explored.

4. TRANSLATIONAL APPLICATIONS OF AI-SELECTED EGYPTIAN PROBIOTICS: FROM FOODBORNE PATHOGEN CONTROL TO PRECISION NUTRITION

Like many other countries, Egypt continues to be challenged by foodborne pathogens, including *Salmonella enterica*, *Listeria monocytogenes*, and *Escherichia coli O157:*H_7. Their resistance to dairy, meat, and ready-to-eat products demands the development of novel countermeasures capable of not only protecting food systems but also offering host-level advantages. Examples of probiotics are known as two-for-one, having antimicrobial properties and promoting systemic health through gastrointestinal and immune support. This part focuses on examining the functional aspects of the Egyptian probiotic strains selected by AI, where replicate testing in well-defined laboratory settings is performed with critical insight into their potential translatability (Abdel Tawab et al., 2023).

Functional Readouts and Methodology

The evaluation was carried out by comparing conventional agar diffusion tests, liquid coculture inhibition assays, and time-kill assays. Agar assays gave initial screening results with inhibition zones indicating the presence of bacteriocin-mediated water. A more dynamic coculture assay demonstrated the progression of suppression over time, and time–kill kinetics were used to analyze the speed and degree of microbe inhibition.

The results showed that *Lactobacillus plantarum* EP1 and *Lactobacillus rhamnosus* EG3 successfully reduced the pathogen to 3–4 collateral cycle units/ml after a span of 12 hours. These results indicate not only bacteriostatic effects but also bactericidal effects, which may indicate that the pathogen is quickly suppressed in a simulated environment. The strains that were positive for CRISPR/Cas had a broader inhibition spectrum, which may indicate increased diversity in the bacteriocin and increased resistance to horizontal gene transfer caused by the phages, which can be considered a sign of functional stability (Mir, 2025).

Evaluation of Bactericidal Evidence with Critical Analysis

Although promising, the use of agar-based assays can increase inhibition, as simplified media may favor the diffusion of bacteriocins. Real food systems have introduced complicating factors—fat globules, protein matrices, and fluctuating pH—that can reduce the effects of antimicrobial agents (Wang, et al., 2023). Additionally, with regard to bacteriocin susceptibility to heat treatment and storage, longitudinal validation is beyond challenging. Without this type of testing, industrial applicability is not certain. In such a manner, the current evidence can be seen as a basis, not as a conclusive result, of food matrix effectiveness (Harahap et al., 2024; Ombarak et al., 2019).

Future Applications to Antiviral Assays: Reasons and Methodological Principles

In addition to the trend of bacterial pathogen research, studies have explored the antiviral aspects of Egyptian strains, as their importance as modulators of host–virus interactions is increasingly recognized with the advent of probiotics. Virus models were utilized to determine the ability of specific strains to prevent entry of viruses into the cells of the host, and this was complemented by the use of cytokine and gene expression analyses to assess the level of immunomodulation by these strains (Liu et al., 2021).

Remarkably, *L. plantarum* EP1 and L. *rhamnosus* EG3 inhibited SARS-CoV-2 infection by almost 50–60%, indicating that the effects were not only due to competition for binding but also to stimulation of interferon signaling as well as activation of mucosal barrier responses. The antiviral activity of the bacteriocin-enriched strains partially overlapped, suggesting that their mode of action is multifactorial, involving the release of metabolites, priming of the immune system, and direct viral interruption (Umair et al., 2022).

Evaluation of Bactericidal Evidence with Critical Analysis

The in vitro antiviral experiments are interesting because they shed some light on the mechanism, but ultimately, they do not reflect the in vivo situation as much. Viral inhibition in true hosts is dependent on microbiome interactions, dietary cofactors, and host genomic variability, which all determine immune responsiveness (González et al., 2021). Additionally, SARS-CoV-2 systems provide an approximation of entry that does not reflect replication dynamics or systemic dissemination. Advanced models, such as gut-on-chip platforms or tests in animals, can be used to validate the findings, but until this is done, the benefits of the findings as far as translation is concerned are low. However, immunological outcomes, such as the upregulation of IFN-beta or stimulation of mucosal IgA, fit the body of literature on the interaction between probiotics and the immune system, leading to ongoing long-term studies designed to elucidate their mechanism of action (Maillard et al., 2023).

CONCLUSION

This study highlights two frontiers: bacterial suppression, which lies in the diversity of bacteriocins, and antiviral regulation, which consists of activating the immune system. Egyptian isolates, especially *L. plantarum* EP1 and L. *rhamnosus* EG3, have particular features in terms of their mechanism and context. However, vital concerns remain, especially with respect to strain-specific clinical trials and restricted investigations of functional stability in advanced food media. Closing these gaps will not only prove their forecasted advantages but also establish Egyptian probiotics firmly in the world of food safety and precision medicine (Al-Madboly et al., 2015).

Applications and implications at the translational level

The combination of antimicrobial and antiviral effects places Egyptian strains at the cutting edge of functional probiotic products. Food systems can incorporate microbial species into traditional dairy products (e.g., *laban rayeb or karish cheese*) to improve microbial safety, harnessing familiarity with fermented foods (Ombarak, et al., 2019). In nutraceutical applications, their immune-modulating property makes them candidates for nutraceuticals used to target immune resilience in individuals with already low defenses against foodborne or novel viral infections. (Di Stadio et al., 2020).

The use of strain selection via artificial intelligence (AI) and multiomic profiling to inform formulations on the basis of host immune status and microbiome composition is an attractive roadmap. These strategies have the ability to turn Egyptian strains of local culture into global biotherapeutics. Nonetheless, strict clinical validation, standard regulatory pipelines, and industry-academic collaborations on how to transform predictive models into safe, evidence-based interventions are needed (Agrawal, et al., 2025).

Application Scenarios: Translating Egyptian Probiotic Research into Functional Outcomes

The functional profiling of Egyptian probiotic strains clearly creates multiple directions leading to new knowledge and practical applications that connect laboratory findings with increased social interest. Many of the current studies in the scholarly field still focus on mechanistic and comparative approaches, but determining practical applications for how these strains can be used in functional foods, dietary supplements, and personalized nutrition platforms is essential. Such application cases provide a connection with the realization of experimental science in the domain of public health so that not only biological insights into the therapeutic and preventive potential of Egyptian strains remain limited to abstract models of reality.

Integration of Functional Food

One of the more imminent uses is the incorporation of Egyptian strains that have been validated into functional foods. The introduction of traditional dairy products such as *laban rayeb* and *karish cheese* in Egypt, which are already being consumed on a wide scale, is an ideal way to provide a source of natural delivery systems for probiotic strains such as *Lactobacillus plantarum* EP1 and *Lactobacillus rhamnosus* (Elkot & Ismail., 2025). By fortifying such dairy products with standardized and high-potency probiotic cultures, such foods could be developed as specific functional foods aimed at stimulating immune resilience and viral protection. Plant-based alternatives to dairy products via fermented legumes and cereals can be explored as further vehicles, mirroring prevailing tendencies toward sustainable and plant-forward eating (Senanayake et al., 2023). This type of integration not only utilizes indigenous eating cultures but also enhances consumer acceptability, which is a major determinant of market growth.

Nutraceutical and supplement development

Another line of research involves the production of diet supplements and probiotic prototypes in the form of encapsulated products. Many firm examples of such probiotic capsules as commercially available goods, such as Linex Adult Probiotec AB Blend 64, have been able to match and produce at a mass scale, which is possible at the level of companies spread across the global market (Bhutto, et al., 2025). These formulations are largely dependent on European strains or North American strains. The construction of supplements with Egyptian isolates would create regionally acclimatized supplements with specific ecological opportunities. Encapsulation technologies, such as microencapsulation and the use of synbiotics containing prebiotics, have the potential to increase the survival of Egyptian probiotics during storage and throughout gastrointestinal transit. These innovations could enable Egyptian organizations to enter both home and international probiotic markets and provide local microbiomes with uniquely appropriate supplements depending on the local microbiome and immune health needs (Abd Elhamid, et al., 2025).

Strategies Specifically Targeted to Egypt

Egypt has particular nutritional and public health issues, such as a high level of infectious disease, gastrointestinal disorders, and an increasing burden of noncommunicable diseases (Khalid et al., 2023; Ministry of Health and Population et al., 2015(. The use of Egyptian probiotics in this context may include the following:

- Traditional dairy products are supplemented with fortification with the desired strains to minimize morbidity related to infections.
- Community-based nutrition programs that integrate accepted probiotic strains into school lunch programs or maternal nutrition programs.
- Food safety interventions include the application of probiotic-derived antimicrobial peptides to address viral contamination in food chains.
- The development of supplement strategies at the national level has advanced the concept of immune resilience in vulnerable individuals, including children, elderly individuals, and immunocompromised individuals.
- Such plans would put probiotic innovation in dialog with the nation's health needs, and the results of this initiative would be scientifically significant as well as very impactful to the surrounding society (Sarfraz et al., 2022). In addition to these national-level strategies, further opportunities lie in the emerging field of personalized nutrition.

Personalized Nutrition Opportunities

One of the furturistic application areas is the manufacture of personalized nutrition systems, wherein Egyptian probiotics may be combined with genomic and microbiomic profiling. Personalized nutrition is the practice of linking dietary treatment to the genetic makeup and intestinal microbiome of an individual, coupled with his or her lifestyle. Since genetic variability in Egyptian populations can alter the efficacy of probiotics in regional populations, Egyptian strains could serve as precision probiotics potentially tailored to geographic populations where populations would have genetic variants in immune or metabolic pathways modifying probiotic efficacy) Roman et al., 2024). Artificial intelligence can additionally optimize these strategies via mechanistic-level models of host–microbe interactions, modeling of therapeutic response, and personal recommendations for probiotics (Asar, et al., 2025). This would place Egypt in the modern fray of AI-aided nutrigenomics, with local strains becoming drivers of precision medicine. Collectively, these tailored solutions auger well into the prospects of overall consideration of how Egypt can connect present-day applications to the future of innovation.

Closing the Gap between the Latest Applications and Future Innovation

Collectively, these application scenarios reveal a spectrum of opportunities between businesses, as we currently know them, and the future of innovation. Although functional foods and supplements are the best track for commercialization, personalized nutrition and AI-designed probiotics embody the nature of future commercialization

(Roman et al., 2024). In the case of Egypt, this two-pronged strategy will enable the country to both quickly implement and remain at the forefront (leading in scientific development terms). Egyptian probiotics can both address local needs and contribute to international knowledge of probiotics by judiciously leveraging the traditional structures of the Egyptian diet and modern biotechnology techniques.

Closures and Future Prospects

This chapter has traced the entire history of Egyptian probiotic research, including the genomic underpinnings as well as their translational promise, not merely as national microbial assets but rather as key assets within the global trend toward AI-based microbiome science and precision medicine. Opening with genomic and metabolomic profiling, progressing into functional immuno-virological tests, and then into regulatory, ethical, and clinical considerations, the discussion highlights the uniting theme of how indigenous microbial biodiversity combined with artificial intelligence and systems-level nutrigenomics may transform the nexus of food, health, and biotechnology (Parvin & Sadras 2024).

One of the main consequences of this story is how Egyptian probiotic strains have moved from describing entities to predictive and actionable instruments. By using AI-driven genomics, operon mapping, and ecological strength modeling, these strains present potential antiviral, immunomodulatory, and metabolic agents with vastly expanded application areas extending well beyond their long-established use in fermented foods. This reframing highlights not only their cultural and nutritional value but also their ability to act as competitive sources of functional food development and therapeutic production worldwide (Asar et al., 2025).

The translational arc presented here shows that the future of Egyptian probiotics is precision and individualization. The ability to encapsulate the strains into culturally honed vehicles, including *laban rayeb, karish cheese,* and fermented grains, as well as explore encapsulation and synbiotic formats, offers accessibility and scalability. Combining AI will also allow such applications to approach individualized nutrition by coupling strain-level capabilities with host genomic variability and other lifestyle considerations, in line with the general vision of precision medicine (Han et al., 2025).

It is also crucial to realize that the future of scientific potential cannot exist without clarity in regulation and a willingness to consider the morality of it all. The situation in Egypt, accompanied by regulatory lapses, reflects both challenges and opportunities. As observed in the examples of the Codex, EFSA, India, and Brazil, hybrid models are feasible—they impose high standards of biosafety and clinical validation but simultaneously incorporate values of fairness, transparency, and fair benefit-sharing. These frameworks not only retain innovation but also ensure that indigenous communities are cobeneficiaries of probiotic commercialization, in which biodiversity is converted into bioeconomic capital, without compromising ethics (Tone-Pah-Hote & Redvers 2022).

The future path forward can be described as multilayered and ambitious: solidifying preclinical reproducibility, advancing region-specific clinical trials, implementing long-term biosafety procedures, and implementing nationwide strain repositories as a part of a coherent microbiome ecosystem. Coinciding with this, investment in AI-based personalization platforms can place Egyptian probiotics at the center of the tumultuous shift toward precision nutrition and functional immunotherapy, transforming them into international, exportable assets rather than mere national ones (Jiang et al., 2024).

In conclusion, the chapter demonstrates a paradigm shift that resonates with readers of this book: Advancing Personalized Medicine with AI-Driven Microbiome and Nutrigenomics. Egyptian probiotics can be seen as a shift toward innovative predictive biotechnologies replacing cultural knowledge and global health innovations instead of local foods. By interconnecting tradition and newness, nature and technology, and control and morality, these streams can ground a future in which microbiome science can help support sustainable health, sustainable food, and equity in innovation. The roadmap is not just a fancy wish list because it is a strategic plan aimed at making Egypt a leader in precision medicine on the basis of the microbiome and at matching the resources related to microorganisms in the country with the worldwide push toward personalizing health through AI technologies.

ACKNOWLEDGMENTS

This work was financially supported by the National Research Centre (NRC), Dokki, Giza, Egypt, under Grant No. MP121202 (2021–2023). The authors sincerely acknowledge the NRC staff, field collaborators, and technical teams for their invaluable contributions, expertise, and dedicated assistance throughout the study.

REFERENCES

Abd Elhaliem, H. S., & Zwain, B. A. (2022). Effect of Lentil, Chickpeas and Okara with probiotic bacteria on the health status of diabetic induced rats. *Egyptian Journal of Nutrition and Health, 17*(2), 103.

Abd Elhamid, A., Habliza, R., & Kholaf, G. (2025). Production of Functional Egyptian Bellila with Probiotic Strains. *Food Technology Research Journal, 8*(1), 17–27. DOI: 10.21608/ftrj.2025.380782.1158

Abdel Tawab, F. I., Abd Elkadr, M. H., Sultan, A. M., Hamed, E. O., El-Zayat, A. S., & Ahmed, M. N. (2023). Probiotic potentials of lactic acid bacteria isolated from Egyptian fermented food. *Scientific Reports, 13*(1), 16601. DOI: 10.1038/s41598-023-43752-0

Abouelnaga, M., & Gouda, A., saad, M., & Ismail, H. (2024). Positive impact of potential probiotic Lacticaseibacillus rhamnosus isolated from local dairy products on raising microbiological quality of white soft cheese. *Egyptian Journal of Food Science, 52*(1), 41–50. DOI: 10.21608/ejfs.2024.285537.1180

Abraham, C., Abreu, M. T., & Turner, J. R. (2022). Pattern Recognition Receptor Signaling and Cytokine Networks in Microbial Defenses and Regulation of Intestinal Barriers: Implications for Inflammatory Bowel Disease. *Gastroenterology, 162*(6), 1602–1616.e6. DOI: 10.1053/j.gastro.2021.12.288

Agrawal, K., Goktas, P., Kumar, N., & Leung, M.-F. (2025). Artificial intelligence in personalized nutrition and food manufacturing: A comprehensive review of methods, applications, and future directions. *Frontiers in Nutrition, 12*, 1636980. DOI: 10.3389/fnut.2025.1636980

Agrawal, K., Goktas, P., Kumar, N., & Leung, M. F. (2025). Artificial intelligence in personalized nutrition and food manufacturing: A comprehensive review of methods, applications, and future directions. *Frontiers in Nutrition, 12*, 1636980. DOI: 10.3389/fnut.2025.1636980

Alexander, M., Upadhyay, V., Rock, R., Ramirez, L., Trepka, K., Puchalska, P., Orellana, D., Ang, Q. Y., Whitty, C., Turnbaugh, J. A., Tian, Y., Dumlao, D., Nayak, R., Patterson, A., Newman, J. C., Crawford, P. A., & Turnbaugh, P. J. (2024). A diet-dependent host metabolite shapes the gut microbiota to protect from autoimmunity. *Cell Host & Microbe*, *43*(1), 1114891. DOI: 10.1016/j.chom.2024.1114891

Alizadeh Behbahani, B., Rahmati-Joneidabad, M., & Taki, M. (2025). Examining the impact of probiotic *Lactiplantibacillus pentosus* 6MMI on inhibiting biofilm formation, adhesion, and virulence gene expression in *Listeria monocytogenes* ATCC 19115. *Biofilm*, *9*, 100255. DOI: 10.1016/j.bioflm.2025.100255

Al-Madboly, L. A., & Abdullah, A. K. (2015). Potent antagonistic activity of Egyptian *Lactobacillus plantarum* against multiresistant and virulent food-associated pathogens. *Frontiers in Microbiology*, *6*, 347. DOI: 10.3389/fmicb.2015.00347

Amin, M. R., Biswas, A. P., Tasnim, M., Islam, M. N., & Azam, M. S. (2025). Probiotics and their applications in functional foods: A health perspective. *Applied Food Research*, *5*(2), 101193. DOI: 10.1016/j.afres.2025.101193

Asar, A. (2024). Microbial Functional Foods as a Magic Secret to Healthy Life Style. *Egyptian Journal of Chemistry*, *67*(2), 161–173. DOI: 10.21608/ejchem.2023.202051.7926

Asar, R., Erenler, S., Devecioglu, D., Ispirli, H., Karbancioglu-Guler, F., Ozturk, H. I., & Dertli, E. (2025). Understanding the Functionality of Probiotics on the Edge of Artificial Intelligence (AI) Era. *Fermentation (Basel, Switzerland)*, *11*(5), 259. DOI: 10.3390/fermentation11050259

Bakkeren, E., Piskovsky, V., & Foster, K. R. (2025). Metabolic ecology of microbiomes: Nutrient competition, host benefits, and community engineering. *Cell Host & Microbe*, *33*(6), 790–807. DOI: 10.1016/j.chom.2025.05.013

Ballerini, M., Galiè, S., & Tyagi, P.. (2025). A gut-on-a-chip incorporating human fecal samples and peristalsis predicts responses to immune checkpoint inhibitors for melanoma. *Nature Biomedical Engineering*, *9*, 967–984. DOI: 10.1038/s41551-024-01318-z

Barba-Ostria, C., Guamán-Bautista, J., Tosi-Vélez, A. A., Puente-Pineda, J. A., Cedeño-Zambrano, M. A., Teran, E., & Guamán, L. P. (2025). Recent Advances in the Therapeutic Potential of Bioactive Molecules from Plants of Andean Origin. *Nutrients*, *17*(11), 1749. DOI: 10.3390/nu17111749

Bhutto, R. A., & Bhutto, N. u. a. H., Mahar, H., Khanal, S., Wang, M., Iqbal, S., Fan, Y., & Yi, J. (2025). Recent trends in coencapsulation of probiotics with prebiotics and their applications in the food industry. *Trends in Food Science & Technology*, *156*, 104829. DOI: 10.1016/j.tifs.2024.104829

Codex Alimentarius Commission. (2003). *Recommended International Code of Practice: General Principles of Food Hygiene* (CAC/RCP 1-1969, Rev 4-2003). Retrieved from www.codexalimentarius.net/download/standards/23/cxp_001e.pdf

Čuljak, N., Bendelja, K., Leboš Pavunc, A., Butorac, K., Banić, M., Savić Mlakar, A., Cvetić, Ž., Hrsan, J., Novak, J., Šušković, J., & Kos, B. (2025). In Vitro Analysis of Probiotic Properties Related to the Adaptation of Levilactobacillus brevis to Intestinal Microenvironment and Involvement of S-Layer Proteins. *International Journal of Molecular Sciences*, *26*(6), 2425. DOI: 10.3390/ijms26062425

Damiani, C., Gaglio, D., Sacco, E., Alberghina, L., Vanoni, M., & Naito, Y. (2020). Systems metabolomics: From metabolomic snapshots to design principles. *Current Opinion in Biotechnology*, *63*, 190–199. DOI: 10.1016/j.copbio.2020.02.013

Di Stadio, A., Ishai, R., Gambacorta, V., Korsch, F., Ricci, G., Della Volpe, A., & Bernitsas, E. (2020). Nutraceuticals as immune-stimulating therapy to fight COVID-19. Combination of elements to improve the efficacy. *European Review for Medical and Pharmacological Sciences*, *24*, 9182–9187.

Dong, Z., Yang, S., Tang, C., Li, D., Kan, Y., & Yao, L. (2025). New insights into microbial bile salt hydrolases: From physiological roles to potential applications. *Frontiers in Microbiology*, *16*, 1513541. DOI: 10.3389/fmicb.2025.1513541

Du Qiao, L. H., Zhang, X., Chen, X., Zhang, J., Zou, J., Zhao, D., Zhu, W., Qian, X., & Li, H. (2025). The convergence of artificial intelligence and microfluidics in drug research and development. *Engineering*. Advance online publication. DOI: 10.1016/j.eng.2025.07.025

Duan, H., Wang, L., Huangfu, M., & Li, H. (2023). The impact of microbiota-derived short-chain fatty acids on macrophage activities in disease: Mechanisms and therapeutic potentials. *Biomedicine and Pharmacotherapy*, *165*, 115276. DOI: 10.1016/j.biopha.2023.115276

Ejtahed, H. S., Parsa, M., & Larijani, B. (2023). Ethical challenges in conducting and the clinical application of human microbiome research. *Journal of Medical Ethics and History of Medicine*, *16*, 5. DOI: 10.18502/jmehm.v16i5.13313

Elhalik, M. A., Mekky, A. E., Khedr, M., & Suleiman, W. B. (2024). Antineoplastic with DNA fragmentation assay and antioxidant, anti-inflammatory with gene expression activity of *Lactobacillus plantarum* isolated from local Egyptian milk products. *BMC Microbiology, 24*(1), 443. DOI: 10.1186/s12866-024-03576-y

Elkot, W. F., & Ismail, H. A. (2025). Technological advances in functional dairy foods and their role in human nutrition: A review. *Aswan Journal of Agricultural and Biological Sciences, 1*(1), 24–33. DOI: 10.21608/AJABS.2025.397996.1005

Fadare, O. S., Enabulele, O. I., Singh, V., & Pradhan, D. (2025). GC–MS based metabolomic profiling of synbiotic action of probiotic lactobacilli with aqueous garlic extract against Salmonella spp. *The Microbe, 8*, 100494. DOI: 10.1016/j.microb.2025.100494

Fentie, E. G., Lim, K., Andargie, Y. E., Azizoglu, U., Shin, J., & Tuganbay, A. (2025). Globalizing the health-promoting potential of fermented foods: A culturomics pathway to probiotics. *Trends in Food Science & Technology, 163*, 105119. DOI: 10.1016/j.tifs.2025.105119

Fitsum, S., Gebreyohannes, G., & Sbhatu, D. B. (2025). Bioactive compounds in fermented foods: Health benefits, safety, and future perspectives. *Applied Food Research, 5*(2), 101097. DOI: 10.1016/j.afres.2025.101097

Ganguly, A., Kirtania, D. K., & Chhetri, M.. (2025). Sustainable development through probiotics: A scientometric analysis of Indian research and innovation. *Discover Sustainability, 6*, 455. DOI: 10.1007/s43621-025-01263-0

González, R., & Elena, S. F. (2021). Interplay between the Host Microbiome and Pathogenic Viral Infections. *mBio, 12*(6), e0249621. DOI: 10.1128/mBio.02496-21

Gunjan, H., Himanshu, Mukherjee, R., et al. (2023). Comparative meta-analysis of antimicrobial resistance from different food sources along with one health approach in Egypt and the UK. *BMC Microbiology, 23*, 291. DOI: 10.1186/s12866-023-03030-5

Hammam, A. R. A., Elfaruk, M. S., Ahmed, M. E., & Sunkesula, V. (2020). Characteristics and technological aspects of the Egyptian cheeses. *International Journal of Current Microbiology and Applied Sciences, 9*(6), 3338–3354. DOI: 10.20546/ijcmas.2020.906.397

Han, X., Liu, Q., Li, Y., Zhang, M., Liu, K., Kwok, L.-Y., Zhang, H., & Zhang, W. (2025). Synergizing artificial intelligence and probiotics: A comprehensive review of emerging applications in health promotion and industrial innovation. *Trends in Food Science & Technology, 159*, 104938. DOI: 10.1016/j.tifs.2025.104938

Harahap, I. A., Moszak, M., Czlapka-Matyasik, M., Skrypnik, K., Bogdański, P., & Suliburska, J. (2024). Effects of daily probiotic supplementation with *Lactobacillus acidophilus* on calcium status, bone metabolism biomarkers, and bone mineral density in postmenopausal women: A controlled and randomized clinical study. *Frontiers in Nutrition, 11*, 1401920. DOI: 10.3389/fnut.2024.1401920

Ikombayev, T., Ospanova, A., Omarova, A., Horáčková, Š., Tuganbay, A., Kassenova, G., Zhamurova, V., Sharipova, S., & Suleimenova, Z. (2025). Functional properties of lactic acid bacteria isolated from raw goat milk and cottage cheese. *Journal of Agriculture and Food Research, 21*, 101822. DOI: 10.1016/j.jafr.2025.101822

Jiang, Y.-H., Xin, W.-G., Zhang, Q.-L., Lin, L.-B., & Deng, X.-Y. (2022). A Novel Bacteriocin Against Shigella flexneri From Lactiplantibacillus plantarum Isolated From Tilapia Intestine: Purification, Antibacterial Properties and Antibiofilm Activity. *Frontiers in Microbiology, 12*, 779315. DOI: 10.3389/fmicb.2021.779315

Jiang, S., Ma, W., Ma, C., Zhang, Z., Zhang, W., & Zhang, J. (2024). An emerging strategy: Probiotics enhance the effectiveness of tumor immunotherapy by mediating the gut microbiome. *Gut Microbes, 16*(1). Advance online publication. DOI: 10.1080/19490976.2024.2341717

Khalid, A., Babry, J. A., Vearey, J., & Zenner, D. (2023). Turning up the heat: A conceptual model for understanding the migration and health in the context of global climate change. *Journal of Migration and Health, 7*, 100172. DOI: 10.1016/j.jmh.2023.100172

Kiran, B. S., Bashir, O., & Pawase, P. A.. (2025). A review of in vitro digestibility models on diverse foods in various segments of human digestive tract. *Discover Food, 5*, 273. DOI: 10.1007/s44187-025-00454-y

Kumar, P. A., Gagana, C. J., Sowmya, T., Lalitha, L., Lavanya, S. N., & Divya, S. (2024). Regulatory requirements for dietary supplements under food safety and standards authority of India (FSSAI) in comparison with Cambodia. *World Journal of Advanced Research and Reviews, 23*(3), 2558–2571. DOI: 10.30574/wjarr.2024.23.3.2829

Li, S.-C., Hsu, W.-F., Chang, J.-S., & Shih, C.-K. (2019). Combination of *Lactobacillus acidophilus* and *Bifidobacterium animalis* subsp. *lactis* Shows a Stronger Anti-Inflammatory Effect than Individual Strains in HT-29 Cells. *Nutrients, 11*(5), 969. DOI: 10.3390/nu11050969

Liu, C., Chu, D., Kalantar-Zadeh, K., George, J., Young, H. A., & Liu, G. (2021). Cytokines: From Clinical Significance to Quantification. Advanced science (Weinheim, Baden-Wurttemberg, Germany), 8(15), e2004433. https://doi.org/DOI: 10.1002/advs.202004433

Liu, H., Huang, S. S., Lingam, G., Kai, D., Su, X., & Liu, Z. (2024). Advances in retinal pigment epithelial cell transplantation for retinal degenerative diseases. *Stem Cell Research & Therapy*, *15*(1), 390.

Liu, M., Fang, Y., Chen, R., Cai, M., Yang, X., Fang, Z., Fang, X., & Dong, S. (2025). Effect of *Lactobacillus plantarum* 1243 fermentation on quality properties and metabolome of Aronia melanocarpa (Michx.) elliott juice. *Food Chemistry: X*, *29*, 102706. DOI: 10.1016/j.fochx.2025.102706

Lone, A., Mottawea, W., Ait Chait, Y., & Hammami, R. (2021). Dual Inhibition of *Salmonella enterica* and *Clostridium perfringens* by New Probiotic Candidates Isolated from Chicken Intestinal Mucosa. *Microorganisms*, *9*(1), 166. DOI: 10.3390/microorganisms9010166

Maillard, F., Meynier, M., Mondot, S., Pepke, F., Galbert, C., Torres Maravilla, E., Kropp, C., Sokol, H., Carvalho, F. A., Jacouton, E., Holowacz, S., Langella, P., Chain, F., & Martín, R. (2023). From in Vitro to in Vivo: A Rational Flowchart for the Selection and Characterization of Candidate Probiotic Strains in Intestinal Disorders. *Microorganisms*, *11*(4), 906. DOI: 10.3390/microorganisms11040906

Maldonado-Reina, A. J., López-Ruiz, R., Romero-González, R., Martínez Vidal, J. L., & Garrido-Frenich, A. (2022). Assessment of Co-Formulants in Marketed Plant Protection Products by LC-Q-Orbitrap-MS: Application of a Hybrid Data Treatment Strategy Combining Suspect Screening and Unknown Analysis. *Journal of Agricultural and Food Chemistry*, *70*(23), 7302–7313. DOI: 10.1021/acs.jafc.2c01152

Manan, A., & Musaalbakri. (2025). Progress in probiotic science: Prospects of functional probiotic-based foods and beverages. *International Journal of Food Science, 2025*, Article 5567567. 34 pages. https://doi.org/DOI: 10.1155/ijfo/5567567

Marco, M. L., Sanders, M. E., & Gänzle, M.. (2021). The International Scientific Association for Probiotics and Prebiotics (ISAPP) consensus statement on fermented foods. *Nature Reviews. Gastroenterology & Hepatology*, *18*, 196–208. DOI: 10.1038/s41575-020-00390-5

Martinez, J., Rodriguez Hovnanian, K. M., & Martinez, E. E. (2023). Biomarkers and Functional Assays of Epithelial Barrier Disruption and Gastrointestinal Dysmotility in Critical Illness-A Narrative Review. *Nutrients*, *15*(18), 4052. DOI: 10.3390/nu15184052

McCoy, R., Oldroyd, S., Yang, W., Wang, K., Hoven, D., Bulmer, D., Zilbauer, M., & Owens, R. M. (2024). In Vitro Models for Investigating Intestinal Host–Pathogen Interactions. *Advancement of Science, 11*(7), 2306727. DOI: 10.1002/advs.202306727

Merenstein, D., Pot, B., Leyer, G., Ouwehand, A. C., Preidis, G. A., Elkins, C. A., Hill, C., Lewis, Z. T., Shane, A. L., Zmora, N., Petrova, M. I., Collado, M. C., Morelli, L., Montoya, G. A., Szajewska, H., Tancredi, D. J., & Sanders, M. E. (2023). Emerging issues in probiotic safety: 2023 perspectives. *Gut Microbes, 15*(1), 2185034. DOI: 10.1080/19490976.2023.2185034

Merrick, B., Sergaki, C., Edwards, L., Moyes, D. L., Kertanegara, M., Prossomariti, D., Shawcross, D. L., & Goldenberg, S. D. (2023). Modulation of the Gut Microbiota to Control Antimicrobial Resistance (AMR)-A Narrative Review with a Focus on Fecal Microbiota Transplantation (FMT). *Infectious Disease Reports, 15*(3), 238–254. DOI: 10.3390/idr15030025

Ministry of Health and Population [Egypt], El-Zanaty and Associates [Egypt], & ICF International. (2015). *Egypt Health Issues Survey 2015*. Ministry of Health and Population & ICF International. https://www.unicef.org/egypt/sites/unicef.org.egypt/files/2018-04/eg_EHIS_2015_en.pdf

Mir, S. S. (2025). Harnessing Microbial Ecosystems in Health and Disease: Innovations in Medical Microbiology for Advancing Precision Medicine and Therapeutic Strategies. *Egyptian Academic Journal of Biological Sciences. C, Physiology and Molecular Biology, 17*(1), 41–67. DOI: 10.21608/EAJBSC.2025.411912

Muhammad, I., Contes, K., Bility, M. T., & Tang, Q. (2025). Chasing Virus Replication and Infection: PAMP-PRR Interaction Drives Type I Interferon Production, Which in Turn Activates ISG Expression and ISGylation. *Viruses, 17*(4), 528. DOI: 10.3390/v17040528

Muthuvelu, K. S., Ethiraj, B., Pramnik, S., Raj, N. K., Venkataraman, S., Rajendran, D. S., Bharathi, P., Palanisamy, E., Narayanan, A. S., Vaidyanathan, V. K., & Muthusamy, S. (2023). Biopreservative technologies of food: An alternative to chemical preservation and recent developments. *Food Science and Biotechnology, 32*(10), 1337–1350. DOI: 10.1007/s10068-023-01336-8

Naito, E., Yoshida, Y., Kunihiro, S., Makino, K., Kasahara, K., Kounoshi, Y., Aida, M., Hoshi, R., Watanabe, O., Igarashi, T., Miyazaki, K., & Ito, H. (2018). Effect of Lactobacillus casei strain Shirota-fermented milk on metabolic abnormalities in obese prediabetic Japanese men: A randomized, double-blind, placebo-controlled trial. *Bioscience of Microbiota, Food and Health, 37*(1), 9–18. DOI: 10.12938/bmfh.17-012

O'Connell Motherway, M., O'Connor, L., Smith, S., Bolton, D., Cormican, M., De Waal, T., Gutierrez, M., Part, M. A. M., Walsh, F., & Duffy, G. (2024). Report of the Scientific Committee of the Food Safety Authority of Ireland on the assessment of the safety of "probiotics" in food supplements. *Food Risk Assess Europe*, *2*, e0050. DOI: 10.2903/fr.efsa.2024.FR-0050

Ombarak, R., Mahrous, H., & Hussien, H. (2019). Antibacterial activity of lactobacilli strains isolated from Laban Rayeb "Egyptian traditional fermented milk" against Pseudomonas fluorescens and Escherichia coli in vitro. *Journal of Current Veterinary Research*, *1*(2), 105–111. DOI: 10.21608/jcvr.2019.57139

Parvin, T., & Sadras, S. R. (2024). Advanced probiotics: Bioengineering and their therapeutic application. *Molecular Biology Reports*, *51*, 361. DOI: 10.1007/s11033-024-09309-8

Ramos Meyers, G., Samouda, H., & Bohn, T. (2022). Short Chain Fatty Acid Metabolism in Relation to Gut Microbiota and Genetic Variability. *Nutrients*, *14*(24), 5361. DOI: 10.3390/nu14245361

Rasika, D. M. D., Vidanarachchi, J. K., Luiz, S. F., Azeredo, D. R. P., Cruz, A. G., & Ranadheera, C. S. (2021). Probiotic Delivery through Non-Dairy Plant-Based Food Matrices. *Agriculture*, *11*(7), 599. DOI: 10.3390/agriculture11070599

Reyed, R. M., & Prabhakar, P. K. (2025a). Mechanisms, Impact, and Therapeutic Strategies for Oxidative Stress in Metabolic Diseases: Understanding and Mitigating Oxidative Stress. In Prabhakar, P. (Ed.), *The Role of Reactive Oxygen Species in Human Health and Disease* (pp. 203–256). IGI Global Scientific Publishing., DOI: 10.4018/979-8-3693-7919-6.ch006

Reyed, R. M., & Prabhakar, P. K. (2025 b). Personalized Antiviral Therapy TailoringTreatments to Genetic Environmental and Lifestyle Factors. In Aslam, M., Yingngam, B., Javed, T., & Riaz, S. (Eds.), *Global Perspectives on Antiviral Drug Development* (pp. 63–138). IGI Global Scientific Publishing., DOI: 10.4018/979-8-3693-9276-8.ch003

Reyed, R. M. (2022). Practical implementation of nutritional serves, gut microbiome, combination therapy, and nutraceuticals supplements. *Microenvironmental and Microbiological Research*, *4*(2), 8. DOI: 10.53388/MMR2022008

Reyed, R. M. (2023). Focusing on individualized nutrition within the algorithmic diet: An in-depth look at recent advances in nutritional science, microbial diversity studies, and human health. *Food Health*, *5*(1), 5. DOI: 10.53388/FH2023005

Roach, J., Mital, R., Haffner, J. J., Colwell, N., Coats, R., Palacios, H. M., Liu, Z., Godinho, J. L. P., Ness, M., Peramuna, T., & McCall, L. I. (2024). Microbiome metabolite quantification methods enabling insights into human health and disease. *Methods (San Diego, Calif.)*, *222*, 81–99. DOI: 10.1016/j.ymeth.2023.12.007

Roman, S., Campos-Medina, L., & Leal-Mercado, L. (2024). Personalized nutrition: The end of the one-diet-fits-all era. *Frontiers in Nutrition*, *11*, 1370595. DOI: 10.3389/fnut.2024.1370595

RRMA. (2025, June 11). *Brazil updates supplement regulation to include new ingredients and labeling rules*. RRMA Global. https://rrma-global.org/news-details/brazil-updates-supplement-regulation-to-include-new-ingredients-and-labelling-rules/MjE0Nw==

Sabino, Y. N. V., Paiva, A. D., Fonseca, B. R., Medeiros, J. D., & Machado, A. B. F. (2025). Deciphering probiotic potential: A comprehensive guide to probiogenomic analyses. *Future Microbiology*, *20*(7–9), 611–622. DOI: 10.1080/17460913.2025.2492472

Sanders, M. E., Merenstein, D. J., Reid, G., Gibson, G. R., & Rastall, R. A. (2019). Author Correction: Probiotics and prebiotics in intestinal health and disease: from biology to the clinic. *Nature Reviews. Gastroenterology & Hepatology*, *16*(10), 642. DOI: 10.1038/s41575-019-0199-6

Sarfraz, M. H., Shahid, A., Asghar, S., Aslam, B., Ashfaq, U. A., Raza, H., Prieto, M. A., Simal-Gandara, J., Barba, F. J., Rajoka, M. S. R., Khurshid, M., & Nashwan, A. J. (2022). Personalized nutrition, microbiota, and metabolism: A triad for eudaimonia. *Frontiers in Molecular Biosciences*, *9*, 1038830. DOI: 10.3389/fmolb.2022.1038830

Sarita, B., Samadhan, D., Hassan, M. Z., & Kovaleva, E. G. (2025). A comprehensive review of probiotics and human health-current prospective and applications. *Frontiers in Microbiology*, *15*, 1487641. DOI: 10.3389/fmicb.2024.1487641

Senanayake, D., Torley, P. J., Chandrapala, J., & Terefe, N. S. (2023). Microbial Fermentation for Improving the Sensory, Nutritional and Functional Attributes of Legumes. *Fermentation (Basel, Switzerland)*, *9*(7), 635. DOI: 10.3390/fermentation9070635

Shahzad, K., Fatima, A., Cadeiras, M., Wisniewski, N., Bondar, G., Cheng, R., Reed, E., & Deng, M. (2012). Challenges and solutions in the development of genomic biomarker panels: A systematic phased approach. *Current Genomics*, *13*(4), 334–341. DOI: 10.2174/138920212800793339

Shankar, V., Gouda, M., Moncivaiz, J., Gordon, A., Reo, N. V., Hussein, L., & Paliy, O. (2017). Differences in Gut Metabolites and Microbial Composition and Functions between Egyptian and U.S. Children Are Consistent with Their Diets. *mSystems*, *2*(1), e00169–e16. DOI: 10.1128/mSystems.00169-16

Shen, Y., Fan, N., Ma, S. X., Cheng, X., Yang, X., & Wang, G. (2025). Gut Microbiota Dysbiosis: Pathogenesis, Diseases, Prevention, and Therapy. *MedComm*, *6*(5), e70168. DOI: 10.1002/mco2.70168

Singh, V. K., Hu, X. H., Singh, A. K., Solanki, M. K., Vijayaraghavan, P., Srivastav, R., Joshi, N. K., Kumari, M., Singh, S. K., Wang, Z., & Kumar, A. (2024). Precision nutrition-based strategy for management of human diseases and healthy aging: Current progress and challenges forward. *Frontiers in Nutrition*, *11*, 1427608. DOI: 10.3389/fnut.2024.1427608

Spacova, I., Ouwehand, A. C., Binda, S., Younes, J. A., Henoud, S., Legrain-Raspaud, S., Dekker, J., Espadaler-Mazo, J., Langella, P., Martín, R., & Pane, M. (2024). Biotics: Applications of functional ingredients in different product categories: The example of probiotics. *Agro Food Industry Hi-Tech*, *35*(1). https://digital.teknoscienze.com/agrofood_industry_hi_tech_35_1_2024/biotics_-_applications_of_functional_ingredients_in_different_product_categories_the_example_of_probiotics

Srivastava, R. (2025). Advancing precision oncology with AI-powered genomic analysis. *Frontiers in Pharmacology*, *16*, 1591696. DOI: 10.3389/fphar.2025.1591696

Sutula, J., Coulthwaite, L. A., Thomas, L. V., & Verran, J. (2013). The effect of a commercial probiotic drink containing Lactobacillus casei strain Shirota on oral health in healthy dentate people. *Microbial ecology in health and disease, 24*, . https://doi.org/DOI: 10.3402/mehd.v24i0.21003

Swanson, K. S., Gibson, G. R., & Hutkins, R.. (2020). The International Scientific Association for Probiotics and Prebiotics (ISAPP) consensus statement on the definition and scope of synbiotics. *Nature Reviews. Gastroenterology & Hepatology*, *17*(11), 687–701. DOI: 10.1038/s41575-020-0344-2

Takada, M., Nishida, K., Kataoka-Kato, A., Gondo, Y., Ishikawa, H., Suda, K., Kawai, M., Hoshi, R., Watanabe, O., Igarashi, T., Kuwano, Y., Miyazaki, K., & Rokutan, K. (2016). The probiotic Lactobacillus casei strain Shirota relieves stress-associated symptoms by modulating the gut–brain interaction in human and animal models. *Neurogastroenterology and Motility*, *28*(7), 1027–1036. DOI: 10.1111/nmo.12804

Tiwari, A., Ika Krisnawati, D., Susilowati, E., Mutalik, C., & Kuo, T. R. (2024). Next-Generation Probiotics and Chronic Diseases: A Review of Current Research and Future Directions. *Journal of Agricultural and Food Chemistry*, *72*(50), 27679–27700. DOI: 10.1021/acs.jafc.4c08702

Tone-Pah-Hote, T., & Redvers, N. (2022). The commercialization of biospecimens from indigenous peoples: A scoping review of benefit-sharing. *Frontiers of Medicine*, *9*, 978826. DOI: 10.3389/fmed.2022.978826

Umair, M., Jabbar, S., Zhaoxin, L., Jianhao, Z., Abid, M., Khan, K. R., Korma, S. A., Alghamdi, M. A., El-Saadony, M. T., Abd El-Hack, M. E., Cacciotti, I., AbuQamar, S. F., El-Tarabily, K. A., & Zhao, L. (2022). Probiotic-Based Bacteriocin: Immunity Supplementation against Viruses. An Updated Review. *Frontiers in Microbiology*, *13*, 876058. DOI: 10.3389/fmicb.2022.876058

Ushkalova, E. A., & Gushchina, Yu. S. (2015). Linex forte v profilaktike i lechenii zheludochno-kishechnykh zabolevanii [Linex forte in the prevention and treatment of gastrointestinal diseases]. *Terapevticheskii Arkhiv*, *87*(12), 138–144. DOI: 10.17116/terarkh20158712138-144

Wang, G., Fan, Y., & Zhang, G.. (2024). Microbiota-derived indoles alleviate intestinal inflammation and modulate the microbiome via microbial crossfeeding. *Microbiome*, *12*, 59. DOI: 10.1186/s40168-024-01750-y

Wang, J., Huang, M., & Du, Y.. (2024). Lactobacillus rhamnosus GG Regulates Host IFN-I Through the RIG-I Signaling Pathway to Inhibit Herpes Simplex Virus Type 2 Infection. *Probiotics and Antimicrobial Proteins*, *16*, 1966–1978. DOI: 10.1007/s12602-023-10137-8

Wang, L., Dekker, M., Heising, J., Zhao, L., & Fogliano, V. (2023). Food matrix design can influence the antimicrobial activity in the food systems: A narrative review. *Critical Reviews in Food Science and Nutrition*, *64*(25), 8963–8989. DOI: 10.1080/10408398.2023.2205937

Wang, Y., Moon, A., Huang, J., Sun, Y., & Qiu, H.-J. (2022). Antiviral Effects and Underlying Mechanisms of Probiotics as Promising Antivirals. *Frontiers in Cellular and Infection Microbiology*, *12*, 928050. DOI: 10.3389/fcimb.2022.928050

Widodo, W., Ariani, A. L., Widianto, D., & Haltrich, D. (2022). Genomic Comparison of Lactobacillus casei AP and *Lactobacillus plantarum* DR131 with Emphasis on the Butyric Acid Biosynthetic Pathways. *Microorganisms*, *9*(2), 425. DOI: 10.3390/microorganisms9020425

Wu, X., Yang, X., Dai, Y., Zhao, Z., Zhu, J., Guo, H., & Yang, R. (2024). Single-cell sequencing to multiomics: Technologies and applications. *Biomarker Research, 12*(1), 110. DOI: 10.1186/s40364-024-00643-4

Yaqub, M. O., Jain, A., Joseph, C. E., & Edison, L. K. (2025). Microbiome-Driven Therapeutics: From Gut Health to Precision Medicine. *Gastrointestinal Disorders (Basel, Switzerland), 7*(1), 7. DOI: 10.3390/gidisord7010007

Yee, C. S., Zahia-Azizan, N. A., Abd Rahim, M. H., Mohd Zaini, N. A., Raja-Razali, R. B., Ushidee-Radzi, M. A., Ilham, Z., & Wan-Mohtar, W. A. A. Q. I. (2025). Smart Fermentation Technologies: Microbial Process Control in Traditional Fermented Foods. *Fermentation (Basel, Switzerland), 11*(6), 323. DOI: 10.3390/fermentation11060323

Yin, B., Tan, G., Muhammad, R., Liu, J., & Bi, J. (2025). AI-Powered Innovations in Food Safety from Farm to Fork. *Foods, 14*(11), 1973. DOI: 10.3390/foods14111973

Yumnam, H., & Sharma, I. (2025). Beneficial microbiota in indigenous fermented foods and beverages. In Tiwari, A., & Sarma, H. (Eds.), *Ethnic and indigenous food technologies. Disaster resilience and green growth.* Springer., DOI: 10.1007/978-981-96-9286-6_14

Zhang, Z., Niu, H., Qu, Q., Guo, D., Wan, X., Yang, Q., & Liu, Z. (2025). Advancements in Lactiplantibacillus plantarum: Probiotic characteristics, gene editing technologies and applications. *Critical Reviews in Food Science and Nutrition*, •••, 1–22. DOI: 10.1080/10408398.2024.2448562

Zimmer, A., Wang, E. R., Choudhary, G., & Zhang, P. (2024). Protocol for simultaneous isolation of high-quality and high-quantity cardiomyocytes and nonmyocyte cells from adult rat hearts. *STAR Protocols, 5*(3), 103174. DOI: 10.1016/j.xpro.2024.103174

Zommara, M., El-Ghaish, S., & Haertle, T.. (2023). Probiotic and technological characterization of selected Lactobacillus strains isolated from different Egyptian cheeses. *BMC Microbiology, 23*, 160. DOI: 10.1186/s12866-023-02890-1

Chapter 9
Ethical and Regulatory Considerations in AI-Driven Healthcare

Ashok Kumar Sah
http://orcid.org/0000-0002-7762-4351
A' Sharqiyah University, Oman

Rahul Kumar
http://orcid.org/0009-0001-0237-3201
Vidya University, India

Sakshi Patel
http://orcid.org/0009-0000-2041-9325
Meerut Institute of Engineering and Technology, India

Prem Shankar Mishra
http://orcid.org/0000-0003-3797-1575
Vidya University, India

Hiren Doshi
Vidya University, India

Rakhi Mishra
Noida Institute of Engineering and Technology Pharmacy Institute, India

ABSTRACT

Artificial Intelligence (AI) is revolutionizing clinical practice by providing previously unheard-of capabilities in patient monitoring, diagnosis, treatment personalization, and healthcare system optimization. Even though artificial intelligence (AI) promises to improve accuracy, efficiency, and scalability, its quick adoption also brings with it difficult moral, legal, and social issues. Using a critical lens based on medical ethics, legal frameworks, and practical implications, this chapter examines the dual reality of AI-driven healthcare—its transformative potential and its associated risks. In the context of algorithmic decision-making, fundamental ethical concepts like beneficence, autonomy, justice, and transparency are reviewed. Data bias, health disparities, patient consent, data privacy, clinical deskilling, and accountability in the case of AI system failures are among the urgent issues covered in this chapter. It assesses new ethical frameworks from international organizations and medical

DOI: 10.4018/979-8-3373-3121-8.ch009

institutions and offers a comparative overview of regulatory environments across the globe. The chapter highlights the importance of collaborative, multidisciplinary approaches to ensure AI remains a tool for equitable and human-centered care by highlighting the roles of key stakeholders, including developers, healthcare professionals, regulators, and patients. In order to integrate technological innovation with the ethical underpinnings of healthcare, it concludes with recommendations for the future that include adaptive regulation, ethical design practices, and ongoing vigilance. This chapter is a call to action as well as a guide for using AI responsibly for the benefit of humankind and health.

1. INTRODUCTION

Artificial Intelligence (AI) is currently revolutionizing the contemporary healthcare paradigm. AI is promising to make the healthcare industry more effective, affordable, and accurate, including such advance discoveries as more predictive diagnostics and procedures that involve using robots as surgeons (Topol, 2019). But along with this transformational capability there are a variety of ethical and regulatory issues to consider. The consequences of using AI in clinical decision-making and patient monitoring, as in other processes traditionally done by humans, become much higher as AI starts to engage in them (Morley et al., 2020). Achieving the implementation of such technologies in a way in which the values of medicine, e.g., beneficence, justice, and autonomy, are observed has become an urgent issue. The assimilation of AI in the healthcare system does not occur technically but as a paradigm that has organically touched on the norms of society, law, and philosophical thinking (Price & Cohen, 2019). It dims assumptions that have been there in relation to human judgment, relationships between patients and doctors and also the accountability within institutions. Therefore, it is important not only to policy-makers and developers, but to all stakeholders in a healthcare ecosystem, to learn what the ethical and regulatory aspects of AI are, and how they can be addressed (Gerke et al., 2020).

1.1. The Rise of AI in Healthcare

There has been a tremendous increase in the use of AI in healthcare in the last 10 years or so. The development of machine learning, natural language processing, and computer vision allowed AI to become an inseparable part of most healthcare functions (Bohr & Memarzadeh, 2020):

- The algorithm could promote early diagnosis by identifying patterns on the medical images with the use of AI.
- The virtual health assistants enhance better patient participation and treatment compliance.
- Predictive models are in real time and are intended to forecast disease outbreaks or patient deterioration.

Also, startups and big tech companies are actively pouring funds into AI-powered solutions, and healthcare systems in various countries are employing this technology to solve the problem of clinician shortage, growing expenses and the necessity of personalized care (Yanamadala et al., 2024). Yet the speed of adoption tends to be faster than the creation of proper ethical principles and regulatory frameworks allowing malpractices, injustices, and unintentional consequences.

1.2. Scope, Purpose, and Key Issues

With Artificial Intelligence (AI) increasingly progressing and entering various domains of healthcare, it should be noted that it is more crucial now than ever to analyze critically this technological transformation on a larger scale (Kumar et al., 2024). This chapter aims at giving a detailed discussion of ethical implications and policy environments that guide the advancement, application, and monitoring of AI-based frameworks in medical settings. The implementations spread over to the entire range of AI-operated usage-based applications-including clinical decision support, diagnostic imaging, end-user digital tools, and administrative efficiency (Malla & Amin, 2023).

This chapter is meant to become the source of information intended to be used as a primer by a varied audience, such as patients and other healthcare professionals, AI developers, as well as ethicists, policymakers, legal scholars, and activists in the field of public health (Fahim, 2024). This piece of work should add to the discussion of a more informed and equitable approach to the design and regulation of AI in healthcare settings by highlighting the major ethical and regulatory issues and using them as the case study through which one can gain insight into those issues (Hirani et al., 2024).

1.3 Why Ethics and Regulation Matter in AI-Driven Healthcare

In contrast to conventional medical technologies, each of the AI systems is highly likely to be an autonomous or semi-autonomous agent and can make or influence decisions with a direct impact upon patient health and wellbeing (Mennella et al.,

2024). The transfer of clinical decision-making process to a machine leaves serious ethical concerns:

- Does patient autonomy and informed decision-making allow algorithms to be respected?
- Do they strengthen the social disparities that prevail besides creating new types of damages?
- When AI goes wrong, who is at fault and how is a redress or accountability possible?

Besides these ethical considerations, the law and regulatory aspects are also of utmost importance (Alkhatib & Darabseh, 2025). Most of the current health data protection regulation, medical device regulation, and liability systems were innovated at a time when AI could not be applied in the clinical setting. This has impaired a checkboard of standards that are only capable of operating at an inadequate speed to keep in stride with innovation that leads to the presence of legal luminescence and dissimilar protection of patients and practitioners (Banerjee et al., 2025).

2. BACKGROUND: AI TECHNOLOGIES AND HEALTHCARE APPLICATIONS

In order to consider the ethical and regulatory perimeters of AI usage in healthcare, we will have to discuss at a first glance, what AI represents, what ways it is currently being used in the healthcare environment, and what the trade-offs are behind the introduction of AI in the healthcare industry (Chen & Decary, 2020). The section will give a basic level of knowledge of the essential AI technologies and the world of their applications with the benefits and dangers it involves (Amann et al., 2020).

2.1. Key AI Technologies in Use

AI in healthcare comprises a suite of technologies designed to mimic or augment human cognitive functions (Belk et al., 2023). These include:

i. **Machine Learning (ML):** ML models are used to provide structured data inputs with the goal of making predictions. They are popular in diagnostic algorithms and risk stratification and are refined and enhanced after receiving new data (Mukhamediev et al., 2022).
ii. **Deep Learning:** It is a type of ML which applies multi-layered neural networks to make sense of multifaceted data like medical images, genetic sequences, and

speeches. It has played a critical role in the developments of radiology, pathology, and ophthalmology (Sarker, 2021).

iii. **Natural Language Processing (NLP):** Natural Language Processing (NLP) permits the computer to comprehend and produce human language. In healthcare it allows one to distill clinical data in forms of unstructured text including physician notes and electronic health records (Wang et al., 2023).

iv. **Computer Vision:** It is an AI subfield that reads and analyses visual data. It is a first among image-based diagnostics, like the presence of a tumor on a radiograph or an unusual area displayed on a dermatology scan (Matsuzaka & Yashiro, 2023).

v. **Reinforcement Learning:** This technique is just starting to come to clinical practice, but it allows AI systems to make decisions by trial and error, and has already been used in adaptive therapy and personalised plans of treatment (Matsuo et al., 2022).

Captively, such technologies are transforming clinical decision-making, operating clinical workflows, as well as patient engagement tactics (Mulla & Patel, 2025).

2.2. Applications, Benefits, and Risks of AI in Healthcare

Artificial Intelligence (AI) has become a fast-growing inseparable component of modern healthcare that increases the accuracy of diagnosis, personalizes and streamlines the systems (Lee & Yoon, 2021). It can be used to process medical images and predict disease, streamline workflows at hospitals, and allow a patient to be monitored remotely using a wearable or mobile app. Clinical Decision Support Systems (CDSS) that uses artificial intelligence (AI) is also transforming clinical practice by providing real-time guidance that supports evidence-based decisions made by the clinicians (Lin et al., 2020). With the processing of large amounts of data genomics, lifestyle, and patient history, AI can facilitate highly personalized treatment and facilitate proactive and data-driven interventions. Its scalability and automation are not only beneficial when it comes to expanding access to care in underserved areas but also on making administrative tasks less time-consuming, thus letting the providers spend more time unwinding with their patients (Pugliese et al., 2021). In addition, the quality of AI to learn and continually improve presents a dynamic tool to the stagnant clinical guidelines (Wasilewski et al., 2024). To better understand how AI is being deployed across different healthcare systems worldwide, *Table 1* outlines adoption trends, primary use cases, and national strategies across five diverse countries.

Table 1. Global Adoption Trends of AI Technologies in Healthcare

Country/ Region	Primary AI Use Cases	Adoption Stage	Notable Initiatives
United States	Diagnostic imaging, predictive analytics, virtual nursing	Mature & expanding	FDA Digital Health Center, Mayo Clinic AI factory
United Kingdom	AI triage, radiology, remote monitoring	Advanced experimentation	NHS AI Lab, AI Award Program
India	Telemedicine AI, triage, tuberculosis screening	Emerging deployment	eSanjeevani AI, NITI Aayog Health AI pilot programs
China	Hospital automation, CT image analysis, pandemic response	Rapid scaling	Ping A Good Doctor, Alibaba AI cloud hospital
Brazil	Mobile health, maternal care AI, triage bots	Pilot phase	SUS Digital Health Strategy, AI for primary care projects

Nevertheless, such developments are associated with multifaceted ethical, technological, and social issues. This can be a problem in such a way that it is challenging to comprehend or explain what many AI models would produce or execute particularly deep learning systems, which can erode credit and responsibility relations (Saraswat et al., 2022). Training bias can provide a foundation to healthcare inequality, and AI over-dependence can cause clinical deskilling. The issue of data privacy and consent is also an acute subject, as AI relies on substantial quantities of extremely sensitive health information. In the meantime, the regulatory systems tend to become out of step with the latest technology advancements, and such essential questions as liability and control are still unanswered. Overall, although AI can be transformative to the sector of healthcare, there is a way it should be adopted through high ethical values, transparent governance, and equity (Younis et al., 2024). The topic of innovating while remembering human-centered care is important to consider so that the AI may be implemented as a means of supplying inclusive and trust-based humane care on a highly effective level.

3. AN ANALYTICAL FRAMEWORK FOR ETHICAL AND REGULATORY GOVERNANCE OF AI IN HEALTHCARE

Artificial intelligence in healthcare is not only a technical tool, but it is a socio-technical system that is integrated into the clinical practice environment, institutional organization, and regulatory framework. The ethical and regulatory issues related to AI will not emerge at one point of use, but will be continually developed throughout the lifecycle of an AI system, including the collection of data and the development of the model, as well as its clinical implementation and subsequent surveillance in the market. In order to express this complexity, a lifecycle-based analytical framework has been adopted in this chapter incorporating both technical processes, ethics, and regulatory controls. On a technical level, healthcare AI systems can be characterized by a series of steps which involve: data acquisition, training of the algorithm, validation, deployment, and ongoing performance appraisal.

There are different risks involved in each stage. Data can be a historical imbalance or underrepresentation of a population, and algorithm models can be predictive and less interpretive or just. In practice, AI systems might be applied in operational clinical environments that are not as similar to those during training, which aggravates the threat of performance drift or accidental damage. Overlapping this technical lifecycle is the ethical aspect which gives normative direction on how AI can be used responsibly. The fundamental ethical principles like autonomy, beneficence, non-maleficence, justice, and transparency have been central though they need to be rethought in the context of algorithm-mediated care. Patient autonomy, in its turn, does not limit itself to the treatment consent, but also awareness of AI use in clinical decision-making. On the same note, justice requires consideration of the possibility that AI systems cause populations to be disproportionately disadvantaged or contribute to the existing healthcare inequalities.

Regulatory dimension converts the ethical expectations into the institutional responsibilities using tools like pre-market assessment, classification of risks, post-market monitoring, and liability models. Nevertheless, most current regulatory frameworks are built around some form of fixed medical technologies and they fail to adapt to adaptive or constantly learning AI systems. It is a malalignment that may leave accountability gaps especially where the accountability responsibility is shared between the developers, healthcare providers, and institutions. These three dimensions are interdependent as opposed to sequential i.e. technical, ethical and regulatory dimensions. The application of ethical principles on the regulatory requirements, the regulatory standards on the technical design options and technical constraints to traditional ethical assumptions may be seen. This combined structure enables a more consistent evaluation of risks, duties and protection by considering AI governance. It also offers the conceptual basis of ethical issues of analysis, regulatory differences, and governance strategies to the following sections.

4. ETHICAL PRINCIPLES IN AI-DRIVEN HEALTHCARE

Triple dependencies More than technical accuracy is needed in the integration of Artificial Intelligence into healthcare system; a critical moral basis that acknowledges the human interests embodied in the medical decision process would be necessary (Li et al., 2023). The use of AI in healthcare has a direct impact on diagnosing, treatment, and outcome unlike conventional technologies, which require minimal human supervision. This entails additional ethical conflicts which cannot be addressed by classical bioethics (Prakash et al., 2022). Although the classical principles of medical ethics have not lost their applicability, they have to be viewed differently namely, through the prism of algorithmic mediation, data dependency, and ML uncertainty.

Figure 1. The AI Ethics Lifecycle in Healthcare. Note: This figure was specifically developed and curated by Reyed M. Reyed (Editor) to provide a tailored, knowledge-based conceptual framework that aligns with the specialized content of this chapter.

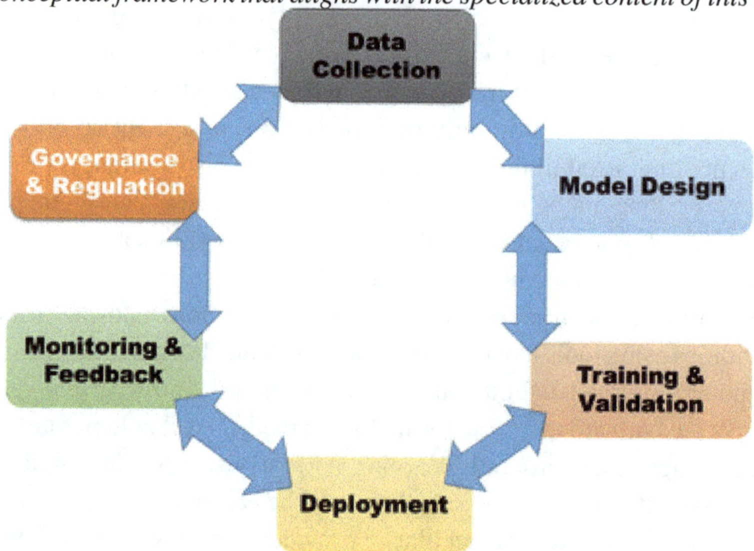

In order to make the case that ethical adequacy in AI is a continuous, dynamic negotiation between technology, clinicians, patients, and society, this section re-examines the fundamental ethical principles—autonomy, beneficence, non-maleficence, justice, and transparency—in the context of AI. Instead of being implemented after the fact, ethical considerations in healthcare AI must be ingrained throughout the system's whole lifecycle (Karimian et al., 2022). Fairness, accountability, and transparency must be considered at every stage, from data collection to governance, as illustrated in *Figure 1*, to guarantee results that are in line with ethical standards.

4.1. Autonomy and Informed Consent in the Algorithmic Age

Ethical medicine advances patient autonomy the ability to make one-self with the knowledge of the events approaching them. AI makes this principle complicated in depth though (Savolainen & Ruckenstein, 2024). A typical informed consent presupposes the ability of the patients (and usually of the clinicians) to reasonably grasp risks and benefits of the medical interventions. However, even though AI systems recommend specific treatment plans using opaque algorithms or probabilistic prediction using uninterpretable models, clinicians and patients might not fully understand how and why a specific decision is made (Pickering, 2021).

Such epistemic paralysis poses relevant questions because the consent is difficult to be informed when the rationale underlying an AI recommendation is not understandable (Astromskė et al., 2021). Is it okay to deny the patients the right to say no to algorithm input? And what about when artificial intelligence systems are so wrapped into the clinical infrastructure that there is no way reasonable to avoid? More models of algorithmic transparency and tiered consent are proposed, in which patients will be informed about both medical risks and the role, limitations, and data dependencies of the AI systems in question (Iserson, 2024). Patients need to be empowered to change their behavioral patterns that allow them to accept AI-encroached care only when they feel they have agency over AI in decision-making.

Most AI-based healthcare systems are based on complex and opaque algorithms the inner processes of which are not comprehensible even by expert clinicians. The challenge presented by this black-box nature is to the main medical ethic of patient autonomy. Unless the patient knows the way AI made a diagnosis or a recommendation, he or she cannot make a fully informed decision.

In response to this, medical professionals need to implement algorithmic transparency measures, including explainable AI (XAI) models, risk scores that are easy to interpret, and disclosure statements that are easy to understand. The argument that AI-assisted care ought to have informed consent is gaining momentum and argues that such consent ought to include:

a. The place of AI in diagnosis or decision support.
b. The nature and origin of the used data.
c. Risks that may arise such as biases or errors.
d. The level of control by humans.

The consent process cannot be complete and autonomy cannot be achieved without such disclosure.

4.2. Beneficence and Non-Maleficence in Data-Driven Care

The other concepts, such as beneficence (or do good) and non-maleficence (or do no harm), will have to be re-considered in the context of the algorithms that would define clinical decisions made more often basing on the historical data (Reddy, 2025). AI can be used to increase beneficence by earlier and accurate risk prediction as well as individualized treatment plans. But it is also that those same systems are capable of causing harm through reinforcing old clinical prejudices, ignoring rare diseases, or being unable to cope with unusual patient profiles (Garno et al., 2025). Most importantly, AI harms are not always obvious or instant. The error less algorithm might be subtle enough, or take sufficient time to develop, that causality is hardly traceable (ZadZiabari & Tabatabaei, 2025). Further, where harm occurs it is ethically and legally problematic: is the culprit the clinician, the developer, the institution or the data itself?

The real beneficence in AI-driven healthcare lies in active harm reduction: stringent pre-production validation, real life testing of the algorithm work in various populations, as well as constant tracking of the performance drifts (Rogers et al., 2021). Similarly non-maleficence requires algorithmic humility, that acknowledges the limitations of existing models and leaves AI as a supplement to human judgment, rather than a replacement (Gashu & Guadie, 2024).

Prejudice in AI systems can manifest itself in two ways: when training datasets do not represent some groups or in historical disparities integrated in the systems. In medical fields, biased AI may cause inequalities in the diagnosis, treatment advice, or care access, which predominantly impact the marginalized (e.g., women, low-income communities, ethnic minorities).

To be ethically responsible design needs:

a. Various and representative data sets.
b. Biases audited on a regular basis over the AI lifecycle.
c. The measurement of outcome disparity in practice clinical settings.
d. Ethics review boards should be included in the development of the model.

Algorithms are offered to support the notion of algorithmic fairness, including equalized odds or demographic parity. Not only is fairness an ethical concern but it is a regulatory expectation in most jurisdictions.

4.3. Justice and Fairness: Beyond Technical Fairness Metrics

The area of justice is most commonly talked about in the context of fairness, i.e. not having AI systems disproportionately harm certain groups. But algorithmic fairness is a more than technical issue of a data representation or model bias (Panarese et al., 2025). It is both a structural and ethical matter that involves other issues of access, power, and social determinants of health. As an example, AI models that used particular data on hospitals largely located in urban areas might be insufficient in the countryside or low-income population and, in extension, contribute to healthcare disparities. Similarly, insurance or triage predictive models can discriminate against an individual on the basis of factors that are socioeconomic proxies of race, geography or disability (Lee et al., 2021).

Also, ethical AI has to supersede abstract measures of fairness and involve taking a stance that concerns the context of justice, meaning not only what the model will do, but at who and why, and at what price. This involves engagement of useful stakeholders, involving marginalized groups into the design of systems, and evaluations that measure long-term and systemic effects (Merriel et al., 2018).

4.4. Transparency and Explainability: From Algorithmic Opacity to Clinical Trust

Transparency is one of the solutions towards the ethical issues of AI, but all too often it has become a magic bullet whose practical meaning is unclear. Conventional ideas about transparency, e.g., posting code or model parameters, might not give clinicians or patients any sense of how an AI system operates, in particular when its results are created using deep neural networks with millions of parameters (Eke & Shuib, 2025). Rather, contextual explainability must be the target: supplying clinicians with plain and actable rationale behind AI suggestions compatible with clinical logic and supplying patients with understandable explanation that would make them engaged. Explainability is not just engineering and not only social but communicative as well. It has to take into account various audiences, mental loads, and clinical situation within which the decision is made (Fernández, 2023).

In addition, the emergence of transparency should not be limited to models but should be applied to data provenance, how the models were developed, and institute accountability. Who did the curating of the information? Whose name put on it? In what assumptions was the model trained? That is only when these questions are answered that AI will have clinical trust and legitimacy among the people (Marey et al., 2024).

When it comes to the ethical use of AI in the health-care sector, then not only will reapplication of extant principles suffice, but a reconsideration of what these principles actually imply in a machine-moderated setting will be needed (Metta et al., 2024). Autonomy will need to be transparent to algorithmic meddling; beneficence will have to counter indeterminacy and intricacy; justice will have to combat bias embedded in systems; and transparency will have to be transformed to fit the era of incomprehensible calculation (Nouis et al., 2025).

Artificial intelligence contributes to making the classical models of clinical accountability more difficult. It is often uncertain who should be held accountable in the event of an AI recommendation leading to a patient harm event because it can be the clinician, developer, institution, or AI vendor.

a. Ethical sound models are based on collective responsibility, where:
b. Clinicians are still the decision-makers.
c. The validation of models and safety tests are left to the developers.
d. The institutions should make sure that AI systems are well trained and integrated.

Liability avenues should be made transparent to avoid blame game and encourage responsible AI construction and safeguard patient privacy.

The importance of explicitly defined ethical guardrails is recognized as artificial intelligence (AI) becomes a part of clinical workflows and decision-making procedures. Ethical considerations are not mere academic postulations only, they are an operational guide to developing, implementing and governing AI technologies in healthcare (Goirand et al., 2021).

Nevertheless, with more autonomy, adaptivity, and obscurity of AI systems, classical biomedical ethics, albeit still the basis, might not be enough to stand on its own. An increasingly number of institutional policies, international declarations and cross-disciplinary initiatives have begun to offer practical mechanisms of ethical analysis in this novel technological frontier.

This segment discusses significant initiatives of various global health organizations, engineering or medical organizations that are making efforts to establish and put in place ethical practices of the AI. It also gives emphasis to real-life case studies that give emphasis on how these frames can (and should) influence the development of AI in medicine (Nasir et al., 2024).

4.5 Global Ethical Frameworks and Guidelines

4.5.1. WHO and UNESCO Guidelines

In 2021, the World Health Organization (WHO) published a ground-breaking report: Ethics and Governance of Artificial Intelligence for Health. This document can be identified as the first major effort to develop ethically acceptable standards, which are health-related and globally applicable (Hoffmann & Nortjé, 2022).

Experience shows that there are six key ethical principles of AI utilization and development in healthcare that should be adhered to according to WHO:

i. **Protect Human Autonomy:** AI does not need to replace the process of human decision-making but rather supplement it. The clinicians should have the final say in care decisions, and the patients should be better informed and empowered rather than coerced and left out.

ii. **Promote Human Well-being and Safety:** The tools used in AI have to be effective and clinically useful. This implies intense pre-market and post-market verification, a method of stopping the unacceptable or unreliable systems.
iii. **Ensure Transparency, Explainability, and Intelligibility:** Artificial intelligence should be transparent to the clinicians and patients using its outputs. Lacking explainability, the trust will dissolve, and the ethical consent will not be possible.
iv. **Foster Responsibility and Accountability:** There should be clearly laid out lines of responsibility, those involved can be the developers, the healthcare institutions themselves or the vendors but this especially in cases of injury or wrong diagnosis.
v. **Inclusion and Equity:** The AI systems should not increase nor reinforce the current health disparities. This parades representative training data sets, localization to particular populations and active marginalized community inclusion.
vi. **Promote Sustainable and Responsive AI:** Ethics has to be dynamic. The evolution of AI systems should be responsible and time-consuming, that is, they will have to adjust to societies, changes in legislation, and new data. The principles are intended to be universal, to cover the resource-limited contexts in which AI may have transformational value, as well those jurisdictions where ethical oversight frameworks may be less robust.

In 2021, 193 nations supported the United Nations Educational, Scientific, and Cultural Organization's (*UNESCO*) adoption of the first international normative framework on AI ethics. In contrast to the WHO, UNESCO's proposal offers a thorough vision for human-centered AI in all fields, going beyond health (Bean et al., 2025). Key tenets include:

i. **Respect for Human Rights and Fundamental Freedoms:** AI must not undermine privacy, freedom of expression, or democratic participation.
ii. **Cultural and Linguistic Diversity:** AI should be designed in a way that respects diverse identities, avoiding "algorithmic monocultures."
iii. **Environmental Sustainability:** The ecological impact of AI—including energy use in model training—must be considered an ethical issue, not merely a technical or economic one.
iv. **Ethical Impact Assessment (EIA):** UNESCO calls for EIA frameworks akin to environmental reviews, wherein AI projects are evaluated for their social and ethical ramifications before deployment.

Together, WHO and UNESCO offer two ethical frameworks: one that is primarily focused on health, while the other offers a foundation based on universal rights.

4.5.2. IEEE, AMA, and Other Institutional Frameworks

The Institute of electrical and electronics engineers (IEEE) the most recognized technical organization in the world formulated a collection of guidelines in a multi-volume publication under the flagship of Ethically Aligned Design (EAD) (Abràmoff et al., 2022). In contrast to purely regulatory or clinical documents, the work of IEEE is addressed to developers and engineers of AI, as well as to persuade them to not work bolt-on ethics, but establish it at the design stage. Such concepts as:

- **Value-Sensitive Design (VSD):** The design of systems must focus on fundamental human values (such as dignity, agency and justice) as opposed to technical specifications.
- **Design Transparency:** There must be traced-based recorders of decisions within the AI systems to allow audit and legal action.
- **Human Oversight Means:** The developers are admonished to provide user-enablement to override them and significant tools around the traceability of decisions.

The design ethics compatibility with the IEEE helps balance the technical development with the rest of society, and as such, it is especially applicable to young innovators.

According to the *AMA* guidance, the Augmented Intelligence in Health Care focuses on the maintenance of the physician-patient relationship and clinical integrity (Adler-Milstein et al., 2022). Its basic tenets are:

- **Enhancement instead of Replacement:** Artificial intelligence should enhance clinicians work, rather than replace clinicians' work.
- **Evidence Based Implementation:** Implementation will not be done until tools are clinically tested or proved through viable forms of research.
- **Fair Access:** AI should not be made an instrument of exclusion, but an instrument that increases access according to both social economic and demographic levels.

The framework proposed by AMA puts physicians in focus of ethical adoption of AI, where professional judgment is imbued in professional education, permanent learning, and cross-disciplinary cooperation. Future of Life Institute encourages long-term safety and existential risk investment in the field of general AI and healthcare automation. Partnership on AI, an engagement between technology firms and the civil society, provides best standards of fairness, bias mitigation, and transparency of algorithms in relation to health as well as humanitarian applications (Tam et al.,

2024). An international harmonization of AI principles has emerged in the OECD Principles on AI (agreed by more than 40 countries) which focuses on inclusive growth, human-centered values, and international cooperation.

4.5.3. Case Studies: When Principles Are Put to the Test

a. Skin Cancer Detection Algorithm (Germany)

An A.I. trained to identify melanoma performed excellently- until an after-deployment audit showed that it was dysfunctional when identifying patients with darker skin tones. The database was predisposed to represent those with fair skin that were prevalent in Europe (Hübner et al., 2023). Applied Ethics:

- The principles of equity and safety stipulated by WHO required retraining on increasingly representative data.
- The right to explanation when it comes to GDPR prompted the need of explainable AI logic in a medical setting.

b. Predictive Risk Scoring in U.S. Hospitals

Hospitals that were using AI system to predict patient deterioration identified that it falsely labelled the black patients as a lower-level risk because they had under-documentation in their records historically (Patel et al., 2023). Applied Ethics:

- Algorithmic audits and recalibration were brought about by AMA principles.
- The IEEE principles of value-sensitive design were retroactively used to enhance greater fairness and accountability.

c. AI Chatbots for Mental Health (Southeast Asia)

Chatbots about mental health were introduced in remote areas with a low level of psychiatric access. The bots worked well to screen the mild cases, but were not so successful at dealing with subtle complexities of language, cultural sensitivity, and could do little to describe the escalation pathways when they encountered suicidal ideation (Algumaei et al., 2025). Applied Ethics:

- The concepts of cultural diversity applied by UNESCO demanded localization and culturally appropriate models of language.
- Human-in-the-loop systems of risky interactions were based on the WHO safety principles.

The spread of AI in healthcare requires more than the nice-to-have ethics. Technical requirements, institutional policies, and regulatory requirements have to be developed out of ethical AI. It is particularly the case in the healthcare industry whose decisions can be life or death. The ethical framework must not be basically fixed, but it must be able to change according to the changing social norms, emerging technologies and observable empirical analysis of positive/negative effects (Algumaei et al., 2025). The way forward is cross-sector communication, an interdisciplinary approach to leadership, and patient-oriented decision-making that will create the path towards a gapless space between what we ought to be doing and what we are constructing. Illustrated in *Table 2* Based on modern literature regarding AI ethics and regulatory practices in healthcare.

Table 2. Case studies illustrating ethical and regulatory challenges in AI-Driven healthcare.

Case Study	Description	Ethical Issue	Regulatory Implication
AI Misdiagnosis Due to Dataset Bias	An AI dermatology diagnostic tool produced significantly lower accuracy for patients with darker skin because the training dataset was predominantly composed of lighter-skinned individuals.	Inequitable access to accurate diagnosis; reinforcement of health disparities.	Necessity for diverse, representative datasets; implementation of mandatory bias audits.
AI-Based Sepsis Prediction System	A hospital deployed an AI model that generated frequent false alarms, causing clinicians to experience alert fatigue and ignore warnings, which resulted in delayed diagnoses.	Patient safety risks; unclear accountability for AI-assisted clinical decisions.	Requirement for post-market monitoring, real-world validation, and adaptive performance auditing.
Privacy Risks in Personalized Nutrition and Microbiome AI	A startup used microbiome samples collected from patients for additional commercial research without obtaining explicit consent.	Violation of patient autonomy; data misuse and erosion of trust.	Need for stricter consent frameworks and well-defined rules for secondary use of sensitive data.
AI for Radiology Decision Support	Radiology departments using AI triage tools saw improved workflow efficiency but clinicians struggled to interpret why certain cases were flagged by the AI.	Lack of explainability; reduced clinician confidence in AI recommendations.	Mandates for explainable AI, especially under high-risk categories defined by regulations such as the EU AI Act.

5. CHALLENGES AND RISKS IN AI-DRIVEN HEALTHCARE

5.1 Bias and Discrimination

A biased decision in the medical field may even be the difference between life and death. When the bias is inherited or amplified by the Artificial Intelligence, it can lead to the spillover effect regarding the whole health systems, that is, the inaccurate diagnosis, treatment delays or the denial of care altogether (Williamson & Prybutok, 2024). The assurances of AI to ensure that healthcare is more objective or effective are greatly flawed when the solutions appear to be fraught with the same inequalities that they are intended to address (Ueda et al., 2024). Ethical risks in AI are not uniform—they vary significantly by application type. *Table 3* compares dominant ethical concerns across key healthcare AI use cases, highlighting where and how harm can emerge.

This section reveals the diagnosed depths of biasness of AI healthcare systems, such as skewed datasets and minuscule performance differences that are still opportunities to uncover unknown areas and solve them by the possible strategies to detect, minimize, and eventually circumvent algorithmic discrimination (Hanna et al., 2025).

Table 3. Comparison of Ethical Concerns by AI Application Type

AI Application	Dominant Ethical Concern	Example of Impact
Diagnostic Imaging	Bias due to underrepresented skin tones	Misdiagnosis in melanoma screening for darker skin tones
Predictive Risk Scoring	Discrimination from socioeconomic proxies	Underestimated care needs in minority populations
AI Chatbots (Mental Health)	Lack of empathy and escalation risk	Suicide risk not detected in NLP-based chatbot
Robotic Surgery Assistance	Accountability in automated actions	Error in surgical assistance with no clear liability
Remote Monitoring Wearables	Privacy and consent complexity	Continuous data tracking without dynamic patient consent

5.1.1. Data Bias in AI Training: Garbage In, Inequity Out

Bias in AI frequently starts even prior to a first line of code has been typed. Medical AI applications use real-world data in the form of electronic health records, medical images, and clinical notes, which were gathered throughout decades (Pot et al., 2021). This data is not neutral at all though. It is an indication of decades-old inequality in access, diagnosis, and treatment. To give one example, machine learning dermatology models that have been trained primarily with light skinned patient images have exhibited troubling degradations in accuracy when applied to darker skin tones. In 2020, a study has found that certain common pulse oximeters inaccurately report the level of oxygen in Black patients and they are likely to be under-treated during the COVID-19 crisis. They are not mere technical blind spots but failing of representation with actual human cost (Norori et al., 2021).

The missing data, under-reporting, and historical underdiagnosis of some populations-particularly, those that are racially, gender, and socioeconomically divided because of structural inequalities- are the cornerstones of biased algorithms. The consequence is a danger to escalate these blind spots unless AI models are formulated with a critical awareness of the problem (Mergen et al., 2025).

5.1.2. Disparities in AI Performance Across Populations

Alternatively, even when the seemingly accurate AI systems might seem to be accurate toward a certain degree, in some groups, its failures even get ignored. A model can have a good average performance but fail Indigenous patients or aged people or individuals with rare cases (Daneshjou et al., 2022). Such differences in performance are especially insidious since aggregate measures of accuracy continue to obscure them. Assume one of the most popular algorithms in the U.S that would help to recognize the patients who can have the additional attention. Scientists found it very discriminative: it vastly underrepresented the medical demands of the Black patients, depending on the healthcare expenditures as an indicator of the health status of individuals, when, in fact, the lower costs translated to the higher health rates. As a matter of fact, black patients usually had fewer means of getting care rather than health needs (Seyyed-Kalantari et al., 2021).

Discrimination does not necessarily appear in the form of open oppression. It may take the form of systematic underestimation, mislabelling or not being sensitive to local context. In addition, since most AI models are closed-source or otherwise commercially safeguarded, healthcare organizations might not even understand how such systems are contributing to the failure of certain groups of people (Jha et al., 2025).

5.1.3. Strategies to Mitigate Bias: From Awareness to Accountability

Partiality is not something that can be eliminated, but it can be identified, reduced and handled responsibly. The most significant one is to recognize the bias as a problem of both society and technology: it is not merely a data science bug (Weiner et al., 2025). It is a result of complicated interpenetrations of historical wrongfulness, institutional framework, and algorithmic reasoning. In order to deal with bias, a number of approaches are coming up:

i. **Bias Audits and Fairness Testing:** Frequent testing of AI models over demographically subgroups to detect the disparities. Some organizations are even developing what is known as the bias dashboard to monitor this with the passage of time (Murikah et al., 2024).
ii. **Representative and Inclusive Data Curation:** Ensuring that a range of populations is included during the data collection process, and address any historical underrepresentation of the population through data augmentation or by re-weighting solutions (Murikah et al., 2024).
iii. **Contextual Design:** Using local clinicians, ethicists, and patient advocates in the design of an AI system so models address the populations in which they will intervene- they will not be merely technically accurate but will be modelled in a manner that is socially aware (Moon & Ahn, 2025).
iv. **Validation Transparency:** Making the details of model-training, data used etc. publicly available as well as the process used to assess fairness. Transparency should be good ethics but it is also good science (Daneshjou et al., 2021).
v. **Regulatory Pressure:** The government and international organizations are starting to mandate bias mitigation into AI certification or approval. As an example, in the suggested EU AI Act, high-risk healthcare applications are categorized, and one will be required to provide evidence on the fairness and safety of the proposed applications (Gray et al., 2024).

Finally, rather than being a stopgap measure, bias reduction needs to become an inherent part of the AI development process. In healthcare, ethical AI raises questions about who should perform well as well as about quality (Nazer et al., 2023).

5.2 Privacy, Data Protection, and Cybersecurity

5.2.1 Privacy, Security, and Data Protection in AI-Driven Healthcare

Data in healthcare is not only numbers; it is identity, vulnerability and trust. A health record tells much more than a diagnosis, it can tell the daily habits, history, genetic tendencies and even mental states of a person. Privacy and security issues are a million times worse when AI comes along in this area (Gawankar et al., 2024).

Contrary to other conventional systems, AI enjoys huge amounts of data, both gathered by wearables, clinical data, genomic databases, and even a presence on social media. But when you have big data then it has its responsibility (Keshta, 2022). In the absence of deep-seated security measures, the potentially life-saving tools might as well be used to subject patients to discrimination, spying, or cyber-related crime.

5.2.2 The Sensitive Nature of Health Data in the AI Era

Healthcare data has always required protection, but AI makes that need more urgent and complex. Unlike a physician who sees a snapshot of your condition, an AI model can synthesize years of data into detailed risk profiles or behavioural predictions. This isn't just sensitive—it's deeply *predictive* and potentially *manipulable* (Murdoch, 2021). Moreover, AI systems increasingly pull from *non-traditional health data*—like fitness trackers, smartphone apps, or online behaviour—to build richer models. These sources often sit outside of traditional health privacy regulations like HIPAA or GDPR. The result? A growing "Gray zone" where deeply personal insights are generated from data the patient never imagined was medical (Trocin et al., 2023).

For example, an AI system might infer depression based on your voice in a call with a telehealth provider—or your risk of diabetes from step count and grocery purchases. These insights are powerful, but they also raise a profound ethical question: Should AI systems know things about us that even we haven't chosen to reveal?

5.2.3 Patient Consent and Data Ownership: Who Really Controls the Data?

The same model of patient consent that applies to most of the traditionally existing systems of consent, e.g., sign a consent form, agree to treatment, is not fit to be applied to the AI environment. The consent, in this case, is either hidden on several pages with a legal jargon or is offered once and then unlimitedly applied, despite the evolving nature of algorithms and re-usage or resale of data (Kostick-Quenet & Rahimzadeh, 2023). Patients rarely know:

- What their data is being used for
- Who is accessing it
- How long it will be stored
- Whether it might be sold or shared with third parties

This opacity is something that is endangering the data sovereignty the idea that individuals have a right to control the uses of their personal data. Because AI is making sharing data more appealing (to innovate) and more dangerous (to privacy), we have to reconsider consent as continuous, situational and retractable (Al-Ani et al., 2024).

There is an emergence of innovative frameworks such as a dynamic consent or data trusts. Such models would enable patients to grant or revoke consent to certain uses, update preferences throughout their lifetimes, and would enable them to understand better what value their data has been contributing. Nevertheless, they demand new online infrastructures and transformations that most healthcare systems are not prepared to address.

5.2.4 Cybersecurity Risks and Mitigation in AI Systems

As artificial intelligence increasingly integrated the healthcare infrastructures, when the cybersecurity field turns into an existential rather than a technical issue. Not only are AI systems vulnerable to attack, but they alter the kind and extent of risk. Most conventional IT systems usually have ample static functions (e.g., storing or accessing data), whereas, when it comes to AI systems, learning, evolving, and anticipating interactions with clinical set-ups are dynamic; hence, AI systems are more difficult to secure and simpler to attack. This is not the only thing a compromised AI system could facilitate; it could clone a disease and make mistakes in diagnosis, offer harmful treatment, or alter pathways of clinical decisions in real-time (Chakraborty et al., 2023). The integration of AI into healthcare infrastructure introduces a range of security vulnerabilities—each of which can threaten patient safety, data integrity, or operational continuity. *Figure 2* illustrates the primary cybersecurity risks, including adversarial attacks, device insecurity, and weak access control systems that collectively demand a robust, multi-layered defence strategy.

Emerging Threat Vectors in Healthcare AI:

- **Membership Inference and Model Inversion:** Adversaries may use small probing attacks against AI models to extract sensitive information about individuals- including whether these individuals were included in the training dataset or even recover medical images of a patient (Bouhaddi & Adi, 2023).
- **Data Poisoning Attacks:** Data poisoning attacks will allow attackers to corrupt training pipelines with minimal actions (such as labelling benign tumors malignant) causing a degradation of performance in a specific manner. It is particularly toxic even in decentralized systems where models constantly learn based on new data (Wymberry & Jahankhani, 2024).
- **Adversarial Inputs:** Small crafted modifications of the inputs to medical systems such as the noise on an X-ray, or the alteration of a single pixel on a photo of a skin lesion, can cause AI systems to make erroneous predictions, but not to human observers. This may in real life cause decisive misdiagnoses without any apparent interference (Sarker, 2023).

- **Supply Chain Attacks:** A large proportion of healthcare organizations apply third-party AI tools or APIs. Cybercriminals are likely to attack these vendors, introducing the malware, or the backdoors to updates or pretrained models, potentially impacting a whole network down the chain (Bécue et al., 2021).
- **Ransomware 2.0:** During the past years, ransomware has grown to be more than the encryption of files. In hospitals where AI plays a leading role, hackers now stand an opportunity to hack into AI-reliant hospital services and render them ineffective, triage processes, diagnostic tests, or even use the robot surgeries systems, and you pay to get them back (Ferdous et al., 2024).

Figure 2. Critical Cybersecurity Vulnerabilities in AI-Driven Healthcare Systems, This figure was specifically developed and curated by Reyed M. Reyed (Editor) to ensure academic rigor and a knowledge-based framework for this chapter

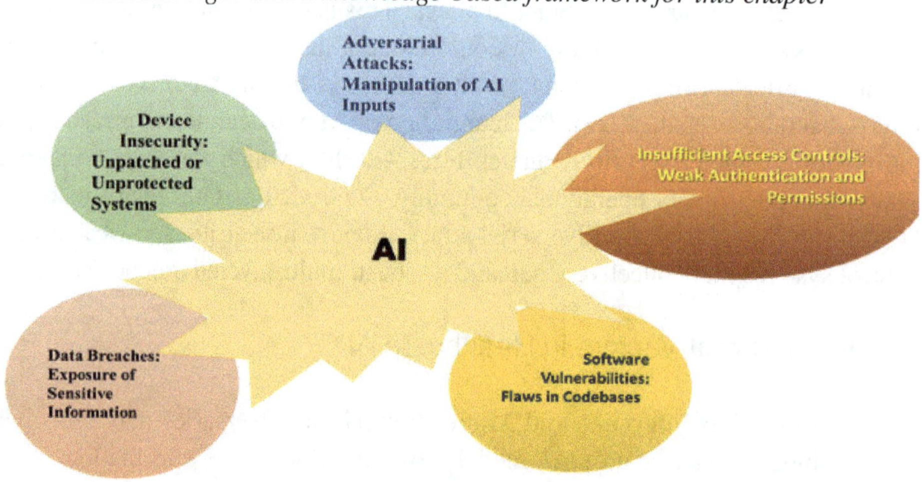

5.3 Consequences: More Than Data Breaches

The effect of cybersecurity breach in AI healthcare is not quantitatively similar to the more conventional one. It is not only a matter of disappeared records but also the matter of mistrust, lost lives, and lost clinical integrity (Nemec Zlatolas et al., 2024). For example:

- An attacker altering diagnostic algorithms could *deliberately cause underdiagnosis of cancers* or overprescription of costly medications.

- A corrupted AI system in emergency departments could disrupt *triage prioritization*, delaying care for critical patients.
- Exposure of mental health or genetic data could lead to *insurance discrimination or social stigma*, even years after the breach.

It is no longer a theoretical danger. Such a breach occurred in 2021, when a large psychotherapy institution in Finland was hacked, and not only the institution, but individual patients themselves, received ransom demands that the confidential notes of the therapy sessions would be published online (Mohsin Khan et al., 2025). With the AI system becoming further and further part of mental health, reproductive care, and personalized medicine, this breach will be even more personal, will cause more harm, and will be even harder to mitigate.

5.4 Mitigation Strategies: Building Resilience by Design

A proactive, AI-aware approach to cybersecurity is necessary to prevent and mitigate these threats; this involves combining organisational policies, technology controls, and regulatory compliance into a cohesive defence plan (Taimoor & Rehman, 2022).

- **Secure AI Model Lifecycle:** Deploy secure model development pipelines, data integrity checks, version control and cryptographic signing; Carry out red-team training and penetration testing during pre-deployment, with a primary concern on adversarial robustness and poisoning resistance (Omran Almagrabi & Khan, 2025).
- **Data Governance and Infrastructure Hardening:** To reduce the amount of raw patient data that is exposed while still allowing model training, use federated learning or differential privacy; Implement encrypted audit trails, role-based access restrictions, and real-time anomaly detection in various data contexts (Aldoseri et al., 2023).
- **Algorithmic Monitoring and Incident Response:** Use AI explainability techniques to identify unusual or out-of-character results, which function as early indicators of model corruption; Establish specialised AI incident response procedures that provide quick isolation or rollback of hacked models without interfering with clinical procedures (Feng et al., 2022).
- **Vendor and Supply Chain Risk Management:** Request clear documentation and security certifications from AI partners and suppliers; Keep an eye out for upgrades and vulnerabilities in third-party libraries and APIs, since many AI attacks take use of poorly maintained dependencies (Kioskli et al., 2025).

- **Regulatory and Ethical Alignment:** Make sure cybersecurity policies align with data protection policies like HIPAA, GDPR and the EU AI Act; Introduce cybersecurity audits into AI ethics committees or boards recognizing the ethical implications of resilience in the systems (Goktas & Grzybowski, 2025).

5.5 The Stakes Are Higher: From Reactive Defence to Proactive Ethics

In AI-driven healthcare, cybersecurity does not only mean preventing malicious parties, but also maintaining the ethical core of healthcare. When AI systems recommend or influence the decisions as to who receives care, when and how, a breach has the characteristics of technical mishaps and is a breach of clinical trust and poses an immediate threat to human dignity. The development of safe, robust, and ethically-balanced AI systems could not be left as a choice (Shu et al., 2025). It is a precondition of responsible innovation, and an experiment of whether we can achieve the serious power of the technologies without breaking the solemn trust between patient and provider.

6. ACCOUNTABILITY AND HUMAN OVERSIGHT IN AI SYSTEMS

With artificial intelligence in healthcare there comes an awkward paradox: clinical power is broadened and the clarity of roles elevated. When an AI-based system should suggest a treatment that results in injuries or cannot identify a serious condition, who is to blame? Was it the doctor that operated the device? The typical developer of the algorithm? The hospital which implemented it? Or the person who consented to use it? In this section, the authors discuss the elasticity of accountability in AI-based medicine and the importance of redefining such responsibility among human and non-human players, legal and moral standards (Smith, 2021).

The application of artificial intelligence in the medical sphere is not only a technological change, but also a change of the society. With machines starting to play a role in decision who should be diagnosed and the treatment prescribed, and what information should form the basis of the medical decision, the question of ethical deployment will need to shift off the shoulders of a particular actor (Rozenblit et al., 2025). In healthcare AI is a collaboration effort vast in the number of stakeholders that influence whether this technology liberates or threatens, cures or kills (Camacho et al., 2025).

6.1. When AI Fails: Who Bears the Liability?

In conventional medical care, the practitioner has both legal and ethical obligation of medical diagnosis and treatment. However, AI is the one that ruins this relationship. Does that mean that when clinician relies upon that advice, never having been informed of the existence of an AI in a radiology application that failed to detect a tumor or incorrectly classified a lesion, can he/she be sued? However, the other side of the coin would be that the clinician had intervened by overriding the AI and the outcome gets worse, is it then the liability that shifts?

This responsibility gap is especially concerning in black-box models, where developers might not even be sure about how the decision makes Stewart Taggart is seen with his head in his hands in a case of responsibility gap that is very worrying in black-box models in that even developers would not be certain of how the decision is made. In the absence of interpretability, accountability is muddy (Schoenherr & Thomson, 2024). The courts and regulators are left with the situation in which no single player had acted negligently but there had been harm either caused due to the concerted effort of a socio-technical system.

The new legal systems are beginning to fill these gaps. The European Union is currently developing the AI Liability Directive that seeks to hold manufacturers responsible of the high-risk AI systems on a presumed-fault basis. But numerous national regulations continue to regard AI as a product, with overall responsibility on the human operator. The model is growing progressively untenable as AI systems develop to be more independent, adaptable, and integrated (Königs, 2022).

6.2. Human Oversight in an AI-Augmented World

In spite of their sophistication, AI systems are to be placed under meaningful human control. Regulatory point 2: this principle as codified in EU AI rules, and invoked by medical associations across the world, acknowledges that machines, however competent, cannot provide ethical and contextual decision making that patient care demands. However, what should the meaningful oversight involve? Clinical decision makers should not be allowed to rubber-stamp algorithmic results (Ahdadou et al., 2024).

The oversight should be an active, informed, and empowered one- needful is that the professionals should be knowledgeable about the strengths of the model, its weaknesses, and any bias. It means that digital literacy in healthcare is going to be new: clinicians will have to be AI-savvy without turning into machine watchers. Furthermore, the health systems should offer the oversight provision conditions (Kim et al., 2022). This includes:

- The time and the resources to challenge or override AI outputs
- Transparent records on which and how AI was used in making decisions
- Overall education and advice not to trust an algorithmic suggestion and when to question it

Supervision is not a sustainable option in settings of workflow-centered stress, algorithmic inscrutability or naive dependence upon technology.

6.3. The Black Box Problem and Traceability

Black box problem the concept of black box is used to describe the intransparency of most AI models particularly deep learning models, who's internal reasonings might not be understood even by the developers of the model. This opaque kind of responsibility is problematic in medicine, in which every decision may lead to life or death (Nasiraee et al., 2023).

In the case that inability to trace the manner in which a decision was made, the latter will be challenging to explain to patients, audit it or stand up in court. More threateningly, black-boxes have the capability of yielding unpredictable results when deployed in the real-world environment, mostly in the presence of underrepresented populations or unrepresentative situations in the training data. To counter this, explainability and decision traceability is becoming emphasized (Fomin & Astromskis, 2023). This includes:

- Using *interpretable models* in high-risk applications
- Providing post-hoc explanations that make outputs intelligible to clinicians
- Logging AI decisions and data inputs for auditability
- Maintaining *version histories* that track changes in model behaviour over time

Traceability in a critical care environment is also not merely a matter of legal defence, it is clinical education. When a diagnosis miscarries, we have to be capable of tracing through the chain and detecting the mistake (in man or in machine) and correcting the machine.

Finally, AI in healthcare results in the hybrid intelligence system: the cooperation of the machine and humans when making instant decisions. In that case, accountability cannot be considered a mere passing of the buck, rather it has to be shared, distributed and be effectively managed. This demands new institutional mechanisms: AI risk governance boards that oversee deployment and performance; Clear role definitions for developers, deployers, and end-users; Responsibility matrices that predefine who is accountable under various failure modes. More to the point, it would need a cultural reorientation, replacing blind trust in innovation with the responsible assimilation, at which all stakeholders would be aware of their responsibilities as per ethical and legal standards and in a framework that is potent, fallible, and ultimately human-powered (Zednik, 2021).

The regulatory approaches to artificial intelligence in healthcare vary significantly across jurisdictions, which is due to the difference in the legal frameworks, maturity of the healthcare systems, and policy priorities. Although the goals of all major governance models are to achieve the balance between innovation and patient safety, they differ in the way they address the issue of risk classification, data security, human supervision, and enforcement policies. As in order to make the effective comparison, *Table 4* will outline the major regulatory aspects of the healthcare AI governance in the United States, the European Union, India, and the World Health Organization, including convergences and vital gaps that are associated with the global implementation and adherence.

Table 4. Comparative Overview of AI Regulation in Healthcare Across Major Jurisdictions

Dimension	United States	European Union	India	WHO
Regulatory Nature	Sector-specific, product-based	Unified, binding legal framework	Policy-driven, emerging	Non-binding global guidance
Primary Authority	FDA (SaMD framework)	GDPR + EU AI Act	Ministry of Health, NITI Aayog	World Health Organization
Risk Classification	Moderate; focus on clinical function	High-risk categorization for healthcare AI	Contextual, use-case dependent	Ethical risk orientation
Data Protection	HIPAA (limited scope)	GDPR (strict consent, transparency)	Draft health data policies	Emphasis on data dignity
Human Oversight	Encouraged, not uniformly mandated	Mandatory for high-risk systems	Policy-recommended	Strongly emphasized
Explainability	Required mainly for CDS tools	Explicit legal requirement	Not clearly defined	Ethical expectation
Post-Market Monitoring	Limited, evolving	Mandatory	Largely absent	Encouraged
Enforcement Strength	Moderate	High	Low–Moderate	Advisory only
Key Limitation	Fragmented oversight	Regulatory complexity	Legal uncertainty	No enforcement power

6.4 Shared Moral Responsibility of Developers and Clinicians

AI creators are no longer software engineers they have become moral creators. Each one of the choices regarding the dataset to be used, features prioritization, and uncertainty treatment has a clinical impact on lives. Instead, engineers should start thinking like ethicists, design with equity and openness in mind. In medicine the old mantra of move fast and break things is done. Instead of this there must be move wisely and heal better. Medical practitioners, too, do not act as a passive consumer of technology anymore (Gundlack et al., 2025). They act as the moral watchdogs of the introduction of AI. Their role is to take the recommendations of algorithmic system with a grain of salt, challenge black box systems, and make sure the clinical judgment never gets contracted to a machine. AI needs to act as a second opinion not a final decision.

7. REGULATORY FRAMEWORKS AND COMPARATIVE GOVERNANCE OF HEALTHCARE AI

Regulatory oversight of artificial intelligence in healthcare has developed unevenly across jurisdictions, reflecting differences in legal traditions, healthcare infrastructures, and policy priorities. This section examines major regulatory frameworks governing healthcare AI, with particular attention to approaches adopted in the United States, the European Union, India, and emerging global guidance.

7.1 United States: FDA, HIPAA, and AI Oversight

The healthcare sector AI regulation in the U.S. is a combination of medical device regulation and data privacy legislation. The Food and Drug Administration (FDA) is in the middle of approving artificial intelligence (AI)-powered tools in order to be used either in diagnosis or therapy, and many are categorized as Software as a Medical Device (SaMD) (Adirim & Rao, 2025).

FDA has assumed a risk-based system. There is even the possibility that AI systems that are providing clinical decision support (CDS) are not even subject to premarket approval so long as they can provide a way for clinicians to review the rationale behind the suggestion on their own. But black-box models or autonomous systems are, however, normally placed under stricter scrutiny. Key recent initiatives include:

- *The FDA's Proposed Regulatory Framework for AI/ML-Based SaMD* (2019): Introduced the concept of a "predetermined change control plan" to manage continuously learning algorithms.

- *The Digital Health Centre of Excellence*: A hub within the FDA focused on accelerating and guiding digital innovation in medicine.

On privacy, there is the Health Insurance Portability and Accountability Act (HIPAA) that acts on how to manage the health information of patients. But HIPAA is not suited to the size and complexity of the AI world, since it does not apply to data in the hands of most third-party health technology companies, or to profiling and inference based on non-traditional data (e.g., social determinants or wearable devices) (Seeram & Kanade, 2024).

AIs are controlled by the U.S. Food and Drug Administration (FDA) with the Software as a Medical Device (SaMD) model. Recent initiatives include:

a. Good practices of Machine Learning (GMLP)
b. Adaptive AI predetermined Change Control Plans (PCCP).
c. Pathways to risk-based evaluation of clinical decision-support tools.

Moreover, HIPAA regulates information privacy, though in most situations, it does not work well with complex AI data flows and dealer with genomic, microbiome, or cross-institutional information. There are still regulatory balls falling through:

a. Machine learning based on artificial intelligence.
b. AI decision explainability conditions.
c. Training of algorithms based on patient data used commercially.

U.S. is transitioning to a more dynamic oversight model though it does not have an extensive AI-focused health regulation.

7.2 European Union: GDPR and the AI Act

The European Union has been more holistic and active in its approach combining data protection, AI regulation and ethical requirements in a unified regulatory framework.

The *General Data Protection Regulation (GDPR)* requires transparency, limited purpose and data minimization in all use of personal data even when used by AI systems. Interestingly, it has a right to explanation to persons who become swept in automated judgment, however, in its effectiveness, this right is debated as well (Meszaros et al., 2022).

GDPR Implications:

a. Strict consent and data minimization requirements
b. Right to explanation for automated decisions
c. Heavy penalties for misuse

The *Artificial Intelligence Act (AI Act)*, which is under process of adoption, would establish a risk-based model that is preset horizontally across all AI systems. Healthcare AI has been broadly classified as a high-risk, which implies that it will be subject to a set of obligations pertaining to the area of transparency, human supervision, and accurateness, robustness, and follow-up (Schmidt et al., 2024).

The European Union also focuses on the concepts of ethical AI, and it is reflected in the guidelines adopted by the High-Level Expert Group, including the so-called lawful, ethical, and robust AI. Although these are not enforceable, they form the expectations and funding priorities of institutions.

7.3 India and Other Emerging Markets

The emergence of AI regulation in healthcare comes at a time of extremely fast digitalization, lack of infrastructure, and significant diversity of the population in countries such as India (Choudhary et al., 2024). There is no particular AI laws established in India yet, though there are a number of policy frameworks under way:

- *Digital Information Security in Healthcare Act (DISHA)* (draft): The bill suggests to govern digital health data and provide it with consent-based sharing.
- *Health Data Management Policy* within the National Digital Health Mission: Concentrates on ownership of A data, anonymization and patron consent.
- The *National Strategy on AI (NITI Aayog)*, though cognizant of the power of AI in healthcare, has demanded scenarios of the regulatory sandbox, or, context sensitive model of regulation, in that, western paradigms of regulation need not be applied verbatim to India.

Other emerging markets (Brazil, South Africa, Southeast Asia) are still in different stages of development with some following the lead and borrowing the mechanisms of the EU model, others focusing on making partnership between the governments and the private sector and making artificial intelligence tools available to everyone (Shvetsova et al., 2023).

7.4 Comparative Regulatory Approaches to AI in Healthcare

Comparative analysis of global regulatory approaches to healthcare AI reveals both convergence and divergence in how jurisdictions address risk, oversight, and accountability. Such asymmetry poses the danger of reproducing the already lingering power disparity, particularly in cases of AI decisions getting markedly opaque or paternalistic (Palaniappan, Lin, Vogel, et al., 2024). In different areas, a number of trends are developing:

- Consensus on risk-based categorizations, human oversight requirements and post-market surveillance.
- Difference of scale: The EU recognizes the need to regulate at the system level whereas the U.S. is still focusing on products; other countries such as India are considering health equity, infrastructural flexibility, and the benefit of people.
- Loopholes in the regulation of continuously learning AI, non-clinical uses (e.g. mental health bots) and cross-border data flows.
- The absence of standardization increases the difficulty of law compliance on the global scale on the developer of AI and could impede the use of safe tools in the markets with low regulation.

Small differences in the approaches denote varying health systems, governing capacity, and social values. Global harmonization might not be realistic, but harmonization of main principles, which are safety, justice, transparency, and accountability, is both possible and far needed. The WHO and OECD are international agencies trying to come up with global guidelines (Balogun & Luetge, 2025). Their efforts might act as the link across the regions, and they might encourage regulatory interoperability without disrespect to the variation in context.

7.5 Policymakers as Stewards of Ethical AI Innovation

Regulators have to replace reactive watchdogs with proactive guardians of ethical innovation. They are left with the twofold responsibility to provide security and not choke development. Static regulation cannot stand dynamic algorithms; regulations have to be dynamic, cyclical, and highly scientific and socially informed. Regulatory frameworks such as the EU AI Act and FDA guidance are emerging and definitely hold promise, but they are and should always be subject to the daily stress tests involving which and how AI is used, and what real-world harms can be produced (Chapardar, 2024). In addition to introducing new regulations, policymakers need to influence the expectations of the masses: What can AI do? Who can collect medical insights and own medical insights generated by an algorithm? What should we do with the erroneous mistakes of AI that impact on the lives of people? They are both political and ethical as well as technical questions (Moilanen et al., 2025).

8. KNOWLEDGE GAPS IN ETHICAL AND REGULATORY GOVERNANCE OF HEALTHCARE AI

Despite the rapid expansion of ethical guidelines and regulatory initiatives for artificial intelligence in healthcare, significant gaps remain between conceptual frameworks and real-world implementation. Much of the existing scholarship emphasizes normative principles or early-stage deployment, while empirical evidence on long-term clinical impact and governance effectiveness remains limited. One critical gap concerns the lack of longitudinal, post-deployment evaluation of AI systems in clinical practice. Many models are assessed at the time of development or regulatory approval, yet there is insufficient monitoring of how performance, bias, and safety evolve once these systems are integrated into complex healthcare environments. This limitation is particularly pronounced for adaptive and continuously learning algorithms.

A second gap relates to equity and inclusiveness. While algorithmic bias is widely recognized, there is limited empirical research examining how healthcare AI systems affect marginalized populations over time, especially in low- and middle-income countries. Regulatory frameworks often articulate fairness as a goal but provide limited guidance on measurement, enforcement, or remediation. Finally, patient-centered governance remains underdeveloped. Patients are frequently positioned as data sources rather than active stakeholders in AI oversight. Questions surrounding meaningful consent, transparency of AI involvement, and accessible mechanisms for redress in cases of harm remain insufficiently addressed in both regulatory policy and empirical research. Addressing these gaps is essential for translating ethical commitments into effective and accountable governance of AI in healthcare (Palaniappan, Lin, & Vogel, 2024).

9. RESEARCH IMPLICATIONS AND FUTURE DIRECTIONS

As AI further roots into the inner parts of healthcare services, including diagnostics, decision-making, resources optimization, and robot surgery, the stakes are higher than ever. The only question that remains is not whether AI will revolutionize healthcare, but how this revolution could be influenced in a way so that it continues as trustworthy, human-centric and ethically justified (Wang et al., 2023). This section proposes major areas of future work and steps that should be followed to ensure AI-driven path leads to AI aiding rather than replacing ethical principles of medicine.

9.1 Longitudinal and Post-Deployment Evaluation

The identified knowledge gaps above justify a number of issues that should be prioritized on future research in order to enhance the ethical and regulatory governance of healthcare AI. To begin with, longitudinal and post-deployment studies are clearly required to research the performance of AI systems after 1-2 years in actual clinical settings. To evaluate emerging bias, safety, and unintended clinical effects, this kind of research must go beyond accuracy measures to determine adaptive bias to changing patient groups and institutional processes.

9.2 Equity-Sensitive and LMIC-Focused Research

Second, the equity-sensitive evaluation frameworks should be addressed in future research. It would involve creating standard ways to assess the data on the differential performance of demographic groups and how AI-based decision-making can be interacting with social determinants of health. A closer focus on empirical research on low-income and middle-income environments is of crucial significance, since the issues of governance in the specified environments are usually poorly represented in existing sources. Future research must also examine how equity-sensitive evaluation frameworks can be operationalized within resource-constrained healthcare systems, ensuring that regulatory aspirations translate into measurable and context-aware outcomes in low- and middle-income settings.

9.3 Generative AI and Adaptive Regulation Challenges

The growth of AI is quicker than the usual policy procedures. Strict regulations that are unified will not help. Instead, what is required is agile, responsive governance such regulatory frameworks which change along with the rate of innovation and are not fixed to very strict principles of ethics. Regulators ought to conduct trials using regulatory sandboxes, live audits and outcome-based systems of approval (Pesapane et al., 2021). International unification of standards (e.g., GDPR, HIPAA, AI Act) might contribute to the uniformity, whereas culture-, financial-, and infrastructural variations should be addressed by contextually-specific adds. More to the point, regulations need to focus on not only what AI does but how AI is constructed, implemented and operational throughout its lifecycle (Li et al., 2024). Anticipating how various healthcare institutions may adapt to ethical AI governance is critical. *Table*

5 provides a forecast of institutional readiness between now and 2030, identifying expected gaps in implementation.

Table 5. Forecasted Ethical-Compliance Readiness of Healthcare Institutions (2025–2030)

Institution Type	Readiness for Ethical AI Governance	Key Gaps Identified	Forecast by 2030
Large Academic Hospitals	Moderate–High	AI training for staff, explainability frameworks	Near full readiness
Regional Public Hospitals	Low–Moderate	Funding, data infrastructure, AI literacy	Partial readiness
Private Digital Health Startups	High tech, Low ethics	Ethical reviews, patient engagement, transparency	Depends on regulatory pressure
Primary Care Clinics	Low	Lack of technical staff and oversight mechanisms	Limited integration expected
Global NGOs (health programs)	Moderate	Cross-border data governance, localization	Moderate but uneven

Most ethical discussions carried out today occur once harm has occurred within the systems. Such reactive nature is not sustainable. The ethical future should be designed first: values, such as justice, autonomy, and safety, must be coded into algorithms that are used to create them. The whole pipeline of AI development needs to be redesigned as an ethical design, including generating inclusive and multicultural datasets, training with equity measurement, validate in subgroup of the population, and deploy with protective measures and consent procedures (Freyer et al., 2024). Ethics has to shift from being a policy appendix to being a technical aspect. In addition, morality must not be used as a roadblock on innovation; rather it must guide it. Proactively ethical AI systems will be stronger, more credible and in the end more powerful (Bekbolatova et al., 2024).

Here we are at a crossroad. The decisions that we make over the next few years on the issue of how to build, utilize and regulate artificial intelligence will define the ethical structure of the health sector in the 21st century. Will AI increase, or decrease inequalities? Does it expand clinical knowledge, or alienate patients? To whom will it be beneficial, many or few? They require vision, vigilance and collective will to answer these questions. What our future can hold with regards to healthcare AI should not just focus on what is possible as an end, but what is right (Tursunbayeva & Renkema, 2023).

9.4 Patient-Centered and Participatory Governance Models

The healthcare is based on trust. However, AI models are marred with either lack of transparency, bias, or are used to exploit personal information that reduces societal trust. The AI of the future in healthcare should be designed to be trustworthy. It implies more than sound forecasts. It is characterized by clear-cut reasoning, comprehensible results and sensitivity in interfaces (Schoenherr et al., 2023). Human-centric AI does not mean to replace the ethical culture of care, though, as it can respect the dignity of patients and assist the judgment of clinicians. Comments Building trust, such as through participation-driven design (collaborative design), explanation structures, and transparent auditing, will be central to adoption in the long run and societal robustness (Seralidou et al., 2025).

Computer scientists are not the only ones who should practice ethical AI in health care. It needs strong interdisciplinary cooperation: clinicians, ethicists, sociologists, patient advocates, public health specialists and legal scholars (Stogiannos et al., 2024). The future of innovation should be created in settings that incorporate these voices in creating solutions, but not just to review them retroactively. Such models will make new institutional models of ethics-integrated labs, interdisciplinary AI councils, and cross-sector partnerships that trade technical innovation and ethical vision. Not only do such alliances address any harm done, but leave the possibility of more culturally competent, context-aware, and more user-friendly AI applications (Stafie et al., 2023).

A change in the perception towards the public is perhaps, most radical. Patients are data; they are also the main beneficiaries and even victims of an AI decision. Their issues cannot be considered like an added matter or PR threat. Real ethical AI needs meaningful, continuous, and responsible public interaction (Agnello et al., 2025). The citizens should also be able to decide what happens to their health information, what risk exposure they have allowed and what degree of algorithmic participation is acceptable to them. Citizen advisory boards, participatory design, and public consultation rooms are not a suggestion anymore; they are the basic tool of democratic governance of technology in healthcare (Geroimenko, 2025).

What comes through this constellation is obvious: no single stakeholder can make it on her/his own. Programmers can construct potent systems; nevertheless, without medical expertise, the systems can end up being useless, or worse still hazardous. Transparency may be required by the healthcare providers, however, without government support, non-transparent models will prevail. Regulators can draft laws but without the backing of people, the application of the laws will appear as illegitimate (Oliver et al., 2025). Patients might demand fairness, but must have institutional proxies, to guarantee it. It is not the job of a single field to maintain ethical AI in healthcare, but rather an assemblage of responsibility, founded on interdisciplinary mutual respect, active communication, and collaboration. What happens to healthcare AI will not be determined by technological virtuosity but more by the success with which we coordinate the works of the builders, operators, regulators, and residents of our collective future (Hofstetter et al., 2021). Third, the studies of the patient-centered governance models are scarce. Research into participatory strategies, including dynamic consent models, patient advisory positions, and open communication methods, may facilitate harmonizing the implementation of AI with societal principles and anticipations. Lastly, disciplinary research that involves technical, legal, and ethical viewpoints is required to enlighten adaptive regulation models that can be responsive to AI systems that constantly learn. All of these research directions could be used in order to facilitate evidence-based policy making, and advance more responsible, inclusive, and more ethically oriented AI use in healthcare.

Embedding participatory and patient-centered governance into healthcare AI is not optional but foundational to ethical legitimacy. Sustained collaboration among patients, clinicians, developers, and regulators enables AI systems to reflect societal values while remaining clinically effective. Treating patients as co-creators rather than passive data sources strengthens trust, accountability, and long-term sustainability of AI-driven healthcare innovation.

10. EXECUTIVE CHECKLIST FOR ETHICAL AI GOVERNANCE IN HEALTHCARE

The following checklist provides a practical governance tool for healthcare institutions, policymakers, and AI developers to support the responsible design, deployment, and oversight of artificial intelligence in clinical settings. It translates ethical principles and regulatory expectations into actionable governance considerations.

A. Data Governance and Privacy

- Ensure patient data are collected, stored, and used in compliance with applicable data protection laws.
- Verify that datasets are representative of the target patient population and reviewed for historical bias.
- Define clear policies for secondary data use, data sharing, and retention.

B. Ethical Design and Bias Mitigation

- Conduct bias audits during model development and at regular intervals post-deployment.
- Document design choices that affect fairness, explainability, and risk exposure.
- Involve multidisciplinary ethics review processes during development.

C. Transparency and Explainability

- Clearly disclose when AI systems are used in clinical decision-making.
- Provide clinicians with interpretable outputs that support, rather than replace, professional judgment.
- Offer patients understandable explanations of AI involvement where relevant.

D. Human Oversight and Accountability

- Maintain meaningful human-in-the-loop or human-on-the-loop oversight for high-risk applications.
- Define responsibility across developers, vendors, healthcare providers, and institutions.
- Establish clear escalation and override procedures when AI recommendations conflict with clinical judgment.

E. Regulatory Compliance and Monitoring

- Classify AI systems according to applicable regulatory risk categories.
- Implement post-market monitoring to detect performance drift, safety issues, or unintended harm.
- Prepare documentation for audits, incident reporting, and regulatory review.

F. Security and System Integrity

- Apply cybersecurity safeguards across the AI lifecycle, including model updates.
- Monitor for adversarial attacks, data poisoning, or unauthorized system access.
- Maintain contingency plans for system failure or compromise.

This checklist is intended to function as a living governance instrument, adaptable to evolving technologies, regulatory requirements, and clinical contexts.

11. CONCLUSION

Already, AI is changing the face of healthcare by changing how we identify, treat, track, and handle disease. However, a more profound change is occurring in the midst of these technological advancements: the transfer of ethical accountability from individual doctors and healthcare facilities to dispersed networks of algorithms, data pipelines, legal frameworks, and international players. This chapter has thoroughly examined that change, exposing both the potential and the risks associated with AI in healthcare.

As we have discussed, in this book, in terms of the technical heart of AI systems, in terms of the regulatory frameworks that hold them all together, in terms of the biases inherent in the data on which they build, and in terms of the dangers of algorithmic inscrutability that they create, there has not been a neutrality to AI, and ethically. The way it is designed and implemented has serious implications on autonomy, fairness, privacy and justice. The principal values of medicine, i.e., do no harm, provide equity, respect autonomy and act with transparency need to be redefined and strengthened in the age when machines also join the human healing process. We explored such ethical principles as well as structural issues: bias that is inherent in training data, the national security risks of AI models, the international disconnect in regulatory system, and the erosion of human responsibility. These are not some hypothetical risks. They are reflected in the differences in real-world rates of diagnosis, violations of confidential data of patients, loss of clinical trust, and discrimination of vulnerable groups.

Although what emerges is this: we cannot decide the future of healthcare AI by only what we can make, but what we consider to be valuable. Intelligent algorithms alone are not the solution to a better future but teamwork, and ethically-aligned systems that make human dignity central to every step, beginning with the datasets we train on, to policies we put in place, to the interfaces to which patients and clinicians ultimately interact. The players, developers, clinicians, regulators, and the populace should not be viewed as independent agents to this reinvention, but as partners in managing this reinvention. AIs could be the most tremendous addition to medicine not only in the future but even now, provided that we deserve it. Ethical due diligence should not be something to be able to check off and something to be able to step over or work around. It has to be a design mindset of responsibility and a habit and it has to be a continuous process of learning as technology is changing and the society is changing. The bigger setback in this new form of algorithmic medicine is not that we may develop machines that can think we need machines that can care. and that is a very human undertaking.

In conclusion, it is important to state that the ethical and responsible introduction of artificial intelligence into healthcare is a deliberate, ethical, and prospective effort that allows keeping patient welfare in the focus of all technological progress. With AI systems further affecting the diagnosis, treatment planning, patient monitoring, and personalized interventions, the ethical-by-design notion needs to permeate the whole AI lifecycle, including data collection and model development, deployment, evaluation, and maintenance. Ethical-by-design schemes guarantee that accountability, fairness, explainability, and patient safety are not merely added features to system development and deployment but rather core concepts on how the system should be developed and implemented. This will be an active solution to the risk of harm, build the confidence of the populace, and promote the sustainable implementation of AI-powered innovations.

One of the most important elements of this ethical basis is the creation of standard AI risk-classification schemes which are specifically applied to healthcare. Since AI applications have wide ranges, such as low-risk administrative tools and risky clinical decision-support systems, standard risk classification facilitates the establishment of suitable oversight, validation and regulatory routes. This type of classification can help policymakers, developers, and medical institutions to determine the extent to which various applications of AI should be scrutinized, so that more risky systems receive a more stringent review. A unified global risk set and a unified approach to risk management among regulators would also help to minimize ambiguity among developers and make AI integration in clinics a safer and more predictable process.

Nevertheless, this step of risk classification is not the end goal. Governance should also be effective in that it requires incessant tracking and practical validation. In live clinical settings, AI models do not always operate as they did in the controlled development setting (particularly when working with different patient groups, different data quality, and different medical practices). These could be continuous performance audits, post-deployment audits and dynamic monitoring systems that assist in detecting model drift or unintended biases or safety concerns that might arise as time goes by. The real-world assessment is more than just a system maintains its reliability, but the performance of AI has to match clinical results and patient expectations as well. The mechanisms secure the safety of patients and give doctors the assurance to rely on AI-based recommendations.

It is also important that patient-centered consent frameworks should be reinforced. Old-fashioned non-dynamism models of consent are not relevant in an age where patient data drives dynamic and continuously learning AI systems. The patients should be fully informed about what they will do with their data, the level of automation regarding the care they receive, and what risks or benefits AI-assisted decisions might have. Autonomy should be respected by means of transparency of communication and availability of explanations and by allowing patients to have meaningful control over their data. Providing patients with the power by educating them on continuous consent and refining their initial consent strengthens ethical conduct and encourages more engagement with the AI-enabled pathways of care.

Finally, there will be a need to move toward the globally uniform standards of governance because AI technologies are cross-border. Consistency, Fragmentation, and Responsible innovation is ensured through principles and interoperable regulations that are alike across the globe. The process of global harmonization promotes collaborative research, cross-border data exchange in a safe mode and promotion of upholding ethical and regulatory protections irrespective of the geographic positioning. The healthcare industry can embrace the transformative possibilities of AI to the fullest by aligning ethical design, uniform risk management, continuous monitoring, patient-centred consent, and international collaboration and protect human dignity, trust, and safety.

ACKNOWLEDGEMENTS

The authors express their sincere gratitude to the Department of Pharmacy, Galgotias University, for providing the necessary support and infrastructure to carry out this work.

CONFLICTS OF INTEREST

The authors declare that there is no conflict of interest regarding the publication of this paper.

FUNDING

This chapter did not receive any specific grant from funding agencies in the public, commercial, or not-for-profit sectors.

REFERENCES

Abràmoff, M. D., Roehrenbeck, C., Trujillo, S., Goldstein, J., Graves, A. S., Repka, M. X., & Silva, E.III "Zeke. (2022). A reimbursement framework for artificial intelligence in healthcare. *NPJ Digital Medicine*, *5*(1), 72. DOI: 10.1038/s41746-022-00621-w

Adirim, T., & Rao, R. (2025). In Adirim, T. (Ed.), *U.S. and Global Regulatory Frameworks For AI in Healthcare BT - Digital Health, AI and Generative AI in Healthcare: A Concise, Practical Guide for Clinicians* (pp. 179–188). Springer Nature Switzerland., DOI: 10.1007/978-3-031-83526-1_14

Adler-Milstein, J., Aggarwal, N., Ahmed, M., Castner, J., Evans, B. J., Gonzalez, A. A., James, C. A., Lin, S., Mandl, K. D., Matheny, M. E., Sendak, M. P., Shachar, C., & Williams, A. (2022). Meeting the Moment: Addressing Barriers and Facilitating Clinical Adoption of Artificial Intelligence in Medical Diagnosis. *NAM Perspectives*, *2022*. Advance online publication. DOI: 10.31478/202209c

Agnello, D. M., Anand-Kumar, V., An, Q., de Boer, J., Delfmann, L. R., Longworth, G. R., Loisel, Q., McCaffrey, L., Steiner, A., & Chastin, S. (2025). Co-creation methods for public health research — characteristics, benefits, and challenges: A Health CASCADE scoping review. *BMC Medical Research Methodology*, *25*(1), 60. DOI: 10.1186/s12874-025-02514-4

Ahdadou, M., Aajly, A., & Tahrouch, M. (2024). Unlocking the potential of augmented intelligence: A discussion on its role in boardroom decision-making. *International Journal of Disclosure and Governance*, *21*(3), 433–446. DOI: 10.1057/s41310-023-00207-2

Al-Ani, A., Rayyan, A., Maswadeh, A., Sultan, H., Alhammouri, A., Asfour, H., Alrawajih, T., Al Sharie, S., Al Karmi, F., Al-Azzam, A. M., Mansour, A., & Al-Hussaini, M. (2024). Evaluating the understanding of the ethical and moral challenges of Big Data and AI among Jordanian medical students, physicians in training, and senior practitioners: A cross-sectional study. *BMC Medical Ethics*, *25*(1), 18. DOI: 10.1186/s12910-024-01008-0

Aldoseri, A., Al-Khalifa, K. N., & Hamouda, A. M. (2023). Re-Thinking Data Strategy and Integration for Artificial Intelligence: Concepts, Opportunities, and Challenges. *Applied Sciences (Basel, Switzerland)*, *13*(12). Advance online publication. DOI: 10.3390/app13127082

Algumaei, A., Yaacob, N. M., Doheir, M., Al-Andoli, M. N., & Algumaie, M. (2025). Symmetric Therapeutic Frameworks and Ethical Dimensions in AI-Based Mental Health Chatbots (2020–2025): A Systematic Review of Design Patterns, Cultural Balance, and Structural Symmetry. *Symmetry*, *17*(7). Advance online publication. DOI: 10.3390/sym17071082

Alkhatib, A. J., & Darabseh, R. S. (2025). Ethical Challenges in AI-driven Healthcare Innovations. *International Journal of New Medicine*, *1*(1), 11–27.

Amann, J., Blasimme, A., Vayena, E., Frey, D., & Madai, V. I., & consortium, the P. (2020). Explainability for artificial intelligence in healthcare: A multidisciplinary perspective. *BMC Medical Informatics and Decision Making*, *20*(1), 310. DOI: 10.1186/s12911-020-01332-6

Astromskė, K., Peičius, E., & Astromskis, P. (2021). Ethical and legal challenges of informed consent applying artificial intelligence in medical diagnostic consultations. *AI & Society*, *36*(2), 509–520. DOI: 10.1007/s00146-020-01008-9

Balogun, E., & Luetge, C. (2025). Gap Analysis on Regulatory Frameworks for AI-embedded Medical Devices. *2025 IEEE Symposium on Trustworthy, Explainable and Responsible Computational Intelligence (CITREx Companion)*, 1–5. DOI: 10.1109/CITRExCompanion65208.2025.10981493

Banerjee, S., Dumka, A., Singh, T., & Patel, A. (2025). *Navigating the Legal Landscape: Ethical and Regulatory Considerations in AI-Driven Healthcare BT - Proceedings of the International Health Informatics Conference* (S. Jain, B. K. Bhargava, D. Kalra, & S. Groppe (eds.); pp. 401–420). Springer Nature Singapore.

Bean, E., Burleigh, C., Haskell, C., Burris-Melville, T., Payne, J., & Pathak, B. (2025). Eavesdropping on UNESCO AI Policy, Leadership, and Ethics. *Journal of Leadership Studies*, *18*(4), 98–110. 10.1002/jls.70007

Bécue, A., Praça, I., & Gama, J. (2021). Artificial intelligence, cyber-threats and Industry 4.0: Challenges and opportunities. *Artificial Intelligence Review*, *54*(5), 3849–3886. DOI: 10.1007/s10462-020-09942-2

Bekbolatova, M., Mayer, J., Ong, C. W., & Toma, M. (2024). Transformative Potential of AI in Healthcare: Definitions, Applications, and Navigating the Ethical Landscape and Public Perspectives. *Health Care*, *12*(2). Advance online publication. DOI: 10.3390/healthcare12020125

Belk, R. W., Belanche, D., & Flavián, C. (2023). Key concepts in artificial intelligence and technologies 4.0 in services. *Service Business*, *17*(1), 1–9. DOI: 10.1007/s11628-023-00528-w

Bohr, A., & Memarzadeh, K. (2020). The rise of artificial intelligence in healthcare applications. In Bohr, A., & Memarzadeh, K. (Eds.), *Artificial Intelligence in Healthcare* (pp. 25–60). Academic Press., 10.1016/B978-0-12-818438-7.00002-2

Bouhaddi, M., & Adi, K. (2023). Mitigating Membership Inference Attacks in Machine Learning as a Service. *2023 IEEE International Conference on Cyber Security and Resilience (CSR)*, 262–268. DOI: 10.1109/CSR57506.2023.10224960

Camacho, M., Perramon, J., Puig-Bosch, X., Dang, V. N., Díaz, O., & Lekadir, K. (2025). Stakeholder engagement: the path to trustworthy AI in healthcare. In Lorenzi, M., & Zuluaga, M. A. (Eds.), *Trustworthy AI in Medical Imaging* (pp. 471–493). Academic Press., 10.1016/B978-0-44-323761-4.00038-9

Chakraborty, A., Biswas, A., & Khan, A. K. (2023). In Biswas, A., Semwal, V. B., & Singh, D. (Eds.), *Artificial Intelligence for Cybersecurity: Threats, Attacks and Mitigation BT - Artificial Intelligence for Societal Issues* (pp. 3–25). Springer International Publishing., DOI: 10.1007/978-3-031-12419-8_1

Chapardar, H. (2024). Industry hybrid regulation: Exploring a model for business-driven circular economy. *Resources, Conservation & Recycling Advances, 21*, 200205. 10.1016/j.rcradv.2024.200205

Chen, M., & Decary, M. (2020). Artificial intelligence in healthcare: An essential guide for health leaders. *Healthcare Management Forum, 33*(1), 10–18. DOI: 10.1177/0840470419873123

Choudhary, M., Gupta, A., Bajpai, A., & Singh, R. (2024). *Exploring Ethical Aspects and Regulatory Framework for India's Sustainable Healthcare Sector in the Era of Artificial Intelligence BT - Transforming Healthcare Sector Through Artificial Intelligence and Environmental Sustainability* (R. Singh, W. Shafik, D. Crowther, & V. Kumar (eds.); pp. 129–147). Springer Nature Singapore. DOI: 10.1007/978-981-97-9555-0_7

Daneshjou, R., Smith, M. P., Sun, M. D., Rotemberg, V., & Zou, J. (2021). Lack of Transparency and Potential Bias in Artificial Intelligence Data Sets and Algorithms: A Scoping Review. *JAMA Dermatology, 157*(11), 1362–1369. DOI: 10.1001/jamadermatol.2021.3129

Daneshjou, R., Vodrahalli, K., Novoa, R. A., Jenkins, M., Liang, W., Rotemberg, V., Ko, J., Swetter, S. M., Bailey, E. E., Gevaert, O., Mukherjee, P., Phung, M., Yekrang, K., Fong, B., Sahasrabudhe, R., Allerup, J. A. C., Okata-Karigane, U., Zou, J., & Chiou, A. S. (2022). Disparities in dermatology AI performance on a diverse, curated clinical image set. *Science Advances, 8*(32), eabq6147. Advance online publication. DOI: 10.1126/sciadv.abq6147

Eke, C. I., & Shuib, L. (2025). The role of explainability and transparency in fostering trust in AI healthcare systems: A systematic literature review, open issues and potential solutions. *Neural Computing & Applications, 37*(4), 1999–2034. DOI: 10.1007/s00521-024-10868-x

Fahim, S. (2024). *Scope of Artificial Intelligence BT - Ethico-Legal Aspect of AI-driven Driverless Cars: Comparing Autonomous Vehicle Regulations in Germany, California, and India* (S. Fahim (ed.); pp. 15–40). Springer Nature Singapore. DOI: 10.1007/978-981-97-6883-7_2

Feng, J., Phillips, R. V., Malenica, I., Bishara, A., Hubbard, A. E., Celi, L. A., & Pirracchio, R. (2022). Clinical artificial intelligence quality improvement: Towards continual monitoring and updating of AI algorithms in healthcare. *NPJ Digital Medicine, 5*(1), 66. DOI: 10.1038/s41746-022-00611-y

Ferdous, J., Islam, R., Mahboubi, A., & Zahidul Islam, M. (2024). AI-Based Ransomware Detection: A Comprehensive Review. *IEEE Access : Practical Innovations, Open Solutions, 12*, 136666–136695. DOI: 10.1109/ACCESS.2024.3461965

Fernández, A. (2023). In Lara, F., & Deckers, J. (Eds.), *Opacity, Machine Learning and Explainable AI BT - Ethics of Artificial Intelligence* (pp. 39–58). Springer Nature Switzerland., DOI: 10.1007/978-3-031-48135-2_3

Fomin, V. V., & Astromskis, P. (2023). *The Black Box Problem*. Brill., DOI: 10.1163/9789004682900_012

Freyer, N., Groß, D., & Lipprandt, M. (2024). The ethical requirement of explainability for AI-DSS in healthcare: A systematic review of reasons. *BMC Medical Ethics, 25*(1), 104. DOI: 10.1186/s12910-024-01103-2

Garno, Z., Elkari, B., & Ourabah, L. (2025). AI-Driven Innovations in Healthcare: Balancing Technology With Ethics. *World Medical \& Health Policy*. 10.1002/wmh3.70031

Gashu, K. D., & Guadie, H. A. (2024). In Gashu, K. D., Mekonnen, Z. A., Chanyalew, M. A., & Guadie, H. A. (Eds.), *Ethics in Public Health Informatics BT - Public Health Informatics: Implementation and Governance in Resource-Limited Settings* (pp. 225–262). Springer Nature Switzerland., DOI: 10.1007/978-3-031-71118-3_7

Gawankar, S., Nair, S., Pawar, V., Vhatkar, A., & Chavan, P. (2024). Patient Privacy and Data Security in the Era of AI-Driven Healthcare. *2024 8th International Conference on Computing, Communication, Control and Automation (ICCUBEA)*, 1–6. DOI: 10.1109/ICCUBEA61740.2024.10775004

Gerke, S., Minssen, T., & Cohen, G. (2020). Ethical and legal challenges of artificial intelligence-driven healthcare. In Bohr, A., & Memarzadeh, K. (Eds.), *Artificial Intelligence in Healthcare* (pp. 295–336). Academic Press., 10.1016/B978-0-12-818438-7.00012-5

Geroimenko, V. (2025). In Geroimenko, V. (Ed.), *Generative AI: From Human–Computer Interaction to Human–Computer Creativity BT - Human-Computer Creativity: Generative AI in Education, Art, and Healthcare* (pp. 3–29). Springer Nature Switzerland., DOI: 10.1007/978-3-031-86551-0_1

Goirand, M., Austin, E., & Clay-Williams, R. (2021). Implementing Ethics in Healthcare AI-Based Applications: A Scoping Review. *Science and Engineering Ethics, 27*(5), 61. DOI: 10.1007/s11948-021-00336-3

Goktas, P., & Grzybowski, A. (2025). Shaping the Future of Healthcare: Ethical Clinical Challenges and Pathways to Trustworthy AI. *Journal of Clinical Medicine, 14*(5). Advance online publication. DOI: 10.3390/jcm14051605

Gray, M., Samala, R., Liu, Q., Skiles, D., Xu, J., Tong, W., & Wu, L. (2024). Measurement and Mitigation of Bias in Artificial Intelligence: A Narrative Literature Review for Regulatory Science. *Clinical Pharmacology \& Therapeutics, 115*(4), 687–697. 10.1002/cpt.3117

Gundlack, J., Negash, S., Thiel, C., Buch, C., Schildmann, J., Unverzagt, S., Mikolajczyk, R., Frese, T., & consortium, P. (2025). Artificial Intelligence in Medical Care – Patients' Perceptions on Caregiving Relationships and Ethics: A Qualitative Study. *Health Expectations, 28*(2), e70216. 10.1111/hex.70216

Hanna, M. G., Pantanowitz, L., Jackson, B., Palmer, O., Visweswaran, S., Pantanowitz, J., Deebajah, M., & Rashidi, H. H. (2025). Ethical and Bias Considerations in Artificial Intelligence/Machine Learning. *Modern Pathology, 38*(3), 100686. 10.1016/j.modpat.2024.100686

Hirani, R., Noruzi, K., Khuram, H., Hussaini, A. S., Aifuwa, E. I., Ely, K. E., Lewis, J. M., Gabr, A. E., Smiley, A., Tiwari, R. K., & Etienne, M. (2024). Artificial Intelligence and Healthcare: A Journey through History, Present Innovations, and Future Possibilities. *Life (Chicago, Ill.), 14*(5). Advance online publication. DOI: 10.3390/life14050557

Hoffmann, W. A., & Nortjé, N. (2022). In Nortjé, N., & Bester, J. C. (Eds.), *The Ethical Principles that Guide Artificial Intelligence Utilization in Clinical Health Care BT - Pediatric Ethics: Theory and Practice* (pp. 403–419). Springer International Publishing., DOI: 10.1007/978-3-030-86182-7_25

Hofstetter, J. S., De Marchi, V., Sarkis, J., Govindan, K., Klassen, R., Ometto, A. R., Spraul, K. S., Bocken, N., Ashton, W. S., Sharma, S., Jaeger-Erben, M., Jensen, C., Dewick, P., Schröder, P., Sinkovics, N., Ibrahim, S. E., Fiske, L., Goerzen, A., & Vazquez-Brust, D. (2021). From Sustainable Global Value Chains to Circular Economy—Different Silos, Different Perspectives, but Many Opportunities to Build Bridges. *Circular Economy and Sustainability, 1*(1), 21–47. DOI: 10.1007/s43615-021-00015-2

Hübner, J., Baltus, H., Eisemann, N., Rohr, M., Schumann, L., Augustin, J., Hagenström, K., Wolf, S., Garbe, C., Augustin, M., & Katalinic, A. (2023). Evaluation of early skin cancer detection in Germany with cancer registry data – challenges, solutions and current trends. *JDDG: Journal Der Deutschen Dermatologischen Gesellschaft, 21*(S5), 13–20. 10.1111/ddg.15171

Iserson, K. V. (2024). Informed consent for artificial intelligence in emergency medicine: A practical guide. *The American Journal of Emergency Medicine*, *76*, 225–230. 10.1016/j.ajem.2023.11.022

Jha, D., Durak, G., Sharma, V., Keles, E., Cicek, V., Zhang, Z., Srivastava, A., Rauniyar, A., Hagos, D. H., Tomar, N. K., Miller, F. H., Topcu, A., Yazidi, A., Håkegård, J. E., & Bagci, U. (2025). A Conceptual Framework for Applying Ethical Principles of AI to Medical Practice. *Bioengineering (Basel, Switzerland)*, *12*(2). Advance online publication. DOI: 10.3390/bioengineering12020180

Karimian, G., Petelos, E., & Evers, S. M. A. A. (2022). The ethical issues of the application of artificial intelligence in healthcare: A systematic scoping review. *AI and Ethics*, *2*(4), 539–551. DOI: 10.1007/s43681-021-00131-7

Keshta, I. (2022). AI-driven IoT for smart health care: Security and privacy issues. *Informatics in Medicine Unlocked*, *30*, 100903. 10.1016/j.imu.2022.100903

Kim, J., Davis, T., & Hong, L. (2022). In Albert, V., Lin, L., Spector, M. J., & Dunn, L. S. (Eds.), *Augmented Intelligence: Enhancing Human Decision Making BT - Bridging Human Intelligence and Artificial Intelligence (M* (pp. 151–170). Springer International Publishing., DOI: 10.1007/978-3-030-84729-6_10

Kioskli, K., Grigoriou, E., Islam, S., Yiorkas, A. M., Christofi, L., & Mouratidis, H. (2025). A risk and conformity assessment framework to ensure security and resilience of healthcare systems and medical supply chain. *International Journal of Information Security*, *24*(2), 90. DOI: 10.1007/s10207-025-01009-z

Königs, P. (2022). Artificial intelligence and responsibility gaps: What is the problem? *Ethics and Information Technology*, *24*(3), 36. DOI: 10.1007/s10676-022-09643-0

Kostick-Quenet, K., & Rahimzadeh, V. (2023). Ethical hazards of health data governance in the metaverse. *Nature Machine Intelligence*, *5*(5), 480–482. DOI: 10.1038/s42256-023-00658-w

Kumar, A., Gupta, A., & Raj, U. (2024). New Era of Intelligent Medicine: Future Scope and Challenges. *2024 11th International Conference on Reliability, Infocom Technologies and Optimization (Trends and Future Directions) (ICRITO)*, 1–6. DOI: 10.1109/ICRITO61523.2024.10522148

Lee, D., & Yoon, S. N. (2021). Application of artificial intelligence-based technologies in the healthcare industry: Opportunities and challenges. *International Journal of Environmental Research and Public Health*, *18*(1), 271.

Lee, M. S. A., Floridi, L., & Singh, J. (2021). Formalising trade-offs beyond algorithmic fairness: Lessons from ethical philosophy and welfare economics. *AI and Ethics, 1*(4), 529–544. DOI: 10.1007/s43681-021-00067-y

Li, F., Ruijs, N., & Lu, Y. (2023). Ethics & AI: A Systematic Review on Ethical Concerns and Related Strategies for Designing with AI in Healthcare. *AI, 4*(1), 28–53. DOI: 10.3390/ai4010003

Li, Y.-H., Li, Y.-L., Wei, M.-Y., & Li, G.-Y. (2024). Innovation and challenges of artificial intelligence technology in personalized healthcare. *Scientific Reports, 14*(1), 18994. DOI: 10.1038/s41598-024-70073-7

Lin, Y., Zhao, X., Miao, Z., Ling, Z., Wei, X., Pu, J., Hou, J., & Shen, B. (2020). Data-driven translational prostate cancer research: From biomarker discovery to clinical decision. *Journal of Translational Medicine, 18*(1), 119. DOI: 10.1186/s12967-020-02281-4

Malla, A. M., & Amin, U. (2023). Scope of technology in health care, special focus on nursing. *Journal of Integrative Nursing, 5*(4).

Marey, A., Arjmand, P., Alerab, A. D. S., Eslami, M. J., Saad, A. M., Sanchez, N., & Umair, M. (2024). Explainability, transparency and black box challenges of AI in radiology: Impact on patient care in cardiovascular radiology. *The Egyptian Journal of Radiology and Nuclear Medicine, 55*(1), 183. DOI: 10.1186/s43055-024-01356-2

Matsuo, Y., LeCun, Y., Sahani, M., Precup, D., Silver, D., Sugiyama, M., Uchibe, E., & Morimoto, J. (2022). Deep learning, reinforcement learning, and world models. *Neural Networks, 152*, 267–275. 10.1016/j.neunet.2022.03.037

Matsuzaka, Y., & Yashiro, R. (2023). AI-Based Computer Vision Techniques and Expert Systems. *AI, 4*(1), 289–302. DOI: 10.3390/ai4010013

Mennella, C., Maniscalco, U., De Pietro, G., & Esposito, M. (2024). Ethical and regulatory challenges of AI technologies in healthcare: A narrative review. *Heliyon, 10*(4). Advance online publication. DOI: 10.1016/j.heliyon.2024.e26297

Mergen, A., Çetin-Kılıç, N., & Özbilgin, M. F. (2025). Artificial Intelligence and Bias Towards Marginalised Groups: Theoretical Roots and Challenges. In Vassilopoulou, J., & Kyriakidou, O. (Eds.), *AI and Diversity in a Datafied World of Work: Will the Future of Work be Inclusive?* (Vol. 12, pp. 17–38). Emerald Publishing Limited., DOI: 10.1108/S2051-233320250000012004

Merriel, S. W. D., Funston, G., & Hamilton, W. (2018). Prostate Cancer in Primary Care. *Advances in Therapy, 35*(9), 1285–1294. DOI: 10.1007/s12325-018-0766-1

Meszaros, J., Minari, J., & Huys, I. (2022). The future regulation of artificial intelligence systems in healthcare services and medical research in the European Union. *Frontiers in Genetics, 13*. Advance online publication. DOI: 10.3389/fgene.2022.927721

Metta, C., Beretta, A., Pellungrini, R., Rinzivillo, S., & Giannotti, F. (2024). Towards Transparent Healthcare: Advancing Local Explanation Methods in Explainable Artificial Intelligence. *Bioengineering (Basel, Switzerland), 11*(4). Advance online publication. DOI: 10.3390/bioengineering11040369

Mohsin Khan, M., Shah, N., Shaikh, N., & Thabet, A. alrabayah, T., & Belkhair, S. (2025). Towards secure and trusted AI in healthcare: A systematic review of emerging innovations and ethical challenges. *International Journal of Medical Informatics, 195*, 105780. 10.1016/j.ijmedinf.2024.105780

Moilanen, T., Suhonen, R., & Kangasniemi, M. (2025). Key professional stakeholders roles in promoting older people's autonomy in residential care. *Nursing Ethics, 32*(2), 575–587. DOI: 10.1177/09697330241247321

Moon, D., & Ahn, S. (2025). Metrics and Algorithms for Identifying and Mitigating Bias in AI Design: A Counterfactual Fairness Approach. *IEEE Access : Practical Innovations, Open Solutions, 13*, 59118–59129. DOI: 10.1109/ACCESS.2025.3556082

Morley, J., Machado, C. C. V., Burr, C., Cowls, J., Joshi, I., Taddeo, M., & Floridi, L. (2020). The ethics of AI in health care: A mapping review. *Social Science & Medicine, 260*, 113172. 10.1016/j.socscimed.2020.113172

Mukhamediev, R. I., Popova, Y., Kuchin, Y., Zaitseva, E., Kalimoldayev, A., Symagulov, A., Levashenko, V., Abdoldina, F., Gopejenko, V., Yakunin, K., Muhamedijeva, E., & Yelis, M. (2022). Review of Artificial Intelligence and Machine Learning Technologies: Classification, Restrictions, Opportunities and Challenges. *Mathematics, 10*(15). Advance online publication. DOI: 10.3390/math10152552

Mulla, S. J., & Patel, M. (2025). Global Regulatory Perspectives on Sensor-Enabled and Biomaterial-Integrated Medical Devices: A Focus on India, Canada, and Australia. *Biomedical Materials & Devices*. Advance online publication. DOI: 10.1007/s44174-025-00327-z

Murdoch, B. (2021). Privacy and artificial intelligence: Challenges for protecting health information in a new era. *BMC Medical Ethics, 22*(1), 122. DOI: 10.1186/s12910-021-00687-3

Murikah, W., Nthenge, J. K., & Musyoka, F. M. (2024). Bias and ethics of AI systems applied in auditing - A systematic review. *Scientific African*, *25*, e02281. 10.1016/j.sciaf.2024.e02281

Nasir, S., Khan, R. A., & Bai, S. (2024). Ethical Framework for Harnessing the Power of AI in Healthcare and Beyond. *IEEE Access : Practical Innovations, Open Solutions*, *12*, 31014–31035. DOI: 10.1109/ACCESS.2024.3369912

Nasiraee, H., Ashouri-Talouki, M., & Liu, X. (2023). Optimal Black-Box Traceability in Decentralized Attribute-Based Encryption. *IEEE Transactions on Cloud Computing*, *11*(3), 2459–2472. DOI: 10.1109/TCC.2022.3210137

Nazer, L. H., Zatarah, R., Waldrip, S., Ke, J. X. C., Moukheiber, M., Khanna, A. K., Hicklen, R. S., Moukheiber, L., Moukheiber, D., Ma, H., & Mathur, P. (2023). Bias in artificial intelligence algorithms and recommendations for mitigation. *PLOS Digital Health*, *2*(6), 1–14. DOI: 10.1371/journal.pdig.0000278

Nemec Zlatolas, L., Welzer, T., & Lhotska, L. (2024). Data breaches in healthcare: Security mechanisms for attack mitigation. *Cluster Computing*, *27*(7), 8639–8654. DOI: 10.1007/s10586-024-04507-2

Norori, N., Hu, Q., Aellen, F. M., Faraci, F. D., & Tzovara, A. (2021). Addressing bias in big data and AI for health care: A call for open science. *Patterns (New York, N.Y.)*, *2*(10). Advance online publication. DOI: 10.1016/j.patter.2021.100347

Nouis, S. C., Uren, V., & Jariwala, S. (2025). Evaluating accountability, transparency, and bias in AI-assisted healthcare decision- making: A qualitative study of healthcare professionals' perspectives in the UK. *BMC Medical Ethics*, *26*(1), 89. DOI: 10.1186/s12910-025-01243-z

Oliver, A. L., Rittblat, R., & Menuhin, J. (2025). From Silos to Synergies: A Nexus Framework for Innovation-Driven Sustainability Ecosystems. *Sustainability*, *17*(14). Advance online publication. DOI: 10.3390/su17146239

Omran Almagrabi, A., & Khan, R. A. (2025). Optimizing Secure AI Lifecycle Model Management With Innovative Generative AI Strategies. *IEEE Access : Practical Innovations, Open Solutions*, *13*, 12889–12920. DOI: 10.1109/ACCESS.2024.3491373

Palaniappan, K., Lin, E. Y. T., & Vogel, S. (2024). Global Regulatory Frameworks for the Use of Artificial Intelligence (AI) in the Healthcare Services Sector. *Health Care*, *12*(5). Advance online publication. DOI: 10.3390/healthcare12050562

Palaniappan, K., Lin, E. Y. T., Vogel, S., & Lim, J. C. W. (2024). Gaps in the Global Regulatory Frameworks for the Use of Artificial Intelligence (AI) in the Healthcare Services Sector and Key Recommendations. *Health Care*, *12*(17). Advance online publication. DOI: 10.3390/healthcare12171730

Panarese, P., Grasso, M. M., & Solinas, C. (2025). Algorithmic bias, fairness, and inclusivity: A multilevel framework for justice-oriented AI. *AI & Society*. Advance online publication. DOI: 10.1007/s00146-025-02451-2

Patel, A. P., Wang, M., Ruan, Y., Koyama, S., Clarke, S. L., Yang, X., Tcheandjieu, C., Agrawal, S., Fahed, A. C., Ellinor, P. T., Tsao, P. S., Sun, Y. V, Cho, K., Wilson, P. W. F., Assimes, T. L., van Heel, D. A., Butterworth, A. S., Aragam, K. G., & Natarajan, P. Program, G. & H. R. T. the M. V. (2023). A multi-ancestry polygenic risk score improves risk prediction for coronary artery disease. *Nature Medicine*, *29*(7), 1793–1803. DOI: 10.1038/s41591-023-02429-x

Pesapane, F., Bracchi, D. A., Mulligan, J. F., Linnikov, A., Maslennikov, O., Lanzavecchia, M. B., Tantrige, P., Stasolla, A., Biondetti, P., Giuggioli, P. F., Cassano, E., & Carrafiello, G. (2021). Legal and Regulatory Framework for AI Solutions in Healthcare in EU, US, China, and Russia: New Scenarios after a Pandemic. *Radiation*, *1*(4), 261–276. DOI: 10.3390/radiation1040022

Pickering, B. (2021). Trust, but Verify: Informed Consent, AI Technologies, and Public Health Emergencies. *Future Internet*, *13*(5). Advance online publication. DOI: 10.3390/fi13050132

Pot, M., Kieusseyan, N., & Prainsack, B. (2021). Not all biases are bad: Equitable and inequitable biases in machine learning and radiology. *Insights Into Imaging*, *12*(1), 13. DOI: 10.1186/s13244-020-00955-7

Prakash, S., Balaji, J. N., Joshi, A., & Surapaneni, K. M. (2022). Ethical Conundrums in the Application of Artificial Intelligence (AI) in Healthcare—A Scoping Review of Reviews. *Journal of Personalized Medicine*, *12*(11). Advance online publication. DOI: 10.3390/jpm12111914

Price, W. N., & Cohen, I. G. (2019). Privacy in the age of medical big data. *Nature Medicine*, *25*(1), 37–43. DOI: 10.1038/s41591-018-0272-7

Pugliese, R., Regondi, S., & Marini, R. (2021). Machine learning-based approach: global trends, research directions, and regulatory standpoints. *Data Science and Management, 4*, 19–29. 10.1016/j.dsm.2021.12.002

Reddy, K. J. (2025). In Reddy, K. J. (Ed.), *Ethical and Privacy Issues in Technology BT - Innovations in Neurocognitive Rehabilitation: Harnessing Technology for Effective Therapy* (pp. 267–284). Springer Nature Switzerland., DOI: 10.1007/978-3-031-88117-6_12

Rogers, W. A., Draper, H., & Carter, S. M. (2021). Evaluation of artificial intelligence clinical applications: Detailed case analyses show value of healthcare ethics approach in identifying patient care issues. *Bioethics*, *35*(7), 623–633. 10.1111/bioe.12885

Rozenblit, L., Price, A., Solomonides, A., Joseph, A. L., Koski, E., Srivastava, G., Labkoff, S., Bray, D., Lopez-Gonzalez, M., Singh, R., deBronkart, D., Barr, P. J., Szolovits, P., Dattani, K., Jaffe, C., Fridsma, D., Baris, R., Leftwich, R., Stolper, R., & Quintana, Y. (2025). Toward responsible AI governance: Balancing multi-stakeholder perspectives on AI in healthcare. *International Journal of Medical Informatics, 203*, 106015. 10.1016/j.ijmedinf.2025.106015

Saraswat, D., Bhattacharya, P., Verma, A., Prasad, V. K., Tanwar, S., Sharma, G., Bokoro, P. N., & Sharma, R. (2022). Explainable AI for Healthcare 5.0: Opportunities and Challenges. *IEEE Access : Practical Innovations, Open Solutions, 10*, 84486–84517. DOI: 10.1109/ACCESS.2022.3197671

Sarker, I. H. (2021). Deep Learning: A Comprehensive Overview on Techniques, Taxonomy, Applications and Research Directions. *SN Computer Science*, *2*(6), 420. DOI: 10.1007/s42979-021-00815-1

Sarker, I. H. (2023). Multi-aspects AI-based modeling and adversarial learning for cybersecurity intelligence and robustness: A comprehensive overview. *Security and Privacy*, *6*(5), e295. 10.1002/spy2.295

Savolainen, L., & Ruckenstein, M. (2024). Dimensions of autonomy in human–algorithm relations. *New Media \& Society, 26*(6), 3472–3490. DOI: 10.1177/14614448221100802

Schmidt, J., Schutte, N. M., Buttigieg, S., Novillo-Ortiz, D., Sutherland, E., Anderson, M., de Witte, B., Peolsson, M., Unim, B., Pavlova, M., Stern, A. D., Mossialos, E., & van Kessel, R. (2024). Mapping the regulatory landscape for artificial intelligence in health within the European Union. *NPJ Digital Medicine*, *7*(1), 229. DOI: 10.1038/s41746-024-01221-6

Schoenherr, J. R., Abbas, R., Michael, K., Rivas, P., & Anderson, T. D. (2023). Designing AI Using a Human-Centered Approach: Explainability and Accuracy Toward Trustworthiness. *IEEE Transactions on Technology and Society*, *4*(1), 9–23. DOI: 10.1109/TTS.2023.3257627

Schoenherr, J. R., & Thomson, R. (2024). When AI Fails, Who Do We Blame? Attributing Responsibility in Human–AI Interactions. *IEEE Transactions on Technology and Society*, *5*(1), 61–70. DOI: 10.1109/TTS.2024.3370095

Seeram, E., & Kanade, V. (2024). In Seeram, E., & Kanade, V. (Eds.), *Ethical and Regulatory Considerations BT - Artificial Intelligence in Medical Imaging Technology: An Introduction* (pp. 151–167). Springer Nature Switzerland., DOI: 10.1007/978-3-031-64049-0_9

Seralidou, E., Kioskli, K., Fotis, T., & Polemi, N. (2025). AI_TAF: A Human-Centric Trustworthiness Risk Assessment Framework for AI Systems. *Computers*, *14*(7). Advance online publication. DOI: 10.3390/computers14070243

Seyyed-Kalantari, L., Zhang, H., McDermott, M. B. A., Chen, I. Y., & Ghassemi, M. (2021). Underdiagnosis bias of artificial intelligence algorithms applied to chest radiographs in under-served patient populations. *Nature Medicine*, *27*(12), 2176–2182. DOI: 10.1038/s41591-021-01595-0

Shu, S., Luo, Q., & Chen, Z. (2025). Proactive vs. passive algorithmic ethics practices in healthcare: The moderating role of healthcare engagement type in patients' responses. *BMC Medical Ethics*, *26*(1), 73. DOI: 10.1186/s12910-025-01236-y

Shvetsova, O., Feroz, M., Salkutsan, S., & Efimov, A. (2023). *Artificial Intelligence Application for Healthcare Industry: Cases of Developed and Emerging Markets BT - Expert Clouds and Applications* (I. Jeena Jacob, S. Kolandapalayam Shanmugam, & I. Izonin (eds.); pp. 419–432). Springer Nature Singapore.

Smith, H. (2021). Clinical AI: Opacity, accountability, responsibility and liability. *AI & Society*, *36*(2), 535–545. DOI: 10.1007/s00146-020-01019-6

Stafie, C. S., Sufaru, I.-G., Ghiciuc, C. M., Stafie, I.-I., Sufaru, E.-C., Solomon, S. M., & Hancianu, M. (2023). Exploring the Intersection of Artificial Intelligence and Clinical Healthcare: A Multidisciplinary Review. *Diagnostics (Basel)*, *13*(12). Advance online publication. DOI: 10.3390/diagnostics13121995

Stogiannos, N., Gillan, C., & Precht, H. sa dos Reis, C., Kumar, A., O'Regan, T., Ellis, V., Barnes, A., Meades, R., Pogose, M., Greggio, J., Scurr, E., Kumar, S., King, G., Rosewarne, D., Jones, C., van Leeuwen, K. G., Hyde, E., Beardmore, C., … Malamateniou, C. (2024). A multidisciplinary team and multiagency approach for AI implementation: A commentary for medical imaging and radiotherapy key stakeholders. *Journal of Medical Imaging and Radiation Sciences, 55*(4), 101717. 10.1016/j.jmir.2024.101717

Taimoor, N., & Rehman, S. (2022). Reliable and Resilient AI and IoT-Based Personalised Healthcare Services: A Survey. *IEEE Access : Practical Innovations, Open Solutions*, *10*, 535–563. DOI: 10.1109/ACCESS.2021.3137364

Tam, T. Y. C., Sivarajkumar, S., Kapoor, S., Stolyar, A. V., Polanska, K., McCarthy, K. R., Osterhoudt, H., Wu, X., Visweswaran, S., Fu, S., Mathur, P., Cacciamani, G. E., Sun, C., Peng, Y., & Wang, Y. (2024). A framework for human evaluation of large language models in healthcare derived from literature review. *NPJ Digital Medicine*, *7*(1), 258. DOI: 10.1038/s41746-024-01258-7

Topol, E. J. (2019). High-performance medicine: The convergence of human and artificial intelligence. *Nature Medicine*, *25*(1), 44–56. DOI: 10.1038/s41591-018-0300-7

Trocin, C., Mikalef, P., Papamitsiou, Z., & Conboy, K. (2023). Responsible AI for Digital Health: A Synthesis and a Research Agenda. *Information Systems Frontiers*, *25*(6), 2139–2157. DOI: 10.1007/s10796-021-10146-4

Tursunbayeva, A., & Renkema, M. (2023). Artificial intelligence in health-care: Implications for the job design of healthcare professionals. *Asia Pacific Journal of Human Resources*, *61*(4), 845–887. 10.1111/1744-7941.12325

Ueda, D., Kakinuma, T., Fujita, S., Kamagata, K., Fushimi, Y., Ito, R., Matsui, Y., Nozaki, T., Nakaura, T., Fujima, N., Tatsugami, F., Yanagawa, M., Hirata, K., Yamada, A., Tsuboyama, T., Kawamura, M., Fujioka, T., & Naganawa, S. (2024). Fairness of artificial intelligence in healthcare: Review and recommendations. *Japanese Journal of Radiology*, *42*(1), 3–15. DOI: 10.1007/s11604-023-01474-3

Wang, C., He, T., Zhou, H., Zhang, Z., & Lee, C. (2023). Artificial intelligence enhanced sensors - enabling technologies to next-generation healthcare and biomedical platform. *Bioelectronic Medicine*, *9*(1), 17. DOI: 10.1186/s42234-023-00118-1

Wang, J. K., Wang, S. K., Lee, E. B., & Chang, R. T. (2023). In Yogesan, K., Goldschmidt, L., Cuadros, J., & Ricur, G. (Eds.), *Natural Language Processing (NLP) in AI BT - Digital Eye Care and Teleophthalmology: A Practical Guide to Applications* (pp. 243–249). Springer International Publishing., DOI: 10.1007/978-3-031-24052-2_17

Wasilewski, T., Kamysz, W., & Gębicki, J. (2024). AI-Assisted Detection of Biomarkers by Sensors and Biosensors for Early Diagnosis and Monitoring. *Biosensors (Basel)*, *14*(7). Advance online publication. DOI: 10.3390/bios14070356

Weiner, E. B., Dankwa-Mullan, I., Nelson, W. A., & Hassanpour, S. (2025). Ethical challenges and evolving strategies in the integration of artificial intelligence into clinical practice. *PLOS Digital Health*, *4*(4), 1–12. DOI: 10.1371/journal.pdig.0000810

Williamson, S. M., & Prybutok, V. (2024). Balancing Privacy and Progress: A Review of Privacy Challenges, Systemic Oversight, and Patient Perceptions in AI-Driven Healthcare. *Applied Sciences (Basel, Switzerland)*, *14*(2). Advance online publication. DOI: 10.3390/app14020675

Wymberry, C., & Jahankhani, H. (2024). In Jahankhani, H., Bowen, G., Sharif, M. S., & Hussien, O. (Eds.), *An Approach to Measure the Effectiveness of the MITRE ATLAS Framework in Safeguarding Machine Learning Systems Against Data Poisoning Attack BT - Cybersecurity and Artificial Intelligence: Transformational Strategies and Disruptive Innovation* (pp. 81–116). Springer Nature Switzerland., DOI: 10.1007/978-3-031-52272-7_4

Yanamadala, V., Bento, V., & Cohen, S. P. (2024). The rise of AI in health care: Transforming the future. *The Journal of the International Society of Physical and Rehabilitation Medicine*, *7*(4).

Younis, H. A., Eisa, T. A. E., Nasser, M., Sahib, T. M., Noor, A. A., Alyasiri, O. M., Salisu, S., Hayder, I. M., & Younis, H. A. (2024). A Systematic Review and Meta-Analysis of Artificial Intelligence Tools in Medicine and Healthcare: Applications, Considerations, Limitations, Motivation and Challenges. *Diagnostics (Basel)*, *14*(1). Advance online publication. DOI: 10.3390/diagnostics14010109

ZadZiabari. A., & Tabatabaei, A. (2025). *Ethics and Regulations in Generative AI BT - Application of Generative AI in Healthcare Systems* (A. Zamanifar & M. Faezipour (eds.); pp. 197–211). Springer Nature Switzerland. DOI: 10.1007/978-3-031-82963-5_8

Zednik, C. (2021). Solving the Black Box Problem: A Normative Framework for Explainable Artificial Intelligence. *Philosophy & Technology*, *34*(2), 265–288. DOI: 10.1007/s13347-019-00382-7

LIST OF ABBREVIATIONS

AI: - Artificial Intelligence
AMA: - American Medical Association
API: - Application Programming Interface
EU: - European Union
FDA: - Food and Drug Administration (USA)

GDPR: - General Data Protection Regulation (EU)
HER: - Electronic Health Record
HIPAA: - Health Insurance Portability and Accountability Act
IEEE: - Institute of Electrical and Electronics Engineers
ML: - Machine Learning
NLP: - Natural Language Processing
PII: - Personally Identifiable Information
UNESCO: - United Nations Educational, Scientific and Cultural Organization
WHO: - World Health Organization

Conclusion

A STRATEGIC FRAMEWORK FOR POLYBIOME–AI SYSTEMS MEDICINE

This volume presents a structured, practice-oriented blueprint for operationalizing a polybiome–AI paradigm in precision health. By integrating conceptual foundations with regionally grounded case studies and translational pipelines—including the NexDi diabetes reclassification model, AI-guided monoclonal antibody discovery, and algorithmic nutrient engineering—it reframes health and disease as emergent properties of interacting biological, nutritional, environmental, and socio-cultural systems. Within this architecture, the polybiome construct advances a functional multi-kingdom systems ontology that extends beyond descriptive taxonomy toward prediction, prevention, and individualized intervention.

Core Contributions

Three interlocking contributions define the work

First, methodological integration and analytical rigor.

The volume establishes a coherent backbone for multi-omic harmonization through interpretable artificial intelligence. Genomic, metabolomic, microbiomic, exposomic, and clinical data streams are synthesized into reproducible and clinically actionable models. By prioritizing explainability and cross-context validation, the framework addresses a central challenge of contemporary biomedicine: converting high-dimensional data into mechanistically grounded insight.

Second, translational operationalization.

The chapters delineate structured pathways from computational modeling to real-world implementation. Digital twins, genotype-aligned nutrient design via precision fermentation, bacteriocin-enriched probiotic development, and AI-assisted immunotherapeutics demonstrate how algorithmic outputs can be translated into scalable interventions. The emphasis remains on operational feasibility, regulatory compatibility, and measurable population relevance rather than technological novelty alone.

Third, ethical governance and systemic responsibility.

Ethical architecture is treated as a foundational design principle rather than a peripheral consideration. Transparency, bias auditing, provenance tracking, and data sovereignty are articulated as non-negotiable requirements for AI deployment in clinical and public health settings. In this way, scientific advancement is aligned with institutional accountability and equitable access.

CIRCULAR MEDICINE: A SYSTEMS REFRAMING

Beyond technical metrics, the volume advances a reframing of medical epistemology. Disease is considered not merely a target for suppression, but an expression of disrupted systemic coherence across host biology, microbiome ecology, environmental exposure, and lived context. Within this perspective—here termed Circular Medicine—intervention seeks restoration of functional equilibrium through calibrated modulation rather than force-based eradication.

Anchored in the polybiome paradigm, this orientation points toward increasingly adaptive healthcare systems in which ecological intelligence and computational analytics evolve in concert. The long-term implication is not technological dominance over biology, but deeper biological alignment supported by digital precision.

POLICY AND IMPLEMENTATION IMPERATIVES

Operational realization requires coordinated structural investment. Priority actions include:

1. Developing interoperable, secure, and standardized multi-omic infrastructures to ensure cross-site reproducibility.
2. Establishing regulatory sandboxes that integrate conventional clinical trials with realworld evidence for AI-driven diagnostics, therapeutics, and nutraceuticals.
3. Mandating explainability standards, systematic bias auditing, and comprehensive documentation of data provenance in clinical decision systems.
4. Strengthening cross-disciplinary workforce capacity across clinical medicine, computational science, and regulatory governance.

5. Implementing culturally contextualized pilot programs to validate polybiome-informed interventions across diverse populations and prevent inequitable diffusion of innovation.

RESEARCH PRIORITIES FOR THE NEXT DECADE

Future research directions logically extend from this integrative model:

- External validation of predictive models across independent and ethnically diverse cohorts.
- Longitudinal life-course studies mapping dynamic polybiome trajectories.
- Harmonized protocols for multi-omic sampling, processing, and reporting.
- Integrated translational pipelines connecting algorithmic discovery to scalable manufacturing and clinical deployment.
- Evaluation frameworks that extend beyond algorithmic accuracy to encompass clinical utility, cost-effectiveness, equity of benefit, and measurable population health outcomes.

GLOBAL COLLABORATION AND FORWARD TRAJECTORY

Contributions from twelve countries provide methodological depth and cultural nuance, strengthening both generalizability and contextual adaptability. The polybiome–AI paradigm articulated here is neither speculative nor rhetorical; it represents a disciplined operational framework positioned for incremental validation and responsible scale-up.

Looking toward 2040, systems medicine informed by polybiome intelligence is expected to become progressively predictive, contextually adaptive, and governance-aligned. Its success will depend not on isolated technological breakthroughs, but on sustained integration across research, clinical practice, industry, and policy domains.

This volume stands not as a declaration of completion, but as a structured point of departure— an invitation to refine, validate, and implement a biologically coherent and computationally accountable model of precision health.

Reyed M. Reyed

Genetic Engineering and Biotechnology Research Institute, Egypt & City of Scientific Research and Technological Applications, Egypt

About the Contributors

Reyed M. Reyed is an Associate Professor of Microbial Biotechnology at the Genetic Engineering and Biotechnology Research Institute (GEBRI), within the City of Scientific Research and Technological Applications (SRTA-City), Egypt. He is internationally recognized as a pioneering innovator in the emerging field of Bio-Digital Polybiomics, which uniquely integrates microbial genomics, AI-driven computational biology, and precision nutrition to transform healthcare paradigms. His visionary leadership has led to the development of groundbreaking frameworks such as GED-PerMed (focused on leveraging big data for precision medicine), and NexDi, a next-generation diagnostic platform designed to translate complex multi-omics data into actionable, personalized therapeutic strategies. Prof. Reyed's academic and research journey is distinguished by prestigious fellowships and competitive grants from top global institutions, including the National Centre for Biological Sciences (NCBS), India and Sandia National Laboratories, USA. These experiences have shaped his expertise in advanced computational methods, systems biology, and integrative data analytics, enabling him to lead innovative projects at the intersection of microbiology, bioinformatics, and clinical nutrition. In addition to his research contributions, Prof. Reyed holds several key editorial leadership roles. He serves as Editor-in-Chief and Academic Editor for multiple internationally peer-reviewed journals, where he upholds the highest standards of scientific rigor and fosters the dissemination of cutting-edge research. His editorial governance is marked by a commitment to multidisciplinary collaboration and the promotion of novel approaches in biotechnology and digital health. Prof. Reyed's strategic vision extends beyond the laboratory and editorial office. He is appointed Chair of the 2025 Global Summit on Climate Change, to be held in Paris, reflecting his dedication to addressing global environmental challenges through sustainable biotechnological innovations. His leadership role in this high-profile event exemplifies his commitment to integrating environmental sustainability with advanced biomedical research. Multidisciplinarily trained as a Registered

Clinical Nutritionist and an expert in bioinformatics, Prof. Reyed bridges critical gaps between fundamental bench science, computational modeling, and practical health applications. He currently leads a dynamic international research consortium spanning Egypt, India, Oman, Turkey, and Japan, focused on advancing national bio-sovereignty and developing transformative healthcare solutions powered by AI-governed biotechnology. This consortium underscores his commitment to ethical AI integration, fostering equitable access to personalized medicine, and promoting scientific collaboration across borders. Through his pioneering work, Prof. Reyed continues to shape the future of precision health by integrating complex biological systems with state-of-the-art artificial intelligence, positioning himself as a global leader at the forefront of the convergence between biotechnology, digital innovation, and sustainable healthcare.

Pranav Kumar Prabhakar Professor, Department of Biotechnology, School of Engineering and Technology Nagaland University, Meriema, Kohima, Nagaland, India. Email: prabhakar.iitm@gmail.com Biographical Profile Prof. Pranav Kumar Prabhakar is a highly distinguished scholar recognized among the World's Top 2% Scientists (Stanford University, USA, 2021-2023). He currently serves as a Professor in the Department of Biotechnology at Nagaland University, India. Prof. Prabhakar earned his PhD in Biotechnology from the prestigious Indian Institute of Technology (IIT) Madras, specializing in the molecular mechanisms of metabolic disorders and advanced strategies for oral insulin delivery. His research integrates the application of natural products with complex signaling pathway modeling, a critical component in the Bio-Digital Polybiomics approach to managing metabolic diseases like diabetes. A fellow of the Royal Society of Chemistry, his prolific academic output includes over 75 research articles, 3 books, and 14 book chapters. Prof. Prabhakar has been honored with numerous prestigious grants, including awards from the Council for Scientific and Industrial Research (CSIR) and the Department of Science and Technology (DST), India. As a seasoned editorial board member for several high-impact journals, he brings extensive expertise in scholarly governance and rigorous peer-review, further strengthening the international scientific authority of this volume.

A. K. Haghi is a retired Professor and a globally prolific scholar with over 1,500 publications, including books, chapters, and refereed journal papers, accumulating more than 4,300 citations with an h-index of 35. He serves as an Honorary Research Associate at the Chemistry Centre, University of Coimbra (Portugal) and the Universidad Autónoma de Coahuila (Mexico). With over 30 years of expertise in engineering and chemoinformatics, Prof. Haghi is the founder and former Editor-in-Chief of the International Journal of Chemoinformatics and Chemical Engineering.

He holds advanced degrees from the University of North Carolina (USA) and the Université de Technologie de Compiègne (France), with a Ph.D. in Engineering Sciences from the Université de Franche-Comté (France). Within this volume, his vast experience in computational modeling and material sciences provides a foundational academic authority to the Bio-Digital Polybiomics paradigm, bridging decades of engineering excellence with next-generation diagnostic innovations.

***

Arshita is a dedicated researcher in the field of bioinformatics at Amity University, Noida, India. Her academic focus lies at the intersection of computational biology, precision medicine, and AI-driven biological data analysis. She is actively engaged in scientific writing and explores innovative methodologies to decode complex biological systems through technological integration. Driven by a passion for data-driven biomedical research, Arshita contributes to the Bio-Digital Polybiomics framework by investigating how computational approaches can advance personalized healthcare. Her work in this volume reflects a commitment to the evolving landscape of bioinformatics, focusing on the application of emerging technologies to address modern challenges in systems biology and precision health.

Sherein Ismail Abd El-Moez is the Head of the Microbiology and Immunology Department at the Veterinary Research Institute, National Research Centre (NRC), Egypt. A distinguished scientist with an H-index of 14 and over 950 citations, she has authored 74 international publications, focusing on microbiology, immunology, and advanced diagnostics. Since 2014, she has served as a Project Officer at the Science and Technology Development Fund (STDF), where she manages national and international research programs and leads scientific selection committees. Prof. Abd El-Moez is a recipient of numerous prestigious honors, including the State Encouragement Prize for Agricultural Science (2015) and First Class Honors from the Presidency of the Arab Republic of Egypt (2017). Additionally, she serves as a Lead and Technical Assessor at the Egyptian Accreditation Council (EGAC), specializing in ISO standards for laboratory proficiency and inspection bodies. Her expertise integrates high-level research leadership with international accreditation standards, contributing significantly to the advancement of veterinary and microbial sciences.

Miral Gamal AbdElwahab is a Lecturer in the Food Industry Technology Program at Thebes Technological University, Luxor, Egypt. She holds a Ph.D. in Veterinary Medical Sciences with a specialization in Food Hygiene from Alexandria University. Dr. AbdElwahab is an expert in food safety, natural antioxidants, and advanced biotechnological applications, with a research portfolio that includes numerous peer-reviewed publications and several registered patents. Beyond her technical research, she has demonstrated significant institutional leadership, serving as a director for specialized units and chairing multiple academic committees. Her work focuses on the intersection of food security and innovative bioprocessing. Within this volume, her expertise in bioactive compounds and food safety frameworks contributes a vital translational dimension to the Bio-Digital Polybiomics paradigm, exploring how technological innovations can enhance the functional properties of food systems for precision nutrition.

Rohini Vipin Adepwar is a faculty member at the Dr. D. Y. Patil School of Allied Health Sciences, Pune, India. An academic and biomedical researcher with over 15 years of experience, she specializes in microbiology, public health surveillance, and biomarker validation. Her distinguished career includes a significant tenure at the ICMR–National AIDS Research Institute, where she focused on molecular biomarkers for HIV and HCV within regulated laboratory environments. Mrs. Adepwar is an expert in laboratory quality systems and outcome-based education, contributing to the advancement of allied health sciences through peer-reviewed research and international collaborations. Within this volume, her expertise in pathogen surveillance and biomarker diagnostic frameworks provides an essential clinical layer for the Bio-Digital Polybiomics paradigm, ensuring that digital health insights are grounded in rigorous laboratory validation and public health standards.

Amr Mohamed Ageez is a Professor of Bioinformatics and Vice Dean for Research and Postgraduate Studies at the Faculty of Biotechnology, MSA University, Egypt. He earned his Ph.D. from the University of Tokyo, Japan, and has held prestigious visiting research positions at both the University of Tokyo and Kobe University. A pioneer in genomic sciences, Prof. Ageez led Egypt's first next-generation sequencing project (the de novo assembly of the Egyptian Buffalo Genome) and has made significant contributions to the genetic enhancement of microbial systems. His current research leverages Next-Generation Sequencing (NGS) to explore probiotic bacterial genomics and the evolutionary dynamics of viral pathogens, including SARS-CoV-2 and H5N1. Within this volume, Prof. Ageez's expertise in comparative genomics and biomarker discovery provides a high-level analytical framework for the Bio-Digital Polybiomics paradigm, integrating strain surveillance with precision health outcomes.

Hina Bansal is an Assistant Professor at the Centre for Bioinformatics and Computational Biology, Amity Institute of Biotechnology, Amity University, India. With over 16 years of academic and research experience, she specializes in the integration of Artificial Intelligence and Machine Learning within the biological sciences. Her extensive research portfolio encompasses network pharmacology, functional genomics, and genome informatics, with a particular focus on Next-Generation Sequencing (NGS) and molecular docking simulations. Dr. Bansal has authored over 40 peer-reviewed publications and serves as an expert reviewer for high-impact journals such as Scientific Reports and BMC Computational Biology. Her technical proficiency in programming languages including Python, C++, and Java allows her to architect complex computational models for biosciences. Within this volume, her expertise contributes to the Bio-Digital Polybiomics framework, specifically in decoding the multi-omics data layers essential for precision healthcare and predictive disease modeling

Hiren Doshi is the Vice Chancellor of Vidya University, Meerut (Delhi NCR), India, where he provides strategic leadership in academic excellence and research innovation. A distinguished chemist and educational strategist, his expertise encompasses drug design, natural product chemistry, chromatography, and bioremediation. Dr. Doshi is a prominent figure in higher education advancement, bridging the gap between traditional chemical sciences and modern industrial applications. His leadership focuses on fostering a culture of research excellence and promoting academic-industry synergy through innovative digital frameworks. Within this volume, Dr. Doshi's expertise in bio-chemical architecture and drug discovery provides a high-level strategic foundation for the Bio-Digital Polybiomics paradigm, integrating chemical innovation with institutional research frameworks for global health solutions.

Osama Ibrahim El Batawy is a Professor at the Food Science Department, Faculty of Agriculture, Ain Shams University, Cairo, Egypt, where he also serves as the Chairman of the Food Safety Committee. With extensive academic tenure since 2020, his research specializes in dairy science and food microbiology, focusing on the microbial dynamics and safety protocols of functional food systems. Prof. El Batawy has contributed significantly to the field through numerous peer-reviewed publications exploring the intersection of food quality and microbial safety. Within this volume, his expertise in dairy-based delivery systems and food safety governance provides a critical framework for the Bio-Digital Polybiomics paradigm, particularly in the standardization of probiotic applications and the management of foodborne microbial interfaces within precision nutrition.

Mostafa Fetoh El-Hosseny is the Head of the Next Generation Sequencing (NGS) Research Unit at the Biodefense Center for Infectious and Emerging Diseases, Ministry of Defense, Egypt. A specialist in Genome Sequencing and Bioinformatics, he possesses extensive expertise in high-throughput molecular diagnostics and multi-platform NGS technologies, including the Illumina iSeq100, MiSeq, and NextSeq 550Dx systems. Dr. El-Hosseny focuses on the integration of infectious disease genomics with advanced bioinformatics analysis to enhance biosafety and biosecurity frameworks. Within the scope of this volume, his expertise in genomic architecture and nucleic acid precision analysis provides a robust computational and molecular foundation for the Bio-Digital Polybiomics paradigm, bridging the gap between raw sequencing data and translational precision medicine.

Mona A. Esawy is a Professor at the Pharmaceutical and Drug Industries Research Institute, National Research Centre (NRC), Egypt. A specialist in enzyme engineering, purification, and immobilization, her career is marked by significant international research tenures in Hong Kong and France (IRD Marcie), where she focused on molecular biology and bacterial gene detection. Prof. Esawy's pioneering discovery of honey-derived bacterial isolates as a novel source of levansucrase has opened new frontiers in probiotic applications and functional food science. Her current research focuses on the development of bio-intelligent anti-plaque agents using dextransucrase, targeting oral pathogenic microorganisms through innovative antimicrobial frameworks. Within this volume, her expertise in enzymatic modulation and microbial isolates provides essential biochemical insights for the Bio-Digital Polybiomics paradigm, particularly in the design of next-generation probiotic and therapeutic formulations.

Ahmed Zaki Ghareeb is an Assistant Professor at the Center of Excellence for Drug Preclinical Studies (CE-DPS), SRTA-City, Egypt. Specializing in experimental surgery and animal modeling, he holds a Ph.D. from Alexandria University and has over two decades of expertise in pharmaceutical development and applied biotechnological research. Dr. Ghareeb's work is pivotal in bridging the gap between in vitro discoveries and clinical applications, focusing on the development of robust animal models to test drug efficacy and safety. Within the scope of this volume, his expertise in preclinical systems provides the essential biological validation layer for the Bio-Digital Polybiomics paradigm, ensuring that AI-driven therapeutic insights are integrated with rigorous in vivo experimental methodologies for advancing precision medicine.

Doaa A. Ghareeb is a Professor of Medical Biotechnology at Alexandria University and the Director of the Bio-Screening and Preclinical Trial Laboratory. A prominent leader in the scientific community, she previously served as the Dean of the Pharmaceutical and Fermentation Industries Development Center at SRTA-City and co-founded the Drug Preclinical Studies Center of Excellence. Prof. Ghareeb is a pioneer in natural product drug design, with a research focus on viral pathogens, oncology, and metabolic disorders. Her prolific career is marked by over 177 peer-reviewed publications and the registration of 11 patents, reflecting her commitment to innovation. A recipient of the ASRT Women's Encouragement Award (2018) and multiple sustainability honors, she has led over 35 national and international research projects. Within this volume, her profound expertise in preclinical validation and biotechnological innovation serves as a foundational pillar for the Bio-Digital Polybiomics framework, bridging the gap between molecular discovery and clinical drug development.

Mohamed Abd Elhakim Kelany is a Senior Researcher and Deputy Technical Manager of the Microbiology Department at the Central Laboratory of Residue Analysis of Pesticides and Heavy Metals in Food (QCAP), Agricultural Research Center, Egypt. A lecturer in Microbiology at the Faculty of Science, Ain Shams University, his research specializes in the intersection of food safety, molecular pathogen typing, and applied microbiology. Dr. Kelany's expertise encompasses the development of reference materials and the implementation of advanced laboratory accreditation and quality management systems. Within this volume, his expertise in molecular biological surveillance and food-borne pathogen diagnostics provides a critical regulatory and safety framework for the Bio-Digital Polybiomics paradigm, bridging the gap between innovative microbial research and standardized food safety protocols.

Irem Kılınç is a researcher specializing in thermal food science and advanced processing technologies. She earned her Bachelor's degree in Food Engineering and a Master's degree in Fishing and Processing from Ege University, Turkey. In 2023, her research excellence was recognized through a JASSO-supported fellowship at the Food Thermal Science Laboratory, Tokyo University of Marine Science and Technology (TUMSAT), Japan. Following her doctoral studies at Izmir Katip Celebi University, she was awarded the prestigious MEXT Scholarship in 2026 to continue her professional career at TUMSAT. Ms. Kılınç's expertise focuses on the intersection of food thermal dynamics and innovative processing methodologies. Her contribution to this volume bridges the gap between traditional food engineering and the emerging requirements of precision health, exploring how processing technologies influence the bio-availability and safety of complex food systems within the polybiome framework

Rahul Kumar is an Assistant Professor at the Faculty of Pharmacy, Vidya University, Meerut, India. A specialist in Computer-Aided Drug Design (CADD) and pharmaceutical chemistry, he holds a Master of Pharmacy from Galgotias University.

Mr. Kumar is a recognized innovator, holding a granted design patent for a MUC-1 based SPR biosensor specifically targeting Triple-Negative Breast Cancer diagnosis. His research bridges the gap between computational modeling and experimental validation, focusing on precision drug discovery and innovative therapeutic technologies. With a prolific record of publications and multiple prestigious awards from Anveshan 2025 and Pharmathon 2024, he provides an essential computational chemistry dimension to the Bio-Digital Polybiomics framework, contributing to the development of bio-intelligent diagnostic and therapeutic platforms.

Prem Shankar Mishra is the Dean of the Faculty of Pharmacy at Vidya University, Meerut, India. A distinguished leader in pharmaceutical sciences, he provides strategic direction in academic innovation, research enhancement, and quality assurance. Prof. Mishra's expertise encompasses drug design, formulation science, and emerging therapeutic technologies, with a prolific record of scholarly contributions that have shaped modern pharmaceutical education. His leadership focuses on fostering institutional growth and mentoring the next generation of healthcare professionals through innovative research frameworks. Within this volume, Prof. Mishra's expertise in formulation dynamics and pharmacological innovation provides a critical translational bridge for the Bio-Digital Polybiomics paradigm, ensuring that computational insights are integrated with robust pharmaceutical development and therapeutic precision.

Rakhi Mishra is a distinguished Professor of Pharmaceutical Chemistry at the Noida Institute of Engineering and Technology (NIET), Greater Noida, India. A leader in pharmaceutical education and interdisciplinary research, Prof. Mishra specializes in the intersection of medicinal chemistry and drug development strategies. Her academic leadership is marked by a strong commitment to fostering innovation-driven learning environments and bridging the gap between industry and academia. With a prolific record of scholarly publications and active engagement in curriculum modernization, her research emphasizes the development of novel therapeutic agents. Within this volume, Prof. Mishra's expertise in pharmaceutical molecular design provides a vital chemical foundation for the Bio-Digital Polybiomics paradigm, integrating chemical informatics with precision pharmacological frameworks for next-generation drug discovery.

Sakshi Patel is an Assistant Professor in the Department of Pharmaceutical Technology at the Meerut Institute of Engineering and Technology (MIET), India. A specialist in pharmaceutical chemistry and computational drug discovery, she holds a Master of Pharmacy from Galgotias University. Ms. Patel's research is at the forefront of innovative diagnostics, evidenced by her registered patent in biosensor design for cancer diagnosis. With a prolific publication record in collaboration with prestigious publishers such as Springer Nature, Elsevier, and IGI Global, her work emphasizes the integration of experimental research with advanced computational methodologies. A recipient of multiple innovation awards, including recognitions at Anveshan 2025 and Pharmathon 2024, she provides a vital technological and diagnostic dimension to the Bio-Digital Polybiomics framework, particularly in the development of bio-intelligent sensors and precision pharmacological screening.

Ashok Kumar Sah is an Assistant Professor in the Department of Medical Laboratory Sciences at A' Sharqiyah University, Oman. A certified Medical Laboratory Scientist (ASCPi) with over 15 years of clinical and research experience, he holds a Ph.D. in Medical Laboratory Science. Dr. Sah is an expert in molecular pathology, cancer biomarkers, and clinical biochemistry, with a prolific publication record of over 55 peer-reviewed articles, including high-impact contributions to The Lancet and NPJ Clean Water. His work focuses on diagnostic innovation and the translational aspects of laboratory medicine in diverse geographic settings. Within this volume, Dr. Sah's expertise in biomarker discovery and clinical pathology provides a critical validation layer for the Bio-Digital Polybiomics paradigm, bridging the gap between advanced diagnostic technologies and global public health outcomes.

Mohamed Gomaa Seadawy is the Head of the Biodefense Center for Infectious and Emerging Diseases, Ministry of Defense, Egypt. With over 15 years of expertise in virology, genomics, and biodefense, he leads national research efforts in infectious disease control and vaccine development. Dr. Seadawy played a pivotal role in Egypt's national pandemic response, overseeing the development of EGYVAX (Egypt's inactivated COVID-19 vaccine) and the whole genome sequencing of over 800 SARS-CoV-2 isolates submitted to the NCBI database. His research excellence extends to DNA vaccine development and cancer genomics using next-generation sequencing (NGS). As a pioneer in establishing BSL-3 laboratory infrastructures and national vaccine research units, Dr. Seadawy's expertise provides a strategic dimension to the Bio-Digital Polybiomics framework, particularly in the areas of genomic surveillance, pathogen-host interactions, and the development of bio-intelligent countermeasures for global health security.

Abdelmohsen M. Soliman is a Professor of Biochemistry at the National Research Centre (NRC), Egypt. A distinguished scientist with extensive expertise in molecular biology, immunology, and advanced drug delivery systems, his work focuses on the development of polymeric products for complex medical and industrial applications. Prof. Soliman has led numerous international and national research projects as a Principal Investigator (PI) and served as the Co-Director of the NATO Advanced Research Workshop (ARW). A recipient of the National Research Centre Award for Scientific Excellence, he holds multiple approved and pending patents in biotechnological innovations. Within this volume, his expertise in biopolymer engineering and molecular diagnostics provides a critical biochemical foundation for the Bio-Digital Polybiomics framework, bridging advanced material science with precision therapeutic strategies.

Kalidindi Divya Srija is a researcher at the Amity Institute of Biotechnology, India, specializing in Bioinformatics with a dual focus on Data Science and computational analytics. Her research expertise lies at the intersection of Artificial Intelligence and healthcare, where she investigates data-driven methodologies to enhance biomedical research and drug discovery. Ms. Srija is dedicated to bridging the gap between computational algorithms and biological insights, with a particular interest in leveraging genomic data analytics to advance precision medicine. Within the framework of this volume, her work contributes to the Bio-Digital Polybiomics paradigm, focusing on the application of data science to decode complex biological systems and optimize therapeutic interventions.

Ashraf Tabll is a Professor at the Egypt Centre for Research and Regenerative Medicine (ECRRM), Cairo. He is an expert in advanced cell culture techniques and the establishment of monoclonal and polyclonal antibody systems for immunodiagnosis and immunotherapy. With over 97 international publications, Prof. Tabll's distinguished career includes postdoctoral fellowships at Yale University and the University of Connecticut Health Center, USA, as well as research stays in Germany supported by DAAD and STDF scholarships. His research excellence was recognized early with the Egyptian State Encouragement Prize in Biological Sciences. Within this volume, Prof. Tabll's expertise in immunometabolic signaling and cell-based diagnostics provides a critical biological foundation for the Bio-Digital Polybiomics framework, bridging the gap between advanced regenerative medicine and precision immunotherapy.

Shweta Thusoo is an Assistant Professor at the Dr. D.Y. Patil School of Allied Health Sciences, Pune, India. She holds a Master's degree in Medical Laboratory Technology and a Bachelor's degree in Zoology (Honours) from the University of Delhi. With over a decade of expertise, her research focuses on hematology, infectious diseases, and antimicrobial resistance (AMR), emphasizing diagnostic innovations and public health applications. Ms. Thusoo is a specialist in integrating AI-driven methodologies into clinical education and precision diagnostics. Her work, published in Scopus-indexed journals, addresses the evolution of healthcare systems through interdisciplinary training and quality management. Within this volume, her expertise in pathology and AI-assisted laboratory workforce training contributes a vital educational and diagnostic dimension to the Bio-Digital Polybiomics paradigm, bridging the gap between traditional clinical pathology and the future of automated precision medicine.

Marwa Ashraf Zayed is a researcher specializing in pharmacology and molecular medicine, currently affiliated with Damanhour University and the City of Scientific Research and Technological Applications (SRTA-City), Egypt. With a strong foundation in tissue culture and pharmaceutical development, her professional experience spans prestigious industrial and research institutions, including Pharco Corporation (EEPI) and the Pharmaceutical and Fermentation Industries Development Center (PFIDC). Dr. Zayed's expertise lies in bridging the gap between molecular diagnostics and therapeutic applications. Beyond her laboratory proficiency, she is an expert in scientific communication and digital content strategy, skills that are vital for the dissemination of complex biological data. Within this volume, her work contributes to the Bio-Digital Polybiomics framework, focusing on the translational aspects of pharmacology and the integration of innovative molecular research into precision medical practice.

Index

A

Adjuvant Therapy 355
Artificial Intelligence (AI) 4, 6, 12, 21, 36, 44, 45, 46, 47, 50, 51, 54, 55, 57, 58, 59, 60, 61, 62, 73, 82, 85, 88, 89, 90, 91, 93, 94, 95, 96, 98, 99, 100, 101, 103, 105, 111, 113, 117, 126, 128, 129, 130, 131, 133, 137, 138, 143, 147, 150, 151, 152, 153, 154, 155, 156, 157, 159, 162, 163, 164, 166, 167, 168, 169, 170, 171, 172, 179, 191, 201, 202, 206, 210, 213, 215, 219, 220, 221, 222, 223, 224, 225, 228, 231, 232, 234, 235, 236, 237, 239, 240, 242, 243, 245, 246, 247, 248, 250, 252, 254, 255, 259, 260, 261, 263, 264, 266, 267, 271, 275, 279, 283, 284, 290, 291, 294, 295, 297, 300, 301, 303, 304, 306, 307, 315, 318, 323, 324, 327, 328, 330, 338, 340, 341, 344, 345, 346, 350, 351, 352, 353, 354, 360, 372, 379, 380, 381, 383, 386, 387, 390, 393, 396, 414, 417, 421, 423, 424, 426, 427, 428, 429, 439, 440, 441, 443, 444, 445, 447, 449, 450, 451, 452, 455, 459, 462, 465, 466, 467, 468, 470, 472, 474, 477, 479, 480, 481, 482, 483, 484, 485, 486, 487, 488, 489, 490, 491, 492, 493

B

bacteriocin clusters 364, 369
Biointelligent Systems 175, 201, 205

C

Chronic Diseases 5, 47, 96, 114, 138, 140, 168, 255, 324, 325, 332, 358, 361, 362, 389, 390, 436
Circular Medicine 173, 179, 182, 185, 191, 195, 197, 202, 203, 204, 205
Clinical Translation 71, 95, 212, 243
Cognitive Oncology (COS) 173, 174, 175, 178, 182, 185, 191, 195, 196, 197, 198, 199, 201, 202, 204, 205
Colorectal Cancer (CRC) 118, 122, 129, 135, 162, 173, 174, 177, 178, 180, 183, 184, 185, 186, 188, 189, 191, 193, 195, 198, 199, 201, 203, 206, 207, 209, 210, 211, 212, 213, 215, 216, 217, 233, 259, 316, 341, 347, 349
CRISPR arrays 358

D

Data Privacy 117, 131, 147, 270, 317, 324, 327, 331, 439, 444, 466
Deep Learning 52, 54, 82, 93, 112, 122, 127, 128, 132, 133, 134, 142, 150, 168, 227, 235, 238, 245, 250, 267, 281, 315, 316, 331, 346, 348, 350, 373, 442, 444, 486, 490
Digital Health Strategy 444
Digital Nutrient Specification 295
Dysbiosis 9, 11, 12, 18, 28, 32, 33, 43, 54, 56, 107, 108, 109, 110, 111, 112, 113, 136, 159, 160, 162, 171, 181, 186, 190, 194, 219, 220, 231, 234, 239, 240, 242, 245, 257, 261, 270, 302, 310, 311, 313, 314, 315, 323, 326, 328, 332, 354, 355, 404, 416, 435

E

Endocrine Disruptors 28, 50, 66, 67, 83
Engineered Bacteria 210, 355
Epigenomics 82, 150, 152, 157, 213, 295, 326

F

Fecal Microbiota Transplantation 29, 309, 310, 319, 330, 331, 335, 338, 344, 345, 346, 354, 355, 432
functional foods 36, 62, 63, 64, 88, 201, 251, 270, 283, 286, 289, 290, 306, 357, 358, 360, 361, 362, 365, 367, 376, 379,

389, 390, 391, 398, 399, 400, 402, 403, 405, 418, 421, 422, 423, 427
Functional Metabolites 287

G

Genetic Polymorphisms 46, 47, 51, 58, 59, 91, 116, 236, 297, 322, 339
Gut-Brain Axis 110, 159, 166, 167
Gut Microbiota 12, 37, 57, 74, 75, 78, 80, 84, 88, 92, 105, 106, 108, 118, 128, 135, 159, 162, 207, 208, 209, 210, 211, 213, 247, 248, 254, 257, 261, 276, 313, 314, 319, 320, 321, 323, 332, 337, 341, 342, 345, 347, 348, 349, 351, 352, 353, 354, 355, 402, 403, 427, 432, 433, 435

H

Hidden Hunger 1, 5, 11, 13, 14, 22, 276
Host–Microbe Interactions 106, 132, 221, 225, 269, 402, 423

I

Immune Tolerance 12, 18, 21, 22, 23, 24, 30, 33, 34, 38, 43, 54, 66, 186, 190, 309, 310, 311, 312, 313, 314, 315, 320, 321, 326, 355

L

Lactobacillus Genomics 381

M

Metabolomics 29, 51, 54, 62, 99, 100, 102, 103, 104, 106, 107, 117, 125, 135, 144, 150, 152, 157, 166, 181, 233, 246, 251, 268, 280, 292, 295, 317, 326, 327, 328, 347, 384, 409, 410, 412, 413, 417, 428
Microbial Interactions 63, 101, 120, 123, 153, 163, 190, 194, 232, 236, 328, 329
Multi-Omics Integration 101, 303, 315, 417

N

Nanoplastics 5, 6, 16, 28, 34, 42, 43, 44, 49, 50, 54, 56, 66, 67, 74, 80
Nutraceuticals 9, 10, 11, 31, 32, 37, 51, 57, 58, 59, 60, 61, 62, 63, 64, 68, 69, 77, 198, 358, 391, 417, 421, 428, 433
Nutrigenomics 1, 4, 6, 9, 10, 12, 19, 29, 33, 36, 38, 42, 45, 46, 47, 48, 49, 50, 54, 55, 59, 61, 63, 64, 65, 71, 74, 75, 77, 81, 89, 93, 94, 95, 96, 97, 98, 99, 101, 110, 111, 112, 113, 114, 115, 116, 121, 123, 124, 126, 130, 131, 132, 140, 144, 147, 158, 159, 161, 163, 164, 165, 166, 168, 169, 170, 172, 194, 201, 203, 205, 214, 216, 219, 220, 221, 222, 223, 224, 227, 231, 233, 236, 244, 246, 248, 251, 255, 257, 265, 270, 279, 286, 294, 295, 323, 324, 325, 326, 347, 348, 351, 361, 376, 379, 397, 399, 412, 414, 423, 424, 425
Nutritional Genomics 78, 99, 208, 210, 221, 287, 288, 290

O

Ontological Shift 174
Organoid Interfaces 173

P

Personalized Intervention 6, 328
Personalized Nutrition 48, 54, 73, 96, 98, 101, 115, 116, 125, 126, 131, 133, 142, 144, 148, 151, 159, 160, 162, 163, 164, 166, 167, 168, 170, 172, 206, 211, 221, 223, 232, 250, 251, 257, 263, 267, 270, 273, 279, 286, 289, 290, 294, 296, 297, 305, 334, 346, 347, 348, 360, 379, 380, 386, 396, 400, 403, 421, 423, 426, 434, 454
Polybiome Dynamics 36, 229
Prebiotics 29, 31, 52, 63, 107, 109, 110, 112, 183, 200, 265, 309, 310, 319, 320, 321, 328, 330, 332, 333, 338, 340, 355, 359, 422, 428, 431, 434, 435

Precision Immunotherapy 219, 244
Precision Nutrition 7, 17, 21, 30, 31, 36, 38, 44, 50, 52, 56, 59, 68, 69, 71, 78, 79, 101, 115, 132, 142, 148, 156, 161, 162, 165, 208, 215, 234, 235, 242, 243, 249, 255, 288, 327, 331, 343, 344, 367, 384, 406, 412, 418, 425, 435
Probiotic Characterization 386
Probiotics 29, 30, 52, 58, 59, 60, 63, 87, 90, 95, 107, 109, 110, 112, 146, 151, 152, 153, 156, 157, 158, 169, 183, 191, 198, 200, 202, 220, 257, 276, 283, 297, 302, 303, 309, 310, 311, 312, 313, 319, 320, 321, 328, 330, 332, 333, 335, 336, 338, 340, 341, 343, 345, 349, 351, 352, 355, 359, 361, 364, 366, 367, 374, 379, 383, 389, 395, 397, 398, 399, 400, 402, 403, 404, 405, 406, 407, 408, 409, 410, 414, 415, 418, 419, 420, 421, 422, 423, 424, 425, 427, 428, 429, 430, 431, 433, 434, 435, 436

R

Regulatory Frameworks 117, 284, 327, 382, 405, 408, 441, 445, 466, 469, 470, 471, 476, 479, 481, 488, 489

S

Safety Assessment 289, 358, 390
Synbiotics 58, 60, 320, 321, 328, 330, 333, 335, 340, 355, 359, 422, 435
Systemic Coherence 173, 204
Systems Biology 44, 83, 120, 174, 204, 268

T

Therapeutic Modeling 202
Type 1 Diabetes 1, 5, 7, 8, 11, 12, 13, 16, 21, 27, 34, 37, 38, 54, 55, 56, 57, 65, 67, 69, 70, 74, 75, 78, 79, 80, 81, 82, 83, 85, 90, 91
Type 5 Diabetes 1, 5, 8, 11, 12, 13, 14, 16, 21, 34, 36, 38, 54, 55, 56, 57, 65, 67, 70, 81, 88

www.ingramcontent.com/pod-product-compliance
Lightning Source LLC
LaVergne TN
LVHW081317300926
849161LV00008B/1791